T0210565

Lecture Notes in Artificial Intelligence 9012

Subseries of Lecture Notes in Computer Science

More information about this series at http://www.springer.com/series/1244

Ngoc Thanh Nguyen · Bogdan Trawiński
Raymond Kosala (Eds.)

Intelligent Information and Database Systems

7th Asian Conference, ACIIDS 2015
Bali, Indonesia, March 23–25, 2015
Proceedings, Part II

 Springer

Editors
Ngoc Thanh Nguyen
Ton Duc Thang University
Ho Chi Minh city
Vietnam

Raymond Kosala
Bina Nusantara University
Jakarta
Indonesia

and

Wroclaw University of Technology
Wroclaw
Poland

Bogdan Trawiński
Wroclaw University of Technology
Wroclaw
Poland

ISSN 0302-9743
Lecture Notes in Artificial Intelligence
ISBN 978-3-319-15704-7
DOI 10.1007/978-3-319-15705-4

ISSN 1611-3349   (electronic)

ISBN 978-3-319-15705-4   (eBook)

Library of Congress Control Number: 2015932661

LNCS Sublibrary: SL7 – Artificial Intelligence

Springer Cham Heidelberg New York Dordrecht London

Printed on acid-free paper

Springer International Publishing AG Switzerland is part of Springer Science+Business Media
(www.springer.com)

Preface

ACIIDS 2015 was the seventh event in the series of international scientific conferences for research and applications in the field of intelligent information and database systems. The aim of ACIIDS 2015 was to provide an internationally respected forum for scientific research in the technologies and applications of intelligent information and database systems. ACIIDS 2015 was co-organized by Bina Nusantara University, Indonesia and Wrocław University of Technology, Poland in cooperation with Ton Duc Thang University, Vietnam and Quang Binh University, Vietnam, and with IEEE Indonesia Section and IEEE SMC Technical Committee on Computational Collective Intelligence as patrons of the conference. It took place in Bali, Indonesia during March 23–25, 2015.

Conferences of series ACIIDS have been well established. The first two events, ACIIDS 2009 and ACIIDS 2010, took place in Dong Hoi City and Hue City in Vietnam, respectively. The third event, ACIIDS 2011, took place in Daegu, Korea, while the fourth event, ACIIDS 2012, took place in Kaohsiung, Taiwan. The fifth event, ACIIDS 2013, was held in Kuala Lumpur in Malaysia while the sixth event, ACIIDS 2014, was held in Bangkok in Thailand.

We received more than 300 papers from about 40 countries all over the world. Each paper was peer reviewed by at least two members of the International Program Committee and International Reviewer Board. Only 117 papers with the highest quality were selected for oral presentation and publication in the two volumes of the ACIIDS 2015 proceedings.

Papers included in the proceedings cover the following topics: semantic web, social networks and recommendation systems, text processing and information retrieval, intelligent database systems, intelligent information systems, decision support and control systems, machine learning and data mining, multiple model approach to machine learning, innovations in intelligent systems and applications, artificial intelligent techniques and their application in engineering and operational research, machine learning in biometrics and bioinformatics with applications, advanced data mining techniques and applications, collective intelligent systems for e-market trading, technology opportunity discovery and collaborative learning, intelligent information systems in security and defense, analysis of image, video and motion data in life sciences, augmented reality and 3D media, cloud-based solutions, Internet of things, big data, and cloud computing.

Accepted and presented papers highlight new trends and challenges of intelligent information and database systems. The presenters showed how new research could lead to new and innovative applications. We hope you will find these results useful and inspiring for your future research.

We would like to express our sincere thanks to the Honorary Chairs, Prof. Harjanto Prabowo (Rector of the Bina Nusantara University, Indonesia) and Prof. Tadeusz Więckowski (Rector of the Wrocław University of Technology, Poland) for their supports.

Our special thanks go to the Program Chairs, Special Session Chairs, Organizing Chairs, Publicity Chairs, and Local Organizing Committee for their work for the conference. We sincerely thank all members of the International Program Committee for their valuable efforts in the review process which helped us to guarantee the highest quality of the selected papers for the conference. We cordially thank the organizers and chairs of special sessions which essentially contributed to the success of the conference.

We also would like to express our thanks to the Keynote Speakers (Prof. Nikola Kasabov, Prof. Suphamit Chittayasothorn, Prof. Dosam Hwang, and Prof. Satryo Soemantri Brodjonegoro) for their interesting and informative talks of world-class standard.

We cordially thank our main sponsors, Bina Nusantara University (Indonesia), Wrocław University of Technology (Poland), Ton Duc Thang University (Vietnam) Quang Binh University (Vietnam), and patrons: IEEE Indonesia Section and IEEE SMC Technical Committee on Computational Collective Intelligence. Our special thanks are due also to Springer for publishing the proceedings, and to other sponsors for their kind supports.

We wish to thank the members of the Organizing Committee for their very substantial work and the members of the Local Organizing Committee for their excellent work.

We cordially thank all the authors for their valuable contributions and other participants of this conference. The conference would not have been possible without their supports.

Thanks are also due to many experts who contributed to making the event a success.

March 2015   Ngoc Thanh Nguyen
                                                               Bogdan Trawiński
                                                               Raymond Kosala

Organization

Honorary Chairs

Harjanto Prabowo   Bina Nusantara University, Indonesia
Tadeusz Więckowski   Wrocław University of Technology,
           Poland

General Chairs

Ngoc Thanh Nguyen   Wrocław University of Technology, Poland
Ford Lumban Gaol   Bina Nusantara University, Indonesia

Program Chairs

Bogdan Trawiński   Wrocław University of Technology, Poland
Raymond Kosala   Bina Nusantara University, Indonesia
Tzung-Pei Hong   National University of Kaohsiung, Taiwan
Hamido Fujita   Iwate Prefectural University, Japan

Organizing Chairs

Harisno   Bina Nusantara University, Indonesia
Suharjito   Bina Nusantara University, Indonesia
Marcin Maleszka   Wrocław University of Technology, Poland

Special Session Chairs

Dariusz Barbucha   Gdynia Maritime University, Poland
John Batubara   Bina Nusantara University, Indonesia

Publicity Chairs

Diana   Bina Nusantara University, Indonesia
Adrianna Kozierkiewicz-
    Hetmańska   Wrocław University of Technology, Poland

Local Organizing Committee

Togar Napitupulu   Bina Nusantara University, Indonesia
Nilo Legowo   Bina Nusantara University, Indonesia
Benfano Soewito   Bina Nusantara University, Indonesia
Fergyanto   Bina Nusantara University, Indonesia
Zbigniew Telec   Wrocław University of Technology, Poland
Bernadetta Maleszka   Wrocław University of Technology, Poland
Marcin Pietranik   Wrocław University of Technology, Poland

Steering Committee

Ngoc Thanh Nguyen (Chair)	Wrocław University of Technology, Poland
Longbing Cao	University of Technology Sydney, Australia
Tu Bao Ho	Japan Advanced Institute of Science and Technology, Japan
Tzung-Pei Hong	National University of Kaohsiung, Taiwan
Lakhmi C. Jain	University of South Australia, Australia
Geun-Sik Jo	Inha University, Korea
Jason J. Jung	Yeungnam University, Korea
Hoai An Le-Thi	Paul Verlaine University – Metz, France
Toyoaki Nishida	Kyoto University, Japan
Leszek Rutkowski	Częstochowa University of Technology, Poland
Suphamit Chittayasothorn	King Mongkut's Institute of Technology Ladkrabang, Thailand
Ford Lumban Gaol	Bina Nusantara University, Indonesia
Ali Selamat	Universiti Teknologi Malaysia, Malyasia

Keynote Speakers

Nikola Kasabov	Auckland University of Technology, New Zealand
Suphamit Chittayasothorn	King Mongkut's Institute of Technology Ladkrabang, Thailand
Dosam Hwang	Yeungnam University, Korea
Satryo Soemantri Brodjonegoro	Indonesian Academy of Sciences, Indonesia

Special Sessions Organizers

1. *Multiple Model Approach to Machine Learning (MMAML 2015)*

Tomasz Kajdanowicz	Wrocław University of Technology, Poland
Edwin Lughofer	Johannes Kepler University Linz, Austria
Bogdan Trawiński	Wrocław University of Technology, Poland

2. *Special Session on Innovations in Intelligent Systems and Applications (IISA 2015)*

Shyi-Ming Chen	National Taiwan University of Science and Technology, Taiwan

3. *Special Session on Innovations in Artificial Intelligent Techniques and Its Application in Engineering and Operational Research (AITEOR 2015)*

Pandian Vasant	Universiti Teknologi PETRONAS, Malaysia
Vo Ngoc Dieu	HCMC University of Technology, Vietnam
Irraivan Elamvazuthi	Universiti Teknologi PETRONAS, Malaysia

Mohammad Abdullah-Al-Wadud	King Saud University, Saudi Arabia
Ahamed Khan	Universiti Selangor, Malaysia
Timothy Ganesan	Universiti Teknologi PETRONAS, Malaysia
Perumal Nallagownden	Universiti Teknologi PETRONAS, Malaysia

4. *Special Session on Analysis of Image, Video and Motion Data in Life Sciences (IVMLS 2015)*

Kondrad Wojciechowski	Polish-Japanese Institute of Information Technology, Poland
Marek Kulbacki	Polish-Japanese Institute of Information Technology, Poland
Jakub Segen	Gest3D, USA
Andrzej Polański	Silesian University of Technology, Poland

5. *Special Session on Machine Learning in Biometrics and Bioinformatics with Application (MLBBA 2015)*

Piotr Porwik	University of Silesia, Poland
Marina L. Gavrilova	University of Calgary, Canada
Rafał Doroz	University of Silesia, Poland
Krzysztof Wróbel	University of Silesia, Poland

6. *Special Session on Collective Intelligent Systems for E-market Trading, Technology Opportunity Discovery and Collaborative Learning (CISETC 2015)*

Tzu-Fu Chiu	Aletheia University, Taiwan
Chia-Ling Hsu	Tamkang University, Taiwan
Feng-Sueng Yang	Aletheia University, Taiwan

7. *Special Session on Intelligent Information Systems in Security & Defence (IISSD 2015)*

Andrzej Najgebauer	Military University of Technology, Poland
Dariusz Pierzchała	Military University of Technology, Poland
Ryszard Antkiewicz	Military University of Technology, Poland
Ewa Niewiadomska-Szynkiewicz	Warsaw University of Technology, Poland
Zbigniew Tarapata	Military University of Technology, Poland
Richard Warner	IIT/Chicago-Kent College of Law, USA
Adam Zagorecki	Cranfield University, UK

8. *Special Session on Advanced Data Mining Techniques and Applications (ADMTA 2015)*

Bay Vo	Ton Duc Thang University, Vietnam
Tzung-Pei Hong	National University of Kaohsiung, Taiwan
Bac Le	Ho Chi Minh City University of Science, Vietnam

9. *Special Session on Cloud-Based Solutions (CBS 2015)*

Ondrej Krejcar	University of Hradec Králové, Czech Republic
Vladimir Sobeslav	University of Hradec Králové, Czech Republic
Peter Brida	University of Žilina, Slovakia
Kamil Kuca	University of Hradec Králové, Czech Republic

10. *Special Session on Augmented Reality and 3D Media (AR3DM 2015)*

Atanas Gotchev	Tampere University of Technology, Finland
Janusz Sobecki	Wrocław University of Technology, Poland
Zbigniew Wantuła	ADUMA S.A., Poland

11. *Special Session on Internet of Things, Big Data, and Cloud Computing (IoT 2015)*

Adam Grzech	Wrocław University of Technology, Poland
Andrzej Ruciński	University of New Hampshire, USA

International Program Committee

Muhammad Abulaish	Jamia Millia Islamia, India
El-Houssaine Aghezzaf	Ghent University, Belgium
Haider M. AlSabbagh	University of Basra, Iraq
Toni Anwar	Universiti Teknologi Malaysia, Malaysia
Ahmad Taher Azar	Benha University, Egypt
Amelia Badica	University of Craiova, Romania
Costin Badica	University of Craiova, Romania
Emili Balaguer- Ballester	Bournemouth University, UK
Zbigniew Banaszak	Warsaw University of Technology, Poland
Dariusz Barbucha	Gdynia Maritime University, Poland
John Batubara	Bina Nusantara University, Indonesia
Ramazan Bayindir	Gazi University, Turkey
Maumita Bhattacharya	Charles Sturt University, Australia
Maria Bielikova	Slovak University of Technology in Bratislava, Slovakia
Veera Boonjing	King Mongkut's Institute of Technology Ladkrabang, Thailand
Mariusz Boryczka	University of Silesia, Poland
Urszula Boryczka	University of Silesia, Poland
Abdelhamid Bouchachia	Bournemouth University, UK
Stephane Bressan	National University of Singapore, Singapore
Peter Brida	University of Žilina, Slovakia
Piotr Bródka	Wrocław University of Technology, Poland

Andrej Brodnik	University of Ljubljana, Slovenia
Grażyna Brzykcy	Poznań University of Technology, Poland
The Duy Bui	VNU University of Engineering and Technology, Vietnam
Robert Burduk	Wrocław University of Technology, Poland
David Camacho	Universidad Autónoma de Madrid, Spain
Frantisek Capkovic	Institute of Informatics, Slovak Academy of Sciences, Slovakia
Oscar Castillo	Tijuana Institute of Technology, Mexico
Dariusz Ceglarek	Poznań School of Banking, Poland
Stephan Chalup	University of Newcastle, Australia
Bao Rong Chang	National University of Kaohsiung, Taiwan
Somchai Chatvichienchai	University of Nagasaki, Japan
Rung-Ching Chen	Chaoyang University of Technology, Taiwan
Shyi-Ming Chen	National Taiwan University of Science and Technology, Taiwan
Suphamit Chittayasothorn	King Mongkut's Institute of Technology Ladkrabang, Thailand
Tzu-Fu Chiu	Aletheia University, Taiwan
Kazimierz Choroś	Wrocław University of Technology, Poland
Dorian Cojocaru	University of Craiova, Romania
Phan Cong-Vinh	NTT University, Vietnam
Jose Alfredo Ferreira Costa	Universidade Federal do Rio Grande do Norte, Brazil
Keeley Crockett	Manchester Metropolitan University, UK
Boguslaw Cyganek	AGH University of Science and Technology, Poland
Ireneusz Czarnowski	Gdynia Maritime University, Poland
Piotr Czekalski	Silesian University of Technology, Poland
Paul Davidsson	Malmö University, Sweden
Roberto De Virgilio	Universita' degli Studi Roma Tre, Italy
Tien V. Do	Budapest University of Technology and Economics, Hungary
Pietro Ducange	University of Pisa, Italy
El-Sayed M. El-Alfy	King Fahd University of Petroleum and Minerals, Saudi Arabia
Vadim Ermolayev	Zaporozhye National University, Ukraine
Rim Faiz	University of Carthage, Tunisia
Victor Felea	Alexandru Ioan Cuza University of Iasi, Romania
Thomas Fober	University of Marburg, Germany
Dariusz Frejlichowski	West Pomeranian University of Technology, Poland
Mohamed Gaber	Robert Gordon University, UK
Patrick Gallinari	LIP6 - University of Paris 6, France
Dariusz Gąsior	Wrocław University of Technology, Poland

Tomasz Kajdanowicz	Wrocław University of Technology, Poland
Nikola Kasabov	Auckland University of Technology, New Zealand
Arkadiusz Kawa	Poznań University of Economics, Poland
Muhammad Khurram Khan	King Saud University, Saudi Arabia
Pan-Koo Kim	Chosun University, Korea
Yong Seog Kim	Utah State University, USA
Attila Kiss	Eötvös Loránd University, Hungary
Frank Klawonn	Ostfalia University of Applied Sciences, Germany
Goran Klepac	Raiffeisen Bank, Croatia
Joanna Kołodziej	Cracow University of Technology, Poland
Marek Kopel	Wrocław University of Technology, Poland
Józef Korbicz	University of Zielona Góra, Poland
Raymond Kosala	Bina Nusantara University, Indonesia
Leszek Koszałka	Wrocław University of Technology, Poland
Adrianna Kozierkiewicz- Hetmańska	Wrocław University of Technology, Poland
Ondrej Krejcar	University of Hradec Králové, Czech Republic
Dalia Kriksciuniene	Vilnius University, Lithuania
Dariusz Król	Wrocław University of Technology, Poland
Marzena Kryszkiewicz	Warsaw University of Technology, Poland
Adam Krzyzak	Concordia University, Canada
Elżbieta Kukla	Wrocław University of Technology, Poland
Marek Kulbacki	Polish-Japanese Institute of Information Technology, Poland
Kazuhiro Kuwabara	Ritsumeikan University, Japan
Halina Kwaśnicka	Wrocław University of Technology, Poland
Helge Langseth	Norwegian University of Science and Technology, Norway
Annabel Latham	Manchester Metropolitan University, UK
Hoai An Le Thi	University of Lorraine, France
Kun Chang Lee	Sungkyunkwan University, Korea
Philippe Lenca	Telecom Bretagne, France
Horst Lichter	RWTH Aachen University, Germany
Sebastian Link	University of Auckland, New Zealand
Rey-Long Liu	Tzu Chi University, Taiwan
Edwin Lughofer	Johannes Kepler University of Linz, Austria
Lech Madeyski	Wrocław University of Technology, Poland
Bernadetta Maleszka	Wrocław University of Technology, Poland
Marcin Maleszka	Wrocław University of Technology, Poland
Yannis Manolopoulos	Aristotle University of Thessaloniki, Greece
Urszula Markowska-Kaczmar	Wrocław University of Technology, Poland
Francesco Masulli	University of Genoa, Italy
Tamás Matuszka	Eötvös Loránd University, Hungary

João Mendes-Moreira	University of Porto, Portugal
Jacek Mercik	Wrocław School of Banking, Poland
Saeid Nahavandi	Deakin University, Australia
Kazumi Nakamatsu	University of Hyogo, Japan
Grzegorz J. Nalepa	AGH University of Science and Technology, Poland
Mahyuddin K.M. Nasution	Universiti Kabangsaan Malaysia, Indonesia
Prospero Naval	University of the Philippines, Philippines
Fulufhelo Vincent Nelwamondo	Council for Scientific and Industrial Research, South Africa
Linh Anh Nguyen	University of Warsaw, Poland
Thanh Binh Nguyen	International Institute for Applied Systems Analysis, Austria
Adam Niewiadomski	Lodz University of Technology, Poland
Toyoaki Nishida	Kyoto University, Japan
Yusuke Nojima	Osaka Prefecture University, Japan
Mariusz Nowostawski	University of Otago, New Zealand
Manuel Núñez	Universidad Complutense de Madrid, Spain
Richard Jayadi Oentaryo	Singapore Management University, Singapore
Shingo Otsuka	Kanagawa Institute of Technology, Japan
Jeng-Shyang Pan	National Kaohsiung University of Applied Sciences, Taiwan
Mrutyunjaya Panda	Gandhi Institute for Technological Advancement, Bhubaneswar, India
Tadeusz Pankowski	Poznań University of Technology, Poland
Marcin Paprzycki	Systems Research Institute, Polish Academy of Sciences, Poland
Jakub Peksiński	West Pomeranian University of Technology, Szczecin, Poland
Danilo Pelusi	University of Teramo, Italy
Xuan Hau Pham	Quang Binh University, Vietnam
Dariusz Pierzchała	Military University of Technology, Poland
Marcin Pietranik	Wrocław University of Technology, Poland
Niels Pinkwart	Humboldt University of Berlin, Germany
Elvira Popescu	University of Craiova, Romania
Piotr Porwik	University of Silesia, Poland
Bhanu Prasad	Florida A&M University, USA
Andrzej Przybyszewski	University of Massachusetts Medical School, USA
Paulo Quaresma	University of Évora, Portugal
Héctor Quintián	University of Salamanca, Spain
Christoph Quix	Fraunhofer FIT, Germany
Monruthai Radeerom	Rangsit University, Thailand
Ewa Ratajczak-Ropel	Gdynia Maritime University, Poland

Rajesh Reghunadhan	Central University of Bihar, India
Przemysław Różewski	West Pomeranian University of Technology, Szczecin, Poland
Leszek Rutkowski	Częstochowa University of Technology, Poland
Henryk Rybiński	Warsaw University of Technology, Poland
Alexander Ryjov	Lomonosov Moscow State University, Russia
Virgilijus Sakalauskas	Vilnius University, Lithuania
Daniel Sanchez	University of Granada, Spain
Juergen Schmidhuber	Swiss AI Lab IDSIA, Switzerland
Bjorn Schuller	Technical University Munich, Germany
Jakub Segen	Gest3D, USA
Ali Selamat	Universiti Teknologi Malaysia, Malaysia
Alexei Sharpanskykh	Delft University of Technology, The Netherlands
Quan Z. Sheng	University of Adelaide, Australia
Andrzej Siemiński	Wrocław University of Technology, Poland
Dragan Simic	University of Novi Sad, Serbia
Gia Sirbiladze	Tbilisi State University, Georgia
Andrzej Skowron	University of Warsaw, Poland
Adam Słowik	Koszalin University of Technology, Poland
Janusz Sobecki	Wrocław University of Technology, Poland
Kulwadee Somboonviwat	King Mongkut's Institute of Technology Ladkrabang, Thailand
Zenon A. Sosnowski	Białystok University of Technology, Poland
Serge Stinckwich	University of Caen Lower Normandy, France
Stanimir Stoyanov	Plovdiv University "Paisii Hilendarski", Bulgaria
Jerzy Świątek	Wrocław University of Technology, Poland
Andrzej Świerniak	Silesian University of Technology, Poland
Edward Szczerbicki	University of Newcastle, Australia
Julian Szymański	Gdańsk University of Technology, Poland
Ryszard Tadeusiewicz	AGH University of Science and Technology, Poland
Yasufumi Takama	Tokyo Metropolitan University, Japan
Pham Dinh Tao	INSA-Rouen, France
Zbigniew Telec	Wrocław University of Technology, Poland
Krzysztof Tokarz	Silesian University of Technology, Poland
Behcet Ugur Toreyin	Çankaya University, Turkey
Bogdan Trawiński	Wrocław University of Technology, Poland
Krzysztof Trawiński	European Centre for Soft Computing, Spain
Maria Trocan	Institut Superieur d'Electronique de Paris, France
Hong-Linh Truong	Vienna University of Technology, Austria
Olgierd Unold	Wrocław University of Technology, Poland
Pandian Vasant	Universiti Teknologi PETRONAS, Malaysia

Joost Vennekens	Katholieke Universiteit Leuven, Belgium
Jorgen Villadsen	Technical University of Denmark, Denmark
Bay Vo	Ton Duc Thang University, Vietnam
Yongkun Wang	University of Tokyo, Japan
Izabela Wierzbowska	Gdynia Maritime University, Poland
Marek Wojciechowski	Poznań University of Technology, Poland
Dong-Min Woo	Myongji University, Korea
Michał Woźniak	Wrocław University of Technology, Poland
Marian Wysocki	Rzeszow University of Technology, Poland
Guandong Xu	University of Technology Sydney, Australia
Xin-She Yang	Middlesex University, UK
Zhenglu Yang	University of Tokyo, Japan
Lean Yu	Chinese Academy of Sciences, AMSS, China
Slawomir Zadrozny	Systems Research Institute, Polish Academy of Sciences, Poland
Drago Žagar	University of Osijek, Croatia
Danuta Zakrzewska	Lodz University of Technology, Poland
Faisal Zaman	Dublin City University, Ireland
Constantin-Bala Zamfirescu	Lucian Blaga University of Sibiu, Romania
Katerina Zdravkova	Ss. Cyril and Methodius University in Skopje, Macedonia
Aleksander Zgrzywa	Wrocław University of Technology, Poland
Jianwei Zhang	National University Corporation Tsukuba University of Technology, Japan
Min-Ling Zhang	Southeast University, China
Zhongwei Zhang	University of Southern Queensland, Australia
Zhi-Hua Zhou	Nanjing University, China

Program Committees of Special Sessions

Multiple Model Approach to Machine Learning (MMAML 2015)

Emili Balaguer-Ballester	Bournemouth University, UK
Urszula Boryczka	University of Silesia, Poland
Abdelhamid Bouchachia	Bournemouth University, UK
Robert Burduk	Wrocław University of Technology, Poland
Oscar Castillo	Tijuana Institute of Technology, Mexico
Rung-Ching Chen	Chaoyang University of Technology, Taiwan
Suphamit Chittayasothorn	King Mongkut's Institute of Technology Ladkrabang, Thailand
José Alfredo F. Costa	Federal University of Rio Grande do Norte, Brazil
Bogusław Cyganek	AGH University of Science and Technology, Poland
Ireneusz Czarnowski	Gdynia Maritime University, Poland

Patrick Gallinari	Pierre et Marie Curie University, France
Fernando Gomide	State University of Campinas, Brazil
Francisco Herrera	University of Granada, Spain
Tzung-Pei Hong	National University of Kaohsiung, Taiwan
Konrad Jackowski	Wrocław University of Technology, Poland
Piotr Jędrzejowicz	Gdynia Maritime University, Poland
Tomasz Kajdanowicz	Wrocław University of Technology, Poland
Yong Seog Kim	Utah State University, USA
Bartosz Krawczyk	Wrocław University of Technology, Poland
Kun Chang Lee	Sungkyunkwan University, Korea
Edwin Lughofer	Johannes Kepler University Linz, Austria
Héctor Quintián	University of Salamanca, Spain
Andrzej Siemiński	Wrocław University of Technology, Poland
Dragan Simic	University of Novi Sad, Serbia
Adam Słowik	Koszalin University of Technology, Poland
Zbigniew Telec	Wrocław University of Technology, Poland
Bogdan Trawiński	Wrocław University of Technology, Poland
Krzysztof Trawiński	European Centre for Soft Computing, Spain
Olgierd Unold	Wrocław University of Technology, Poland
Pandian Vasant	Universiti Teknologi PETRONAS, Malaysia
Michał Woźniak	Wrocław University of Technology, Poland
Zhongwei Zhang	University of Southern Queensland, Australia
Zhi-Hua Zhou	Nanjing University, China

Special Session on Innovations in Intelligent Systems and Applications (IISA 2015)

I-Cheng Chang	National Dong Hwa University, Hualien, Taiwan
Shyi-Ming Chen	National Taiwan University of Science and Technology, Taipei, Taiwan
Po-Hung Chen	St. John's University, New Taipei City, Taiwan
Shou-Hsiung Cheng	Chienkuo Technology University, Changhua, Taiwan
Mong-Fong Horng	National Kaohsiung University of Applications, Taiwan
Wei-Lieh Hsu	Lunghwa University of Science and Technology, Taoyuan, Taiwan
Feng-Long Huang	National United University, Miaoli, Taiwan
Pingsheng Huang	Ming Chuan University, Taoyuan County, Taiwan
Bor-Jiunn Hwang	Ming Chuan University, Taoyuan County, Taiwan
Huey-Ming Lee	Chinese Culture University, Taipei, Taiwan
Li-Wei Lee	De Lin Institute of Technology, New Taipei City, Taiwan
Chung-Ming Ou	Kainan University, Taoyuan County, Taiwan

Jeng-Shyang Pan	Harbin Institute of Technology, China
Victor R.L. Shen	National Taipei University, New Taipei City, Taiwan
Chia-Rong Su	Chang Gung University, New Taipei City, Taiwan
An-Zen Shih	Jinwen University of Science and Technology, Taiwan
Chun-Ming Tsai	Taipei Municipal University of Education, Taiwan
Wen-Chung Tsai	Chaoyang University of Technology, Taichung, Taiwan
Cheng-Fa Tsai	National Pingtung University of Science and Technology, Taiwan
Cheng-Yi Wang	National Taiwan University of Science and Technology, Taipei, Taiwan
Chih-Hung Wu	National Taichung University of Education, Taiwan

Special Session on Innovations in Artificial Intelligent Techniques and Its Application in Engineering and Operational Research (AITEOR 2015)

Gerhard-Wilhelm Weber	Middle East Technical University, Turkey
Junzo Watada	Waseda University, Japan
Kwon-Hee Lee	Dong-A University, South Korea
Hindriyanto Dwi Purnomo	Satya Wacana Christian University, Indonesia
Charles Mbohwa	University of Johannesburg, South Africa
Gerardo Maximiliano Mendez	Instituto Tecnológico de Nuevo León, Mexico
Leopoldo Eduardo Cárdenas Barrón	Tecnológico de Monterry, Mexico
Petr Dostál	Brno University of Technology, Czech Republic
Erik Kropat	Universität der Bundeswehr München, Germany
Timothy Ganesan	Universiti Teknologi PETRONAS, Malaysia
Vedpal Singh	Universiti Teknologi PETRONAS, Malaysia
Gerrit Janssens	Hasselt University, Belgium
Suhail Qureshi	UET Lahore, Pakistan
Monica Chis	SIEMENS Program and System Engineering, Romania
Michael Mutingi	National University of Singapore, Singapore
Ugo Fiore	University of Naples Federico II, Italy
Utku Kose	Uşak University, Turkey
Nuno Pombo	University of Beira Interior, Portugal
Kusuma Soonpracha	Kasetsart University, Thailand
Armin Milani	Islamic Azad University, Iran

Leo Mrsic	University College of Law and Finance Effectus Zagreb, Crotia
Igor Litvinchev	Nuevo Leon State University, Mexico
Goran Klepac	Raiffeisen Bank, Croatia
Herman Mawengkang	University of Sumetera Utara, Indonesia

Special Session on Analysis of Image, Video and Motion Data in Life Sciences (IVMLS 2015)

Aldona Drabik	Polish-Japanese Institute of Information Technology, Poland
Leszek Chmielewski	Warsaw University of Life Sciences, Poland
André Gagalowicz	Inria, France
David Gibbon	AT&T, USA
Celina Imielinska	Vesalius Technologies, USA
Ryszard Klempous	Wrocław University of Technology, Poland
Ryszard Kozera	Warsaw University of Life Sciences, Poland
Marek Kulbacki	Polish-Japanese Institute of Information Technology, Poland
Aleksander Nawrat	Silesian University of Technology, Poland
Lyle Noakes	The University of Western Australia, Australia
Jerzy Paweł Nowacki	Polish-Japanese Institute of Information Technology, Poland
Eric Petajan	Directv, USA
Gopal Pingali	IBM, USA
Andrzej Polański	Polish-Japanese Institute of Information Technology, Poland
Andrzej Przybyszewski	University of Massachusetts, USA
Jerzy Rozenbilt	University of Arizona, Tucson, USA
Jakub Segen	Gest3D, USA
Aleksander Sieroń	Medical University of Silesia, Poland
Konrad Wojciechowski	Polish-Japanese Institute of Information Technology, Poland

Special Session on Machine Learning in Biometrics and Bioinformatics with Application (MLBBA 2015)

Marcin Adamski	Białystok Technical University, Poland
Ryszard Choraś	Uniwersytet Technologiczno-Przyrodniczy, Poland
Nabendu Chaki	University of Calcutta, India
Rituparna Chaki	University of Calcutta, India
Rafał Doroz	University of Silesia, Poland
Marina Gavrilova	University of Calgary, Canada
Phalguni Gupta	Indian Institute of Technology Kanpur, India

Anil K. Jain	Michigan State University, USA
Hemant B. Kekre	NMIMS University, India
Vic Lane	University of London, UK
Davide Maltoni	Università di Bologna, Italy
Ashu Marasinghe	Nagaoka University of Technology, Japan
Nobuyuki Nishiuchi	Tokyo Metropolitan University, Japan
Javier Ortega-García	Universidad Autónoma de Madrid, Spain
Giuseppe Pirlo	Università degli Studi di Bari, Italy
Piotr Porwik	University of Silesia, Poland
Arun Ross	Michigan State University, USA
Khalid Saeed	Białystok Technical University, Poland
Refik Samet	Ankara University, Turkey
Michał Woźniak	Wroclaw University of Technology, Poland
Krzysztof Wróbel	University of Silesia, Poland

Special Session on Collective Intelligent Systems for E-market Trading, Technology Opportunity Discovery and Collaborative Learning (CISETC 2015)

Ya-Fung Chang	Tamkang University, Taiwan
Peng-Wen Chen	Oriental Institute of Technology, Taiwan
Kuan-Shiu Chiu	Aletheia University, Taiwan
Tzu-Fu Chiu	Aletheia University, Taiwan
Chen-Huei Chou	College of Charleston, USA
Chia-Ling Hsu	Tamkang University, Taiwan
Fang-Cheng Hsu	Aletheia University, Taiwan
Kuo-Sui Lin	Aletheia University, Taiwan
Min-Huei Lin	Aletheia University, Taiwan
Yuh-Chang Lin	Aletheia University, Taiwan
Pen-Choug Sun	Aletheia University, Taiwan
Leuo-Hong Wang	Aletheia University, Taiwan
Ai-Ling Wang	Tamkang University, Taiwan
Henry Wang	Chinese Academy of Sciences, China
Feng-Sueng Yang	Aletheia University, Taiwan
Ming-Chien Yang	Aletheia University, Taiwan

Special Session on Intelligent Information Systems in Security & Defence (IISSD 2015)

Ryszard Antkiewicz	Military University of Technology, Poland
Leon Bobrowski	Białystok University of Technology, Poland
Urszula Boryczka	University of Silesia, Poland
Mariusz Chmielewski	Military University of Technology, Poland
Rafał Kasprzyk	Military University of Technology, Poland
Jacek Koronacki	Polish Academy of Sciences, Poland

Leszek Kotulski	AGH University of Science and Technology, Poland
Krzysztof Malinowski	Warsaw University of Technology, Poland
Andrzej Najgebauer	Military University of Technology, Poland
Ewa Niewiadomska-Szynkiewicz	Warsaw University of Technology, Poland
Dariusz Pierzchała	Military University of Technology, Poland
Jarosław Rulka	Military University of Technology, Poland
Zenon A. Sosnowski	Białystok University of Technology, Poland
Zbigniew Tarapata	Military University of Technology, Poland
Richard Warner	IIT/Chicago-Kent College of Law, USA
Marek Zachara	AGH University of Science and Technology, Poland
Adam Zagorecki	Cranfield University, UK

Special Session on Advanced Data Mining Techniques and Applications (ADMTA 2015)

Bay Vo	Ton Duc Thang University, Vietnam
Tzung-Pei Hong	National University of Kaohsiung, Taiwan
Bac Le	Ho Chi Minh City University of Science, Vietnam
Chun-Hao Chen	Tamkang University, Taiwan
Chun-Wei Lin	Harbin Institute of Technology Shenzhen Graduate School, China
Wen-Yang Lin	National University of Kaohsiung, Taiwan
Guo-Cheng Lan	Industrial Technology Research Institute, Taiwan
Yeong-Chyi Lee	Cheng Shiu University, Taiwan
Le Hoang Son	Ha Noi University of Science, Vietnam
Le Hoang Thai	Ho Chi Minh City University of Science, Vietnam
Vo Thi Ngoc Chau	Ho Chi Minh City University of Technology, Vietnam
Van Vo	Ho Chi Minh University of Industry, Vietnam

Special Session on Cloud-Based Solutions (CBS 2015)

Ana Almeida	Porto Superior Institute of Engineering, Portugal
Oliver Au	The Open University of Hong Kong, Hong Kong
Zoltan Balogh	Univerzita Konštantína Filozofa v Nitre, Slovakia
Jorge Bernardino	Polytechnical Institute of Coimbra, Portugal
Peter Brida	University of Žilina, Slovakia
Jozef Bucko	Technical University of Košice, Slovakia
Tanos Costa	Franca Military Institute of Engineering, Brazil
Bipin Desai	Concordia University, Canada
Ivan Dolnak	University of Žilina, Slovakia
Elsa Gomes	Porto Superior Institute of Engineering, Portugal

Alipio Jorge	University of Porto, Portugal
Ondrej Krejcar	CBAR, University of Hradec Králové, Czech Republic
Kamil Kuca	Biomedical Research Center, University Hospital of Hradec Králové, Czech Republic
Juraj Machaj	University of Žilina, Slovakia
Norbert Majer	The Research Institute of Posts and Telecommunications (VUS), Slovakia
Goreti Marreiros	Porto Superior Institute of Engineering, Portugal
Peter Mikulecký	University of Hradec Králové, Czech Republic
Marek Penhaker	VSB – Technical University of Ostrava, Czech Republic
Teodorico Ramalho	University of Lavras, Brazil
Maria Teresa Restivo	Universidade do Porto, Portugal
Tiia Ruutmann	Tallinn University of Technology, Lativia
Ali Selamat	Universiti Teknologi Malaysia, Malaysia
José Salmeron	Universidad Pablo de Olavide, Seville, Spain
Vladimir Sobeslav	University of Hradec Králové, Czech Republic
Vassilis Stylianakis	University of Patras, Greece
Jan Vascak	Technical University of Košice, Slovakia

Special Session on Augmented Reality and 3D Media (AR3DM 2015)

Jędrzej Anisiewicz	ADUMA, Poland
Robert Bregovic	Tampere University of Technology, Finland
Piotr Chynał	Wrocław University of Technology, Poland
Bogusław Cyganek	AGH University of Science and Technology, Poland
Irek Defee	Tampere University of Technology, Finland
Piotr Hrebeniuk	Cohesiva, Poland
Marcin Sikorski	Polish-Japanese Institute of Information Technology, Poland
Mårten Sjöström	Mid Sweden University, Sweden
Jarmo Viteli	University of Tampere, Finland
Jakub Wójnicki	ADUMA MOBILE, Poland

Special Session on Internet of Things, Big Data and Cloud Computing (IoT 2015)

Jorgi Mongal Batalla	National Institute of Telecommunications, Poland
Adam Grzech	Wrocław University of Technology, Poland
Jason Jeffords	DeepIS, USA
Krzysztof Juszczyszyn	Wrocław University of Technology, Poland

Piotr Krawiec	National Institute of Telecommunications, Poland
Jacek Lewandowski	Coventry University, UK
Marek Natkaniec	AGH University of Science and Technology, Poland
Andy Rindos	IBM Research, USA
Joel Rodrigues	Univeristy of Beira Interior, Portugal
Andrzej Ruciński	University of New Hampshire, USA
Henry Salveraj	University of Nevada-Las Vegas, USA
Edward Szczerbicki	University of Newcastle, Australia
Paweł Świątek	Wrocław University of Technology, Poland
Halina Tarasiuk	Warsaw University of Technology, Poland
David Zydek	Idaho State University, USA

Contents – Part II

Augmented Reality and 3D Media

Cloud Based Solutions

Internet of Things, Big Data and Cloud Computing

Artificial Intelligent Techniques and Their Application in Engineering and Operational Research

XXX   Contents – Part II

Contents – Part I

Intelligent Database Systems

Intelligent Information Systems

Decision Support and Control Systems

Machine Learning and Data Mining

Multiple Model Approach to Machine Learning (MMAML 2015)

Innovations in Intelligent Systems and Applications

Bio-inspired Optimization Techniques and Their Applications

Swarm Based Mean-Variance Mapping Optimization for Solving Economic Dispatch with Cubic Fuel Cost Function

Khoa H. Truong[1(✉)], Pandian Vasant[1], M.S. Balbir Sing[1], and Dieu N. Vo[2]

[1] Department of Fundamental and Applied Sciences, Universiti Teknologi
PETRONAS, Tronoh, Malaysia
{trhkhoa89,pvasant}@gmail.com, balbir@petronas.com.my
[2] Department of Power Systems, HCMC University of Technology,
Ho Chi Minh City, Vietnam
vndieu@gmail.com

Abstract. In power generation system, the economic dispatch (ED) is used to allocate the real power output of thermal generating units to meet required load demand so as their total operating cost is minimized while satisfying all units and system constraints. This paper proposes a novel swarm based mean-variance mapping optimization ($MVMO^S$) for solving the ED problem with the cubic fuel cost function. The special feature of the proposed algorithm is a mapping function applied for the mutation based on the mean and variance of n-best population. This method has been tested on 3, 5 and 26 units and the obtained results are compared to those from genetic algorithm (GA), particle swarm optimization (PSO) and firefly algorithm (FA). Test results have indicated that the proposed method is efficient for solving the ED problem with cubic fuel cost function.

Keywords: Economic dispatch · Cubic fuel cost function · Mean-variance mapping optimization · Swarm based Mean-variance mapping optimization

1 Introduction

Economic dispatch (ED) is one of the important optimization tasks in the power generation system. Its objective is to determine the real power output of thermal generating units to meet required load demand at minimum total fuel cost while satisfying all unit and system equality and inequality constraints[1,2].

Traditionally, the cost function objective of the ED problem is approximated as a quadratic function. However, a cubic function is more realistic than a quadratic function to express the operating cost. The fuel cost function becomes more nonlinear when cubic function is considered. Liang and Glover [3] proposed an iterative dynamic programming (DP) method for solving ED problem with cubic fuel cost function considering power transmission losses along with a clear description of modifying generator cost functions. Also, Jiang and Ertem [4] suggested the Newton

© Springer International Publishing Switzerland 2015
N.T. Nguyen et al. (Eds.): ACIIDS 2015, Part II, LNAI 9012, pp. 3–12, 2015.
DOI: 10.1007/978-3-319-15705-4_1

approach to solve the ED problem with cubic fuel cost functions. They developed a linear transmission loss model and included it into the ED problem. In the past decade, the ED problem with cubic cost function has been successfully solved by using artificial intelligence techniques. Kumaran and Mouly [5] used genetic algorithm (GA) to solve the cubic cost function ED problem. Adhinarayanan and Sydulu [6] utilized particle swarm optimization (PSO) for solving ED problem with cubic cost function. They showed that PSO obtaines better results and efficiency than GA. Adhinarayanan and Sydulu [7,8] presented a very fast and effective non-iterative λ-logic based algorithm for solving ED problem with cubic cost function. They concluded that this algorithm archieves a near optimal solution for large-scale systems. The power transmission loss was ignored in this case. Amoli et al. [9] have successfully applied firefly algorithm (FA) for solving the ED problem with cubic cost function.

Recently, a new meta-heuristic search algorithm, namely Mean-variance mapping optimization (MVMO), has been developed by István Erlich in 2010 [10]. This algorithm falls into the category of the so-called "population-based stochastic optimization technique". The similarities between MVMO and the other known stochastic algorithms are in three evolutionary operators including selection, mutation and crossover. The extensions of MVMO has been developed which named swarm based mean-variance mapping optimization (MVMO^S) [11]. Unlike the single particle MVMO, the search process of MVMO^S is started with a set of particles. In addition, two parameters of MVMO including the scaling factor and variable increment parameters have been extended to enhance the mapping. Hence, the ability for global search of MVMO^S is found to be more powerful than the original version. In this paper, MVMO^S is proposed as a novel optimization technique for solving the ED problem with cubic fuel cost function.

2   Problem Formulation

The optimization problem of the ED is to minimize the total fuel cost F_T, which be written as:

$$\text{Minimize } F_T = \sum_{i=1}^{N} F_i(P_i) \quad i = 1, 2, 3, ..., N \tag{1}$$

The solution of ED can be highly improved by introducing higher order generator cost functions. Cubic cost function displays the actual response of thermal generators more accurately. The cubic fuel cost function of a thermal generating unit is presented as follows [4]:

$$F_i(P_i) = a_i + b_i P_i + c_i P_i^2 + d_i P_i^3 \tag{2}$$

The constraints of the ED problem must be satisfied during the optimization process are presented as follows:

1. *Real power balance equation:* The total active power output of generating units must be equal to total active power load demand P_D plus power loss P_L:

$$\sum_{i=1}^{N} P_i = P_D + P_L \tag{3}$$

The power loss P_L is calculated by the below formulation [2]:

$$P_L = \sum_{i=1}^{N}\sum_{j=1}^{N} P_i B_{ij} P_j + \sum_{i=1}^{N} B_{0i} P_i + B_{00} \tag{4}$$

2. *Generator capacity limits:* The active power output of generating units must be within the allowed limits:

$$P_{i,min} \leq P_i \leq P_{i,max} \tag{5}$$

3   Swarm Based Mean - Variance Mapping Optimization

The search process of the MVMO^S starts with a set of particles. Initial variables is normalized to the range [0,1] as follows:

$$x\_normalized = rand(n\_par, n\_var). \tag{6}$$

However, the function evaluation is carried out always in the original scales of the problem space. The de-normalization of optimization variables is carried by using (7):

$$P_i = P_{i,min} + Scaling. \, x\_normalized(\iota,:) \tag{7}$$

where

$$Scaling = P_{i,max} - P_{i,min}$$

MVMO^S utilizes swarm implementation to enhance the power of global searching of the classical MVMO by starting the search with a set of n_p particles, each having its own memory and represented by the corresponding archive and mapping function. At the beginning of the optimization process, each particle performs m steps independently to collect a set of reliable individual solutions. Then, the particles start to communicate and to exchange information.

The key feature of MVMO is a special mapping function which applied for mutating the offspring based on mean-variance of the solutions stored in the archive.

The mean $\overline{x_i}$ and variance v_i are calculated as follows [10]:

$$\overline{x_i} = \frac{1}{n}\sum_{j=1}^{n} x_i(j) \tag{8}$$

$$v_i = \frac{1}{n}\sum_{j=1}^{n}(x_i(j)-\overline{x}_i)^2 \qquad (9)$$

where $j = 1, 2, ..., n$ (n is archive size).

The transformation of x_i^* to x_i via mapping function is calculated in (10) and depicted in Fig. 1. The transformation mapping function, h, is calculated in (12) by the mean $\overline{x}$ and shape variables s_{i1} and s_{i2} [10]:

$$x_i = h_x + (1 - h_1 + h_o).x_i^* - h_o. \qquad (10)$$

where h_x, h_1, h_0 are the outputs of transformation mapping function (12) based on different inputs given by:

$$h_x = h(x = x_i^*) , \quad h_o = h(x = 0), \quad h_1 = h(x = 1) \qquad (11)$$

$$h(x_i, s_{i1}, s_{i2}, x) = \overline{x}_i.(1 - e^{-x.s_{i1}}) + (1 - \overline{x}_i).e^{-(1-x).s_{i2}} \qquad (12)$$

where,

$$s_i = -\ln(v_i).f_s \qquad (13)$$

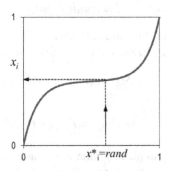

Fig. 1. Variable mapping

The scaling factor f_s in (13) is a MVMO parameter which allows for controlling the search process during iteration. In the MVMO^S algorithm, this factor is extended for the need of exploring the search space at the beginning more globally whereas, at the end of the iterations, the focus should be on the exploitation. It is determined by [11]:

$$f_s = f_s^*.(1 + rand()) \qquad (14)$$

where,

$$f_s^* = f_{s\_ini}^* + \left(\frac{i}{i_{final}}\right)^2 \left(f_{s\_final}^* - f_{s\_ini}^*\right) \qquad (15)$$

rand () is a random number in the range [0, 1]. In (15), the variable *i* represents the iteration number.

The shape variable s_{i1} and s_{i2} in (12) are determined by an algorithm in [11].

4   Implemention of MVMO^S to ED

4.1   Handing of Constraints

Neglecting the power transmission losses, the equality constraint (3) is rewritten by:

$$\sum_{i=1}^{N} P_i = P_D \tag{16}$$

By using the slack variable method [12] to guarantee that the equality constraint (16) is always satisfied. The power output of the slack unit is calculated as follows:

$$P_s = P_D - \sum_{\substack{i=1 \\ i \neq s}}^{N} P_i \tag{17}$$

The fitness function for the proposed MVMO^S will include the objective function (1) and penalty terms for the slack unit if inequality (5) is violated. The fitness function is as follows:

$$F_T = \sum_{i=1}^{N} F_i(P_i) + K \times \left[\left(\max(0, P_s - P_{s,\max}) \right)^2 + \left(\max(0, P_{s,\min} - P_s) \right)^2 \right] \tag{18}$$

4.2   Implementation of MVMO^S to ED

The steps of procedure of MVMO^S for the ED problem are described as follows:

Step 1: Setting the parameters for MVMO^S including $iter_{max}$, n_var, n_par, $mode$, d_i, Δd_0^{ini}, Δd_0^{final}, archive zize, $f_{s\_ini}^*$, $f_{s\_final}^*$, n_randomly, n_randomly_min, indep.runs(m), D_{min}
Set $i = 1$, i denotes the function evaluation

Step 2: Normalize initial variables to the range [0,1] (i.e. swarm of particles).
$x\_normalized = rand(n\_par, n\_var)$

Step 3: Set $k = 1$, k denotes particle counters.

Step 4: De-normalized variables using (7), calculate power output for the slack generator using (17) to evaluate fitness function in (18), store f_{best} and x_{best} in archive.

Step 5: Increase $i = i+1$. If $i < m$ (independent steps), go to Step 6. Otherwise, go to Step 7.

Step 6: Check the particles for the global best, collect a set of reliable individual solutions. The *i*-th particle is discarded from the optimization process if the distance D_i is less than a certain user defined threshold D_{min}. If the particle is deleted, increase $k = k+1$, decrease $n_p = n_p - 1$ and go to step 4. Otherwise, go to Step 7.

Step 7: Create offspring generation through three evolutionary operators: selection, mutation and crossover.3

Step 8: if $k < n_p$, increase $k = k+1$ and go to step 4. Otherwise, go to step 9.

Step 9: Check termination criteria. If stoping criteria is satisfied, stop. Otherwise, go to step 3. The algorithm of the proposed MVMO^S is terminated when the maximum number of iterations $iter_{max}$ is reached.

5   Numerical Results

The proposed MVMO^S has been tested on 3 test systems including 3, 5 and 26 thermal generation units where the cost function is cubic form. For each case, the algorithm of MVMO^S is run 50 independent trials on a Core i5 CPU 3.2 GHz PC with 4GB of RAM. The implementation of the proposed MVMO^S is coded in the Matlab R2013a platform.

Since different parameters of the proposed method have effects on the performance of MVMO^S. Hence, it is important to determine an optimal set of parameters of the proposed methods for dealing with ED problems. For each problem, the selection of parameters is carried out by varying only one parameter at a time and fixing the others. The parameter is first fixed at the low value and then increased. Multiple runs are carried out to choose the suitable set of parameters.

5.1   Case 1: 3-Unit System

The input data for 3-generating units are given in [2]. The total demanded load P_D of this case is 2500 MW, the transmission power loss is neglected. The obtained results by the MVMO^S are compared to those from genetic algorithm (GA) [6], particle swarm optimization (PSO) [6] and firefly algorithm (FA) [9], which are presented in Table 1.

Table 1. Results and comparisons for 3-unit system

Unit	Power outputs P_i			
	GA	PSO	FA	MVMO^S
1	725.02	724.99	729.0682	704.1826
2	910.19	910.15	906.8021	881.8226
3	864.88	864.85	864.1315	913.9944
Total power	2500.00	2500.00	2500.0000	2500.0000
Total Cost ($/h)	22730.14	22729.35	22728	22569.2239

The parameters for MVMO^S for this case are as follows: $iter_{max}$ = 1000, n_var (generators) = 3, n_p = 5, $archive\ size$ = 5, $mode$ = 4, $indep.runs$ (m) = 100, $n\_randomly$ = 2, $n\_randomly\_min$ = 2, $f^*_{s\_ini}$ = 0.95 , $f^*_{s\_final}$ = 3 , d_i = 1 , Δd_0^{ini} = 0.4 , Δd_0^{final} = 0.02 , D_{min} = 0.

Table 1 shows that the power output obtained by the MVMO^S is always satisfy the constraints and the MVMO^S provides the total cost less than GA, PSO and FA.

5.2   Case 2: 5-Unit System

The test system is from [5] including 5 generating units supplying to a power load demand of 1800 MW. The transmission power loss is also neglected in this case. The obtained results by the MVMO^S are compared to those from genetic algorithm (GA) [6], particle swarm optimization (PSO) [6] and firefly algorithm (FA) [9], which are shown in Table 2.

The parameters for MVMO^S for this case are as follows: $iter_{max}$ = 7000, n_var (generators) = 5, n_p = 5, $archive\ size$ = 5, $mode$ = 4, $indep.runs$ (m) = 100, $n\_randomly$ = 3, $n\_randomly\_min$ = 2, $f^*_{s\_ini}$ = 0.95 , $f^*_{s\_final}$ = 3 , d_i = 1 , Δd_0^{ini} = 0.4 , Δd_0^{final} = 0.02 , D_{min} = 0.

Table 2. Results and comparisons for 5-unit system

Unit	Power outputs P_i			
	GA	PSO	FA	MVMO^S
1	320.00	320.00	327.8004	320.000
2	343.74	343.70	341.9890	315.9476
3	472.60	472.60	460.4127	528.7392
4	320.00	320.00	327.8004	320.000
5	343.74	343.70	341.9890	315.3132
Total power	1800.00	1800.00	1800.00	1800.00
Total Cost ($/h)	18611.07	18610.40	18610	18519.5822

In Table 2, the power output obtained by the MVMO^S is always satisfy the constraints and the MVMO^S provides the total cost less than GA, PSO and FA.

5.3   Case 3: 26-Unit System

The data of the test system including 26 thermal generating units with cubic fuel cost function is from [8]. The system load demand for this case is 2000MW neglecting transmission power loss. The obtained results by the MVMO^S are compared to those from genetic algorithm (GA) [6], particle swarm optimization (PSO) [6] as given in Table 3.

The parameters for MVMO^S for this case are as follows: $iter_{max}$ = 40000, n_var (generators) = 26, n_p = 10, $archive\ size$ = 5, $mode$ = 4, $indep.runs$ (m) = 400,

$n\_randomly = 12$, $n\_randomly\_min = 10$, $f^*_{s\_ini} = 0.95$, $f^*_{s\_final} = 3$, $d_i = 1$, $\Delta d_0^{ini} = 0.4$, $\Delta d_0^{final} = 0.02$, $D_{min} = 0$.

In Table 3, the power output obtained by the MVMO^S is always satisfy the constraints and the MVMO^S provides the total cost less than GA and PSO.

Table 3. Results and comparisons for 26-unit system

Unit	Power outputs P_i		
	GA	PSO	MVMO^S
1	2.40	2.40	2.3937
2	2.40	2.40	2.400
3	2.40	2.40	2.400
4	2.40	2.40	2.400
5	2.40	2.40	2.400
6	4.00	4.00	4.000
7	4.00	4.00	4.000
8	4.00	4.00	4.000
9	4.00	4.00	4.000
10	15.20	15.20	15.7337
11	15.20	15.20	23.0998
12	15.20	15.20	17.7768
13	15.20	15.20	15.4274
14	25.00	25.00	25.0000
15	25.00	25.00	25.0000
16	25.00	25.00	25.0000
17	129.71	129.69	123.3045
18	124.71	124.69	116.6811
19	120.42	120.40	115.5890
20	116.72	116.70	116.4675
21	68.95	68.95	68.9500
22	68.95	68.95	68.9500
23	68.95	68.95	68.9500
24	337.76	337.85	346.0765
25	400.00	400.00	400.000
26	400.00	400.00	400.000
Total power	2000.00	2000.00	2000.00
Total Cost ($/h)	27671.2441	27671.2276	27267.3334

5.4   Robustness Analysis

The convergence of heuristic methods may not obtain exactly same solution because these methods initialize variables randomly at each run. Hence, their performances could not be judged by the results of a single run. Many trials should be carry out to reach a impartial conclusion about the performance of the algorithm. Therefore, in this study, the proposed algorithm is run 50 independent trials. The mean cost, max cost, average cost and standard deviation obtained by the proposed method to evaluate the robustness characteristic of the proposed method for ED problems. The robustness analysis of three test cases are presented in Table 4.

Table 4. Robustness analysis

	Case 1	Case 2	Case 3
Min total cost ($/h)	22569.2234	18519.5822	27267.3334
Average total cost ($/h)	22569.2234	18520.3270	27318.2241
Max total cost ($/h)	22569.2234	18524.4878	27392.0571
Standard deviation ($/h)	0.0	0.9345	29.3559
Ratio (%)	0	0.005	0.108
Average CPU time (s)	0.547	2.942	21.648

As seen in Table 4, the difference between the maximum and minimum costs obtained the proposed MVMO^S is small. The ratio between the standard deviation and the minimum cost is less than 0.108%. It shows that the performance the proposed MVMO^S is robust.

6   Conclusion

This paper has presented an application of new method for solving ED problem with cubic fuel cost function. Three test cases were carried out to demonstrate the effectiveness and efficiency of the proposed MVMO^S. The numerical results and robustness analysis show that the proposed method has better solution the than GA, PSO and FA and its performance is robust. Therefore, the proposed MVMO^S could be favorable for solving ED problems.

Acknowledgment. This research work is sponsored by Graduate Assistant Scheme of Universiti Teknologi PETRONAS.

Nomenclature

N	total number of generating units
a_i, b_i, c_i, d_i	fuel cost coefficients of generator i
B_{ij}, B_{0i}, B_{00}	B-matrix coefficients for transmission power loss
P_i	power output of generator i
$P_{i,max}$	maximum power output of generator i
$P_{i,min}$	minimum power output of generator i
K	the penalty factor for the slack unit
P_s	power output of slack unit
n_var	number of variable (generators)

n_par	number of particles
$mode$	variable selection strategy for offspring creation
$archive\ zize$	n-best individuals to be stored in the table
d_i	initial smoothing factor
Δd_0^{ini}	initial smoothing factor increment
Δd_0^{final}	final smoothing factor increment
$f_{s\_ini}^*$	initial shape scaling factor
$f_{s\_final}^*$	final shape scaling factor
D_{min}	minimum distance threshold to the global best solution
$n\_randomly$	initial number of variables selected for mutation
$indep.runs$	m steps independently to collect a set of reliable individual solutions

References

1. Xia, X., Elaiw, A.: Optimal dynamic economic dispatch of generation: a review. Electric Power Systems Research **80**(8), 975–986 (2010)
2. Wood, A.J., Wollenberg, B.F.: Power generation, operation, and control. Wiley (2012)
3. Liang, Z.-X., Glover, J.D.: A zoom feature for a dynamic programming solution to economic dispatch including transmission losses. IEEE Transactions on Power Systems **7**(2), 544–550 (1992)
4. Jiang, A., Ertem, S., Subir, S., Kothari, D.: Economic dispatch with non-monotonically increasing incremental cost units and transmission system losses. Discussion. IEEE transactions on power systems **10**(2), 891–897 (1995)
5. Kumaran, G., Mouly, V.S.R.K.: Using evolutionary computation to solve the economic load dispatch problem. In: Proceedings of the 2001 Congress on Evolutionary Computation, vol. 291, pp. 296–301 (2001)
6. Adhinarayanan, T., Sydulu, M.: Particle swarm optimisation for economic dispatch with cubic fuel cost function. In: TENCON 2006, IEEE Region 10 Conference 2006, pp. 1–4 (2006)
7. Adhinarayanan, T., Sydulu, M.: Fast and effective algorithm for economic dispatch of cubic fuel cost based thermal units. In: First International Conference on Industrial and Information Systems 2006, pp. 156–160. IEEE (2006)
8. Theerthamalai, A., Maheswarapu, S.: An effective non-iterative " λ logic based" algorithm for economic dispatch of generators with cubic fuel cost function. International Journal of Electrical Power & Energy Systems **32**(5), 539–542 (2010)
9. Amoli, N.A., Jalid, S., Shayanfar, H.A., Barzinpour, F.: Solving economic dispatch problem with cubic fuel cost function by firefly algorithm. In: 8th International Conference on "Technical and Physical Problems of Power Engineering" (ICTPE) Fredrikstad, Norway, 5–7 September (2012)
10. Erlich, I., Venayagamoorthy, G.K., Worawat, N.: A mean-variance optimization algorithm. In: IEEE Congress on Evolutionary Computation (CEC), pp. 1–6, 18–23 July 2010
11. Rueda, J.L., Erlich, I.: Evaluation of the mean-variance mapping optimization for solving multimodal problems. In: IEEE Symposium on Swarm Intelligence (SIS) 2013, pp. 7–14 (2013)
12. Kuo, C.-C.: A novel coding scheme for practical economic dispatch by modified particle swarm approach. IEEE Trans on Power Syst **23**, 1825–1835 (2008)

Multiobjective Optimization of Bioactive Compound Extraction Process via Evolutionary Strategies

Timothy Ganesan[1(✉)], Irraivan Elamvazuthi[2],
Pandian Vasant[3], and Ku Zilati Ku Shaari[1]

[1] Department of Chemical Engineering,
Universiti Teknologi Petronas, 31750 Tronoh, Perak, Malaysia
tim.ganesan@gmail.com
[2] Department of Electrical and Electronics Engineering,
Universiti Teknologi Petronas, 31750 Tronoh, Perak, Malaysia
[3] Department of Fundamental and Applied Sciences,
Universiti Teknologi Petronas, 31750 Tronoh, Perak, Malaysia

Abstract. Systematic and simultaneous optimization of a collection of objectives is called multiobjective or multicriteria optimization. These sorts of optimization procedures are becoming commonplace in fields involving engineering design, process and system optimization. In this work, the multiobjective (MO) optimization of the bioactive compound extraction process was carried out. Using the Normal Boundary Intersection (NBI) approach the MO optimization problem is transformed into a weighted form called the beta-subproblem. This subproblem is then solved using two evolutionary strategies (differential evolution (DE) and genetic algorithm (GA)). Using these evolutionary strategies, the solutions to the extraction process which form the efficient Pareto frontier was generated. The Hypervolume Indicator (HVI) was applied to the solutions to rank the strategies based on the solution quality. Critical analyses and comparative studies were then carried out on the strategies employed in this work and that from the previous work.

Keywords: Multi-objective optimization · Extraction process · Normal Boundary Intersection approach · Differential evolution (DE) · Genetic algorithm (GA) · Hypervolume Indicator (HVI)

1 Introduction

The problem formulation of the extraction process was done by Shashi *et al*, (2010) [1]. This formulation involves the modeling of the objective functions and the identification of the range of the extraction process parameters. The primary target was yield optimization of specific extracted chemical products from the *Gardenia Jasminoides Ellis* fruit. The process yields three bioactive compounds; crocin, geniposide and total phenolic compounds. Identifying a series of optimal process yields that generate an efficient Pareto frontier is critical. Gauging solution quality in MO optimization can be very difficult and tricky. Ideas involving solution properties like diversity and convergence have become popular in recent times [2], [3]. The utility of these ideas have proved useful for

© Springer International Publishing Switzerland 2015
N.T. Nguyen et al. (Eds.): ACIIDS 2015, Part II, LNAI 9012, pp. 13–21, 2015.
DOI: 10.1007/978-3-319-15705-4_2

developing metrics for evaluating specific aspects of the generated solutions. Hence, these metrics provide the decision maker with some useful information regarding the technique's effectiveness [4]. However, due to the local nature of these metrics, an absolute ranking of the solutions is not attainable. One effective approach that can be utilized for the overall ranking of solution sets is the Hypervolume Indicator (HVI) [5] which is based on the idea of Pareto dominance. This metric measures the Hypervolume (multidimensional) enclosed by a Pareto front approximation with respect to a reference set (see [6], [7], and [8]). This metric ensures strict compliance to monotonicity related to Pareto dominance [9], [10]. This makes the ranking of solution sets and hence algorithms feasible for any given MO problem. The techniques introduced in this work is directed to generating a series of solutions (with the associated weights) which efficiently approximates the Pareto frontier.

This problem was attempted using the particle swarm optimization (PSO) technique (within a weighted sum framework) in Shashi *et al.,* (2010) [1]. In that work, more emphasis was given on the modeling works as compared to the optimization procedures. Although an individual solution optimal solution is attained, the approximate Pareto frontier of the solutions was not constructed and rigorous solution evaluation (for ranking purposes) was not conducted.   In this work, the bioactive compound extraction process optimization problem was tackled using Differential Evolution (DE) [11] and Genetic Algorithm (GA) [12]. This was carried out within the basis of the Normal Boundary Intersection (NBI) framework [13]. The ranking of the techniques of the Pareto frontiers produced by the algorithms were carried out using the HVI metric [5], [6]. Comparative analyses were then conducted on the individual best solutions and the frontiers obtained in this work against those obtained in Shashi *et al.,* (2010) [1].

Genetic Algorithms (GA) were the earliest form of evolutionary algorithms introduced by Holland, (1992) [12]. These algorithms contain the fundamental components that make up an evolutionary algorithm such as the cross-over operator, mutation operator and fitness evaluation mechanisms (which aids the algorithm to successively improve the population's fitness during execution). Differential Evolution (DE) is a population-based evolutionary algorithm that has been derived from Genetic Algorithms (GA) [12]. DE was developed in the nineties by Storn and Price [11]. DE has been used extensively to solve problems which are non-differentiable, non-continuous, non-linear, noisy, flat, multidimensional, have many local minima, constraints or high degree of stochasticity. Lately, DE has been applied to a variety of areas including optimization problems in chemical and process engineering [14],[15],[16]. This paper is organized as follows: Section 2 of this paper presents an overview of the evolutionary strategies. In Section 3, the process formulation is described followed by Section 4 which discusses the computational results. Finally, this paper ends with the concluding remarks in Section 5.

2   Evolutionary Strategies

Evolutionary intelligence originate from the idea presented in Holland's [12] genetic algorithm. The central theme of all evolutionary algorithms is that survival of fittest (natural selection) which acts on a population of organisms under environmental stress. Thus, a fitness distribution is formed in the mentioned population of organisms (individuals). In the case of optimization, let there be an objective function to be optimized. Then, various random candidate solutions (individuals) to the problem can be gauged by applying a fitness function to each of these solutions (where the higher the fitness function, the better the solution quality with respect to the objective function). Each evaluated individual with respect to the best fitness values are then chosen for producing the next generation of potential solution vectors through the process of cross-over and mutation. The cross-over operator crosses the parent individuals from the previous generation (fittest individuals) and produces the next generation of individuals. The mutation operator on the other hand, perturbs the gene pool of the population to generate new optimization capabilities in the new individual offspring. Thus, each individual competes in their population as they achieve higher fitness values. The repeated execution of this cycle is meta-heuristic since the solution vectors in the population are subsequently improved as the iteration proceeds. In this work, Genetic algorithm (GA) and Differential Evolution (DE) algorithms are used.

2.1   Differential Evolution (DE)

DE is a class of evolutionary techniques first introduced in 1995 by Storn and Price [11]. This class of techniques pioneered the development of perturbative evolutionary algorithms. DE begins by randomly initializing a population (P) in the first generation of at least four individuals. These individuals are real-coded vectors with some size, N. A single principal parent (x^p_i) and three auxiliary parents denoted (x^a_i) is randomly selected from the population, P. In DE, every individual, I in the population, P would become a principle parent, x^p_i at one generation or the other. Therefore each individual has a chance to breed with the auxiliary parents, x^a_i. The three auxiliary parents then be mutated (via differential mutation) to generate a mutated vector, V_i.

$$V_i = x^a_1 + F(x^a_2 - x^a_3) \qquad (1)$$

where $F \in [0,1]$ is the real-valued mutation amplification factor. Next V_i is then recombined (or exponentially crossed-over) with x^p_i to generate a child trial vector, x^{child}_i. The cross-over probability (CR) is set by the user. In DE, 'knock-out competition' is used to select the survivors which will then be inserted into the next population's gene pool. This fitness selection procedure required the principle parent, x^p_i and the child trial vector, x^{child}_i to engage in direct competition:

$$x_i(gen+1) = \begin{cases} x_i^{child}(gen) \leftrightarrow & f(x_i^{child}) \ better \ than \ f(x_i^p) \\ x_i^p(gen) \leftrightarrow & otherwise \end{cases} \qquad (2)$$

The parameters initialized for the DE algorithm is given in Table 1. The algorithm of the DE method is shown in Execution Scheme 1.

Table 1. DE Parameter Setting

Initialized Parameters	Values
Individual Size, N	6
Population Size, P	7
Mutation amplification factor, F	0.3
Cross-over Probability, CR	0.667

Execution Scheme 1: Differential Evolution (DE)

Step 1: Initialize input parameters
Step 2: Set random population vectors, x^G_i
Step 3: Select one random principal parents, x^P_i
Step 4: Select three random auxilary parents, x^a_i
Step 5: Implement differential mutation to generate a mutated vector, V_i
Step 6: Merge V_i with x^P_i to produce a child trial vector, x^{child}_i
Step 7: Perform fitness evaluation for the next generation
Step 8: IF the fitness condition is fulfilled and t= T_{max} , stop and print solutions
        ELSE repeat step 3

2.2   Genetic Algorithm (GA)

GA is one of the first evolutionary search and optimization techniques [12]. This population-based approach uses an N-point crossover operator to create new offspring for successive generations. To avoid algorithmic stagnation at some local minima, mutation operators are usually employed to diversify the search. In this work a bit flip-type mutation operator was utilized. The GA scheme applied in this work is provided in Execution Scheme 2. The parameter settings initialized prior to the execution of the GA used in this work are shown in Table 2. The flow of the GA algorithm is shown in Execution Scheme 2:

Table 2. Genetic Algorithm Parameter Setting

Parameters	Values
Define individual string length	5 bits
Define amount of individuals in the population	6
Mutation Probability	0.3333
Recombination Probability	½
Cross-over type	N-point
Mutation type	N-bit flip
Selection type	Tournament

3   Process Formulation

The formulation of the extraction process was developed in Shashi *et al.*, (2010) [1]. Through this work the model describing the yields (of specific chemical products) which are extracted from the *Gardenia Jasminoides Ellis* fruit was attained. Equipped with this, the complete MO optimization model for the extraction process of bioactive compounds from gardenia with respect to the constraints was successfully developed. The MO optimization model was established to maximize the yields which consists of three bioactive compound; crocin (f_1), geniposide (f_2) and total phenolic compounds (f_3). This process extraction MO system is given as follows:

$$Maximize \rightarrow \quad Yields\ (f_1, f_2, f_3)$$

$$subject\ to\ process\ constraints. \tag{3}$$

The objective functions (yields of each of the bioactive compound in the units of mg/g of dry powder) are modeled with respect to the constrained decision variables. This model is given as follows:

$$f_1 = 3.84 + 0.06978158X_1 + 0.0208399X_2 + 0.0417941X_3 - 0.00121613X_1^2$$
$$+ 0.000094209X_2^2 - 0.00031697X_3^2 + 0.000575415X_1X_2 - 0.000910268X_2X_3 \tag{4}$$
$$+ 0.000770328X_1X_3$$

$$f_2 = 46.2468 + 0.68221X_1 + 0.428799X_2 + 1.0X_3 - 0.016162X_1^2 - 0.0034969X_2^2 \tag{5}$$
$$- 0.0116497X_3^2 + 0.0121561X_1X_2 - 0.0095057X_2X_3 + 0.00889058X_1X_3$$

$$f_3 = -6.8 + 0.41185X_1 + 0.339428X_2 + 0.34154X_3 - 0.00538109X_1^2 - 0.00211038X_2^2 \tag{6}$$
$$- 0.00419323X_3^2 + 0.00164593X_1X_2 - 0.00125152X_2X_3 + 0.000783732X_1X_3$$

$$X_1 \in [19.5, 80.5],\ X_2 \in [27.1, 72.9],\ X_3 \in [7.1, 52.9] \tag{7}$$

4   Computational Results and Discussion

All techniques employed in this work were developed using the C++ programming language on a personal computer (PC) with an Intel dual core processor (running at 2 GHz). The solution sets which construct the Pareto frontier was generated using the two evolutionary techniques (GA and DE). The HVI was utilized to evaluate and absolutely rank these solutions in terms of dominance. In these evaluations, the nadir point which is the most non-dominated point is utilized as a reference point for the computation of the hypervolume. The nadir point used in this work is $(r_1, r_2, r_3) = (0, 0, 0)$. Along with the construction of the entire Pareto frontier, the individual solutions were classified into the best, median and worst solutions were identified. These solutions generated by both the algorithms with their respective rankings are shown in Table 3.

Table. 3. Individual Solutions Generated by the DE and GA Algorithms

Technique	Description	Best	Median	Worst
Differential Evolution (DE)	Objective Function (f_1,f_2,f_3)	(8.5265, 109.414,24.7106)	(8.2968, 106.016, 24.3831)	(7.4067, 92.6152, 20.5962)
	Decision Variable (x_1,x_2,x_3)	(54.9127, 72.8872, 40.4613)	(42.5321, 72.599, 35.4178)	(23.9009, 72.5093, 27.9649)
	HVI	23052.89406	21447.08821	14128.51344
	Weights (w_1,w_2,w_3)	(0.2,0.7,0.1)	(0.1, 0.2, 0.7)	(0.5, 0.2, 0.3)
Genetic Algorithm (DE)	Objective Function (f_1,f_2,f_3)	(8.0301, 95.1255 19.8195)	(7.7289, 91.1258, 18.606)	(7.6231, 88.7974, 17.7027)
	Decision Variable (x_1,x_2,x_3)	(29.1553, 72.7404, 9.2592)	(23.8437, 72.6443, 9.9997)	(21.8074, 72.7049, 8.37861)
	HVI	15139.5241	13104.1788	11983.0976
Particle Swarm Optimization by Shashi et al., (2010)[1]	Objective Function (f_1,f_2,f_3)	(8.52, 108.761, 24.67)	(7.7289, 91.1258, 18.606)	(7.6231, 88.7974, 17.7027)
	Decision Variable (x_1,x_2,x_3)	(50.11, 72.23, 28.72)	-	-
	HVI	22860.3006	-	-
	Weights (w_1,w_2,w_3)	(0.33, 0.33, 0.33)	-	-

The Pareto frontier was constructed approximately using 28 solutions from various weights. The weights were varied at an interval of 0.1 for the range of 0 to 1. The Pareto frontiers generated using the DE and GA techniques are presented in Figure 1 respectively:

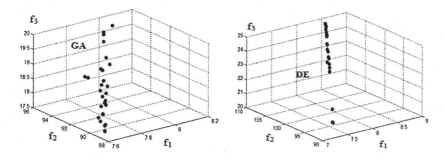

Fig. 1. Pareto frontier spread obtained using the GA and DE methods

Using the HVI, the dominance ranking of the solutions produced by each of the algorithms employed in this work is possible. This way the HVI of the entire Pareto frontier was evaluated. The execution time for constructing the entire frontier by the DE technique is higher than the GA technique by 71.719% (where DE takes 4.4414 seconds and the GA takes 18.2913 seconds). The DE technique outperforms the GA approach in terms of dominant frontier generation. It can be observed that the DE approach produces a more dominant frontier as compared to the GA approach by 54.439%. In terms of best individual solution by dominance ranking, the DE approach outperforms both the GA and the PSO[1] techniques by 52.2696% and 0.8425% respectively. Thus, the DE approach has produced a new individual optima as compared to the previous work [1].

It can be observed in Table 3 that the DE and PSO[1] techniques share a similar feature during the optimization of the individual objectives. Both methods tend to emphasize on maximizing the second objective f_2 as compared to the other objectives. In most computational optimization techniques, the factor of execution time usually conflicts with the notion of solution quality. This is because better solution quality is usually achieved when the technique spends more time searching the objective space thoroughly. However, in this case, the DE technique seems to produce highly dominant solutions at the frontier and it simultaneously manages to do this with minimal execution time as compared to the GA approach. This is may be attributed to the low algorithmic complexity of the DE method as compared to the GA and most other metaheuristic approaches. In addition, it can be observed in Figure 1 that the solutions constructing the Pareto frontier generated by the DE technique is very narrow and specific in certain regions of the objective space as compared to the one produced by the GA approach. Thus, the reason the DE technique is more effective is because it manages to limit the search to the region in the objective space with the highest fitness and not get deviated into other less fruitful search trajectories. This property gives the DE technique more capability in finding more optimal solutions in the objective space. The HVI is a reliable tool for ranking the solution spreads for a MO optimization problem. However, this metric is very dependent on the selection of the nadir point. If the correlation between the nadir point and the hypervolume value is irregular, then the HVI should be tested with multiple values of the nadir point to attain a trustworthy dominance ranking. Both evolutionary techniques employed in this work performed stable computations. Search stagnation, solution divergence or halting problems did not occur during the numerical experimentations. The solutions constructing the Pareto frontier were all within the specified ranges and thus realistically feasible. In this work, a new optimal set of solutions (see Pareto frontier in Figure 1) has been achieved using the DE approach within the NBI framework. As compared to the weighted sum approach in Shashi *et al.*, (2010) [1], the NBI framework utilized in this work seem to be more effective. This may be due to the geometrical aspect in the NBI framework which equips the metaheuristic with enhanced search capabilities as compared to the weighted sum approach which merely acts as a conventional scalarization system. Since DE is an enhanced evolutionary-type algorithm (perturbative improvement), the diversification and the rigorousness of the search is high as compared with the GA and PSO [1] methods.

5   Conclusions and Future Research Directions

Using the DE technique, a new local maximum for the individual objectives was achieved. In addition, a more dominant approximation of the Pareto frontier was constructed by using the DE method. It was observed that among evolutionary strategies, perturbative techniques such as DE are computationally inexpensive and effective in solving industrial MO optimization problems. When gauged with the HVI metric, the DE approach produced the most dominant approximate of the Pareto frontier as compared to the GA and the PSO [1] methods.

In the future, more thorough investigations using an optimization framework should be applied to this problem [17]. In addition, other algorithmic enhancement mechanisms such as (chaos-based improvements [18]) could be incorporated into the DE approach. For future works, other meta-heuristic algorithms such as hybrid approaches and PSO variant approaches (e.g. Hopfield-PSO [19] and Binary-PSO [20]) could be applied to extraction process problem. Besides the HVI, convergence, diversity and spacing metrics could also be employed to provide more insight regarding the solution properties.

References

1. Deep, S.K., Katiyar, V.K.: Extraction optimization of bioactive compounds from gardenia using particle swarm optimization. In: Proceedings of Global Conference on Power Control and Optimization (2010)
2. Li, X., Branke, J., Kieley, M.: On performance metrics and particle swarm methods for dynamic multiobjective optimization problems. In: IEEE Congress on Evolutionary Computation, pp 576–583 (2007)
3. Li, X.: Better spread and convergence: particle swarm multiobjective optimization using the maximin fitness function. In: Deb, K., Tari, Z. (eds.) GECCO 2004. LNCS, vol. 3102, pp. 117–128. Springer, Heidelberg (2004)

4. Stanikov, R.B., Matusov, J.B.: Multicriteria Optimization and Engineering. Chapman and Hall, New York (1995)
5. Auger, A., Bader, J., Brockhoff, D., Zitzler, E.: Theory of the hypervolume indicator: optimal μ-distributions and the choice of the reference point. In: Proceedings of the Tenth ACM SIGEVO Workshop on Foundations of Genetic Algorithms, pp 87–102 (2009)
6. Grosan, C.: Performance metrics for multiobjective optimization evolutionary algorithms. In: Proceedings of Conference on Applied and Industrial Mathematics (CAIM), Oradea (2003)
7. Zitzler, E., Thiele, L.: Multiobjective optimization using evolutionary algorithms - a comparative case study. In: Conference on Parallel Problem Solving from Nature (PPSN V), pp 292–301 (1998)
8. Knowles, J., Corne, D.: Properties of an Adaptive Archiving Algorithm for Storing Non-dominated Vectors. IEEE Transactions on Evolutionary Computation 7(2), 100–116 (2003)
9. Sandgren, E.: Multicriteria design optimization by goal programming. In: Adeli, H. (ed.) Advances in Design Optimization, pp. 225–265. Chapman & Hall, London (1994)
10. Igel, C., Hansen, N., Roth, S.: Covariance Matrix Adaptation for Multi-objective Optimization. Evolutionary Computation 15(1), 1–28 (2007)
11. Storn, R., Price, K.V.: Differential evolution – a simple and efficient adaptive scheme for global optimization over continuous spaces, ICSI, Technical Report TR-95-012 (1995)
12. Holland, J.H.: Adaptation in Natural and Artificial Systems: An Introductory Analysis with Applications to Biology, Control and Artificial Intelligence. MIT Press, USA (1992)
13. Das, I., Dennis, J.E.: Normal-boundary intersection: A new method for generating the Pareto surface in nonlinear multicriteria optimization problems. SIAM Journal of Optimization 8(3), 631–657 (1998)
14. Babu, B.V., Munawar, S.A.: Differential evolution for the optimal design of heat exchangers. In: Proceedings of All-India seminar on Chemical Engineering Progress on Resource Development: A Vision 2010 and Beyond, Bhuvaneshwar (2000)
15. Babu, B.V., Singh, R.P.: Synthesis & optimization of heat integrated distillation systems using differential evolution. In: Proceedings of All- India seminar on Chemical Engineering Progress on Resource Development: A Vision 2010 and Beyond, Bhuvaneshwar (2000)
16. Angira, R., Babu, B.V.: Optimization of non-linear chemical processes using modified differential evolution (MDE). In: Proceedings of the 2nd Indian International Conference on Artificial Intelligence, Pune, India, pp. 911–923 (2005)
17. Ganesan, T., Elamvazuthi, I., Shaari, K.Z.K., Vasant, P.: An Algorithmic Framework for Multiobjective Optimization. The Scientific World Journal 2013 (2013)
18. Ganesan, T., Elamvazuthi, I., Shaari, K.Z.K., Vasant, P.: Multiobjective Optimization of Green Sand Mould System Using Chaotic Differential Evolution. In: Gavrilova, M.L., Tan, C., Abraham, A. (eds.) Transactions on Computational Science XXI. LNCS, vol. 8160, pp. 145–163. Springer, Heidelberg (2013)
19. Elamvazuthi, I., Ganesan, T., Vasant, P.: A comparative study of HNN and Hybrid HNN-PSO techniques in the optimization of distributed generation (DG) power systems. In: 2011 International Conference on Advanced Computer Science and Information System (ICACSIS), pp 195–200. IEEE (2011)
20. Mirjalili, S., Lewis, A.: S-shaped versus V-shaped transfer functions for binary Particle Swarm Optimization. Swarm and Evolutionary Computation 9, 1–14 (2013)

Hybrid Swarm Intelligence-Based Optimization for Charging Plug-in Hybrid Electric Vehicle

Imran Rahman[1(✉)], Pandian Vasant[1], Balbir Singh Mahinder Singh[1], and M. Abdullah-Al-Wadud[2]

[1] Department of Fundamental and Applied Sciences,
Universiti Teknologi PETRONAS, Tronoh, Malaysia
{imran.iutoic,pvasant}@gmail.com, balbir@petronas.com.my
[2] Department of Software Engineering,
College of Computer and Information Sciences, King Saud University, Riyadh, KSA
mwadud@ksu.edu.sa

Abstract. Plug-in hybrid electric vehicle (PHEV) has the potential to facilitate the energy and environmental aspects of personal transportation, but face a hurdle of access to charging system. The charging infrastructure has its own complexities when it is compared with petrol stations because of the involvement of the different charging alternatives. As a result, the topic related to optimization of Plug-in hybrid electric vehicle charging infrastructure has attracted the attention of researchers from different communities in the past few years. Recently introduced smart grid technology has brought new challenges and opportunities for the development of electric vehicle charging facilities. This paper presents Hybrid particle swarm optimization Gravitational Search Algorithm (PSOGSA)-based approach for state-of-charge (SoC) maximization of plug-in hybrid electric vehicles hence optimize the overall smart charging.

Keywords: Smart charging · State-of-charge · Plug-in hybrid electric vehicle · PSOGSA · Swarm intelligence

1 Introduction

The vehicular network recently accounts for around 25% of CO_2 emissions and over 55% of oil consumption around the world [1]. Carbon dioxide is the primary greenhouse gas emitted through human activities like combustion of fossil fuels (coal, natural gas, and oil) for energy and transportation. Several researchers have proved that a great amount of reductions in greenhouse gas emissions and the increasing dependence on oil could be accomplished by electrification of transport sector [2]. Indeed, the adoption of hybrid electric vehicles (HEVs) has brought significant market success over the past decade. Vehicles can be classified into three groups: internal combustion engine vehicles (ICEV), hybrid electric vehicles (HEV) and all- electric vehicles (AEV) [3]. Plug-in hybrid electric vehicles (PHEVs) which is very recently introduced promise to boost up the overall fuel efficiency by holding a higher capacity battery system, which can be directly charged from traditional power grid system,

N.T. Nguyen et al. (Eds.): ACIIDS 2015, Part II, LNAI 9012, pp. 22–30, 2015.
DOI: 10.1007/978-3-319-15705-4_3

that helps the vehicles to operate continuously in "all-electric-range" (AER) All-electric vehicles or AEV is a vehicle using electric power as only sources to move the vehicle [4]. Plug-in hybrid electric vehicles with a connection to the smart grid can possess all of these strategies. Hence, the widely extended adoption of PHEVs might play a significant role in the alternative energy integration into traditional grid systems [5]. There is a need of efficient mechanisms and algorithms for smart grid technologies in order to solve highly heterogeneous problems like energy management, cost reduction, efficient charging infrastructure etc. with different objectives and system constraints [6].

According to a statistics of Electric Power Research Institute (EPRI), about 62% of the entire United States (US) vehicle will comprise of PHEVs within the year 2050 [7]. Large numbers of PHEVs have the capability to threaten the stability of the power system. For example, in order to avoid interruption when several thousand PHEVs are introduced into the system over a short period of time, the load on the power grid will need to be managed very carefully. One of the main targets is to facilitate the proper interaction between the power grid and the PHEV. For the maximization of customer satisfaction and minimization of burdens on the grid, a complicated control mechanism will need to be addressed in order to govern multiple battery loads from a numbers of PHEVs appropriately [8]. The total demand pattern will also have an important impact on the electricity industry due to differences in the needs of the PHEVs parked in the deck at certain time [9]. Proper management can ensure strain minimization of the grid and enhance the transmission and generation of electric power supply. The control of PHEV charging depending on the locations can be classified into two groups; household charging and public charging. The proposed optimization focuses on the public charging station for plug-in vehicles because most of PHEV charging is expected to take place in public charging locations [10].

Charging stations are needed to be built at workplaces, markets/shopping malls and home. In [11], authors proposed the necessity of building new smart charging station with effective communication among utilities along with sub-station control infrastructure in view of grid stability and proper energy utilization. Furthermore, assortment of charging stations with respect to charging characteristics of different PHEVs traffic mobility characteristics, sizeable energy storage, cost minimization; Quality of Services (QoS) and optimal power of intelligent charging station are underway [12].

One of the important constraints for accurate charging is State-of-Charge (SoC). Charging algorithm can accurately be managed by the precise State of charge estimation. The performance of PHEV depends upon proper utilization of electric power which is solely affected by the battery state-of-charge (SoC). In Plug-in hybrid electric vehicles (PHEVs), a key parameter is the state-of-charge (SoC) of the battery as it is a measure of the amount of electrical energy stored in it. It is analogous to fuel gauge on a conventional internal combustion (IC) car [13]. There is a need of in-depth study on maximization of average SoC in order to facilitate intelligent energy allocation for PHEVs in a charging station. Hybrid PSOGSA was developed by Seyedali Mirjalili [14] at soft computing research lab of Universiti Teknologi Malaysia (UTM) in 2010 in order to integrate the ability of exploitation in PSO with the ability of exploration in GSA.

PSOGSA-based optimization has already been used by the researchers for economic load dispatch [15], optimal static state estimation [16], dual channel speech enhancement [17], training feed-forward neural networks [18] and multi-distributed generation planning [19]. Specifically, we are investigating the use of the Hybrid particle swarm optimization Gravitational Search Algorithm (PSOGSA) method for developing real-time and large-scale optimizations for allocating power.

The remainder of this paper is organized as follows: Next section will describe the specific problem that we are trying to solve. We will provide the optimization objective and constraints, flowchart of PSOGSA algorithm as well as describe how the algorithm works for our optimization problems. The simulation results and analysis are presented then with an extensive analysis. Finally, conclusions and future directions are drawn.

2   Problem Formulation

The idea behind smart charging is to charge the vehicle when it is most beneficial, which could be when electricity price, demand is lowest, when there is excess capacity [20].

Suppose, there is a charging station with the capacity of total power P. Total N numbers of PHEVs need to serve in a day (24 hours). The proposed system should allow PHEVs to leave the charging station before their expected leaving time for making the system more effective. It is worth to mention that, each PHEV is regarded to be plugged-in to the charging station once. The main aim is to allocate power intelligently for each PHEV coming to the charging station. The State-of-Charge is the main parameter which needs to be maximized in order to allocate power effectively. For this, the objective function considered in this paper is the maximization of average SoC and thus allocate energy for PHEVs at the next time step. The constraints considered are: charging time, present SoC and price of the energy.

The objective function is defined as:

$$\max J\left(k\right) = \sum_i w_i\left(k\right) SoC_i\left(k+1\right) \tag{1}$$

$$w_i\left(k\right) = f\left(C_{r,i}\left(k\right), T_{r,i}\left(k\right), D_i\left(k\right)\right) \tag{2}$$

$$C_{r,i}\left(k\right) = \left(1 - SoC_i\left(k\right)\right) * C_i \tag{3}$$

where $C_{r,i}(k)$ is the battery capacity (remaining) needed to be filled for i no. of PHEV at time step k; C_i is the battery capacity (rated) of the i no. of PHEV; remaining time for charging a particular PHEV at time step k is expressed as $T_{r,i}(k)$; the price difference between the real-time energy price and the price that a specific customer at the i no. of PHEV charger is willing to pay at time step k is presented by $D_i(k)$; $w_i(k)$ is the charging weighting term of the i no. of PHEV at time step k (a function of charging time, present SoC and price of the energy); $SoC_i(k + 1)$ is the state of charge of the i no. of PHEV at time step $k + 1$.

Here, the weighting term indicates a bonus proportional to the attributes of a specific PHEV. For example, if a PHEV has a lower initial SoC and less charging time (remaining), but the driver is eager to pay a higher price, the system will provide more power to this particular PHEV battery charger:

$$w_i(k) \alpha \left[Cap_{r,i}(k) + D_i(k) + \frac{1}{T_{r,i}}(k) \right] \tag{4}$$

The charging current is also assumed to be constant over Δt.

$$\left[SoC_i(k+1) - SoC_i(k) \right].Cap_i = Q_i = I_i(k)\Delta t \tag{5}$$

$$SoC_i(k+1) = SoC_i(k) + I_i(k)\Delta t / Cap_i \tag{6}$$

Where the sample time Δt is defined by the charging station operators, and $I_i(k)$ is the charging current over Δt.

The battery model is regarded as a capacitor circuit, where C_i is the capacitance of battery (Farad). The model is defined as:

$$C_i \cdot \frac{dV_i}{dt} = I_i \tag{7}$$

Therefore, over a small time interval, one can assume the change of voltage to be linear,

$$C_i \cdot \left[V_i(k+1) - V_i(k) \right] / \Delta t = I_i \tag{8}$$

$$V_i(k+1) - V_i(k) = I_i \Delta t / C_i \tag{9}$$

As the decision variable used here is the allocated power to the PHEVs, by replacing $I_i(k)$ with $P_i(k)$ the objective function finally becomes:

$$J(k) = \sum w_i \cdot \left[SoC_i(k) + \frac{2P_i(k)\Delta t}{0.5.C_i \cdot \left[\sqrt{\frac{2P_i(k)\Delta t}{C_i} + V^2_i(k)} + V_i(k) \right]} \right] \tag{10}$$

Power obtained from the utility ($P_{utility}$) and the maximum power ($P_{i,max}$) absorbed by a specific PHEV are the primary energy constraints being considered in this paper.

The overall charging efficiency of a particular charging infrastructure is described by η. From the system point of view, charging efficiency is supposed to be constant at any given time step. Maximum battery SoC limit for the i no. of PHEV is $SoC_{i,max}$. When SoC_i reaches the values close to $SoC_{i,max}$, the i no. of battery charger shifts to a standby mode. The state of charge ramp rate is confined within limits by the constraint ΔSoC_{max}. The overall control system is changed the state when i) system utility data updates; ii) a new PHEV is plugged-in; iii) time periodΔt has periodically passed.

3   The Hybrid PSOGSA Algorithm

In this paper, a new hybrid population-based algorithm (PSOGSA) [14] is proposed with the combination of Particle Swarm Optimization (PSO) and Gravitational Search Algorithm (GSA). The basic idea is to fit in the exploitation ability in PSO with the

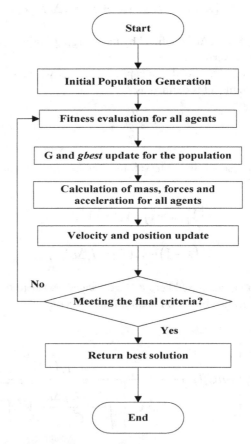

Fig. 1. Hybrid PSOGSA Algorithm Flowchart

exploration ability in GSA to synthesize both algorithms' strength. The basic idea of PSOGSA is to combine the ability of social thinking (*gbest*) in PSO with the local search capability of GSA. In order to combine these two algorithms, velocity update is proposed as

$$v_i(t+1) = w \times v_i(t) + \alpha' \times rand \times ac_i(t) + \beta' \times rand \times (gbest - x_i(t)) \quad (11)$$

where $v_i(t)$ is the velocity of agent i at iteration t, w is a weighting factor, *rand* is a random number between 0 and 1, $ac_i(t)$ is the acceleration of agent at iteration t, and *gbest* is the best solution so far. The position of the particle $x_i(t + 1)$ in each iteration is updated using the equation

$$x_i(t+1) = x_i(t) + v_i(t+1) \quad (12)$$

The flowchart of hybrid PSOGSA method is shown in fig 1.

4   Simulation Results and Analysis

The Hybrid PSOGSA algorithm were applied to find out best fitness of the objective function. All the simulations were run on a Core™ i5-3470M CPU@ 3.20 GHz processor, 4.00 GB RAM and MATLAB R2013a.

The parameter settings for Hybrid PSOGSA are demonstrated in Table 1. The size of swarm is set to the standard value which is 100 and values for C1 and C2 were taken as 0.5 and 1.5 [21]. Other parameters are set from the experiences of previous research articles [22, 23, 24].

Table 1. PSOGSA parameter settings

Parameters	Values
Size of the swarm	100
Maximum Iteration	100
PSO parameter, C_1	0.5
PSO parameter, C_2	1.5
Gravitational Constant, G_0	1
GSA Constant parameter, α	23
Number of runs	20

Table 2 summarizes the simulation results for 50 and 100 plug-in hybrid electric vehicles (PHEVs) respectively for finding the maximum fitness value of objective function J (k). In order to evaluate the performance and show the efficiency and superiority of the proposed algorithm, we ran each scenario total 20 times.

Table 2. Average best fitness and Computational time for PSOGSA

Number of PHEVs	Average Best Fitness	Average Computational Time
50	144.838	4.248 Sec.
100	183.094	7.877 Sec.

The average best fitness increases when number of vehicles are more in number from 144.838 to 183.094. Moreover, the computational time increases for 100 PHEVs. The average computational time for 50 PHEVs is 4.248 seconds while for 100 PHEVs, it becomes 7.877. Computational complexity of hybrid algorithm can be controlled by rigorous attempts of parameter tuning which will be our further research concern.

Fig. 2 and Fig. 3 shows the convergence behavior (iteration vs. Best fitness) for both 50 and 100 numbers of PHEVs. From the figures, it is clear that, the convergence occurs at the same pattern hence prove the stability of this hybrid optimization. It can be apparently seen that although the algorithm has been set to run for maximum 100 iterations, the fitness value converges before 10 iterations and become stable. So, there is an early convergence which may cause the fitness function to trap into local minima. This can be avoided by increasing the size of swarm hence the computational time will also be increased as well. As a result, a trade-off should be taken into consideration between the proper convergence and computational time.

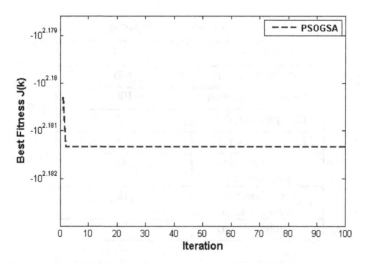

Fig. 2. Best Fitness vs. Iteration (50 PHEVs)

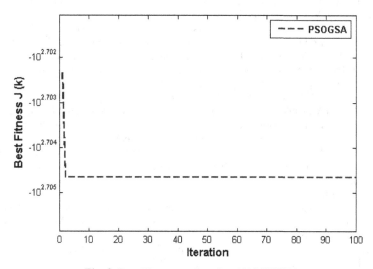

Fig. 3. Best Fitness vs. Iteration (100 PHEVs)

5   Conclusion and Future Works

In this paper, Hybrid particle swarm optimization Gravitational Search Algorithm (PSOGSA)-based optimization was performed in order to optimally allocate power to each of the PHEVs entering into the charging station. A sophisticated controller will need to be designed in order to allocate power to PHEVs appropriately. For this wake, the applied algorithm in this paper is a step towards real-life implementation of such controller for PHEV Charging Infrastructures. Here, two (02) different numbers of PHEVs were considered for MATLAB Simulation. The researchers should try to develop efficient control mechanism for charging infrastructure in order to facilitate upcoming PHEVs penetration in highways. In future, more vehicles should be considered for intelligent power allocation strategy as well as should be applied to ensure higher fitness value and low computational time.

Acknowledgement. The authors would like to thank Universiti Teknologi PETRONAS (UTP) for supporting the research under UTP Graduate Assistantship (GA) scheme.

References

1. Transport, Energy and C02-Moving Towards Sustainability, Paris (2009). http://www.iea.org/newsroomandevents/pressreleases/2009/october/name,20274,en.html
2. Holtz-Eakin, D., Selden, T.M.: Stoking the fires? CO2 emissions and economic growth. Journal of public economics **57**(1), 85–101 (1995)
3. Tie, S.F., Tan, C.W.: A review of energy sources and energy management system in electric vehicles. Renewable and Sustainable Energy Reviews **20**, 82–102 (2013)
4. Environmental assessment of plug-in hybrid electric vehicles. Volume 1: Nationwide greenhouse gas emissions, Electric Power Research. Institute (EPRI), Palo Alto, CA, Tech. Rep. 1015325 (2007)

5. Lund, H., Kempton, W.: Integration of renewable energy into the transport and electricity sectors through V2G. Energy policy **36**, 3578–3587 (2008)
6. Hota, A.R., Juvvanapudi, M., Bajpai, P.: Issues and solution approaches in PHEV integration to the smart grid. Renewable and Sustainable Energy Reviews **30**, 217–229 (2014)
7. Soares, J., Sousa, T., Morais, H., Vale, Z., Canizes, B., Silva, A.: Application-Specific Modified Particle Swarm Optimization for energy resource scheduling considering vehicle-to-grid. Applied Soft Computing **13**(11), 4264–4280 (2013)
8. Su, W., Chow, M.-Y.: Computational intelligence-based energy management for a large-scale PHEV/PEV enabled municipal parking deck. Applied Energy **96**, 171–182 (2012)
9. Su, W., Chow, M.-Y.: Performance evaluation of a PHEV parking station using particle swarm optimization. In: 2011 IEEE Power and Energy Society General Meeting, pp. 1-6 (2011)
10. Su, W., Chow, M.-Y.: Performance evaluation of an EDA-based large-scale plug-in hybrid electric vehicle charging algorithm. IEEE Transactions on Smart Grid **3**, 308–315 (2012)
11. Boyle, G.: Renewable Electricity and the Grid: The Challenge of Variability. Earth scan Publications Ltd (2007)
12. Hess, A., Francesco, M., Reinhardt, M., Casetti, C.: Optimal deployment of charging stations for electric vehicular networks. In: Proceedings of the First Workshop on Urban Networking, pp. 1-6. ACM, New York (2012)
13. Chang, W.-Y.: The State of Charge Estimating Methods for Battery: A Review. ISRN Applied Mathematics, **2013**, Article ID 953792 (2013)
14. Mirjalili, S., Hashim, S.Z.M.: A new hybrid PSOGSA algorithm for function optimization. In: IEEE International Conference on Computer and Information Application (ICCIA), pp. 374-377 (2010)
15. Dubey, H.M., Pandit, M., Panigrahi, B., Udgir, M.: Economic Load Dispatch by Hybrid Swarm Intelligence Based Gravitational Search Algorithm. International Journal of Intelligent Systems & Applications **5** (2013)
16. Mallick, S., Ghoshal, S.P., Acharjee, P., Thakur, S.S.: Optimal static state estimation using improved particle swarm optimization and gravitational search algorithm. International Journal of Electrical Power & Energy Systems **52**, 254–265 (2013)
17. Kunche, P., Rao, G.S.B., Reddy, K.V.V.S., Maheswari, R.U.: A new approach to dual channel speech enhancement based on hybrid PSOGSA. International Journal of Speech Technology 1-12 (2014)
18. Mirjalili, S., Mohd Hashim, S.Z., Moradian Sardroudi, H.: Training feedforward neural networks using hybrid particle swarm optimization and gravitational search algorithm. Applied Mathematics and Computation **218**, 11125–11137 (2012)
19. Tan, W.S., Hassan, M.Y., Rahman, H.A., Abdullah, M.P., Hussin, F.: Multi-distributed generation planning using hybrid particle swarm optimisation-gravitational search algorithm including voltage rise issue. IET Generation, Transmission & Distribution **7**, 929–942 (2013)
20. Mayfield, D.: Site Design for Electric Vehicle Charging Stations, ver.1.0, Sustainable Transportation Strategies (2012)
21. Ganesan, T., Vasant, P., Elamvazuthy, I.: A hybrid PSO approach for solving non-convex optimization problems. Archives of Control Sciences **22**(1), 87–105 (2012)
22. Ganesan, T., Elamvazuthi, I., Ku Shaari, K.Z., Vasant, P.: Swarm intelligence and gravitational search algorithm for multi-objective optimization of synthesis gas production. Applied Energy **103**, 368–374 (2013)
23. Vasant, P.: Hybrid Evolutionary Optimization Algorithms: A Case Study in Manufacturing Industry. Smart Manufacturing Innovation and Transformation: Interconnection and Intelligence **59** (2014)
24. Rahman, I., Vasant, P.M., Singh, B.S.M., Abdullah-Al-Wadud, M.: Intelligent energy allocation strategy for PHEV charging station using gravitational search algorithm. AIP Conference Proceedings **1621**, 52–59 (2014)

A Genetic Algorithm with Grouping Selection and Searching Operators for the Orienteering Problem

Pawel Zabielski$^{(\boxtimes)}$, Joanna Karbowska-Chilinska$^{(\boxtimes)}$, Jolanta Koszelew, and Krzysztof Ostrowski

Faculty of Computer Science, Bialystok University of Technology, Białystok, Poland
p.zabielski@pb.edu.pl

Abstract. In the Orienteering Problem (OP), a set of linked vertices, each with a score, is given. The objective is to find a route, limited in length, over a subset of vertices that maximises the collective score of the visited vertices. In this paper, we present a new, efficient genetic algorithm (nGA) that solves the OP. We use a special grouping during selection, which results in better-adapted routes in the population. Furthermore, we apply a searching crossover to each generation, which uses the common vertices between distinct routes in the population; we also apply a searching mutation. Computer experiments on the nGA are conducted on popular data sets. In some cases, the nGA yields better results than well-known heuristics.

Keywords: Genetic algorithm · Orienteering problem · Grouping selection · Searching operators

1 Introduction

The OP can be modelled as a weighted complete graph problem. Let G be a graph with n vertices, where each vertex has a profit $p_i \geq 0$, and each edge between vertices i and j has a cost t_{ij}. The objective is to determine a path from a given starting point s to a given ending point e, that maximises the total profit. In addition, the total cost of the edges on this path must be less than the constraint t_{max}, and any vertex on the path can only be visited once. The mathematical formulation of the OP could be found in [20].

In the literature, the OP is also known as the Traveling Salesman Problem with Profits [5], the Selective Traveling Salesperson Problem [9], [12] or the Maximum Collection Problem [11].

In addition to numerous applications in logistics [15] and planning production [2], there is a rapid growth of OP applications in tourism and culture. For instance, the OP models the problem of generating the most rewarding route of museum visits, as measured by the preferences expressed by museum visitors [8]. Variants of the OP, such as the Team Orienteering Problem (TOP) [1], and the Orienteering Problem with Time Windows (OPTW) [18], are the basis of

© Springer International Publishing Switzerland 2015
N.T. Nguyen et al. (Eds.): ACIIDS 2015, Part II, LNAI 9012, pp. 31–40, 2015.
DOI: 10.1007/978-3-319-15705-4_4

electronic tourist guides. These guides generate routes that maximise tourist satisfaction, taking into account, for example, personal preferences, time constraints, available budget, and the opening and closing hours of locations [19].

OP is NP-hard [7]. Exact solutions (branch-and-bound and branch-and-cut [4], [7] methods) are time-consuming and not applicable in practice. Therefore, in many systems, meta-heuristic approaches are usually used: genetic algorithms [16], local search methods [3], [17], tabu search [14] and the ant colony optimisation approach [13].

This article presents the genetic algorithm nGA, which is the improved version of our previous algorithm (GA), described in [10]. To obtain well-diversified individuals, we propose modifications to the previous algorithm in the selection, mutation and crossover phases. The numerous experiments are carried out on the benchmark instances and in some cases, the nGA yields better results than well-known heuristics.

The paper is organised as follows. In Section 2, we describe the structure of the nGA in detail. The results of computational experiments run on benchmark datasets are discussed in Section 3. Conclusions are drawn and further work is suggested in Section 4.

2   Description of nGA

The method presented in this paper is an improved version of the GA algorithm described in [10]. The main differences between the nGA and the GA are clearly visible in the selection, crossover and mutation steps. First, the tournament selection step for the nGA divides the population into k parts, and selects the best individuals, not in the whole population but separately for each part. This approach results in individuals after selection that are more diverse and genetic operators (crossover and mutation) that are more effective than those in the GA. Second, the crossover operator in the nGA takes into account common vertices between paths, not just a single, randomly selected pair of vertices as in the GA. Third, the mutation phase is far more effective in the nGA than in the GA because multiple attempts are made to insert or delete an existing point in the best position.

During improving population the vertices can be repeated on the path as the result of the exchange of paths fragments in the crossover. However due to a deletion mutation, in the final population duplicated vertices are very rare. If duplicated vertices are in the final solution profits of the repeated vertices are not taken into account.

Route determination in the nGA is a multi-stage process that is described in detail in the following subsections.

2.1   Initialisation

First, an initial population of P_{size} solutions is generated. Tours are encoded into chromosomes as a sequence of vertices, which is the most natural way of adapting a genetic algorithm to the OP. During the initialization step, we simply add to the generating tour, new random, adjacent, unvisited vertices v_1, v_2, ..., v_j to the tour until $t_{sv_1} + t_{v_1v_2} + ... + t_{v_je}$ does not exceed t_{max}. At every step we remove the previously visited vertex from the set of unvisited vertices.

2.2 Evaluation

We next calculate the value of the fitness function F for each tour (chromosome) in P. The fitness function should estimate the quality of individuals, according to the sum of profits $TotalProfit$ and the length of the tour $TravelLength$. Although in the process of improving population each vertex can be visited more than once in the tour, the $TotalProfit$ is only increased when the vertex is visited for the first time. In the nGA the fitness function is equal to $TotalProfit^3/TravelLength$, which empirically gives the best results, such that the final population contains a tour with the highest total profit.

2.3 Improving the Population

After generating the initial population, the nGA improves the population through the iterative application of selection, crossover and mutation steps. The algorithm stops after n_g generations or earlier if have not the profit improvements in the last 100 generations. The result of the nGA is the path with the highest total profit from the final generation. If the final path contains repeated vertices they are removed.

Tournament Grouping Selection. P_{size} individuals are divided into k groups, where k is the parameter of nGA. The first P_{size}/k tournaments are then carried out, not on the overall population but on individuals 1 to P_{size}/k. Next, P_{size}/k tournaments are carried out on individuals $P_{size}/k+1$ to $2P_{size}/k$, etc. For each of k groups, we randomly select t_{size} individuals from the each group, and the best chromosome in this subgroup (with the highest value of $TotalProfit^3/TravelLength$) is copied to the next population. Then, the whole tournament group is returned to the old population. Finally, after P_{size} repetitions of this step, a new population exists.

Searching Crossover. In the crossover process, two random individuals are initially selected as parents from all population. Then, we determine the set of intersections, i.e., the genes that are common to both parents. If there are no common genes, crossover does not occur, and the parents remain unchanged. If there are common genes, we generate children by crossing the parents between successive pairs in the set of intersections. The viable children has a higher fitness than either of its parents and does not exceed t_{max} limit. If there are many crossover options we choose the one which maximises fitness of the fitter child. If one of the children does not preserve t_{max} constraint, the fitter parent from the new population replaces it. If both children do not preserve this constraint for each pair of successive crossing points, the parents replace them in the new population (no changes applied).

Searching Mutation. After the selection and crossover phases, the population undergoes a heuristic mutation. First, a random individual is selected for mutation. Then, 2-opt algorithm, standard for Traveling Salesman Problem is run on the selected individual. Next, the improving individual is mutated n_m times. There are two possible types of mutation: inserting a new gene or deleting an existing gene (the probability of each is 0.5).

Inserting mutation: The standard version of an insertion mutation, in which the new gene is randomly inserted into an individual, does not adequately improve the fitness of the individual. Therefore, we consider all possible insertions into the chromosome of a new available gene u that is not already present. The location with the highest value of $(p_u)^2/TravelLengthIncrease_u$ is chosen for the insertion, where $TravelLengthIncrease_u$ is the increased travel length of the chromosome after u is inserted and p_u denotes a profit of the vertex u . If $TravelLengthIncrease_u$ is less than 1, we must consider the highest value of $(p_u)^2$. For example, in Figure 1, there are four ways to insert a new gene into the chromosome (1, 4, 3, 5, 1) without exceeding t_{max}. In this case, the insertion of gene $u=2$ between genes 1 and 4 gives the highest value of $(p_u)^2/TravelLengthIncrease_u$. The gene u is not inserted when its addition causes exceeding t_{max} limit.

Deletion mutation: In the deletion mutation, first only genes that appear in the chromosome more than once are considered for deletion, except for the first and last genes. If any candidates exist we choose the gene whose deletion shortens the travel length the most. If no such candidate and the travel length exceeds $0.9 \cdot t_{max}$ we remove the gene u which minimises $(p_u)^2/TravelLengthDecrease_u$, where $TravelLengthDecrease_u$ is decreased travel length of the chromosome

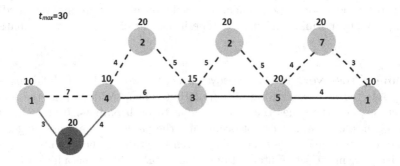

Fig. 1. Insertion mutation (with $t_{max} = 30$), where the inserted gene is first vertex with number 2

after u is removed. For example, in Figure 2, we see only one candidate suitable for deletion: gene 5 (either between genes 1 and 3 or between genes 4 and 1). If we delete gene 5 from the first position, then the travel length shortens by 4, but if we delete gene 5 from the second position, then the travel length shortens by 2. Thus, we delete the gene 5 from between genes 1 and 3.

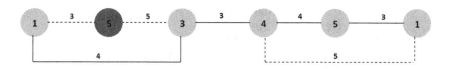

Fig. 2. Deletion mutation (with $t_{max} = 30$), where the removed gene is the first vertex with number 5

3   Experiments and Results

The algorithm was implemented in C++ and run on an Intel Core i7, 1.73 GHz CPU (turbo boost to 2.93 GHz). The computational experiments were conducted on the popular datasets Tsiligirides [15], Fischetti [4] and Chao [3], which consist of 29, 7 and 38 test instances, respectively.

The nGA parameters were determined experimentally, and Table 1 lists their values. The nGA was run 15 times, and the best score and the total time of the 15 runs were recorded. The experiment was repeated 31 times. A median of the observed best scores and the total time of 15 runs is presented in Tables 2-6.

The nGA results are compared in Tables 2-6 to the best known solutions from the literature: the stochastic algorithm of [15] (TS), the heuristics method of [3] (CH), the branch-and-cuts exact solution of [4] (OPT), the guided local search method of [17] (GLS), and the genetic algorithm of [16] (TGA). In these tables, for all cases, gaps are given as a percentage. The gap is the difference between the score obtained by the previous method and the score obtained by the nGA, divided by the score obtained by the previous method. Minus before the gap value denotes that the result nGA exceeds result of the comparable method by

Table 1. nGA parameters

name	explanation	value
P_{size}	initial population size	100 for $n < 100$, 200 otherwise
t_{size}	number of individuals chosen in a tournament selection	2
k	number of groups in a tournament selection	10
n_g	maximum number of nGA generations	300 for $n < 100$, 600 otherwise
n_m	number of mutations on a selected individual	10

Table 2. Comparison of results for the Tsiligirides test problems ($n=32$)

t_{max}	our solution		other heuristics				%gap nGA with			
	nGA	time (s)	TS	CH	TGA	GLS	TS	CH	TGA	GLS
5	10	0.1	10	10	10	10				
10	15	0.1	15	15	15	15				
15	45	0.2	45	45	45	45				
20	65	0.2	65	65	65	55				-18.18
25	90	0.2	90	90	90	90				
30	110	0.2	110	110	110	80				-37.50
35	135	0.2	135	135	135	135				
40	155	0.2	150	155	155	145	-3.33			-6.90
46	175	0.2	170	175	175	175	-2.94			
50	190	0.3	185	190	190	180	-2.70			-5.56
55	205	0.3	195	205	205	200	-5.13			-2.50
60	225	0.3	220	220	225	220	-2.27	-2.27		-2.27
65	240	0.3	235	240	240	240	-2.13			
70	260	0.3	255	260	260	260	-1.96			
73	265	0.3	260	265	265	265	-1.92			
75	270	0.3	265	275	270	270	-1.89	-1.82		
80	280	0.4	270	280	280	280	-3.70			
85	285	0.3	280	285	285	285	-1.79			
Avg.	167.8	0.3	164.2	167.8	167.8	163.9	-1.65	-0.03	0.00	-4.05

Table 3. Comparison of results for the Tsiligirides test problems ($n=21$)

t_{max}	our solution		other heuristics				%gap nGA with			
	nGA	time (s)	TS	CH	TGA	GLS	TS	CH	TGA	GLS
15	120	0.2	120	120	120	120				
20	200	0.2	190	200	200	200	-5.26			
23	210	0.2	205	210	210	210	-2.44			
25	230	0.2	230	230	230	230				
27	230	0.2	230	230	230	220				-4.55
30	265	0.2	250	265	265	260	-6.00			-1.92
32	300	0.2	275	300	300	300	-9.09			
35	320	0.2	315	320	320	305	-1.59			-4.92
38	360	0.2	355	360	360	360	-1.41			
40	390	0.2	395	395	395	380	1.27	1.27	1.27	-2.63
45	450	0.2	430	450	450	450	-4.65			
Avg.	279.5	0.2	272.3	259.1	279.8	275.9	-2.65	0.12	0.12	-1.27

a calculated value (expressed as a percentage). An empty cell denotes that the gap equals 0. Tests for which our method gave better results than the comparison methods are marked in bold. The last rows in Tables 2-6 show the average values in comparison to the other methods.

Table 4. Comparison of results for the Chao test problems (n=66)

t_{max}	our solution		other heuristics		% gap nGA with	
	nGA	time (s)	CH	GLS	CH	GLS
15	120	0.2	120	120		
20	**205**	**0.2**	**195**	**175**	**-4.88**	**-14.63**
25	290	0.3	290	290		
30	400	0.4	400	400		
35	465	0.4	460	465	-1.08	
40	575	0.5	575	575		
45	650	0.5	650	640		-1.54
50	730	0.6	730	710		-2.74
55	825	0.6	825	825		
60	915	0.6	915	905		-1.09
65	980	0.7	980	930		-5.10
70	1070	0.7	1070	1070		
75	1140	0.7	1140	1140		
80	1215	0.7	1215	1195		-1.65
85	1270	0.8	1270	1265		-0.39
90	1340	0.8	1340	1300		-2.99
95	**1395**	**0.8**	**1380**	**1385**	**-1.08**	**-0.72**
100	**1465**	**0.8**	**1435**	**1445**	**-2.05**	**-1.37**
105	1520	0.9	1510	1505	-0.66	-0.99
110	1560	0.9	1550	1560	-0.64	
115	1595	0.9	1595	1580		-0.94
120	1635	0.9	1635	1635		
125	1665	0.9	1655	1665	-0.6	
130	1680	0.9	1680	1680		
Avg.	1029.38	0.66	1025.63	1019.17	- 0.46	-1.42

It should be noted that the results of the nGA scored higher than the comparison methods in 6 of the 74 test problems. For the 29 Tsiligirides test problems listed in Tables 2 - 3, the nGA produced scores that exceeded those of the TS on 18 problems, exceeded those of the GLS on 10 problems, and exceeded those of the CH method on 2 problems. The comparison of the results yielded by the nGA to those yielded by the CH and GLS methods for the Chao test problems are presented in Table 4 (the locations of the points in the test problems [3] take on a square or a diamond shape, respectively). The experimental results showed that on these benchmarks, the nGA gave better solutions than the comparison heuristics in less than 0.9 s, in 5 of the 38 problems (marked in bold in Tables 4-5). In Table 6, the comparison of the results from the nGA to those from the GLS and OPT methods are shown for the selected instances of Fischetti data sets. Notably, the nGA outperformed the GLS method for the problems containing more than 52 locations (the average gap for these instances is above 8%). From one of the Fischetti test problems (n=101, t_{max}=3955) the nGA algorithm determines a better solution than the result published by [4]. It should be noted that [19] confirmed that Fischetti solutions are not always optimal because of rounding the distances to the nearest integer. The optimal route generated by our solution is published in Table 7.

Table 5. Comparison of results for the Chao test problems ($n=64$)

t_{max}	our solution		other heuristics		% gap nGA with	
	nGA	time (s)	CH	GLS	CH	GLS
15	96	0.2	96	96		
20	294	0.2	294	294		
25	390	0.3	390	390		
30	474	0.4	474	474		
35	**576**	**0.5**	**570**	**552**	**-1.05**	**-4.35**
40	714	0.5	714	702		-1.71
45	816	0.6	816	780		-4.62
50	900	0.6	900	888		-1.35
55	984	0.7	984	972		-1.23
60	1062	0.7	1044	1062	-1.72	
65	1116	0.7	1116	1110		-0.54
70	1188	0.8	1176	1188	1.02	
75	1236	0.8	1224	1236	-0.98	
80	**1278**	**0.8**	**1272**	**1260**	**-0.47**	**-1.43**
Avg.	794.6	0.6	790.7	786.0	-0.37	-1.09

Table 6. Comparison of results for the Fischetti test problems

n	t_{max}	our solution		other heuristics		% gap nGA with	
		nGA	time (s)	GLS	OPT	GLS	OPT
52	213	1707	0.4	1702	1707	-0.29	0.12
101	**3955**	**3442**	**2.4**	**3265**	**3359**	**-5.42**	**-2.47**
101	10641	3165	2.3	3165	3212		1.46
201	14684	6478	3.5	5428	6547	-19.34	1.05
300	24096	8770	4.5	8088	9161	-8.43	4.27
319	21045	10498	5.2	9145	10900	-14.79	3.69
401	7641	12417	6.5	11362	13648	-9.29	8.98
Avg.	-	6639.6	3.5	6022.1	6932.6	-8.22	2.43

Table 7. New, optimal routes for the Fischetti dataset

n	t_{max}	length of the route	profit	route
101	3955	3947.767	3399	1-60-8-90-51-83-65-80-81-5-61-22-41-24-25-43-42-7- -34-39-100-44-45-16-98-66-76-59-95-77-93- -28-37-56-46-50-73-92-27-10-33- -20-78-32-12-85-82-75-21-49-86-97-63-15-1

4 Conclusions and Future Work

Computer experiments has shown that the nGA performs better than the comparable methods GLS, TS, CH in the quality of the results without increasing the execution time. The results nGA and comparable to TGA. The execution time of nGA could be decreased by the simple parallelization - each execution of the algorithm could be run parallel on a separate processor. Moreover the tournament grouping selection could be executed parallel in each group.

In the future, we intend to extend the nGA to solve two more complicated, but more applicable, versions of the OP: the Orienteering Problem with Time Windows (OPTW) [18], and the Time Dependent Orienteering Problem with Time Windows (TDOPTW) [6]. We would like to apply the algorithm in the mobile application of the touristic planner, which ideally create personalised routes that maximise tourists satisfaction.

Acknowledgments. The authors gratefully acknowledge support from the Polish Ministry of Science and Higher Education at the Bialystok University of Technology (grant S/WI/1/2014, S/WI/2/13 and W/WI/2/2013).

References

1. Archetti, C., Hertz, A., Speranza, M.: Metaheuristics for team orienteering problem. Journal of Heuristics **13**, 49–76 (2007)
2. Balas, E.: The prize collecting traveling salesman problem. Networks **19**, 797–809 (1989)
3. Chao, I.M., Golden, B.L., Wasil, E.A.: A fast and effective heuristic for the orienteering. European Journal of Operational Research **88**, 475–489 (1996)
4. Fischetti, M., Salazar, J.J., Toth, P.: Solving the orienteering problem through branch-and-cut. INFORMS Journal on Computing **10**, 133–148 (1998)
5. Feillet, D., Dejax, P., Gendreau, M.: Traveling Salesman Problems With Profits: An Overview. Transportation Science **38**, 188–205 (2001)
6. Garcia, A., Arbelaitz, O., Vansteenwegen, P., Souffriau, W., Linaza, M.T.: Hybrid approach for the public transportation time dependent orienteering problem with time windows. In: Corchado, E., Graña Romay, M., Manhaes Savio, A. (eds.) HAIS 2010, Part II. LNCS, vol. 6077, pp. 151–158. Springer, Heidelberg (2010)
7. Gendreau, M., Laporte, G., Semet, F.: A branch-and-cut algorithm for the undirected selective traveling salesman problem. Networks **32**(4), 263–273 (1998)
8. Jaen, J., Mocholi, J.A., Catala, A.: Digital ants as the best cicerones for museum visitors. Applied Soft Computing **11**, 111–119 (2011)
9. Laporte, G., Martello, S.: The selective travelling salesman problem. Discrete Applied Matheuristics **26**, 193–207 (1990)
10. Karbowska-Chilinska, J., Koszelew, J., Ostrowski, K., Zabielski, P.: Genetic algorithm solving orienteering problem in large networks. Frontiers in Artificial Intelligence and Applications **243**, 28–38 (2012)
11. Kataoka, S., Morito, S.: An algorithm for single constraint maximum collection problem. Journal of the Operations Research Society of Japan **31**(4), 515–31 (1988)

12. Piwońska, A., Koszelew, J.: A memetic algorithm for a tour planning in the selective travelling salesman problem on a road network. In: Kryszkiewicz, M., Rybinski, H., Skowron, A., Raś, Z.W. (eds.) ISMIS 2011. LNCS, vol. 6804, pp. 684–694. Springer, Heidelberg (2011)
13. Sevkli, Z., Sevilgen, E.: Discrete particle swarm optimization for the orienteering Problem. In: Evolutionary Computation (CEC) IEEE Congress, pp. 1–8 (2010)
14. Tang, H., Miller-Hooks, E.: A tabu search heuristic for the team orienteering problem. Computers & Operational Research 32(6), 1379–1407 (2005)
15. Tsiligirides, T.: Heuristic methods applied to orienteering. Journal of the Operational Research Society 35(9), 797–809 (1984)
16. Tasgetiren, M.F.: A genetic algorithm with an adaptive penalty function for the orienteering problem. Journal of Economic and Social Research 4(2), 20–40 (2002)
17. Vansteenwegen, P., Souffriau, W., Van Oudheusden, D.: A guided local search metaheuristic for the team orienteering problem. European Journal of Operational Research. 196, 118–127 (2009)
18. Vansteenwegen, P., Souffriau, W., Van Oudheusden, D.: Iterated local search for the team orienteering problem with time windows. Computers & Operational Research 36, 3281–3290 (2009)
19. Vansteenwegen, P., Souffriau, W., Van Oudheusden, D.: The City Trip Planner: An expert system for tourists. Expert Systems with Applications 38(6), 6540–6546 (2011)
20. Vansteenwegen, P., Souffriau, W., Van Oudheusden, D.: The Orienteering Problem: A survey. European Journal of Operational Research 209(1), 1–10 (2011)

Solving the Set Covering Problem
with a Shuffled Frog Leaping Algorithm

Broderick Crawford[1,2,3]($\boxtimes$), Ricardo Soto[1,4,5], Cristian Peña[1], Wenceslao
Palma[1], Franklin Johnson[1,6], and Fernando Paredes[7]

[1] Pontificia Universidad Católica de Valparaíso, Valparaíso, Chile
{broderick.crawford,ricardo.soto,wenceslao.palma}@ucv.cl,
cristian.pena.v@mail.pucv.cl
[2] Universidad Finis Terrae, Providencia, Chile
[3] Universidad San Sebastián, Santiago, Chile
[4] Universidad Autónoma de Chile, Santiago, Chile
[5] Universidad Central de Chile, Santiago, Chile
[6] Universidad de Playa Ancha,Valparaíso, Chile
franklin.johnson@upla.cl
[7] Escuela de Ingeniería Industrial, Universidad Diego Portales, Santiago, Chile
fernando.paredes@udp.cl

Abstract. In this paper we design and evaluate a shuffled frog leaping
algorithm that solves the set covering problem. The shuffled frog leaping algorithm is a novel metaheuristic inspired by natural memetics. It
consists of an individual memetic evolution and a global memetic information exchange between a population of virtual frogs representing possible solutions of a problem at hand. The experimental results show the
effectiveness of our approach which produces competitive results solving
a portfolio of set covering problems from the OR-Library.

Keywords: Shuffled frog leaping algorithm · Set covering problem ·
Metaheuristics · Artificial and computational intelligence

1 Introduction

The Set Covering Problem (SCP) is a weel-known NP-hard problem in the strong
sense [17]. The SCP is a class of representative combinatorial optimization problem
that has been applied to many real world problems such as production planning
in industry [27], the facility location problem [28] and crew scheduling problems
in airlines [19]. SCP can be formally defined as follows. Let $A = (a_{ij})$ be an m-
row, n-column, zero-one matrix. We say that a column j covers a row i if $a_{ij} = 1$.
Each column j is associated with a nonnegative real cost c_j. Let $I = \{1, ..., m\}$
and $J = \{1, ..., n\}$ be the row set and column set, respectively. The SCP calls for
a minimum cost subset $S \subseteq J$, such that each row $i \in I$ is covered by at least one
column $j \in S$. A mathematical model for the SCP is

$$Minimize \quad f(x) = \sum_{j=1}^{n} c_j x_j \tag{1}$$

© Springer International Publishing Switzerland 2015
N.T. Nguyen et al. (Eds.): ACIIDS 2015, Part II, LNAI 9012, pp. 41–50, 2015.
DOI: 10.1007/978-3-319-15705-4_5

subject to

$$\sum_{j=1}^{n} a_{ij}x_j \geq 1, \quad \forall i \in I \tag{2}$$

$$x_j \in \{0,1\}, \quad \forall j \in J \tag{3}$$

The goal is to minimize the sum of the costs of the selected columns, where $x_j = 1$ if the column j is in the solution, 0 otherwise. The restrictions ensure that each row i is covered by at least one column.

The SCP has been solved using Branch-and-bound and branch-and-cut based algorithms [2,16], classical greedy algorithms [10] and heuristics based on Lagrangian relaxation with subgradient optimization [7,9]. However, top-level general search strategies such as tabu search [8], simulated annealing [6], genetic algorithms [4], ant colony optimization (ACO) [11,25], electromagnetism (uni-cost SCP) [24] and gravitational emulation search [1] have been also applied to solve the SCP finding good or near-optimal solutions within a reasonable amount of time. In this paper, we propose a discrete Shuffled Frog Leaping Algorithm (SFLA) to solve the SCP. To the best of our knowledge, this is the first work proposing a binary coded SFLA to solve the SCP. The SFLA is a metaheuristic method proposed in [13] that combines the advantages of memetic evolution of genetic-based memetic algorithms [18] and the social behavior of particle swarm optimization (PSO) algorithms [20]. Accurate results, few parameters adjustment, great capability in global search and a fast convergence speed are the most distinguished characteristics and benefits of SFLA. Thus, in recent years the SFLA metaheuristic has been successfully applied to several optimization problems such as the 0/1 knapsack problem [5], traveling salesman problem [22], the vehicle routing problem [21], water resource distribution [13], the unit commitment problem [12] and the economic load dispatch problem [26], the resource-constrained project scheduling problem [15].

The main contribution of this paper is the design of a novel approach based on a binary coded SFLA to solve the SCP. Extensive experiments have been performed on a portfolio of SCPs from the Beasley's OR-Library showing that our approach can generate good quality solutions. The rest of this paper is organized as follows. In Section 2, we survey the SFLA and we describe our approach to solve the SCP. The Section 3 presents the experimental results obtained when applying the algorithm for solving instances of SCP contained in the OR-Library. Finally, in Section 4 we conclude.

2   The Shuffled Frop Leaping Algorithm

In the SFLA, a population of virtual frogs is partitioned into subsets referred to as memeplexes where each frog acts as host or carrier [14] of a unit of cultural evolution (a meme). Local search is performed simultaneously in each memeplex and in order to provide global exploration virtual frogs are reorganized into new memeplexes. After a defined number of memetic evolution time loops, the memeplexes are forced to mix using a shuffling process in order to avoid that the memetic evolution is free from memeplexes bias. Once a maximum number of shuffling iterations is achieved, the local search and the shuffling process are stopped. More precisely, an initial population of P frogs is created randomly. For a problem with n decision variables, a frog i is represented as a vector $X_i = (x_i^1, x_i^2, ..., x_i^n)$. Then, the initial P frogs are sorted in descending order w.r.t their fitness. Afterwards, all the frogs are partitioned into m memeplexes each containing f frogs (i.e. $P = m \times f$). The strategy to generate the partitions is as follows: the frog ranked first goes to the first memeplex, the second frog goes to the second memeplex, the mth frog goes to the mth memeplex, the $(m + 1)$th frog goes to the $(m + 1)$th memeplex, and so forth. The frog with the global best fitness is identified as X_g and for each memeplex, the frog with the best fitness X_b and the worst fitness X_w are also identified. Then, within each memeplex the worst frog X_w changes all its decision variables w.r.t the best frog X_b as is showed in Eqs. 4 and 5. Where $rand$ is a random number between 0 and 1, d_{min}^j and d_{max}^j is the minimum and maximum allowed change in a frog's decision variable, respectively. However, X_w is replaced only if Eqs. (4) and (5) generate a better solution. Otherwise, this process is repeated but X_b is replaced by X_g. If the obtained value is not better than the old one, a new frog is randomly generated to replace X_w. This process continues for a fixed number of iterations. Afterwards, the frog population is re-evaluated, sorted and repartitioned into memeplexes generating a global information exchange among the frogs. This evolutive process is carried out until a terminate criteria is satisfied.

$$d_w^j = rand \times (x_b^j - x_w^j), \quad 1 \le j \le n \tag{4}$$

$$x_{new}^j = x_w^j + d_w^j, \qquad d_{min}^j < d_w^j < d_{max}^j \tag{5}$$

Algorithm 1. Shuffled Frog Leaping Algorithm
---
    initialize parameters
    generate population of frogs randomly
    evaluate fitness of each frog
    while stop criteria is not satisfied **do**
        sort frogs w.r.t. their fitness
        construct memeplexes
        for each memeplex **do**
            local search using eq. 4 and eq. 5
        end for
        shuffle all frogs
    end while

A Binary Coded Shuffled Frog Leaping Algorithm to Solve the SCP.
The SFLA cannot handle directly the SCP because the Eqs. (4) and (5) are not able to generate values belonging to $\{0, 1\}$. Thus, the original SFLA must be

modified to solve the SCP. In this section we present our approach to solve the
SCP using a binary coded SFLA.

In order to speed up the proposed algorithm, we preprocess the tested SCP
instances using two preprocessing methods which have proved to be the most
effective [25][16]: column domination and column inclusion. In column domination, if the rows covered by a column j can be covered by other columns with
a total cost lower than c_j, the column j is safely removed. The aforementioned
situation is a NP-complete problem itself, thus is impractical to check all the
possible solutions. However, column domination can be used in a limited way
putting all the columns in increasing order of cost. To break ties, we use the
number of rows that they cover in decreasing order. In column inclusion, a column is included in an optimal solution when a row is covered by only one column
after applying column domination.

Each frog is represented by a n-bit binary string, where n is the number of
decision variables of the problem. The initial population is created randomly.
We use transfer functions [23] (see Table 1) and discretization methods in order
to map a continuous search space to a discrete search space as follows. When
the worst frog X_w of each memeplex changes all its decision variables x_w^j w.r.t
the best frog X_b using eq. 4 the value d_w^j is sent to a transfer function. Thus,
a transfer function calculates the probability $(T(d_w^j))$ that the j-th bit of the
vector representing a solution of the worst frog change from 0 to 1 and vice
versa. The transfer functions shown in Table 1 are classified in S-shaped and
V-shaped functions.

Table 1. S-shaped and V-shaped transfer functions

S-shaped family		V-shaped family	
Name	Transfer function	Name	Transfer function
S1	$T(d_w^j) = \frac{1}{1+e^{-2d_w^j}}$	V1	$T(d_w^j) = \left\| erf\left(\frac{\sqrt{2}}{\pi}d_w^j\right)\right\| = \left\|\frac{\sqrt{2}}{\pi}\int_o^{\frac{\sqrt{2}}{\pi}d_w^j} e^{-t^2} dt\right\|$
S2	$T(d_w^j) = \frac{1}{1+e^{-d_w^j}}$	V2	$T(d_w^j) = \|\tanh(d_w^j)\|$
S3	$T(d_w^j) = \frac{1}{1+e^{-d_w^j/2}}$	V3	$T(d_w^j) = \left\|\frac{d_w^j}{\sqrt{1+d_w^{j^2}}}\right\|$
S4	$T(d_w^j) = \frac{1}{1+e^{-d_w^j/3}}$	V4	$T(d_w^j) = \left\|\frac{2}{\pi}\arctan(\frac{\pi}{2}d_w^j)\right\|$

Then, we use the following two discretization methods in order to assign a 0
or a 1 to the j-th bit of the vector of the worst frog X_w:

- Elitist Selection: We use it to take the value of the j-th bit from the best
frog X_b and assign it to the j-th bit of X_w.

$$x_{new}^j = \begin{cases} x_b^j & \text{if } rand < T(d_w^j) \\ 0 & \text{otherwise} \end{cases}$$

- Roulette Wheel Selection: is the most common selection strategies used in
metaheuristics. It assigns to each frog a selection probability that is proportional to its fitness. In our approach we use it to choose the frog from which
to take the value of the j-th bit and assign it to the j-th bit of X_w.

When a solution is infeasible we apply a repairing method [4] based on a heuristic operator which also provides a local optimisation step that removes any redundant column in the repaired solution. Basically, to repair an unfeasible solution all the uncovered rows must be identified and then a greedy heuristic is applied in order to consider first the columns with a low cost ratio ($\frac{\text{cost of a column}}{\text{number of uncovered rows which it covers}}$) and to drop first whenever possible the columns with high costs.

Putting it All Together. We propose a binary coded SFLA to solve the SCP, initially the instances of the problem are preprocessed in order to speed up the algorithm and a random population of P binary strings is created. Then, the solutions are sorted in ascending order w.r.t. their fitness and the population is partitioned into memeplexes. Afterwards, a local search procedure is called within each memeplex. This procedure replaces the worst solution using the best solution or the global best solution. A repairing procedure is called when unfeasible solution are generated. In order to map a continuous search space to a discrete search space we use a transfer function and a discretization method. Once the local search is ended the memeplexes are shuffled in order to ensure global exploration. The algorithm continues until a termination criterion is achieved. Finally, the best solution to the SCP is obtained.

3   Experimental Evaluation

The effectiveness of our proposal is tested using SCP test instances from OR-Library [3]. The proposed algorithm was implemented using Java language and conducted on a 2.4 GHz Intel Core i7 with 8GB RAM running Windows 8.1.

$$RPD = (Z - Z_{opt})/Z_{opt} \times 100 \qquad (6)$$

The relative percentage deviation (RPD) is calculated in order to evaluate the quality of a solution. The RPD value quantifies the deviation of the objective value Z from Z_{opt} which in our case is the best known value (BKV) for each instance (see the second column). We report the minimum, maximum, and average of the obtained solutions. To compute RPD we use $Z = Min$. This measure is computed as is showed in Eq. 6. In all experiments, the binary SFLA is executed 30 times over each of the chosen SCP test instances. We test all the combinations of transfer functions and discretization methods over all these instances. However, because the quality of solutions have the same behavior over all the tested instances, we report the results obtained over the following instances: 4.1, 5.1, 6.1, A.1, B.1, C.1, D.1, NRE.1 (Tables 2 to 9). We used a population of 200 frogs ($P = 200$), 20 memeplexes ($m = 20$) and 20 iterations within each memeplex ($it = 20$). These parameters were selected empirically after a large number of tests over all the SCP instances. Tables 2 to 9 show that V-shaped transfer functions show better results in 75% of instances. V-shaped transfer functions show high exploration when they become saturated early because they tend to modify the variables more frequently. By other side, the exploitation of the SFLA is improved by the use of elitism in the discretization method. This ensures a good balance between exploration and exploitation.

Table 2. Experimental results over the instance 4.1 (BKV=429) of SCP

Discretization Method	Transfer Function	Min.	Max.	Avg.	RPD
Roulette	S1	436	449	443.20	1.63
Roulette	S2	437	449	443.13	1.86
Roulette	S3	436	449	442.37	1.63
Roulette	S4	435	449	440.63	1.40
Roulette	V1	439	449	443.53	2.33
Roulette	V2	438	449	444.90	2.10
Roulette	V3	437	449	444.50	1.86
Roulette	V4	440	449	444.73	2.56
Elitist	S1	437	449	443.30	1.86
Elitist	S2	435	448	442.17	1.40
Elitist	S3	433	449	444.13	0.93
Elitist	S4	436	449	443.80	1.63
Elitist	V1	433	446	441.40	0.93
Elitist	V2	435	446	440.93	1.40
Elitist	V3	431	446	439.20	0.47
Elitist	V4	432	447	439.67	0.70

Table 3. Experimental results over the instance 5.1 (BKV=253) of SCP

Discretization Method	Transfer Function	Min.	Max.	Avg.	RPD
Roulette	S1	257	275	270.60	1.58
Roulette	S2	269	277	271.97	6.32
Roulette	S3	268	278	272.00	5.93
Roulette	S4	267	280	272.30	5.53
Roulette	V1	266	280	275.07	5.14
Roulette	V2	272	280	275.10	7.51
Roulette	V3	269	280	274.70	6.32
Roulette	V4	271	280	275.10	7.11
Elitist	S1	268	280	273.67	5.93
Elitist	S2	268	280	274.40	5.93
Elitist	S3	265	280	273.37	4.74
Elitist	S4	261	280	272.43	3.16
Elitist	V1	257	277	271.03	1.58
Elitist	V2	262	275	271.40	3.56
Elitist	V3	261	276	271.03	3.16
Elitist	V4	267	279	271.53	5.53

Table 4. Experimental results over the instance 6.1 (BKV=138) of SCP

Discretization Method	Transfer Function	Min.	Max.	Avg.	RPD
Roulette	S1	145	152	148.80	5.07
Roulette	S2	146	152	148.23	5.80
Roulette	S3	145	152	148.37	5.07
Roulette	S4	145	152	148.10	5.07
Roulette	V1	145	152	148.83	5.07
Roulette	V2	147	152	149.43	6.52
Roulette	V3	147	152	149.60	6.52
Roulette	V4	145	152	149.57	5.07
Elitist	S1	145	152	149.00	5.07
Elitist	S2	147	152	149.87	6.52
Elitist	S3	146	152	149.87	5.80
Elitist	S4	146	152	149.87	5.80
Elitist	V1	146	152	148.03	5.80
Elitist	V2	144	152	147.63	4.35
Elitist	V3	145	152	148.00	5.07
Elitist	V4	141	152	148.63	2.17

Table 5. Experimental results over the instance A.1 (BKV=253) of SCP

Discretization Method	Transfer Function	Min.	Max.	Avg.	RPD
Roulette	S1	258	261	260,13	1.98
Roulette	S2	257	261	260,17	1.58
Roulette	S3	257	261	260,23	1.58
Roulette	S4	257	261	260,03	1.58
Roulette	V1	259	261	260,43	2.37
Roulette	V2	259	261	260,40	2.37
Roulette	V3	258	261	260,50	1.98
Roulette	V4	258	261	260,60	1.98
Elitist	S1	257	261	260.77	1.58
Elitist	S2	260	261	260.77	2.77
Elitist	S3	257	261	260.60	1.58
Elitist	S4	257	261	260.33	1.58
Elitist	V1	257	261	260.40	1.58
Elitist	V2	257	261	260.17	1.58
Elitist	V3	256	261	259.83	1.19
Elitist	V4	257	261	259.97	1.58

Table 6. Experimental results over the instance B.1 (BKV=69) of SCP

Discretization Method	Transfer Function	Min.	Max.	Avg.	RPD
Roulette	S1	77	86	81.60	11.59
Roulette	S2	79	86	82.57	14.49
Roulette	S3	75	86	81.63	8.70
Roulette	S4	76	86	80.93	10.14
Roulette	V1	79	86	83.80	14.49
Roulette	V2	79	86	84.17	14.49
Roulette	V3	79	86	83.70	14.49
Roulette	V4	78	86	82.10	13.04
Elitist	S1	79	86	83.47	14.49
Elitist	S2	79	86	83.03	14.49
Elitist	S3	77	86	82.83	11.59
Elitist	S4	76	86	82.73	10.14
Elitist	V1	79	86	81.33	14.49
Elitist	V2	76	86	81.67	10.14
Elitist	V3	76	86	80.93	10.14
Elitist	V4	76	86	81.17	10.14

Table 7. Experimental results over the instance C.1 (BKV=227) of SCP

Discretization Method	Transfer Function	Min.	Max.	Avg.	RPD
Roulette	S1	233	235	234.70	2.64
Roulette	S2	233	235	234.57	2.64
Roulette	S3	234	235	234.77	3.08
Roulette	S4	234	235	234.70	3.08
Roulette	V1	234	235	234.60	3.08
Roulette	V2	234	235	234.73	3.08
Roulette	V3	234	235	234.77	3.08
Roulette	V4	233	235	234.57	2.64
Elitist	S1	233	235	234.83	2.64
Elitist	S2	234	235	234.53	3.08
Elitist	S3	233	235	234.80	2.64
Elitist	S4	233	235	234.67	2.64
Elitist	V1	233	235	234.60	2.64
Elitist	V2	233	235	234.37	2.64
Elitist	V3	233	235	234.57	2.64
Elitist	V4	233	236	234.33	2.64

Table 8. Experimental results over the instance D.1 (BKV=60) of SCP

Discretization Method	Transfer Function	Min.	Max.	Avg.	RPD
Roulette	S1	61	62	61.63	1.67
Roulette	S2	61	62	61.40	1.67
Roulette	S3	61	62	61.57	1.67
Roulette	S4	60	62	61.43	0.00
Roulette	V1	61	62	61.70	1.67
Roulette	V2	61	62	61.43	1.67
Roulette	V3	61	62	61.63	1.67
Roulette	V4	61	62	61.63	1.67
Elitist	S1	61	62	61.87	1.67
Elitist	S2	61	62	61.60	1.67
Elitist	S3	61	62	61.77	1.67
Elitist	S4	61	62	61.73	1.67
Elitist	V1	61	62	61.70	1.67
Elitist	V2	61	66	61.80	1.67
Elitist	V3	61	67	61.90	1.67
Elitist	V4	61	63	61.67	1.67

Table 9. Experimental results over the instance NRE.1 (BKV=29) of SCP

Discretization Method	Transfer Function	Min.	Max.	Avg.	RPD
Roulette	S1	30	30	30.00	3.45
Roulette	S2	29	30	29.93	0.00
Roulette	S3	30	30	30.00	3.45
Roulette	S4	30	30	30.00	3.45
Roulette	V1	30	30	30.00	3.45
Roulette	V2	29	30	29.93	0.00
Roulette	V3	29	30	29.97	0.00
Roulette	V4	30	30	30.00	3.45
Elitist	S1	30	30	30.00	3.45
Elitist	S2	29	30	29.93	0.00
Elitist	S3	29	30	29.93	0.00
Elitist	S4	30	30	30.00	3.45
Elitist	V1	30	30	30.00	3.45
Elitist	V2	29	33	30.03	0.00
Elitist	V3	29	35	30.07	0.00
Elitist	V4	30	36	30.20	3.45

4 Conclusions

In this work, we have proposed a binary coded Shuffled Flog Leaping Algorithm to solve the Set Covering Problem. The experimental results show that, regardless of the transfer function employed, the proposed binary coded SFLA has high quality near optimal solutions when the roulette and elitist discretization methods are applied. The discretization methods employed are basically elitists which allow the frogs to move forward the best position. This process can generate a bias in the beginning of the search process causing a premature convergence and a loss of diversity. However, at the same time the shuffling process of the SFLA introduces a better exploration of the search space leading to good quality results.

Acknowledgments. The author Broderick Crawford is supported by grant CONICYT/FONDE-CYT/REGULAR/1140897, Ricardo Soto is supported by grant CONICYT /FONDECYT/INICIACION/11130459 and Fernando Paredes is supported by grant CONICYT/FONDECYT/REGULAR/1130455.

References

1. Balachandar, S.R., Kannan, K.: A meta-heuristic algorithm for set covering problem based on gravity. 4(7), 944–950 (2010)
2. Balas, E., Carrera, M.C.: A dynamic subgradient-based branch-and-bound procedure for set covering. Operations Research 44(6), 875–890 (1996)
3. Beasley, J.E.: A lagrangian heuristic for set-covering problems. Naval Research Logistics (NRL) 37(1), 151–164 (1990)
4. Beasley, J., Chu, P.: A genetic algorithm for the set covering problem. European Journal of Operational Research 94(2), 392–404 (1996)
5. Bhattacharjee, K.K., Sarmah, S.: Shuffled frog leaping algorithm and its application to 0/1 knapsack problem. Applied Soft Computing 19, 252–263 (2014). http://www.sciencedirect.com/science/article/pii/S1568494614000799
6. Brusco, M., Jacobs, L., Thompson, G.: A morphing procedure to supplement a simulated annealing heuristic for cost- and coverage-correlated set-covering problems. Annals of Operations Research 86, 611–627 (1999)
7. Caprara, A., Fischetti, M., Toth, P.: A heuristic method for the set covering problem. Operations Research 47(5), 730–743 (1999)
8. Caserta, M.: Tabu search-based metaheuristic algorithm for large-scale set covering problems. In: Doerner, K., Gendreau, M., Greistorfer, P., Gutjahr, W., Hartl, R., Reimann, M. (eds.) Metaheuristics, Operations Research/Computer Science Interfaces Series, vol. 39, pp. 43–63. Springer, US (2007)
9. Ceria, S., Nobili, P., Sassano, A.: A lagrangian-based heuristic for large-scale set covering problems. Mathematical Programming 81(2), 215–228 (1998)
10. Chvatal, V.: A greedy heuristic for the set-covering problem. Mathematics of Operations Research 4(3), 233–235 (1979)
11. Crawford, B., Soto, R., Monfroy, E., Castro, C., Palma, W., Paredes, F.: A hybrid soft computing approach for subset problems. Mathematical Problems in Engineering, Article ID 716069, 1–12 (2013)
12. Ebrahimi, J., Hosseinian, S.H., Gharehpetian, G.B.: Unit Commitment Problem Solution Using Shuffled Frog Leaping Algorithm. IEEE Transactions on Power Systems 26(2), 573–581 (2011)
13. Eusuff, M.M., Lansey, K.E.: Optimization of Water Distribution Network Design Using the Shuffled Frog Leaping Algorithm. Journal of Water Resources Planning and Management 129(3), 210–225 (2003)
14. Eusuff, M., Lansey, K., Pasha, F.: Shuffled frog-leaping algorithm: a memetic meta-heuristic for discrete optimization. Engineering Optimization 38(2), 129–154 (2006). http://dx.doi.org/10.1080/03052150500384759
15. Fang, C., Wang, L.: An effective shuffled frog-leaping algorithm for resource-constrained project scheduling problem. Computers & OR 39(5), 890–901 (2012)
16. Fisher, M.L., Kedia, P.: Optimal solution of set covering/partitioning problems using dual heuristics. Management Science 36(6), 674–688 (1990)
17. Garey, M.R., Johnson, D.S.: Computers and Intractability: A Guide to the Theory of NP-Completeness. W. H. Freeman & Co., New York (1990)
18. Glover, F.W., Kochenberger, G.A.: Handbook of Metaheuristics (International Series in Operations Research & Management Science). Springer, January 2003
19. Housos, E., Elmroth, T.: Automatic optimization of subproblems in scheduling airline crews. Interfaces 27(5), 68–77 (1997)
20. Kennedy, J., Eberhart, R.: Particle swarm optimization. In: IEEE International Conference on Neural Networks, pp. 1942–1948 (1995)

21. Luo, J., Chen, M.R.: Improved shuffled frog leaping algorithm and its multi-phase model for multi-depot vehicle routing problem. Expert Systems with Applications **41**(5), 2535–2545 (2014)
22. Luo, X., Yang, Y., Li, X.: Solving tsp with shuffled frog-leaping algorithm. In: ISDA (3), pp. 228–232. IEEE Computer Society (2008)
23. Mirjalili, S., Lewis, A.: S-shaped versus v-shaped transfer functions for binary particle swarm optimization. Swarm and Evolutionary Computation **9**, 1–14 (2013)
24. Naji-Azimi, Z., Toth, P., Galli, L.: An electromagnetism metaheuristic for the unicost set covering problem. European Journal of Operational Research **205**(2), 290–300 (2010)
25. Ren, Z.G., Feng, Z.R., Ke, L.J., Zhang, Z.J.: New ideas for applying ant colony optimization to the set covering problem. Computers & Industrial Engineering **58**(4), 774–784 (2010)
26. Roy, P., Roy, P., Chakrabarti, A.: Modified shuffled frog leaping algorithm with genetic algorithm crossover for solving economic load dispatch problem with valve-point effect. Appl. Soft Comput. **13**(11), 4244–4252 (2013)
27. Vasko, F.J., Wolf, F.E., Stott, K.L.: Optimal selection of ingot sizes via set covering. Oper. Res. **35**(3), 346–353 (1987)
28. Vasko, F.J., Wilson, G.R.: Using a facility location algorithm to solve large set covering problems. Operations Research Letters **3**(2), 85–90 (1984)

Machine Learning in Biometrics
and Bioinformatics with Applications

Automatic Evaluation of Area-Related Immunogold Particles Density in Transmission Electron Micrographs

Bartłomiej Płaczek[1]([✉]), Rafał J. Bułdak[2,3], Andrzej Brenk[4], and Renata Polaniak[3]

[1] Institute of Computer Science, University of Silesia, Sosnowiec, Poland
`Placzek.Bartlomiej@gmail.com`
[2] Department of Physiology, Faculty of Medicine with the Division of Dentistry, Silesian Medical University, Zabrze, Poland
[3] Departament of Human Nutrition, Faculty of Public Health, Silesian Medical University, Zabrze, Poland
[4] Department of Genaral Biochemystry, Faculty of Medicine with the Division of Dentistry, Silesian Medical University, Zabrze, Poland

Abstract. Immunogold particles are used in electron microscopy to determine sub-cellular location of biological relevant macromolecules, such as proteins, lipids, carbohydrates, and nucleic acids. In this paper an algorithm is proposed which enables automatic evaluation of the immunogold particles density in transmission electron micrographs. The introduced algorithm combines two different feature localization approaches. Coarse locations of the immunogold particles are recognized by image convolution with a Gaussian prototype and a multi-scale filtering is used to refine the locations. This algorithm was evaluated by using micrographs of human colorectal carcinoma cells. A higher accuracy of the immunogold particles detection was achieved in comparison with a state-of-the-art method. The improved detection accuracy enables a more precise evaluation of the area-related immunogold particles density.

Keywords: Image processing · Immunogold labeling · Electron microscopy

1 Introduction

Immunoelectron microscopy is a powerful tool for localization and quantification of selected macromolecules (proteins, lipids, carbohydrates, or nucleic acids) at sub-cellular level [8,12]. In this technique, the molecules of interest (antigens) are detected by using colloidal nanogold conjugated antibodies. The antibody is adsorbed onto the surface of a colloidal gold, and then colloidal gold is carried by the antibody to the positions of the corresponding target antigen in cells. The gold particles are used due to their spherical shapes and high electron density, which increases electron scatter to give dark circular markers.

© Springer International Publishing Switzerland 2015
N.T. Nguyen et al. (Eds.): ACIIDS 2015, Part II, LNAI 9012, pp. 53–61, 2015.
DOI: 10.1007/978-3-319-15705-4_6

Area-related numerical density of the immunogold particles is important information, which is useful for various research purposes in medicine [2,8,13]. Usually, the area-related density has to be calculated for a given internal cell compartment, e.g., cytoplasm or nucleus, and expressed in units of immunogold particles per μm^2. To this end, the immunogold particles in micrographs have to be accurately detected, localized and counted. Manual localization and counting of the immunogold particles is biased, poorly reproducible, and time-consuming [14,15]. These issues cannot be effectively addressed by the existing general-purpose image processing software [3]. Automatic evaluation of the area-related immunogold particles density can be achieved by using dedicated digital image processing methods [9]. In this paper an algorithm is proposed that detects the immunogold particles and calculates the area-related density based on feature localization in digital images.

The paper is organized as follows. Section 2 briefly reviews related works. Details of the proposed algorithm are presented in Section 3. Section 4 describes experiments on immunogold particles density evaluation for micrographs of human colorectal carcinoma cells. Finally, conclusions and future research directions are given in Section 5.

2   Related Works

There are several works in the literature that deal with automatic localization and counting of the immunogold particles in transmission electron micrographs. The state-of-the-art methods can be categorized as based on image segmentation or feature localization. In case of the segmentation approaches, image regions are selected where the immunogold particles appear. A single segment may contain more than one particle, thus additional operations are necessary to count the particles. When using the feature localization methods, centers of the circular particles are recognized in micrographs and the subsequent determination of particle number is straightforward.

A segmentation method, which applies edge detection and region growing for the immunogold particles recognition was introduced in [14]. In that early approach the region growing parameters have to be tuned manually for individual micrographs, moreover, the segmentation results need to be verified and corrected by an operator. Another semi-automated method was proposed by Lebonvallet et al. [5]. According to that method, the recognition of immunogold particles is based on morphological operations and segmentation with interactive (user-assisted) threshold selection. A simple binary thresholding was used in [7] to recognize foreground regions, which are then categorized based on their shape and size as small markers, large markers, or clusters of markers.

The feature localization approach underlies the immunogold particle detection methods that were proposed in [1] and [15]. According to the first method [1], Gaussian kernel is used as a synthetic prototype of the immunogold particle. A correlation image is computed using the prototype and then a hysteresis thresholding is applied on the correlation image for marker recognition. Detected candidate markers with a low circularity coefficient are removed to reduce the false positive error rate. The recent algorithm reported in [15] utilizes a multi-scale Difference-of-Gaussians (DoG) image representation to detect and categorize the immunogold particles in micrographs. Locations of the target particles are detected as local maxima in the multi-scale image representation. In order to remove false detections that can occur at strong edges or ridges, an analysis of Hessian matrix is performed in that method.

The algorithm presented in this paper enables automatic evaluation of the area-related immunogold particles density in images obtained from electron microscopy. Two different feature localization approaches are combined in this algorithm to improve accuracy of the immunogold particle recognition and counting. According to the introduced algorithm, coarse positions of the immunogold particles and clusters of these particles are recognized by using the Gaussian prototype. The precise locations of individual immunogold particles are then determined based on the DoG filtering.

3   Proposed Approach

In order to evaluate the area-related immunogold particles density in a micrograph, the particles have to be accurately detected. Fig. 1 shows a micrograph of a human colorectal carcinoma cell, in which the immunogold particles appear as small dark circles. It should be noted that the particles can be separated or clustered into regions of varying size and shape. The proposed approach uses a two-stage image processing procedure to localize and count the immunogold particles.

At the first stage, candidate (coarse) particle locations are recognized by using the Gaussian kernel G as a prototype of the immunogold particle:

$$G(x, y, \sigma) = \frac{1}{2\pi\sigma^2} \exp\left(-\frac{x^2 + y^2}{2\sigma^2}\right) \tag{1}$$

where $\sigma = 0.6 \cdot r$, and r is the known radius of the circular particle (in pixels).

The prototype is convolved with negative of an input image I. An example of this operation is presented in Fig. 2 b. At the next step, a threshold τ is used to get a binary image B, in which the foreground (white) regions correspond to the candidate particle locations:

$$B(x, y) = \begin{cases} 1, & G(x, y, \sigma) * (1 - I(x, y)) > \tau, \\ 0, & \text{else,} \end{cases} \tag{2}$$

where $*$ denotes the convolution operation, and $I(x, y) \in [0, 1]$ is the intensity of pixel (x, y) in input image.

False candidate locations of the particles are obtained at this stage due to presence of large dark non-particle objects in the input image (Fig. 2 c). To eliminate the false locations, areas of the foreground regions in binary image B

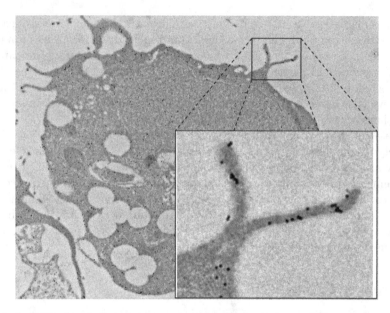

Fig. 1. Individual and clustered immunogold particles in transmission electron micrograph

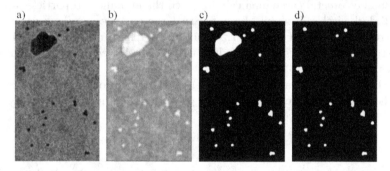

Fig. 2. Recognition of immunogold particles based on Gaussian prototype: a) input image I, b) result of convolution filtering, c) binary image B, d) candidate particle locations

are analyzed and regions with the area above a given maximum are removed (Fig. 2 d). The maximum area parameter can be determined on the basis of an observation that a maximum number of immunogold particles clustered into one region is usually below 10. Area of a region that corresponds to one particle is estimated based on the known radius r.

During second processing stage, the proposed algorithm determines precise locations of the detected immunogold particles within the regions of candidate locations (Fig 3 c). This operation is performed by using a multi-scale image representation, which is based on the DoG filtering [6]. According to the method

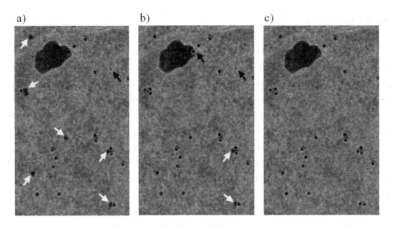

Fig. 3. Examples of the immunogold particles detection results: a) Algorithm 1 ($\rho = 2$), b) Algorithm 1 ($\rho = 4$), c) Algotithm 2 ($\tau = 0.75$)

proposed in [15], the immunogold particle locations correspond to local maxima at a scale level σ of the image representation R:

$$R(x,y,\sigma) = G(x,y,\sigma) * (1 - I(x,y)) - G(x,y,\sqrt{2}\sigma) * (1 - I(x,y)) \qquad (3)$$

where $\sigma = 0.6 \cdot r$.

An element $R(x,y,\sigma)$ of the multiscale image representation is recognized as a local maximum if its value is higher than the values of eight neighboring elements at the same scale level (σ) and higher than the values of two elements with the same location (x,y) at scale levels $\sigma + 0.6$, $\sigma - 0.6$. A local maximum found at (x,y,σ) is taken into consideration if the value of $R(x,y,\sigma)$ is above a given threshold α and the pixel (x,y) was selected as candidate particle location at the first stage. It means that the local maxima are ignored if they do not coincide with the candidate locations. Moreover, distances between the local maxima that correspond to the detected locations of immunogold particles have to be equal or greater than $2r$.

Finally, the area-related immunogold particles density d (in particles per μm^2) is calculated for a region of interest by using the following formula:

$$d = \frac{c}{a \cdot s^2}, \qquad (4)$$

where c is the count of immunogold particles detected in region of interest, a denotes area, i.e., number of pixels in the region of interest, and s (in μm per pixel) is a known scale coefficient, which depends on the microscope magnification and the camera resolution. The region of interest corresponds to a selected cell compartment. It is determined by an input mask of the same size as the input image.

4   Experiments

The goal of the performed experiments was to evaluate accuracy of the proposed algorithm and compare it against the recent method from the literature. Micrographs of human colorectal HCT-116 carcinoma cells with immunogold particles were used for the experimental evaluation. The cells were obtained from the American Type Culture Collection (ATCC). An immunogold staining method for transmission electron microscopy was applied to analyze sub-cellular localization of visfatin [2]. The test micrographs were acquired from FEI Tecnai G2 BioTWIN transmission electron microscope (FEI, Netherlands) at 120 kV and 16000x magnification using Morada CCD camera (Olympus, Hamburg, Germany). Radius of the immunogold particles that appear in micrographs corresponds to 3 pixels ($r = 3$). The scale coefficient s is 0.004 μm per pixel.

Accuracy of the area-related immunogold particles density evaluation depends on correct detection and localization of individual immunogold particles in micrographs. Therefore, the false positive and false negative rates of the particles detection are analyzed in this study for the two compared algorithms.

The false detection rates for the examined algorithms were evaluated using a set of test micrographs that include over 3 500 immunogold particles. The micrographs were divided into 20 test images containing between 150 and 200 particles each. True positions of the particles were determined for the micrographs by an expert. A detection is considered as correct if the distance between the center of the detected particle and center of true particle is less than the radius r.

4.1   Compared Algorithms

Two algorithms for immunogold particles detection were taken into consideration during the experiments. Algorithm 1 is equivalent to the state-of-the-art method, which was proposed by Wang et al. in [15]. The proposed algorithm is referred to as Algorithm 2.

According to Algorithm 1, candidate locations of the immunogold particles are recognized as local maxima in the multi-scale DoG image representation (3). After that, a local Hessian analysis is performed to eliminate false detections. A candidate particle location at pixel (x, y) is discarded, if the following condition is satisfied:

$$\frac{\text{Tr}^2(\mathbf{H})}{\text{Det}(\mathbf{H})} < \frac{(\rho + 1)^2}{\rho}, \tag{5}$$

where ρ is a parameter of the algorithm and $\mathbf{H}$ is the Hessian matrix calculated for pixel (x, y) of the image representation R at scale level σ:

$$\mathbf{H} = \begin{bmatrix} R_{xx} & R_{xy} \\ R_{xy} & R_{yy} \end{bmatrix} \tag{6}$$

In case of $\rho = 1$, the condition (5) is satisfied only for ideal, radially symmetric objects. When a higher value of ρ is used then the condition holds also for more irregular, elongated structures.

According to Algorithm 2 (the proposed approach), coarse particle locations are determined using the Gaussian prototype and then the multi-scale DoG representation is computed to find the precise locations, as discussed in Sect. 3.

4.2   Results

Based on preliminary results, the threshold parameter α of local maxima detection in the multi-scale image representation R was set to 0.4 for both algorithms. The maximum area parameter for Algorithm 2 was determined by taking into account the limit of 10 particles: $10\pi r^2 \approx 283$ pixels ($r = 3$). Examples of the experimental results are presented in Fig. 3. The false positive and false negative detections are indicated, respectively, by black and white arrows. For Algorithm 1 with parameter $\rho = 2$, there are seven false negative detections (Fig. 3 a). In this example a false detection also occurs. The false negative detections can be eliminated by increasing the value of ρ, however for higher ρ values the number of false positive detections expands. This effect is illustrated in Fig. 3 b for $\rho = 4$, where five of the false negatives are eliminated, but one new false positive detection arises at edge of the large dark object. For the analyzed micrograph example, correct detection of all immunogold particles was obtained by using Algorithm 2 with threshold $\tau = 0.75$ (Fig. 3 c).

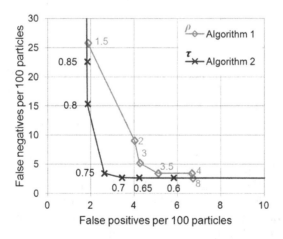

Fig. 4. Average false negative and false positive detection rates for the compared algorithms

Figure 4 shows averaged results of the experiments for all test images. The lowest false positive and false negative rates were achieved for Algorithm 2 with threshold $\tau = 0.75$. The results obtained for Algorithm 2 are clearly better than those of Algorithm 1. In case of Algorithm 2, the false positive rate is significantly reduced at the first processing stage by using the Gaussian prototype. The superiority of Algorithm 2 is evident especially in case of micrographs that

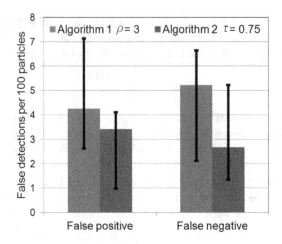

Fig. 5. Range of detection error rates for the compared algorithms

include dark objects as well as low contrast and highly textured regions. Range of the error rates for the compared algorithms is illustrated in Fig. 5. The error bars in Fig. 5 correspond to the maximum and minimum rate of false detections obtained from the set of 20 test images. The columns in the chart show average number of the false positive and false negative detections per 100 true particles.

5 Conclusions

The proposed algorithm combines two different feature localization approaches to improve the evaluation of area-related immunogold particles density for transmission electron micrographs. Coarse locations of the immunogold particles are recognized by image convolution with the Gaussian prototype. The coarse locations are then refined by multi-scale DoG filtering. Experiments were conducted on a set of micrographs that show immunogold particles in human colorectal carcinoma cells. The proposed approach was compared against state-of-the-art method. Results of the experiments confirm that the immunogold particles can be detected with higher accuracy when using the introduced algorithm. The improved detection accuracy enables a more precise evaluation of the area-related immunogold particles density. Further research challenges include automatic calibration of the algorithm parameters as well as improvement of the clustered particles recognition by application of Hough Transform [11] and fuzzy descriptors of image attributes [4,10].

Acknowledgments. The authors thank Prof. Romuald Wojnicz, Dr. Natalia Matysiak, and Łukasz Mielańczyk for providing immunoelectron micrographs of HCT-116 cells.

References

1. Brandt, S., Heikkonen, J., Engelhardt, P.: Multiphase method for automatic alignment of transmission electron microscope images using markers. Journal of structural biology **133**(1), 10–22 (2001)
2. Bułdak, R.J., Skonieczna, M., Matysiak, N., Wyrobiec, G., Kukla, M., Michalski, M., Żwirska-Korczala, K.: Changes in subcellular localization of visfatin in human colorectal HCT-116 carcinoma cell line after cytochalasin B treatment. European Journal of Histochemistry **58**(3), 239–246 (2014)
3. Kołodziejczyk, A., Ładniak, M., Piórkowski, A.: Constructing software for analysis of neuron, glial and endothelial cell numbers and density in histological NISSL-stained rodent brain tissue. Journal of Medical Informatics Technologies **23**, 77–86 (2014)
4. Kudłacik, P., Porwik, P.: A new approach to signature recognition using the fuzzy method. Pattern Analysis and Applications **17**(3), 451–463 (2014)
5. Lebonvallet, S., Mennesson, T., Bonnet, N., Girod, S., Plotkowski, C., Hinnrasky, J., Puchelle, E.: Semi-automatic quantitation of dense markers in cytochemistry. Histochemistry **96**(3), 245–250 (1991)
6. Lowe, D.G.: Distinctive image features from scale-invariant keypoints. International journal of computer vision **60**(2), 91–110 (2004)
7. Monteiro-Leal, L.H., Troster, H., Campanati, L., Spring, H., Trendelenburg, M.F.: Gold finder: a computer method for fast automatic double gold labeling detection, counting, and color overlay in electron microscopic images. Journal of Structural Biology **141**, 228–239 (2003)
8. Nejatbakhsh, R., Kabir-Salmani, M., Dimitriadis, E., et al.: Subcellular localization of L-selectin ligand in the endometrium implies a novel function for pinopodes in endometrial receptivity. Reprod Biol Endocrinol **10**, 46 (2012)
9. Płaczek, B., Bułdak, R.J., Polaniak, R., Matysiak, N., Mielanczyk, Ł., Wojnicz, R.: Detection of immunogold markers in images obtained from transmission electron microscopy. Journal of Medical Informatics Technologies **23**, 111–118 (2014)
10. Płaczek, B.: A real time vehicle detection algorithm for vision-based sensors. In: Bolc, L., Tadeusiewicz, R., Chmielewski, L.J., Wojciechowski, K. (eds.) ICCVG 2010, Part II. LNCS, vol. 6375, pp. 211–218. Springer, Heidelberg (2010)
11. Porwik, P., Sosnowski, M., Wesolowski, T., Wrobel, K.: A computational assessment of a blood vessel's compliance: a procedure based on computed tomography coronary angiography. In: Corchado, E., Kurzyński, M., Woźniak, M. (eds.) HAIS 2011, Part I. LNCS, vol. 6678, pp. 428–435. Springer, Heidelberg (2011)
12. Saluja, R., Jyoti, A., Chatterjee, M., Habib, S., et al.: Molecular and biochemical characterization of nitric oxide synthase isoforms and their intracellular distribution in human peripheral blood mononuclear cells. Biochimica et Biophysica Acta (BBA)-Molecular Cell Research **1813**(10), 1700–1707 (2011)
13. Shimizu, Y., Kabir-Salmani, M., Azadbakht, M., Sugihara, K., Sakai, K., Iwashita, M.: Expression and localization of galectin-9 in the human uterodome. Endocrine journal **55**(5), 879–887 (2008)
14. Starink, J.P., Humbel, B.M., Verkleij, A.J.: Three-dimensional localization of immunogold markers using two tilted electron microscope recordings. Biophysical journal **68**(5), 2171–2180 (1995)
15. Wang, R., Pokhariya, H., McKenna, S.J., Lucocq, J.: Recognition of immunogold markers in electron micrographs. Journal of structural biology **176**(2), 151–158 (2011)

Fusion of Granular Computing and k–NN Classifiers for Medical Data Support System

Marcin Bernas$^{(\boxtimes)}$, Tomasz Orczyk, and Piotr Porwik

Institute of Computer Science, University of Silesia, Bedzinska 39,
41–200 Sosnowiec, Poland
{marcin.bernas,tomasz.orczyk,piotr.porwik}@us.edu.pl

Abstract. The medical data and its classification should be particularly treated. The data can not be modified or altered, because this could lead to overestimation or false decisions. Some classifiers, using random factors, can generate false, higher overall accuracy of diagnosis. Medical support systems should be trustworthy and reliable even at the cost of system complexity. In this paper fusion of two classifiers has been proposed, where k–NN classifier and classifier based on a justified granulation paradigm were employed. Additionally, proposed solution allows to visualize obtained classification results. Accuracy of the proposed solution has been compared with various classifiers. All methods presented in this work were tested on real medical data coming from three medical datasets. Finally, some remarks for further research have been proposed.

Keywords: Medical data support system · Justified granular paradigm · Classifiers fusion

1 Introduction

Medical diagnosis support systems are increasingly used in medical practice and help in daily work of medical doctors. These systems, as software, are often built in many medical devices. Well–known and reliable classifiers tend to fail when faced with new problems such as an atypical data. Therefore, new methods must be developed to deal with the challenges arising and improve the quality of real–life decision support systems [1]. Presented research has been carried out separately on archival datasets which contained observations on changes in breast tissue, heart disease and liver patients.

Human decision support system, in the medical practice, can be considered, inter alia, as diagnosis support on the basis of the patient's medical data. In this paper diagnosis support consists of medical data classification, where fusion of granulation paradigms [2] and k–NN classifier [3] have been applied.

The first method is based on a granular computing paradigm [4]. Using this paradigm the data can be aggregated into many formal representations of information granules: intervals [5], fuzzy sets [6], rough sets [7], shadowed sets [8], or probabilistic sets [9]. There are several works which prove the usefulness of

© Springer International Publishing Switzerland 2015
N.T. Nguyen et al. (Eds.): ACIIDS 2015, Part II, LNAI 9012, pp. 62–71, 2015.
DOI: 10.1007/978-3-319-15705-4_7

this concept [10,11]. The granulation method tends to generalize the information at cost of losing the details. Therefore, as a second method, the k–nearest neighborhood (k–NN) classifier has been selected [12,24]. This classifier tends to find the several closest solutions for a given data. The accuracy of proposed method was compared with various classifiers: perceptron based neural network [13], fuzzy rules classifier [14], random trees [15] and Bayes network classifier [24]. The remainder of the work is organized as follows. Section 2 presents a proposed method. The classification results present Section 3. Section 4 includes conclusion with remarks on possibilities of further development of the method.

2   Description of the Method

The proposed method bases on the observation of physicians. The medical diagnosis is often taken on the basis of the patient's parameter analysis. If observed parameters fall into appropriate interval, then parameters favor a given diagnosis. On the other hand the similarity between single cases can be analyzed [17]. These methods of diagnosing, among many others, are commonly used in the medical environment.

The first method is implemented as the justified granulation paradigm [18] based classifier. Granulation technique offers an interval representation of attributes. The second method uses the k–NN classifier and allows to find the similar, but rare cases – patients with similar parameters marked. Finally, the voting scheme based on the weights was used to obtain the final decision. The main principles of the proposed classifiers connection presents Fig. 1. Both the

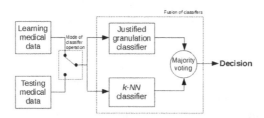

Fig. 1. The diagram of the medical data flow in the proposed classification process

k–NN based classifiers and granulation methods are well–known and are widely described in literature [3,12]. Therefore, only the authors' contribution and modifications of these methods will be presented.

2.1   Description of the Analyzed Data

For the purpose of experiments, three datasets containing real medical data, have been used. The first database contains heart disease diagnosis (Hungarian Institute of Cardiology. Budapest: Andras Janosi, M.D.; University Hospital, Zurich,

Switzerland: William Steinbrunn, M.D.; University Hospital, Basel, Switzerland: Matthias Pfisterer, M.D.; V.A. Medical Center, Long Beach and Cleveland Clinic Foundation: Robert Detrano, M.D., Ph.D.) [19]. The second database contains electrical impedance measurements of freshly excised tissue samples from breast [20]. The third database contains 416 liver patient records and 167 non liver patient records. The data set was collected from north east of Andhra Pradesh, India [23]. These datasets have no missing values. Mentioned above datasets can be characterized as follows:

- Heart disease: 13 parameters, 270 cases, 2 groups of diagnoses (angiographic disease status), group distribution: ">50% narrowing" – 44%, "<50% narrowing" – 56%;
- Breast tissue: 9 parameters, 106 cases, 6 groups of diagnoses (tissue type), group distribution: "Connective" – 13%, "Fibro–adenoma" – 14%, "Glandular" – 15%, "Mastopathy" – 17%, "Carcinoma" – 20%, "Adipose" – 21%;
- Liver patients: 10 parameters, 583 cases, 2 groups of diagnoses (patient type), group distribution: "liver patient" – 71%, "non-liver patient" – 29%.

Data statistics are presented in Table 1.

2.2 Information Mining and Classification Using Justified Granulation Method

The proposed method is universal and can be applied to datasets analysis without modifications. Each dataset consists of medical parameters treated as an input data of the medical support systems. In these systems various classifiers can work. It will be explained in the next sections of the paper.

As was mentioned above the three medical databases have been tested. In the first and second database $m = 13$ and $m = 9$ parameters have been extracted, while in the third database $m = 10$ parameters are analyzed.

The first and third dataset is description of patients with two stages of angiographic disease of hearth and two types of liver patient, so only two classes of the diagnosis can by determined – $n = 2$. In case of a second dataset the diagnosis of six types of changes in breast tissue can be observed, hence $n = 6$.

The proposed method processes every $i^{th}, i = 1, ..., m$, input parameter separately. Values of the ith parameter through all patients are assigned to the appropriate set, which represent a class of the diagnosis. In our case only restricted number of classes are defined according to the properties of the medical datasets, but in practice number of classes is not limited, therefore $n \in N$.

The set is representing a class and will be denoted as $X_{i,j}$, where $i = 1, ..., m$ and $j = 1, ..., n$. It means that any set contains values of elements representing j^{th} class (diagnosis) for i^{th} parameter.

Because any sets include series of not ordered numeric values, these values can be gathered into some intervals $\overline{X_{i,j}}$. Searching of optimal interval is not trivial, therefore the justified granularity method have been adopted for this task. The granulation method has been thoroughly described in [18]. The justified

Table 1. Data statistics for datasets used in the research

	Parameter	Mean	Std. dev.	Parameters description
Heart Disease Dataset	age	54.43	9.11	age – age in years, sex – sex (1 = male; 0 = female), cp – chest pain type (1 = typical angina, 2 = atypical angina, 3 = non–anginal pain, 4 = asymptomatic), trestbps – resting blood pressure (in mm Hg on admission to the hospital), chol – serum cholestoral in mg/dl, fbs – fasting blood sugar >120 mg/dl (1 = true; 0 = false), restecg – resting electrocardiographic results (0 = normal, 1 = having ST–T wave abnormality, 2 = showing probable or definite left ventricular hypertrophy by Estes' criteria), thalach – maximum heart rate achieved, exang – exercise induced angina (1 = yes; 0 = no), oldpeak – ST depression induced by exercise relative to rest, slope – the slope of the peak exercise ST segment (1 = upsloping, 2 = flat, 3 = downsloping), ca – number of major vessels (0–3) colored by flourosopy, thal – 3 = normal; 6 = fixed defect; 7 = reversable defect.
	sex	0.68	0.47	
	cp	3.17	0.95	
	trestbps	131.34	17.86	
	chol	249.66	51.69	
	fbs	0.15	0.36	
	restecg	1.02	1.00	
	thalach	149.68	17	
	exang	0.33	0.47	
	oldpeak	1.05	1.15	
	slope	1.59	0.61	
	ca	0.67	0.94	
	thal	4.70	1.94	
Breast Tissue Dataset	I0	784.25	754.0	I0 – Impedance (ohm) at zero frequency, PA500 – phase angle at 500 kHz, HFS – high–frequency slope of phase angle, DA – impedance distance between spectral ends, Area – area under spectrum, A/DA – area normalized by DA, Max IP – maximum of the spectrum, DR – distance between I0 and real part of the maximum frequency point, P – length of the spectral curve.
	PA500	0.12	0.1	
	HFS	0.11	0.1	
	DA	190.57	190.8	
	Area	7335.16	18580.3	
	A/DA	23.47	23.4	
	Max IP	75.38	81.3	
	DR	166.71	181.3	
	P	810.64	763.0	
Liver Patients Dataset	Age	44.75	16.19	Age – Age of the patient, Gender – Gender of the patient, TB – Total Bilirubin, DB – Direct Bilirubin, Alkphos – Alkaline Phosphotase, Sgpt – Alamine Aminotransferase, Sgot – Aspartate Aminotransferase, TP – Total Protiens, ALB – Albumin, AGR – Albumin and Globulin Ratio.
	Gender	0.76	0.43	
	TB	3.30	6.21	
	DB	1.49	2.81	
	Alkphos	290.58	242.94	
	Sgpt	80.71	182.62	
	Sgot	109.91	288.92	
	TP	6.48	1.09	
	ALB	3.14	0.80	
	AGR	0.95	0.32	

granulation G allows to find an interval representation $\overline{X_{i,j}} = G(X_{i,j}, \alpha) = [x_a, x_b]$ as a balance between generality and specificity, according to the $\alpha \in [0, 1]$ attribute. The x_a and x_b are left and right bound of interval, respectively. Values of the α attribute significantly affects the selection of interval bounds. Fig. 2 illustrates the basic principles of the interval searching. For $\alpha = 0$, all elements

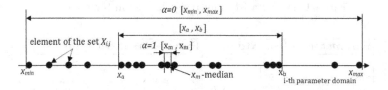

Fig. 2. Illustration of the different intervals inside the set $X_{i,j}$ for the ith parameter

of the set $X_{i,j}$ lie inside the created interval, while for $\alpha = 1$ interval is restricted to the median value only. The α values between 0 and 1 select interval bounds within the homogeneous elements of $X_{i,j}$ set. Optimal attribute for every i^{th} parameter will be computed for all $X_{i,j}$ sets by means of the two formulas. The first formula defines the overlapping of the two intervals within a given i^{th} parameter, where $i = 1, ..., m, j, k \in [1, n]$:

$$\lambda(\overline{X_{i,j}}, \overline{X_{i,k}}) = \lambda([x_a, x_b], [x'_a, x'_b]) =$$

$$= \begin{cases} \frac{x_b - x'_a}{x'_b - x_a} & \text{if } x_a \leq x'_a \wedge x'_a < x_b \wedge x_b \leq x'_b \\ \frac{x_b - x_a}{x_b - x'_a} & \text{if } x'_a \leq x_a \wedge x_a < x'_b \wedge x'_b \leq x_b \\ \frac{x_b - x_a}{x_b - x_a} & \text{if } x'_a > x_a \wedge x'_b < x_b \\ \frac{x_b - x_a}{x'_b - x'_a} & \text{if } x_a > x'_a \wedge x_b < x'_b \\ 0 & \text{otherwise} \end{cases} . \quad (1)$$

The formula minimizes overlapping ratio of the intervals representing separated classes, and thus favors smaller intervals only, which are closer to the median. Therefore the second formula was proposed. In the formula δ, number of elements inside the interval $\overline{X_{i,j}}$ is taken into consideration. The value of the formula δ is a ratio, which determines number of elements that lie inside the interval $\overline{X_{i,j}}$ to the cardinality of the set $X_{i,j}$:

$$\delta(\overline{X_{i,j}}, X_{i,j}) = \frac{\sum_{x \in X_{i,j}} f(x, \overline{X_{i,j}})}{card(X_{i,j})}, \quad (2)$$

where: $f(x, \overline{X_{i,j}}) = f(x, [x_a, x_b]) = \begin{cases} 1 & \text{if } x \geq x_a \wedge x \leq x_b \\ 0 & \text{otherwise} \end{cases}$, $card$ – cardinality of a set, f – membership function of the element x to the interval $[x_a, x_b]$.

Ultimately, the optimal value of attribute for i^{th} parameter is formulated as combination of formulas (1) and (2) as follows:

$$\alpha_i = \underset{\alpha \in [0,1]}{\arg\max} \left(\sum_{k=1}^{n} \sum_{j=1}^{n} \frac{1 - \lambda(\overline{X_{i,j}}, \overline{X_{i,k}})}{n^2} - p \right) + \left(p - \sum_{j=1}^{n} \frac{\delta(\overline{X_{i,j}}, X_{i,j})}{n} \right) , \quad (3)$$

where: $p \in [0, 1]$ is a constant calculated in the classifier's learning mode,

$$\overline{X_{i,j}} = G(X_{i,j}, \alpha)$$

Changes of the p value can improve the classifier's overall accuracy. The constant p is optimized on the basis of the learning dataset.

We have proposed weighted classifiers because granulation techniques provide simpler models for the classes. For this technique, the analyzed input data can belong at the same time to different classes. For this reason, class importance should be established. It means that each jth class is pointed out with the appropriate weight:

$$w_g(j) = \frac{\sum_{i=1}^{m} f\left(x_i, \overline{X_{i,j}}\right)}{\sum_{k=1}^{n} \sum_{=1}^{m} f\left(x_i, \overline{X_{i,j}}\right)} \quad , j = 1, ..., n , \quad (4)$$

where:
x_i – the classified i^{th} parameter, m – number of attributes, n – number of classes.

The function f returns 1 if classified parameters value belongs to the interval and returns 0 in other case. The sum of weights $w_g(j), j = 1, ..., n$ must be equal to 1.

Changes of the p value can improve the classifier's overall accuracy. The constant p is optimized on the basis of the learning dataset.

As a result of classification process a class with maximal weight is selected:

$$C_g = \underset{j}{\arg\max}\{w_g(j) : j = 1, ..., n\} , \quad (5)$$

where: C_g – the result class for a granular classifier.

The calibration process was performed on learning set by means of comparing the classifier's result (C_g) with a declared class to calculate overall accuracy. The optimal value of p constant is the value for which the maximal overall accuracy is achieved. The result obtained for various p values depicts Fig. 3. Processed datasets differ in number of classes and parameters, however selection of the constant p shows that the best value $p = 0.9$ was achieved for every analysed dataset.

2.3 K–NN Clasifier

As a second classifier, the k–NN technique has been selected as a tool to find special, single cases among patients. The k–NN classifier was implemented in the KNIME environment [3], where the optimal number of nearest neighbors (k)

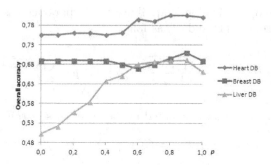

Fig. 3. The classifier's overall accuracy changes for different p values separately for different medical datasets

has been established on the basis of a learning set. In practice, the most suitable value of k should be estimated in simulation experiments, so in our approach this condition has been fulfilled. The experiments conducted shows that parameter k should be set up to $k = 7$ for the first and third dataset and $k = 5$ for the second medical dataset, respectively. The results of the k neighbor's estimation for overall accuracy k–NN classifier improvement present Fig. 4. In classical app-

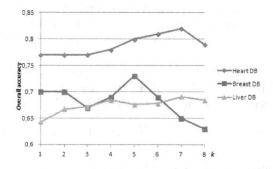

Fig. 4. The classifier's overall accuracy changes for different k values – separately for different medical datasets

roach discriminant functions of k–NN classifier with nearest neighbor estimator of the element x can be described as follows:

$$g_j(x) = \frac{k_j}{k}, j = 1, ..., n \,, \qquad (6)$$

where: k_j is a number of elements, of the jth class, which belong to the k nearest neighbors of the element x, while sum of $\sum_{j=1}^{n}(g_j(x)) = 1$.

The formula (6) estimates a posteriori probability of the jth class for element x.

Hence k–NN classifier C_k realizes the rule:

$$C_k(x) = j \Leftrightarrow \forall_{\substack{i=1,\dots,n \\ j\neq i}} g_j(x) > g_i(x) \,. \tag{7}$$

The estimator (6) can also be treated as weighted support function of the k–NN classifier for the class j^{th}:

$$w_k(j) = g_j(x) \,. \tag{8}$$

It is obvious that sum of the weights $w_k(j)$ is equal to 1.

2.4 Classifiers Fusion

The final classification is obtained by majority voting of the two classifiers. Idea of such classification presents also Fig. 1. Weights are appropriately calculated for granular and k–NN classifiers. Ultimately, class with the largest sum of weights is selected:

$$C_f = \arg\max_{j}\{w_k(j) + w_g(j) \ : \ j = 1, \dots, n\} \,, \tag{9}$$

where: C_f – points out the class number (classifier's final decision)
The principle of the final decision through classifiers fusion presents Fig. 5.

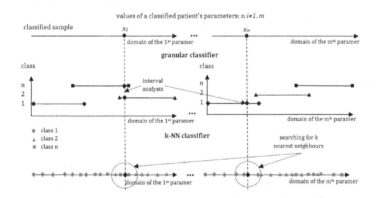

Fig. 5. Illustration of proposed classification method for exemplary three classes

3 Obtained Results

The proposed method was verified using 10–fold cross validation, separately for three aforementioned medical datasets. Proposed classifiers fusion was compared with other well–known single classification methods using the same 10-fold cross validation method - namely the Random Forest[15], Instance Based kNN[3], Bayes Network[21] and PNN[13]. All the classifier were working in their out of the box settings as implemented in the KNIME[3]/WEKA[22] environments. The classifiers comparison result presents Table 3, where as quality of classification the overall accuracy was stated. For the Heart Disease database the proposed method was the second best, outperforming the Random Forest, kNN and PNN classifiers, while for the Breast Tissue and Liver Patients databases the Proposed Method obtained the best overall accuracy, outperforming all compared classifiers. The proposed method gave the best result for the breast tissue dataset,

Table 2. The overall accuracy (%) of various classifiers

Method / Database	Proposed method	Random Forest	k–NN	Bayes Network	PNN
Heart disease database	81.6	79.8	80.1	83.1	63.8
Breast tissue database	75.5	69.8	69.1	67.0	53.8
Liver patient database	72.1	68.3	65.7	66.2	71.4

where six distinctive classes of disease can be performed. The Bayes network gave slightly better result for heart disease dataset, but this method indicated the greater misclassifications for breast tissue dataset.

The proposed solution offers a stable classification results. It is obvious that carried out investigation should be much more extended in the future.

4 Summary

The paper proposes a fusion of two classifiers, where one generalize the information while the second analyzes specific cases. As it was proved classifiers fusion offers higher overall accuracy compared to single classifiers. These researches show that fusion and adaptation of granular computing paradigm and k–NN classifier creates an effective classification method. Performed investigations are promising and can be included to the specialized medical expert systems. In the future the proposed solution will be extended and compared with other classifiers and tested by means of various medical datasets.

References

1. Foster, K.R., Koprowski, R., Skufca, J.D.: Machine learning, medical diagnosis, and biomedical engineering research - commentary. BioMedical Engineering OnLine **13**, 94 (2014)
2. Song, M., Wang, Y.: Human centricity and information granularity in the agenda of theories and applications of soft computing. Applied Soft Computing (2014). doi:10.1016/j.asoc.2014.04.040
3. Aha, D., Kibler, D., Albert, M.: Instance-Based Learning Algorithms. Machine Learning **6**(1), 37–66 (1991)

4. Zhang, Y., Zhang, L., Xu, C.: The Property of Different Granule and Granular Methods Based on Quotient Space. Information Granularity, Big Data, and Computational Intelligence Studies in Big Data **8**, 171–190 (2015)
5. Huang, B., Zhuang, Y., Li, H.: Information granulation and uncertainty measures in interval-valued intuitionist fuzzy information systems. European Journal of Operational Research **231**, 162–170 (2013)
6. Kudlacik, P., Porwik, P.: A New Approach To Signature Recognition Using The Fuzzy Method. Pattern Analysis And Applications **17**(3), 451–463 (2014)
7. Cao, Y., Liu, S., Zhang, L., Qin, J., Wang, J., Tang, K.: Prediction of protein structural class with Rough Sets. BMC Bioinformatics **7**, 20 (2006)
8. Pedrycz, W.: Interpretation of clusters in the framework of shadowed sets. Pattern Recognition Letters **26**(15), 2439–2449 (2005)
9. Hirota, K.: Concepts of probabilistic sets. Fuzzy Sets and Systems **5**(1), 31–46 (1981)
10. Mago, V., Morden, H., Fritz, C., Tiankuang, W., Namazi, S., Geranmayeh, P., Chattopadhyay, R., Dabbaghian, V.: Analyzing the impact of social factors on homelessness: a Fuzzy Cognitive Map approach. BMC Medical Informatics and Decision Making **13**, 94 (2013)
11. Emam, K., Dankar, F., Neisa, A., Jonker, E.: Evaluating the risk of patient re-identification from adverse drug event reports. BMC Medical Informatics and Decision Making **13**, 114 (2013)
12. Bernas, M., Placzek, B., Porwik, P., Pamula, T.: Segmentation of vehicle detector data for improved k-nearest neighbours-based traffic flow prediction. IET Intelligent Transport Systems. doi:10.1049/iet-its.2013.0164
13. Berthold, M., Diamond, J.: Constructive training of probabilistic neural networks. Neurocomputing **19**(1–3), 167–183 (1998)
14. Berthold, M.: Mixed fuzzy rule formation. International Journal of Approximate Reasoning **32**(2–3), 67–84 (2003)
15. Breiman, L.: Random Forests. Machine Learning **45**(1), 5–32 (2001)
16. John, G., Langley, P.: Estimating continuous distributions in Bayesian classifiers. In: Besnard, P., Hanks, S. (eds.) Proceedings of the Eleventh Conference on Uncertainty in Artificial Intelligence, pp. 338–345 (1995)
17. Pantazi, S.V., Arocha, J.F., Moehr, J.R.: Case-based medical informatics. BMC Medical Informatics and Decision Making **4**, 19 (2004)
18. Pedrycz, W., Gomide, F.: Fuzzy Systems Engineering: Toward Human-Centric Computing. John Wiley, Hoboken (2007)
19. Aha, D., Kibler, D.: Instance-based prediction of heart-disease presence with the Cleveland database. University of California (1988)
20. Jossinet, J.: Variability of impedivity in normal and pathological breast tissue. Med. & Biol. Eng. & Comput **34**, 346–350 (1996)
21. Friedman, N., Geiger, D., Goldszmidt, M.: Bayesian Network Classifiers. Machine Learning **29**, 131–163 (1997)
22. Hall, M., Frank, E., Holmes, G., Pfahringer, B., Reutemann, P., Witten, I.: The WEKA Data Mining Software: An Update; SIGKDD Explorations, 11(1) (2009)
23. Ramana, B.V., Prasad Babu, M., Venkateswarlu, N.: A Critical Study of Selected Classification Algorithms for Liver Disease Diagnosis. International Journal of Database Management Systems (IJDMS) **3**(2), 101–114 (2011)
24. Porwik, P., Doroz, R., Orczyk, T.: The k-NN classifier and self-adaptive Hotelling data reduction technique in handwritten signatures recognition, Pattern Analysis and Applications. doi:10.1007/s10044-014-0419-1

Lip Print Recognition Method Using Bifurcations Analysis

Krzysztof Wrobel[(✉)], Rafał Doroz, and Malgorzata Palys

Institute of Computer Science, University of Silesia,
ul. Bedzinska 39, 41-200 Sosnowiec, Poland
{krzysztof.wrobel,rafal.doroz,malgorzata.palys}@us.edu.pl
http://zsk.tech.us.edu.pl, http://biometrics.us.edu.pl

Abstract. The paper presents a method of automatic personal identification on the basis of an analysis of lip prints. The method is based on a new approach, in which each lip print is described by bifurcations. In order to extract bifurcations, a method for lip print pre-processing was proposed. The bifurcations obtained were compared with each other using the similarity coefficient developed for the needs of this study. The effectiveness of this coefficient was verified experimentally.

Keywords: Biometrics · Lip print · Image pre-processing · Bifurcations

1 Introduction

Currently, the need for implementing modern systems for personal identification or verification is constantly growing. Such systems are often based on biometric methods. The most popular biometric identification methods include: signature recognition, iris recognition, fingerprint recognition and profiling [1,3,6,7]. There are also other methods, e.g. cheiloscopy, which are becoming more popular.

The term *cheiloscopy* is derived from Greek words: *cheilos* – lips, *skopeo* – to observe. Cheiloscopy is a domain of criminology which deals with the examination of lip prints and the personal identification based on lip prints [10,12]. The vermilion border (Latin: *rubor labiorum*) is a demarcation between the outer layer of a lip and its inner part. In other words, it is an intermediate part of a lip.

In cheiloscopy, the identification of a given person takes place on the basis of an analysis of characteristic features located in the vermilion border. A single lip print contains, on an average, 1145 individual features forming a unique pattern different for each person [8]. This is a very large number as compared with a fingerprint, in which approx. 100 individual features can be identified. There is no doubt that such a huge potential of lip prints can be used for personal identification [15–17]. Thank to this, in recent years, apart from applications of cheiloscopy in criminology, research are conducted on its use for biometric personal identification and verification.

© Springer International Publishing Switzerland 2015
N.T. Nguyen et al. (Eds.): ACIIDS 2015, Part II, LNAI 9012, pp. 72–81, 2015.
DOI: 10.1007/978-3-319-15705-4_8

So far, there are a few methods for lip print examinations. These methods use, inter alia, statistical analyses [8], similarity coefficients [4], lip shape analyses [5], Dynamic Time Warping [13], segmentation [9], Hough Transform [17]. In these methods, only single features, such as lines forming the lip print pattern, are analysed and compared with each other. A new approach was proposed in this study. It involves not only analyses of single features, but also the determination and investigation of the relationships between them.

2 Proposed Method

The method proposed in this study is used to determine the similarity between two lip prints compared with each other. The comparison takes place by designating a set of features in every image and then comparing them with each other. As the result of such a comparison, a set of the most similar features in both images is obtained. On the basis of the set obtained in this way, the coefficient of similarity between images is determined. The similarity coefficient allows deciding whether the lip prints compared belong to the same person. Steps of this method are shown in the block diagram (Fig. 1).

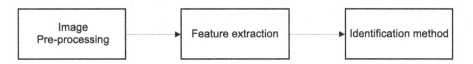

Fig. 1. Block diagram of the proposed method

2.1 Image Pre-processing

Accurate and correct acquisition of the feature set is one of the factors determining the effectiveness of the method presented. Similarly to fingerprints, lip prints may have a poor quality or have a low contrast and be noised. This causes that the extraction of features from such a print is not an easy task. An example of a poor quality print is shown in Figure 2.

In order to expose the characteristic features better, the image is pre-processed. In the first stage of pre-processing, each print from the test database was subjected

Fig. 2. An example of a poor quality lip print

to the process of linear contrast stretching [2]. This process consists in increasing the differences in the brightness level between pixels of the image. If the lip print is characterized by a low contrast, it becomes clearer and therefore easier for analysing. The effect of linear contrast stretching is shown in Figure 3a.

Unfortunately, during the acquisition of the mouth image, the areas around the mouth, such as the cheeks or chin, may also be registered. In further steps of the method, they are interpreted as elements of the lip print, which may cause errors in recognition. Therefore, the next step is to locate the area of the mouth (upper and lower lips), and then to assign the black colour to all the pixels that do not belong to this area.

The mouth area is located by analysing the brightness of pixels in the image. As shown in Figure 3a, the lip print area is darker. The colour of the pixels with brightness lower than a certain threshold is set to black. In this study, the threshold value was determined using the method proposed by Ridler-Calvard [14]. The Ridler's-Calvard's algorithm chooses an initial threshold and then iteratively calculates the next one by taking the mean of the average intensities of the background and foreground pixels determined by the first threshold, repeating this until the threshold converges. Figure 3b shows the effect of the described operations.

The image prepared in this way is subjected to the binarization process. The pixel examined in the image will become white, if the level of its brightness is higher than the average brightness level in the area around it. It was assumed that the size of the analysed area is 7x7 pixels. Otherwise, the pixel will become black. Then, the image is subjected to the operation of negation. As a result, an image is obtained, in which black pixels form the lip print pattern. Such an image is shown in Figure 3c.

As a result of the use of the methods described so far, the lines in the image that form the lip print pattern have a different thickness. This hinders the extraction of bifurcations occurring in the image, which are used in a further part of the study for determining the similarity between lip prints. In order to eliminate this inconvenience, the image is subjected to the process of skeletonization. As a result, an image is obtained, in which all lines have a thickness of one pixel. The Pavlidis method [11] was used in the studies. The result of using this method is shown in Figure 3d. The features to be compared are extracted from the image prepared in this way.

2.2   Feature Extraction

Under the method presented, lower and upper bifurcations were compared in order to determine the similarity of lip prints. Such bifurcations belong to the most frequently occurring features that can be extracted in the easiest way. The set of bifurcations forms a unique pattern for each person, on the basis of which it is possible to identify a given person. For this reason, in relation to a single bifurcation, the features describing relationships between a given bifurcation and the remaining bifurcations were analysed in addition to the determination of its coordinates and the orientation angle.

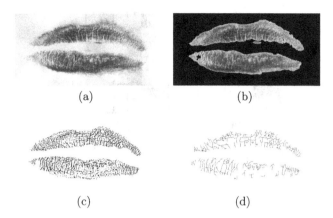

Fig. 3. (a) Image after the linear contrast stretching, (b) image after assign the black colour to all the pixels that do not belong to this mouth area, (c) image after the binarization process, (d) image after the process of skeletonization

In order to find bifurcations, all black pixels in the lip print are analysed. If an analysed pixel is to be considered as the centre of a bifurcation, two conditions must be fulfilled. Firstly, bifurcations are characterized by the presence of three black pixels around the pixel being analysed. This condition is checked by placing a 3x3 mask on the analysed pixel. The sum of black pixels in the mask must equal four. An example of an identified bifurcation is shown in Figure 4.

The second condition is that the minimum length of components of a bifurcation is equal to four pixels, including the pixel being analysed. A lack of the above condition resulted in the detection of pixel clusters that were not bifurcations. An example of such clusters is presented in Figure 4a. The analysis of the length of components of the bifurcation was carried out by placing a mask (with the size of 7x7 pixels) on the pixel being analysed. Then, the black pixels located only on the edges of the mask (Fig. 4b) are summed up. If the sum equals three, the coordinates of the pixel being analysed will be regarded as the coordinates of the bifurcation. A pixel with the coordinates (x, y) which satisfies the above two conditions is treated as the centre of the bifurcation.

The orientation angle is determined for the bifurcation in accordance with the assumptions of the method. In order to determine this angle, values of three angles (Ω, β, α) between the arms of the bifurcation need to be found. Examples of such angles are shown in Figure 5a. The arm, which together with the remaining arms forms two largest angles, determines the direction of the bifurcation. The θ angle between this direction and the X axis is called the bifurcation orientation angle (Fig. 5b). The arrangement of bifurcations in a lip print in relation to each other is unique to each person. The coordinates and orientation angles of each bifurcation in relation to adjacent bifurcations form systems of bifurcations. It was assumed that the two bifurcations described by the coordinates

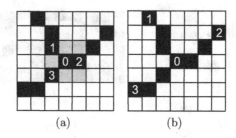

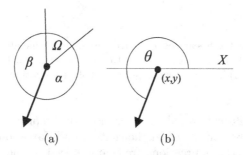

Fig. 4. An example of a bifurcation. The analysed pixel is designated with the number 0.

Fig. 5. (a) The Ω, β and α angles between arms of the bifurcation. The direction of the bifurcation is marked with a bold line, (b) the bifurcation orientation angle determined.

(x_i, y_i), (x_j, y_j) and the θ_i and θ_j angles are adjacent, if the following conditions are fulfilled:

$$(x_i - dist < x_j < x_i + dist) \quad \text{and} \quad (y_i - dist < y_j < y_i + dist), \qquad (1)$$

$$|\theta_i - \theta_j| < ang, \qquad (2)$$

where:
$dist$ – the maximum accepted distance between centres of bifurcations,
ang – the admissible difference between the bifurcation orientation angles.

Therefore, the lip print Lip can be described with the use of a set containing p bifurcation systems:

$$Lip = \{s_1, s_2, ..., s_p\}. \qquad (3)$$

Bifurcation systems consist of a set containing three features (d, α, β) that describe the relationships between the bifurcations. A single relationship is described by the following features:
d – the Euclidean distance between the centre of the analysed bifurcation and the centre of an adjacent bifurcation,
α – the angle between the direction of the analysed bifurcation and the section connecting the centre of the bifurcation with the centre of an adjacent bifurcation,

β – the angle between the orientation angle of the analysed bifurcation and the orientation angle of an adjacent bifurcation.

An example of a bifurcation system with features marked on it is shown in Figure 6.

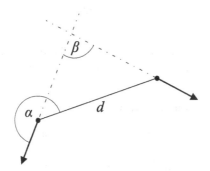

Fig. 6. A bifurcation system with features marked on it

The relationships determined for each bifurcation can be presented in a tabular form. Examples for the three bifurcations are shown in Table 1.

Table 1. Examples of relationships between the three bifurcationsand adjacent bifurcations

Analysed bifurcation			Adjacent bifurcation			
no.	coordinates (x, y)	θ	no.	d	α	β
1	(127, 210)	51	2	68	178	103
			3	47	77	87
			4	51	114	129
			11	39	270	38
			14	71	15	137
2	(86, 148)	163	1	68	116	103
			4	59	135	124
			5	57	192	141
			8	57	276	17
			9	70	47	8
3	(102, 198)	347	1	47	177	87
			4	39	118	83
			5	71	106	39
			6	67	34	82
			14	73	265	169

A single bifurcation system can also be presented in the form of a bifurcation matrix:

$$s_i = \begin{bmatrix} d_1^i & \alpha_1^i & \beta_1^i \\ d_2^i & \alpha_2^i & \beta_2^i \\ \vdots & \vdots & \vdots \\ d_n^i & \alpha_n^i & \beta_n^i \end{bmatrix}, \tag{4}$$

where:
s_i – the matrix determined for the i-th system of bifurcations,
n – the number of triple features describing the system.

2.3   Lip Prints Identification Method

In practice, many characteristic features can be found on lip prints. It often happens that similar characteristic features are present in the same places of lip prints obtained from different people, which makes the identification more difficult. This fact was used for developing a new coefficient of similarity between lip prints. This coefficient is based on the observation that the probability of occurrence of two identical bifurcation systems in different individuals is smaller than the probability of occurrence of two identical bifurcations. Therefore, the method presented here analyses not only single bifurcations, but also bifurcation systems.

The lip print identification method consists in comparing a given lip print with lip prints of other people contained in the database. In the first step of this method, the similarity between bifurcation matrices is determined for two lip prints being compared. Each row of the matrix includes three features (d, α, β). In order to compare two bifurcation matrices s_k and s_l derived from two lip prints being compared (Lip_1 and Lip_2), the following similarity coefficient was developed:

$$sim(s_k, s_l) = \frac{1}{n} \sum_{i=1}^{n} \min_{j=1,\ldots,m} \{|d_i^k - d_j^l| + |\alpha_i^k - \alpha_j^l| + |\beta_i^k - \beta_j^l|\}, \tag{5}$$

where:
$s_k = (d_i^k, \alpha_i^k, \beta_i^k) \in Lip_1$ – i-th row of the bifurcation matrix s_k,
$s_l = (d_j^l, \alpha_j^l, \beta_j^l) \in Lip_2$ – j-th row of the bifurcation matrix s_j,
n – number of rows in the bifurcation matrix s_k,
m – number of rows in the bifurcation matrix s_l.

The advantage of the coefficient given by the formula (5) is that it can compare the bifurcation systems described by a matrix with different numbers of rows.

The similarity between the compared lip prints is determined on the basis of the values of the similarity between individual bifurcation systems. Let the first lip print be represented by the set of bifurcation systems $Lip_1 = \{s_1^1, s_2^1, \ldots, s_p^1\}$, while the second image by the set $Lip_2 = \{s_1^2, s_2^2, ..., s_q^2\}$. The similarity between the lip prints is determined using the following formula:

$$SIM(Lip_1, Lip_2) = \frac{1}{p} \sum_{i=1}^{p} \min_{j=1,...,q} \{sim(s_i^1, s_j^2)\},\qquad(6)$$

where:
p – number of bifurcation systems in the lip print Lip_1,
q – number of bifurcation systems in the lip print Lip_2.

The lower the value of the similarity coefficient SIM, the more similar the lip prints. After a comparison of the lip print analysed with all the lip prints in the database, the lip print for which the lowest value of the similarity coefficient SIM was obtained indicates the person who provided the lip print.

3   Experiments and Results

The effectiveness of the method was verified experimentally. The database used for the studies consisted of 120 lip prints obtained from 30 people (4 prints per person). Each lip print was adequately prepared using the methods described in previous sections. The prints were compared with each other using the round-robin method. As a result of the comparison of two prints, the value of the similarity between them was obtained. During the studies, various values of the $dist$ and ang parameters used at the stage of feature extraction were analysed. These parameters were changed in the following ranges:

  – $dist$ parameter from 10 to 45, with the step 5,
  – ang parameter from 5 to 25, with the step 5.

The total number of the parameter sets created was 40. Value of EER was determined for each set of parameters. The parameter sets, for which 10 lowest values of EER were obtained, are shown in Table 2.

Table 2. The EER values obtained for different combinations of values of the parameters used in the identification method

Parameter		EER [%]
dist	*ang*	
20	5	23.04
15	5	23.88
15	10	24.05
20	10	24.40
10	15	24.74
15	15	25.39
25	15	25.48
20	20	25.56
10	25	25.63
10	20	25.66

When analysing the results presented in Table 2, it can be seen that the best value of EER 23% was obtained for the values $dist = 20$, $ang = 5$. Then the impact of the ang parameter on the results obtained was analysed. The analyses were performed for the following values of the $dist$ parameter: 10, 15, 20, 25. The results are presented in the Figure 7.

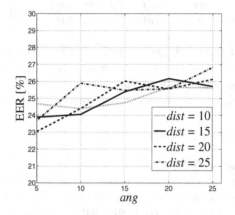

Fig. 7. The impact of the ang parameter on the results obtained

As it appears from the diagram, an increase in the value of the ang parameter results in an increase in the value of EER, regardless of the $dist$ value. The analysis indicates that in order to achieve the best possible results, only the bifurcations with similar orientation angles should be compared. This is consistent with intuition.

4   Conclusions

The paper presents a new method of personal identification based on lip prints. The solution presented here can be easily implemented, while the modular design facilitates future modifications. During the studies, the values of the parameters used in this method were analysed. The best result of EER was at a level of approx. 23%. Considering the fact that there is a small number of solutions for automatic lip print identification, the result obtained can be considered as satisfactory.

During the studies, there were observed situations where some pixels imitated bifurcations that in reality did not exist in the lip print pattern. Such bifurcations may have a negative impact on the identification results. In connection with this fact, further work will focus on developing other methods for pre-processing and extraction of bifurcations, which will eliminate the problem in question. Weights of individual features will be introduced, which will allow determining their impact on the research results.

References

1. Agarwal, G., Ratha, N., Bolle, R.M.: Biometric verification: looking beyond raw similarity scores. In: Workshop on Multibiometrics (CVPR), New York, pp. 31–36 (2006)
2. Al-amri, S.S., Kalyankar, N.V., Khamitkar, S.D.: Linear and Non-linear Contrast Enhancement Image. International Journal of Computer Science and Network Security (IJCSNS) **10**(2), 139–143 (2010)
3. Bhatnagar, J., Kumar, A.: On some performance indices for biometric identification system. In: Lee, S.-W., Li, S.Z. (eds.) ICB 2007. LNCS, vol. 4642, pp. 1043–1056. Springer, Heidelberg (2007)
4. Cha, S.: Comprehensive Survey on Distance/Similarity Measures between Probability Density Functions. International Journal of Mathematical Models and Methods in Applied Sciences **1**(4), 300–307 (2007)
5. Choras, M.: The lip as a biometric. Pattern Analysis And Applications (Springer) **13**, 105–112 (2010)
6. Doroz, R., Wrobel, K.: Method of signature recognition with the use of the mean differences. In: Proceedings of the 31st International IEEE Conference on Information Technology Interfaces (ITI 2009), Croatia, pp. 231–235 (2009)
7. Kasprowski, P.: The impact of temporal proximity between samples on eye movement biometric identification. In: Saeed, K., Chaki, R., Cortesi, A., Wierzchoń, S. (eds.) CISIM 2013. LNCS, vol. 8104, pp. 77–87. Springer, Heidelberg (2013)
8. Kasprzak, J., Leczynska, B.: Cheiloscopy. Human identification on the basis of lip trace (in Polish). KGP, Warsaw, Poland (2001)
9. Koprowski, R., Wrobel, Z.: The cell structures segmentation. In: 4th International Conference on Computer Recognition Systems (CORES 05), pp. 569–576 (2005)
10. Newton, M.: The Encyclopedia of Crime Scene Investigation. Facts on File, New York (2008)
11. Pavlidis, T.: A Thinning Algorithm For Discrete Binary Images. Computer Graphics And Image Processing **13**, 142–157 (1980)
12. Petherick, W.A., Turvey, B.E., Ferguson, C.E.: Forensic Criminology. Elsevier Academic Press, London (2010)
13. Porwik, P., Orczyk, T.: DTW and voting-based lip print recognition system. In: Cortesi, A., Chaki, N., Saeed, K., Wierzchoń, S. (eds.) CISIM 2012. LNCS, vol. 7564, pp. 191–202. Springer, Heidelberg (2012)
14. Ridler, T.W., Calvard, S.: Picture Thresholding Using An Iterative Selection Method. IEEE Transactions On Systems, Man, And Cybernetics **8**(8), 630–632 (1978)
15. Suzuki, K., Tsuchihashi, Y.: Personal identification by means of lip prints. Journal of Forensic Medicine **17**, 52–57 (1970)
16. Tsuchihashi, Y.: Studies on personal identification by means of lip prints. Forensic Science, pp. 127–231 (1974)
17. Wrobel, K., Doroz, R., Palys, M.: A method of lip print recognition based on sections comparison. In: IEEE Int. Conference on Biometrics and Kansei Engineering (ICBAKE 2013), Akihabara, Tokyo, Japan, pp. 47–52 (2013)

Detecting the Reference Point
in Fingerprint Images with the Use
of the High Curvature Points

Rafał Doroz$^{(\boxtimes)}$, Krzysztof Wrobel, and Malgorzata Palys

Institute of Computer Science, University of Silesia, ul. Bedzinska 39,
41-200 Sosnowiec, Poland
{rafal.doroz,krzysztof.wrobel,malgorzata.palys}@us.edu.pl
http://zsk.tech.us.edu.pl,http://biometrics.us.edu.pl

Abstract. A new method for finding a reference point in fingerprints is presented in this paper. The method proposed in this study is based on the IPAN99 algorithm used to detect high curvature points on the contour of a graphical object. This algorithm was modified to detect high curvature points on friction ridges in order to allow locating a reference point on a fingerprint. The IPAN99 algorithm requires that the contour being analysed should have a thickness of one pixel, so each fingerprint was properly prepared before starting an analysis with the use of the IPAN99 algorithm. In order to assess the efficiency of the method, distances between the coordinates of reference points determined with the use of the proposed method and those indicated by an expert were compared. This method was compared with other algorithms for determination of reference points.

Keywords: Biometrics · Fingerprint · Reference point · IPAN99

1 Introduction

Biometrics is a science dealing with personal identification or verification based on an analysis of physiological or behavioral features of a given person [5,12]. Thanks to the continuous technological development in recent years, a gradual increase in the interest in biometric techniques can be observed. The experience gained during years of research causes that biometric techniques are no longer treated as a replacement for passwords - they became primarily a part of security systems. The most commonly used biometric identifiers include fingerprints [4, 6,11]. The biometric identification of persons and criminology constitute two main areas of application of the fingerprint analysis. Up to now, many automatic personal identification methods based on fingerprints have been developed. Many of these methods rely on locating the reference point. The reference point is a point on a fingerprint, in which friction ridges have the highest curvature [5,8].

This study presents a new method for locating the reference point in fingerprint images. It consists of two stages. In the first stage, all friction ridges are

© Springer International Publishing Switzerland 2015
N.T. Nguyen et al. (Eds.): ACIIDS 2015, Part II, LNAI 9012, pp. 82–91, 2015.
DOI: 10.1007/978-3-319-15705-4_9

described by chains of points. In the second stage, the reference point is located on the basis of an analysis of the curvature of the chains of points. Values of the curvature of individual chains are calculated using the IPAN99 algorithm. As it results from the research, it is now one of the best algorithms for finding high curvature points in digital images [1].

This paper is continuation of the research presented in [13,14]. New process of extracting and recording chains of points in a digital fingerprint image has been proposed. This method allows to localize the reference point more precisely.

2   Proposed Method

The proposed method consists of a number of stages:

1. Image pre-processing. Preparation of a digital image of a fingerprint for further stages of the analysis.
2. Identification of chains of points. The process of extracting and recording chains of points in a digital fingerprint image.
3. Application of the IPAN99 algorithm for searching for a reference point in the chains of points extracted earlier.

2.1   Image Pre-Processing

A fingerprint image, in which a reference point is determined, may be characterized by poor quality, low contrast and may be noised or blurred (Fig. 1a). This makes it difficult to extract the friction ridges necessary to determine such a point. For this reason, in the proposed method, the fingerprint image was subjected to a quality improvement process (Fig. 1b). The method described in [3] was used for this purpose. Then the operation of skeletonization was performed. It consists in the use of thinning algorithms that allow reducing the thickness of lines in the image. After the application of the thinning algorithm, the line thickness is equal to 1 pixel. This facilitates creation of chains of points. Figure 1c presents an image after the skeletonization. The thinning algorithm described by Pavlidis [7] was used in the method presented here.

(a)   (b)   (c)

Fig. 1. Image pre-processing: (a) a low-quality fingerprint image, (b) the image after quality improvement, (c) the image after skeletonization

2.2   Identification of Chains of Points

After the completion of the preprocessing, the next step is to determine in the
fingerprint image the chains of points describing the friction ridges. Chain is a
sequence of points $L = (\mathbf{p}_1, \ldots, \mathbf{p}_n)$ forming a friction ridges in the image, where
n is the number of points in the chain. Each point $\mathbf{p}_i = (x_i, y_i)$ belonging to the
chain is characterized by two parameters:

- i - the number of the point in the chain $i = 1, 2, \ldots, n$,
- x_i, y_i - coordinates of the i-th point.

The point $\mathbf{p}_1$ will be the start of the chain, while $\mathbf{p}_n$ will be the end of the chain.
Two points with the coordinates (x_i, y_i) and (x_{i+1}, y_{i+1}) will belong to the chain,
if they are located in the neighbourhood, i.e. when the following dependence is
fulfilled:

$$\forall_{\mathbf{p}_i = (x_i, y_i) \in L} \quad (|x_i - x_{i+1}| \leq 1 \wedge |y_i - y_{i+1}| \leq 1). \tag{1}$$

A sample chain of points is shown in Figure 2.

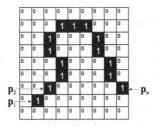

Fig. 2. An example of graphical representation of a chain of points

As seen in Figure 2, a chain of points has only one start point and one end
point. Bifurcations on friction ridges cause that the lines may have multiple
starts or ends, which makes it difficult to represent them as a chain of points.
The algorithm proposed in this paper determines chains both for friction ridges
with bifurcations and without them.

In the presented approach, the points forming friction ridges are divided into
three types: bifurcation points, end points and other points. In the first stage
of determining the chains of points, the image is analysed and the type of each
point is specified. Depending on the type, an appropriate number is assigned to
it. If only one point of a friction ridge is located in the neighbourhood of the
analysed point, it means that this point is the end point. In such a case, this
point is designated with number 1. This point is treated as a bifurcation of the
friction ridge, if at least three other points of the friction ridge are located in
its neighbourhood. Such a point is designated with number 3. All other points
located on friction ridges were designated with number 2. In addition, all the
points outside of the friction ridges are designated with number 0. An example
of a friction ridge with point values marked on it is shown in Figure 3.

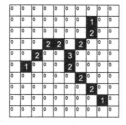

Fig. 3. An example of a friction ridge with point values marked on

In order to define a chain, its end points must be found. For this purpose, the algorithm analyses the values of points in individual columns of the image. The image analysis method is shown in Figure 4. The search starts from the point with coordinates (1,1) and ends in the point with the coordinates (w, h), where: w is the image width and h is the image height.

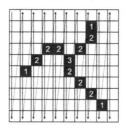

Fig. 4. The sequence of searching for the first pixel with value 1

Then, all possible chains starting at each point designated with the number 1 and ending at another point designated with the number 1 are found. After a point designated with the number 1 is found, it becomes the currently examined point and its coordinates (x_i, y_i) are stored in the computer memory. The number describing this point is changed to 0, which means that it is not taken into account in subsequent iterations. This point is the first point of the chain. Then the neighbourhood of this point is checked in the sequence shown in Figure 5.

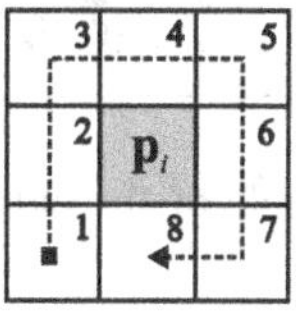

Fig. 5. The sequence of checking the points in the neighbourhood of the point p_i

If a neighbouring point is designated with the number 1 or 2, it is added to the chain of points. In addition, if a point designated as 2 is found, it becomes the currently examined point and the process of searching for its neighbours starts again. The number describing this point is changed to 0. If a point designated with the number 1 is found, this means that it is the last element of the chain and the algorithm starts to search for the next chain.

It is more difficult to analyse a case where the point found is designated with the number 3, i.e. is located on a bifurcation of a friction ridge. In such a case, the coordinates of the analysed point and neighbouring points are stored, while the numbers describing these points are changed to 0. The friction ridge is divided into parts as a result of this operation. This is presented in Figure 6.

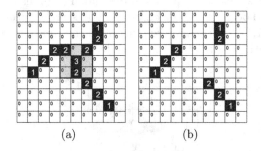

Fig. 6. Friction ridge (a) before the division, (b) after the division

After the division, each part of the friction ridge is analysed separately. New ends are determined for the friction ridge and then the chains based on them. The chains obtained in this way are connected with each other. The chains resulting from this operation have only one start point and one end point. An example of the obtained chains is shown in Figure 7.

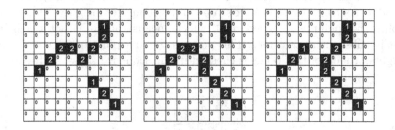

Fig. 7. An example of the extracted chains

In the case of friction ridges with a larger number of bifurcations, the procedure described here is repeated for each point of a bifurcation. Along with an increase in the number of bifurcations occurring in a friction ridge, the number of chains of points that can be identified in this ridge increases too. Chains determined for friction ridges with two bifurcations are shown in Figure 8.

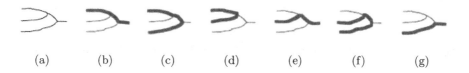

(a)   (b)   (c)   (d)   (e)   (f)   (g)

Fig. 8. (a) A fragment of a friction ridge, (b)-(g) chains extracted from it

The above procedure is repeated until all possible chains of points describing the friction ridges are found. This operation ends when only the pixels designated with the number 0 are present in the image. An example of fingerprints with identified chains of points are shown in Figure 9.

Fig. 9. Fingerprints with identified chains of points

Such prepared chains are analysed using the IPAN99 algorithm in order to determine the high curvature points.

2.3   Application of the IPAN99 Algorithm

Detection of the highest curvature points with the use of the IPAN99 algorithm takes place in two stages [1]. In the first stage, the i-th point $\mathbf{p}_i$ of the chain of points is taken as a corner, if it is possible to inscribe a triangle with a specific opening angle and different lengths of sides $(\mathbf{p}_{i-1}, \mathbf{p}_i, \mathbf{p}_{i+1})$ in this chain of points (Fig. 10).

Triangles are constructed according to the following conditions:

$$d_{min}^2 \leq |\mathbf{p}_i - \mathbf{p}_{i+1}|^2 \leq d_{max}^2, \tag{2}$$

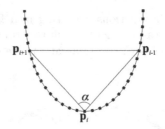

Fig. 10. Detection of the highest curvature points based on the IPAN99 algorithm

$$d_{\min}^2 \leq |\mathbf{p}_i - \mathbf{p}_{i-1}|^2 \leq d_{\max}^2, \tag{3}$$

$$\alpha \leq \alpha_{\max}, \tag{4}$$

where:
$d_{\min}$ - parameter specifying the minimum length of triangle sides,
$d_{\max}$ - parameter specifying the maximum length of triangle sides,
$\alpha_{\max}$ - critical angle, which determines the value of the angle of a triangle inscribed in the chain of points in a given point in order to classify this point as a candidate for a corner.
$a = |\mathbf{p}_i - \mathbf{p}_{i+1}|$ – distance between $\mathbf{p}_i$ and $\mathbf{p}_{i+1}$ points,
$b = |\mathbf{p}_i - \mathbf{p}_{i-1}|$ – distance between $\mathbf{p}_i$ and $\mathbf{p}_{i-1}$ points,
$c = |\mathbf{p}_{i+1} - \mathbf{p}_{i-1}|$ - distance between $\mathbf{p}_{i+1}$ and $\mathbf{p}_{i-1}$ points,
$\alpha \in [-\pi, \pi]$ - opening angle of a triangle, defined as follows:

$$\alpha = \arccos \frac{a^2 + b^2 - c^2}{2ab}. \tag{5}$$

Inscribing a triangle at any point $\mathbf{p}_i$ is started with determining the smallest possible lengths of triangle sides. Then, next triangles are created by increasing the lengths of their sides. The algorithm is stopped, if a triangle does not meet one of the conditions 2-4. From among all acceptable triangles in a given point $\mathbf{p}_i$, the triangle with the smallest opening angle $\alpha(\mathbf{p}_i)$ is selected.

In the second stage, the point $\mathbf{p}_i$ is rejected, if in its neighbourhood there is a point $\mathbf{p}_\nu$, which has a smaller opening angle:

$$\alpha(\mathbf{p}_i) > \alpha(\mathbf{p}_\nu). \tag{6}$$

The point $\mathbf{p}_\nu$ belongs to the neighbourhood of the point $\mathbf{p}_i$, if it fulfills the condition $|\mathbf{p}_i - \mathbf{p}_\nu|^2 \leq d_{\min}^2$.

Depending on the values of the $d_{\min}, d_{\max}, \alpha_{\max}$ parameters, the IPAN99 algorithm detected a different number of high curvature points on each friction ridge. The result of operation of the algorithm was the n set of $\{\mathbf{p}_1, \mathbf{p}_2, \ldots, \mathbf{p}_n\}$ points for each friction ridges on a single fingerprint. For the needs of this study it has been accepted that the $\mathbf{p}_r$ reference point is defined as follows:

$$\mathbf{p}_r = \min_{i=1,\ldots,n} (\alpha(\mathbf{p}_i)). \tag{7}$$

The reference points on fingerprint images determined with the use of the IPAN99 algorithm (marked as dot), as well as the reference points indicated by an expert (marked as cross) are shown in Fig. 11.

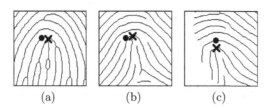

(a)   (b)   (c)

Fig. 11. (a)-(c) Example of operation of the IPAN99 algorithm on fingerprint images

3 Experiments and Results

The aim of the studies was to determine the effectiveness of the proposed method for determining a reference point. The results were compared with those obtained with the use of other methods known from the literature.

The studies were conducted using three different fingerprint databases [2]. The databases contained grey-scale fingerprint images with different quality, resolution and number of images. Examples of fingerprints from the test databases are shown in Table 1.

Table 1. Example fingerprint images from the used databases

Database	A	B	C
Resolution	240x320 px	256x256 px	256x364 px
Number of images in database	80	168	78
Exemplary image			

A reference point was determined for each test image with the use of the method proposed here and every method compared with it. In addition, the coordinates of the reference points identified with the use of the methods analysed here were compared with the coordinates of the reference points determined by a fingerprinting expert. For this purpose, the distance d between the reference points determined was calculated using the following formula:

$$d = \sqrt{(x_m - x_e)^2 + (y_m - y_e)^2},\tag{8}$$

where:

(x_m, y_m) - coordinates of the point indicated by a given method [px],

(x_e, y_e) - coordinates of the point indicated by a fingerprinting expert [px].

The following parameters were selected experimentally for the IPAN99 algorithm: $d_{\min} = 17, d_{\max} = 19, \alpha_{\max} = 150$. Each database was tested separately. The distance d was calculated for each image from the test database. Then, the results obtained for each database were averaged. The results are presented in Table 2.

Table 2. The average distances between coordinates of the reference points identified with the use of the compared methods and with the use of a fingerprinting expert

Database	Average distance d [px]			
	Proposed method	**Method 1** [10]	**Method 2** [9]	**Method 3** [4]
A	19.82	21.19	26.06	44.70
B	20.52	25.64	26.31	99.27
C	27.64	36.70	35.40	36.24

As it can be seen in Table 2, an average distance of a point indicated by the proposed method from a point indicated by the expert is shorter for each database than in the case of other methods being compared.

For the database A, the difference between the proposed method and Method 1 is not large, i.e. is less than 2 pixels. In the case of other methods, the difference is larger – approx. 6 and 24 pixels, respectively. For the fingerprint database B, the difference is greater – even more than 78 pixels. For the database C, the methods being compared indicate a similar average distance that is greater by approx. 9 pixels than the distance calculated using the proposed method.

During the studies, there were situations where the expert had difficulties with locating a reference point in the images explicitly. Examples of such images are shown in Figure 12. In the first image, there is no reference point. In the case of the second image, it was difficult to select one of the two possible reference points (marked in the Figure 12b as x). For these reasons, such images could not be analysed.

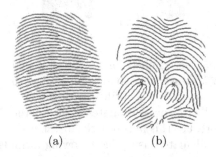

(a)   (b)

Fig. 12. (a)-(b) Images where the expert had difficulties with locating a reference point

4   Conclusions

The presented method for searching for a reference point in a fingerprint image proved to be useful in practical applications. The results obtained are comparable with other methods and differ only slightly from expert's indications.

Further studies will be associated with the use of other image pre-processing methods and an analysis of the time required by the method. Other parameters of the IPAN99 algorithm will be tested too. The set of test images will also be expanded by images contained in other databases.

References

1. Chetverikov, D., Szabo, Z.: Detection of high curvature points in planner curves. In: 23rd Workshop Of The Austrian Pattern Recognition Group, pp. 175–184 (1999)
2. Fingerprint Verification Competition: http://Bias.Csr.Unibo.It/Fvc2006
3. Greenberg, S., Aladjem, M., Kogan, D., Dimitrov, I.: Fingerprint image enhancement using filtering techniques. In: Proceedings Of The 15th International Conference On Pattern Recognition, vol. 3, Barcelona, Spain, pp. 322–325 (2000)
4. Jain, A.K., Prabhakar, S., Jonh, L., Pankanti, S.: Filterbank-Based Fingerprint Matching. IEEE Trans. on Image Processing 9(5), 846–859 (2000)
5. Maltoni, D., Maio, D., Jain, A.K., Prabhakar, S.: Handbook Of Fingerprint Recognition. Springer Professional Computing Series, NY (2003)
6. Park, U., Pankanti, S., Jain, A.K.: Fingerprint verification using sift features. In: SPIE Defense And Security Symposium, paper 6944–19, Orlando, USA (2008)
7. Pavlidis, T.: A Thinning Algorithm For Discrete Binary Images. Computer Graphics And Image Processing 13, 142–157 (1980)
8. Porwik, P., Wieclaw, L.: A new approach to reference point location in fingerprint recognition. IEICE Journal Electronics Express 1(18), 575–581 (2004)
9. Porwik, P., Wieclaw, L.: Fingerprint reference point detection using neighbourhood influence method. In: Kurzyński, M., Puchała, E., Woźniak, M., żołnierek, A. (eds.) Computer Recognition Systems 2. AISC, vol. 45, pp. 786–793. Springer, Heidelberg (2007)
10. Porwik, P., Wrobel, K.: The New Algorithm Of Fingerprint Reference Point Location Based On Identification Masks. In: Kurzyński, M., Puchała, E., Woźniak, M., żołnierek, A. (eds.) Computer Recognition Systems. AISC, vol. 30, pp. 807–814. Springer, Heidelberg (2005)
11. Wang, R., Bhanu, B.: Predicting Fingerprint Biometrics Performance From A Small Gallery. Pattern Recognition Letters 28(1), 40–48 (2007)
12. Wrobel, K., Doroz, R., Palys, M.: A method of lip print recognition based on sections comparison. In: IEEE Int. Conference on Biometrics and Kansei Engineering (ICBAKE 2013), Akihabara, Tokyo, Japan, pp. 47–52 (2013)
13. Wrobel, K., Doroz, R.: New method for finding a reference point in fingerprint images with the use of the IPAN99 algorithm. Journal of Medical Informatics & Technologies 13, 59–63 (2009)
14. Wrobel, K., Doroz, R.: The method for finding a reference point in fingerprint images basing on an analysis of characteristic points. In: Proc. of the Third World Congress on Nature and Biologically Inspired Computing (NaBIC'11), pp. 504–508. IEEE Press, Salamanca (2011)

Kinect-Based Action Recognition in a Meeting Room Environment

Faisal Ahmed[1]([✉]), Edward Tse[2], and Marina L. Gavrilova[1]

[1] Department of Computer Science, University of Calgary, Calgary, AB, Canada
faahmed@ucalgary.ca
[2] SMART Technologies ULC, Calgary, AB, Canada

Abstract. In recent years, rapid development of biometric technologies opened the door to a new class of fast and reliable identity management solutions. Gait is one of the few biometrics that can be recognized unobtrusively from a distance. This paper presents a new gait and action recognition approach based on the Kinect sensor. The proposed method utilizes the joint angles formed by different body parts during walking in order to construct a scale and view-invariant feature set for gait recognition. Next, we develop a new Kinect-based "walking" and "sitting" action recognition method in a meeting room environment. The method is the first step towards the biometric-based interactive meeting room system, which is capable not only to perform user authentication, but also to conduct action recognition during collaborative activities. The proposed method achieves promising results in empirical evaluation compared to some other Kinect-based approaches.

Keywords: Gait recognition · Activity recognition · Kinect for windows v2 · Smart meeting room

1 Introduction

With the increasing development of computing technologies, there has been a growing demand from industries for new generations of interactive technologies to support high productivity and to optimize time spent during meetings. In addition, secure real-time user authentication and access right management solutions are being actively researched and deployed in both universities and corporate settings [1]. In response to these demands, researchers are actively seeking new communication and collaboration technologies, which will reduce meeting attendees' enrollment time and provide the meeting organizer with the tools for efficient real-time attendance tracking and access resource management. Traditional approaches to the problem involve the use of IDs, smart cards, passwords, coupled with user identity management solutions, all of which require a significant amount of time and additional resources [2]. Rapid development of biometric technologies opened the door to a new class of fast and reliable identity management solutions, and changed the research landscape. A biometric system

© Springer International Publishing Switzerland 2015
N.T. Nguyen et al. (Eds.): ACIIDS 2015, Part II, LNAI 9012, pp. 92–101, 2015.
DOI: 10.1007/978-3-319-15705-4_10

can be defined as a pattern recognition system that can recognize individuals based on the characteristics of their physiology or behavior [3]. Human biometric traits can roughly be divided into two categories: physiological and behavioral. Physiological biometric systems utilize certain physical characteristics, such as face, iris, ear, fingerprint, palm, etc. for individual recognition. On the other hand, behavioral biometric systems rely on human behavior-mediated activities, such as gait, voice, handwriting, signature, etc.

This paper introduces for the first time the idea to utilize biometric gait and action recognition for developing a smart meeting room system. It is based on the premises that, physiological and behavioral biometrics can be seamlessly integrated with technologies enabling meeting room setup, and, in addition to individual access management, can provide highly efficient group authentication capabilities. The objective is to enable meeting organizer to immediately start the meeting, keep track of attendance, and allow all participants to seamlessly access shared resources without compromising secure contents. In addition, the developed system should support archiving meeting statistics, determining input of each of the participants by recognizing certain activities. This will help to create more conductive collaborative environment and use time more effectively.

Due to the unobtrusive nature of data acquisition and the ability to recognize individuals at a distance, biometric gait recognition has attracted much attention in the recent years [4]. We propose a simple, yet effective gait recognition method using an inexpensive consumer-level sensor, namely the Microsoft Kinect v2. We also present a real-time "walk" and "sit" activity detection method using the Kinect. The proposed system tracks the distance of the individual's head joint from the floor plane, which can effectively be used to detect if a person is walking or sitting on a chair. Empirical evaluation shows promising performance compared to some recent Kinect-based methods.

2   Related Work

2.1   Gait Recognition

The release of the Kinect sensor and the software development kit (SDK) by Microsoft has opened a new window of research directions in model-based gait recognition. In model-based approaches, explicit models are used to represent human body parts (legs, arms, etc.) [5, 6]. Parameters of these models are estimated in each frame and the change of the parametric values over time is used to represent gait signature. However, the computational cost involved with model construction, model fitting, and estimating parameter values for earlier methods made most of the model-based approaches time-consuming and computationally expensive [6]. As a result, these approaches were infeasible for many real-world applications. BenAbdelKader et al. [7] estimated two spatiotemporal parameters of gait, namely stride length and cadence for biometric authentication of a person from video. Urtasun and Fua [8] proposed a gait analysis method that relies on fitting 3-D temporal motion models to synchronized video sequences. Recovered motion parameters from the models are then used to characterize individual gait signature. Yam et al. [9] modeled human leg structure and motion in order to discriminate between gait signatures obtained from walking and running. Although the method presents an effective way to view and scale independent gait recognition, it is computationally expensive and sensitive to the quality of the gait sequence [10].

On the other hand, Kinect is a low-cost consumer-level device made up of an array of sensors, which includes i) a color camera, ii) a depth sensor, and iii) a multi-array microphone setup. In addition, Kinect sensor can track and construct a 3D virtual skeleton model from human body in real-time with a relatively low computation cost [11]. All these functionalities of Kinect have led to its application in different real-world problems, such as home monitoring, healthcare, surveillance, etc. The low-computation real-time skeleton tracking feature has encouraged some recent gait recognition methods that extract features from the tracked skeleton model. Ball et al. [4] used Kinect for unsupervised clustering of gait samples. Features were extracted only from the lower body part. Preis et al. [12] presented a Kinect skeleton-based gait recognition method based on 13 biometric features: height, the length of legs, torso, both lower legs, both thighs, both upper arms, both forearms, step-length, and speed. However, these features are mostly static and represent individual body structure, while gait is considered to be a behavioral biometric, defined as the pattern of the movement of body parts during locomotion. Gabel et al. [13] used the difference in position of these skeleton points between consecutive frames as their feature. However, their proposed method was only evaluated for gait parameter extraction rather than person identification.

2.2   Activity Recognition

Individual activity recognition is a challenging task that has attracted much attention due to its potential applicability in different scenarios, such as video surveillance, health monitoring, etc. However, only a few researchers have addressed the problem of activity recognition in a meeting room context. Nait-Charif and McKenna [14] developed a head tracker system, which track head position of meeting room participants in order to detect different activities, such as entering, exiting, going to the whiteboard, getting up and sitting down. However, this system has two potential limitations. Firstly, it is heavily dependent on scene-specific constraints [14]. Secondly, the tracking takes place in a 2D space, which limits its applicability greatly. Another approach proposed by Mikic et al. [15] considered 3 basic activities: i) a person located in front of the whiteboard, ii) a lead presenter speaking, and iii) other participants speaking. The objective of the activity recognition was only to select the best-view camera.

3   Proposed Method

3.1   Gait Recognition Using Kinect v2

We propose a model-based gait recognition approach based on the skeleton data provided by the Kinect. To our knowledge, the work presented in this paper is the first one that uses the Kinect v2 sensor for gait recognition. Released in mid-July 2014, Kinect v2 offers a greater overall precision, responsiveness, and intuitive capabilities than the previous version [16]. The v2 sensor has a higher depth fidelity that enables it to see smaller objects more clearly, which results in a more accurate 3D object perception [16]. In addition, while the skeleton tracking range is broader, the tracked joints are more accurate and stable than the previous version [16]. Our proposed gait recognition method works in three steps: 1) Detection of gait cycle from video sequences, 2) Feature vector construction, and 3) Classifier training and testing.

The first component of the proposed gait recognition method is to isolate a complete gait cycle so that salient features can be extracted from it. Regular human walking is considered to be a cyclic motion, which repeats in a relatively stable frequency [6]. Therefore, features extracted from a single gait cycle can represent the complete gait signature. A gait cycle is composed of a complete cycle from rest (standing) position-to-right foot forward-to-rest-to-left foot forward-to rest or vice versa (left food forward followed by a right foot forward) [17]. In order to identify gait cycles, the horizontal distance between the left and right ankle joints was tracked over time, as shown in Fig. 1(a). A moving average filter was used to smooth the distance vector. During the walking motion, the distance between the two ankle joints will be the maximum when the right and the left legs are farthest apart and will be the minimum when the legs are in the rest (standing) position. Therefore, by detecting three subsequent minima, it is possible to find the three subsequent occurrences of the two legs in the rest position, which corresponds to the beginning, middle, and ending points of a complete gait cycle, respectively [18]. The process is illustrated in Fig. 1 (b).

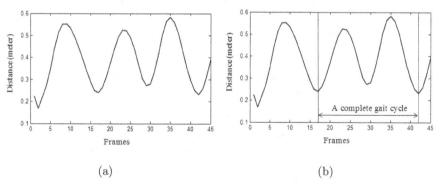

(a)   (b)

Fig. 1. (a) Distance between the left and right ankle joints of a person walking, (b) Detection of a complete gait cycle

The Kinect v2 can detect and track 25 different skeletal joints of an individual. We propose to use different angles formed by these skeletal joints as the gait feature. We argue that, while human locomotion involves similar order of limb movement among different individuals, the poses in which they move their limbs differ. Therefore, we try to capture these differences in pose by the joint angles formed by different limbs during walking. The advantages of using joint angle features are two-fold: firstly, the computed joint angle features are view and scale independent. This means that, the feature values will not be affected by the variation of the distance of the subject from the camera or the direction of subject's walking. This makes joint angle features highly potential for real-world applications. Secondly, according to [4], joint distance-based features proposed in recent works [12, 13] are found to vary over time significantly. As a result, consistent feature extraction is difficult in some cases. On the other hand, although the distances of the joints vary over time, angles formed by the joints remain unaffected. This property makes joint angles a very good candidate for gait feature representation. Given the coordinates of 3 joints A, B, and C in a 3-D space, the angle Θ formed by $A \to B \to C$ using the right hand rule from B can be calculated as:

$$\Theta = \cos^{-1} \frac{\overrightarrow{AB}.\overrightarrow{BC}}{||\overrightarrow{AB}||||\overrightarrow{BC}||} \tag{1}$$

Here $\overrightarrow{AB} = B - A$, $\overrightarrow{BC} = C - B$, the dot(.) represents dot product between two vectors, and $||\overrightarrow{AB}||$ and $||\overrightarrow{BC}||$ represent the length of $\overrightarrow{AB}$ and $\overrightarrow{BC}$, respectively. Using this formula, a set of 9 joint angles was calculated. Table 1 lists the selected joint angles in this study. These joint angle values change continuously with the change of pose when a person walks. In order to quantify these changes over time as features, we calculate the mean, standard deviation, mode, maximum, and mini-mum of all the selected joint angles. These measures are then used as the final feature set for the proposed gait recognition method. We used a support vector machine (SVM) for the classification task. There are different types of kernel available for SVM classification. In our work, we use a radial basis function (RBF) kernel.

Table 1. Selected joint angle features for gait recognition (angle formed by $A \to B \to C$ using the right hand rule from B)

No.	A	B	C
1	ShoulderRight	ElbowRight	WristRight
2	ShoulderLeft	ElbowLeft	WristLeft
3	HipRight	KneeRight	AnkleRight
4	HipLeft	KneeLeft	AnkleLeft
5	ShoulderRight	HipRight	KneeRight
6	ShoulderLeft	HipLeft	KneeLeft
7	ElbowRight	ShoulderRight	HipRight
8	ElbowLeft	ShoulderLeft	HipLeft
9	Neck	SpineBase	SpineMid

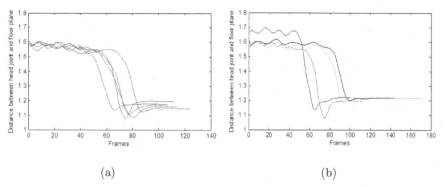

(a)   (b)

Fig. 2. (a) Distance between the head joint and the floor in 5 video clips of the same person, (b) distance between the head joint and the floor in 5 video clips of different persons

3.2 Activity Recognition

We present an effective "walk" and "sit" action detection and tracking method us-ing the Kinect sensor. The fundamental assumption made here is that, a person enter-ing the meeting room is walking. Based on this assumption, it is possible to track whether a person is standing/walking or sitting on a chair by tracking the distance of the head joint from floor plane. Kinect sensor can detect and provide the clip plane of the floor in the form of a four-component vector: (x, y, z, w). Based on this floor clip plane and the 3-D coordinate of the head joint (h_x, h_y, h_z), the Euclidean distance D between the head joint and the floor plane can be calculated as:

$$D = \frac{x \times h_x + y \times h_y + z \times h_z + w}{\sqrt{x^2 + y^2 + z^2}} \qquad (2)$$

We observed that, when a "sit" action is initiated (a person starts to move his body to sit on a chair), the distance between the head joint and the floor plane starts to get shorter since the head is moving downwards with the whole body, as shown in Fig. 2. This sharp continuous decrease in the distance of the head joint from the floor plane can potentially be used to detect a transition from "walk/stand" to a "sit" action. To achieve that, we continuously track the distance between the head joint and the floor plane when a person enters the meeting room (walking). When the distance goes below a certain threshold, a candidate for the sit event is found. After a person completes the sit action, the distance between the head joint and the floor plane becomes stable (almost parallel to the x-axis) after a short bump. This behavior is consistent for all the participants, as demonstrated in Fig. 2(b). We take advantage of this fundamen-tal property to track the "sit" action. A flowchart for detecting a transition from a "walk/stand" to a "sit" action is shown in Fig. 3. In our experiments, we found that, a good estimation of the thresholds T1 and T2 is 0.3 and 0.7, respectively.

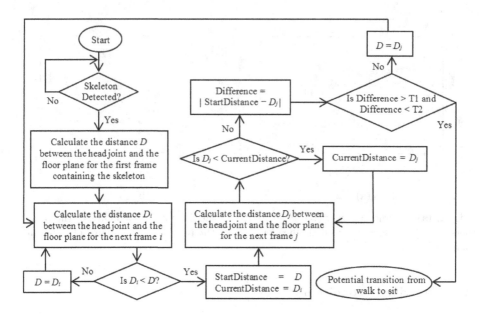

Fig. 3. Flowchart for detecting a transition from "walk" to "sit" action

The same process can be used to detect a transition from a "sit" action to a "walk/stand" action by simply replacing the ">" signs with "<" and vice versa.

4   Experiments and Results

In order to evaluate the effectiveness of the proposed gait and action recognition method, we created a 9 person Kinect video database. The videos were recorded in a meeting room environment. For each person, we recorded a series of 5 videos. The position of the Kinect was fixed throughout the recording session. Each of the video scenes contains a person entering the meeting room, walking toward a chair, and then sitting on the chair. As a result, this database can be used to evaluate both the gait recognition and the activity tracking methods.

We conducted a 5-fold cross-validation in order to evaluate the effectiveness of the proposed method. In a 5-fold cross-validation, the whole dataset is randomly divided into 5 subsets, where each subset contains an equal number of samples from each category. The classifier is trained on 4 subsets, while the remaining one is used for testing. The average classification rate is calculated after repeating the above process for 5 times. Since our database contains 5 videos per person, in each fold, we trained the SVM with 4 instances of videos per person and tested with the remaining one. We compare the proposed joint angle based gait recognition method with two other recent Kinect-based gait recognition approaches. Details about these two methods can be found in [4] and [12]. Table 2 shows the 5-fold cross-validation result of the proposed method against the other two. The confusion matrix for the proposed method is shown in Table 3. From the experimental results, it can be said that, gait feature representation based on joint angular information is more robust and achieves higher recognition rate than some of the existing gait recognition approaches. The superiority of the proposed method is due to the utilization of view and distance invariant joint angle features, which are more robust than the joint-distance based feature representations.

Table 2. Gait recognition rate for 5-fold cross validation

Method	Recognition Rate (%)
Proposed joint angle based method	88.89
Ball et al. [4]	57.78
Preis et al. [12]	86.67

Table 3. Confusion matrix for the 5-fold cross validation using the proposed method (Rows represent true class and columns represent classification. The participants were labeled as P1, P2, P3, ..., P9).

	P1	P2	P3	P4	P5	P6	P7	P8	P9
P1	5	0	0	0	0	0	0	0	0
P2	0	4	1	0	0	0	0	0	0
P3	0	0	4	0	1	0	0	0	0
P4	0	0	0	5	0	0	0	0	0
P5	0	0	0	0	4	0	0	1	0
P6	0	0	0	0	0	4	0	1	0
P7	0	0	0	0	0	0	5	0	0
P8	0	0	0	0	1	0	0	4	0
P9	0	0	0	0	0	0	0	0	5

Based on the proposed activity detection approach, we have built a "sit" and "walk" activity tracker. The system tracks any user who enters the meeting room and shows the current status of the user ("walk/stand" or "sit") at the top of his head. In our experiments, we found that, the detection of walk and sit actions are quite accurate and consistent over the entire video sequence. We considered two different scenarios for evaluating the tracker. In the first scene, a single subject enters the room and sits on a chair. In the second scene, two subjects enter the room, sit on chairs for some time and then leave the room. Figure 4 shows the tracking results for a sample video.

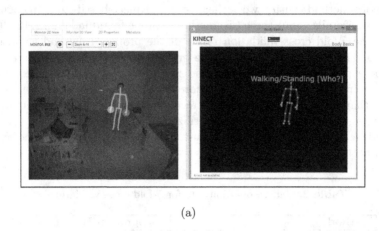

(a)

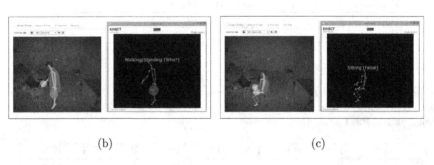

(b)   (c)

Fig. 4. Activity tracking in a 1 person scene, (a) Kinect detects the presence of a person and starts to track the person (the tracked activity is shown in the right side window), (b) the person is pulling the chair to sit on it, (c) the person is sitting on the chair (notice that, based on the walk action, Kinect also recognized the person by analyzing the gait)

5   Conclusion

In this paper, we present our current research efforts toward building a biometric-based authentication and activity detection system for collaborative environment. We propose a robust Kinect-based gait recognition method that utilizes joint angular information in order to construct a view and scale independent feature vector. In addition, we also develop a "walk" and "sit" activity tracker for meeting room participants. Experimental analysis show promising results for both the proposed gait recognition and the activity detection methods. In future, we plan to incorporate recognition of emotions and gestures while sited, speech analysis, and group-activity detection, such as group discussion, group note-taking, monologue, etc. in order to make the system more effective for group authentication, collaborative activity detection, and individual contribution analysis.

Acknowledgments. The authors would like to thank NSERC DISCOVERY program grant RT731064, URGC, NSERC ENGAGE, AITF, and SMART Technologies ULC, Canada for partial support of this project.

References

1. Deutschmann, I., Nordstrom, P., Nilsson, L.: Continuous Authentication Using Behavioral Biometrics. IT Professional **15**(4), 12–15 (2013)
2. Jeon, I.S., Kim, H.S., Kim, M.S.: Enhanced Biometrics-based Remote User Authentication Scheme Using Smart Cards. Journal of Security Engg. **8**(2), 237–254 (2011)
3. Prabhakar, S., Pankanti, S., Jain, A.K.: Biometric recognition: security and privacy concerns. IEEE Security and Privacy **1**(2), 33–42 (2003)
4. Ball, A., Rye, D., Ramos, F., Velonaki, M.: Unsupervised Clustering of People from 'Skeleton' Data. In: Proc. ACM/IEEE Intl. Conf. on HRI, pp. 225–226 (2012)
5. Zhang, R., Vogler, C., Metaxas, D.: Human gait recognition. In: Proc. IEEE Conference on CVPR Workshops, pp. 18–18 (2004)
6. Wang, J., She, M., Nahavandi, S., Kouzani, A.: A review of vision-based gait recognition methods for human identification. In: Proc. IEEE Intl. Conf. on Digital Image Computing: Techniques and Application, pp.320–327 (2010)
7. BenAbdelkader, C., Cutler, R., Davis, L.: Stride and cadence as a biometric in automatic person identification and verification. In: Proc. IEEE Intl. Conf. on Automatic Face and Gesture Recognition, pp. 372–377 (2002)
8. Urtasun, R., Fua, P.: 3D Tracking for Gait Characterization and Recognition. In: Proc. IEEE Automatic Face and Gesture Recognition, pp. 17–22 (2004)
9. Yam, C., Nixon, M.S., Carter, J.N.: Automated person recognition by walking and running via model-based approaches. Pattern Recognition **37**, 1057–1072 (2004)
10. Sinha, A., Chakravarty, K., Bhowmick, B.: Person Identification using Skeleton Information from Kinect. In: Proc. Intl. Conf. on Advances in Computer-Human Interactions, pp. 101–108 (2013)
11. Shotton, J., Fitzgibbon, A., Cook, M., Sharp, T., Finocchio, M., Moore, R., Kipman, A., Blake, A.: Real-time human pose recognition in parts from single depth image. In: Proc. IEEE Conf. on CVPR, pp. 1297–1304 (2011)
12. Preis, J., Kessel, M., Linnhoff-Popien, C., Werner, M.: Gait Recognition with Kinect. In: Proc. Workshop on Kinect in Pervasive Computing (2012)
13. Gabel, M., Gilad-Bachrach, R., Renshaw, E., Schuster, A.: Full body gait analysis with Kinect. In: Proc. Annual Intl. Conf. of the IEEE Engg. in Medicine and Biology Society, pp. 1964–1967 (2012)
14. Nait-Charif, H., McKenna, S.J.: Head Tracking and Action Recognition in a Smart Meeting Room. In: Proc. IEEE Intl. Workshop on PETS (2003)
15. Mikic, I., Huang, K., Trivedi, M.: Activity monitoring and summarization for an intelligent meeting room. In: Proc. Workshop on Human Motion, pp. 107–112 (2000)
16. Kinect for windows features. http://www.microsoft.com/en-us/kinectforwindows/meetkinect/features.aspx (2014). Accessed on 10 October 2014
17. Kale, A., Sundaresan, A., Rajagopalan, A.N., Cuntoor, N.P., Roy-Chowdhury, A.K., Kruger, V., Chellapa, R.: Identification of Humans Using Gait. IEEE Trans. on Image Processing **13**(9), 1163–1173 (2004)
18. Sarkar, S., Phillips, P.J., Liu, Z., Vega, I.R., Grother, P., Bowyer, K.W.: The humanID gait challenge problem: data sets, performance, and analysis. IEEE Trans. on Pattern Analysis and Machine Intelligence **27**(2), 162–177 (2005)

Advanced Data Mining Techniques and Applications

Analyzing Users' Interests with the Temporal Factor Based on Topic Modeling

Thanh Ho[1(✉)] and Phuc Do[2(✉)]

[1] Faculty of Information System, University of Economics and Law,
Vnu-Hcm, Vietnam
thanhht@uel.edu.vn
[2] University of Information Technology, Vnu-Hcm, Vietnam
phucdo@uit.edu.vn

Abstract. In this paper, we proposed a Temporal-Author-Recipient-Topic (TART) model which can simultaneously combine authors' and recipients' interests and temporal dynamics of social network. TART model can discover topics related authors and recipients for different time periods and show how authors and recipients interests are changed over time. All parts of model are integrated on social network analysis system based on topic modeling. The model is experimented on the collection of Vietnamese texts from a student forum containing: 13,208 messages of 2,494 users. The system finds out many useful authors' and recipients' interests in particular topics over time and opened new research and application directions.

Keywords: Topic modeling · TART model · Authors and recipients interests · Temporal dynamics

1 Introduction

Social network analysis (SNA) is the study of mathematical models for analyzing the interactions among actors. Normally, analyzing methods often focus on studying the structure of social network and rarely mentioning about the content of messages among actors. However, due to the practical needs, content-based research works are increasing rapidly [3][4][14][15]. With the purpose of analyzing a social network based on contents, we would like to discover topics to be discussed among authors and recipients and then find out the communities interested in some specific topics. We choose a suitable analytical method and build a high education support system through social network [16][17][19]. In this method, the input is forum data of a social network and the output is a list of labeled topics and users with similar interests.

There are several research papers to study social network analysis. In this work, we focus on LDA model [1], ART model[4], Gibbs Sampling [2] and automatically labeling topics [6][8]. From these research papers, we have successfully built a system with three main parts: term extraction, detection and automatic labeling topics. We combined temporal factor with authors' and recipients' interests from the idea of ART model [4][18] and propose a new model that is called Temporal-Author-Recipient

© Springer International Publishing Switzerland 2015
N.T. Nguyen et al. (Eds.): ACIIDS 2015, Part II, LNAI 9012, pp. 105–115, 2015.
DOI: 10.1007/978-3-319-15705-4_11

Topic (TART). TART focuses on interests and relationships with respect to time and can provide promising answers to the questions about authors' and recipients' interests during each time period.

The novelty of work described in this paper lies on the formalization of the analyzing users' interests with the temporal factor based on topic model, we experimented with the model on the collection of Vietnamese texts from a student forum of our university. With these results, our purpose is to apply this research into education fields on social networks. We hope that this research is able to support other fields such as Marketing, Society, etc...To solve this problem, we have studied many models and social network analysis methods in [3][4][18][19]. After that we develop a TART model based on LDA, ART models. The rest of this paper is organized as follows: section 2 introduces related works; section 3 introduces the model is proposed; section 4 presents experiment, results and discussion; section 5 includes conclusions and future works.

2   Related Works

In this section, we introduce two generative models for documents based-on topic modeling: (1) documents model as a mixture of topics, and (2) model authors- recipients-documents. All t models use the same notations such as a document d is a vector of N_d words, w_d, where each w_{id} is chosen from a vocabulary of size V, and a vector of A_d authors a_d, chosen from a set of authors of size A. A collection of D documents is defined by $D = \{(w_1, a_1),...,(w_D, a_D)\}$. Besides, there are a lot of notations using for each specific model and these notations are introduced in next sections.

2.1   Latent Dirichlet Allocation

Latent Dirichlet Allocation (LDA) model is a generative probabilistic model for collections of discrete text data [1][2][7][11]. In general, LDA is a three-level hierarchical Bayesian model [2][3][4][11], in which each document is described as a random mixture over a latent set of topics. Each topic is modeled as a discrete distribution of a set of words. LDA is suitable for the set of corpus and the set of grouped discrete data. LDA can be used to model the document so as to discover the underlying topics of that document. The generative process of a set of corpus consists of three steps [1][2][7][11] (see Fig 1). 1) Each document has a probabilistic distribution of its topics; this distribution is estimated as the Dirichlet distribution, 2) For each word in a document, a specific topic is chosen based on the distribution of the topics of that document, 3) Each keyword will be chosen from the multinomial distribution of the keywords according to the chosen topic.

The purpose of LDA [1][11] is to detect each word belonging to a specific topic, from that we can guess the label of that topic. The importance of topic model is the posterior distribution. This can be seen as the generative process and the posterior inference for the latent set of variables which are the keywords of topic. In LDA, this process is calculated by the equation:

$$p(\theta, \phi, z \mid w, \alpha, B) = \frac{p(\theta, \phi, z, w \mid \alpha, \beta)}{p(w \mid \alpha, \beta)} \tag{1}$$

In the equation (1), we have the variables $z, \theta, \emptyset$. For each θ_j which is a vector of topics of document j, z_i is the topic of word w_i, $\emptyset^{(k)}$ is the matrix $K \times V$ with $\emptyset_{i,j} = p(w_i, z_j)$. However, in equation (1), we can't calculate precisely the normal factor $p(w \mid \alpha, \beta)$. Therefore, in [2] proposed Gibbs Sampling in the generative model of LDA. Details of Gibbs Sampling can be found in [1][2].

2.2   Author-Recipient-Topic Model

ART model [4] is a development for LDA that simultaneously models message content, as well as the directed social network in which the messages are sent. In its generative process for each message, an author a_d and a set of recipients r_d will be observed. To generate each word, a recipient x is chosen from the set r_d; and then, a topic z is chosen from a multinomial topic distribution $\theta_{a_d,x}$ (see Fig 2). This distribution is specific with the author-recipient pair (a_d, x). Finally, the word w is generated by choosing samples from a defined-topic multinomial distribution $\emptyset_z$.

The process of choosing samples is based on the Gibbs sampling algorithm [2][18]. The final result is the discovery of topics in a social network where the messages are created. In ART, given hyperparameters α and β, the author α_d and the set of recipients r_d, the joint distribution of an author mixture θ, a topic mixture $\emptyset$, a set of recipients x_d (belonging to X_d), a set of topics z_d (belonging to N_d), and a set of words w_d (belonging to N_d) is given by:

$$p(\theta, \emptyset, x_d, z_d, w_d \mid \alpha, \beta, a_d, r_d)$$
$$= p(\theta \mid \alpha)p(\emptyset \mid \beta)\prod_{n=1}^{N_d} p(x_{dn} \mid r_d)p(z_{dn} \mid \theta_{a_d,x_{dn}})p(w_{dn} \mid \emptyset_{z_{dn}}) \qquad (2)$$

By using the rule of total probability, we can integrate out θ and $\emptyset$, summing over x_d and z_d, we get the marginal distribution of a document as follows:

$$p(w_d \mid \alpha, \beta, a_d, r_d) = \iint p(\theta \mid \alpha)p(\emptyset \mid \beta)$$
$$\prod_{n=1}^{N_d}\sum_{x_{dn}}\sum_{z_{dn}} p(x_{dn} \mid r_d)p(z_{dn} \mid \theta_{a_d,x_{dn}})p(w_{dn} \mid \emptyset_{z_{dn}})d\emptyset d\theta \qquad (3)$$

2.3   Gibbs Sampling for Generative Models

The point of Gibbs sampling [2] is that given a multivariate distribution it is simpler to sample from a conditional distribution than to marginalize by integrating out a joint distribution. By applying Gibbs sampling (see Gilks, Richardson, & Spiegelhalter, 1996), the model that we proposed based-on distribution on z and then use the results to infer Θ and Φ [2][3][4][5] and t (temporal distributions Ψ) as well as the latent variables corresponding to the assignments of individual words to topics z and time which authors and recipients discuss about that topics. The details will be introduced in the following.

3   Temporal-Author-Recipient-Topic model

3.1   Introduction

We proposed a Temporal-Author-Recipient-Topic model in the field of social net-work analysis and information extraction based on topic modeling. Our key ideas focus on extracting words, discovering and labeling topics, and analyzing topics with authors, recipients and temporal factor. We develop TART model based-on LDA and ART models, we construct a distributions Ψ based on topics z, temporal t generated from a Dirichlet(μ) distribution. TART can discover and analyze how relationships between authors and recipients. Besides, the authors' and recipients' interests on so-cial network via discussing were analyzed to show that authors and recipients having the interests of degree for each time period t_i. The ideas of TART (see Fig 3) are: (1) at various times t_i, the authors and recipients can discuss the different topics or the same topic which can at time t_1, the user u_1 is the most discussed but at time t_2, user u_2 is the most discussed; (2) determining probability based-on the interests and discus-sion of users on time period and each topic; (3) discovering the experts in each topic over time and spreading influence of users on social networks; (4) discovering the network structures based on topics, authors and recipients over time.

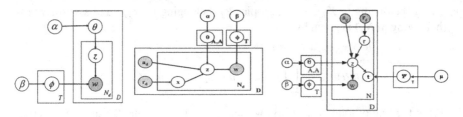

Fig. 1. LDA model [1]   **Fig. 2.** Author-Recipient-   **Fig. 3.** The Temporal-Author-
                                  Topic model [4]   Recipient-Topic model

3.2   Parameter Estimation of TART Model

According to TART model, with the first T is temporal (time period) and Ψ based on topics z, temporal t generated from a Dirichlet(μ) distribution, the total probability of the model is:

$$P(P(T \mid Z, \mu) = \int P(\Psi \mid \mu) \, P(T \mid \Psi, Z) \, d\Psi = \prod_{t=1}^{T} [P(\Psi_t \mid \mu) \, P(T \mid \Psi_t)] \, d\Psi = \prod_{t=1}^{T} [P(\Psi_t \mid \mu)] \prod_{y=1}^{Y} P(T_y \mid \Psi_t)$$
$$= \prod_{t=1}^{T} [P(\Psi_t \mid \mu) \prod_{t=1}^{T} \prod_{y=1}^{Y} \Psi_{ty} \, d\Psi \tag{4}$$

Using Gibbs sampling [2] for generative model is proposed for TART model but in a more complex form as follows:

$$P(z_{di}, a_{di}, r_{di}, t_{di} \mid z_{-di}, a_{-di}, r_{-di}, t_{-di} w, \alpha, \beta, \mu)$$

$$\propto \frac{m_{a_{di}z_{di}}^{-di} + \alpha}{\sum_z (m_{a_{di}z}^{-di} + \alpha)} \frac{n_{z_{di}w_{di}}^{-di} + \beta}{\sum_v (n_{z_{di}w}^{-di} + \beta)} \frac{n_{z_{di}t_{di}}^{-di} + \mu}{\sum_t (n_{z_{di}t}^{-di} + \mu)} \frac{n_{r_{di}z_{di}}^{-di} + \alpha}{\sum_t (n_{r_{di}z}^{-di} + \alpha)} \tag{5}$$

where z_{di} a_{di} r_{di} t_{di} are represented in order topic, author, recipient, temporal. $n_{z_{di}w_{di}}$ is number of times that the topic z_i using the word w_i, $m_{a_{di}z_{di}}$ is number of times that the author a_i giving the topic z_i. During parameter estimation for TART model, the system will keep track 4 matrices to analyze users' interests, including: T (topic) x W (word), A (author) x T (topic), R (recipient) x T (topic) and T (topic) x T (temporal). Based-on these matrices, topic and temporal distribution Φ_{zw}, topic and temporal distribution Ψ_{zt}, author and topic distribution Θ_{az}, recipient and topic distribution Θ_{rz}. The matrices are given by:

$$\theta_{az} = \frac{m_{az} + \alpha}{\sum_z (m_{az} + \alpha)} \qquad \psi_{zt} = \frac{n_{zt} + \mu}{\sum_t (n_{zt} + \mu)} \tag{6)(7}$$

$$\phi_{zw} = \frac{n_{zw} + \beta}{\sum_w (n_{zw} + \beta)} \qquad \theta_{rz} = \frac{m_{rz} + \alpha}{\sum_z (m_{rz} + \alpha)} \tag{8)(9}$$

The Algorithm below is used for analyzing users' interests with the temporal factor based on topic modeling of TART model:

Algorithm. The generative process of TART is as follows

```
1.  Input: topics, profile of users, corpus of messages
2.  Output: Users' interests on each topic over time
3.  Initialization
4.  Repeat:
5.  For each author a =1, ... , A of document d
6.      draw Θₐ from Dirichlet (α);
7.  For each topic z =1, ... , T;
8.      draw Θₐ from Dirichlet (α);
9.      draw Θᵣ from Dirichlet (α);
10.     draw Φᵤ from Dirichlet (β);
11.     draw Ψₜ from Dirichlet (μ);
12. For each word w =1, ... , Nd of document d
13.     draw an author a uniformly from all authors ad;
14.     draw a topic z from multinomial (Θₐ) conditioned  on a;
15.     draw a recipient r from multinomial (Θᵣ) conditioned on x;
16.     draw a word w from multinomial (Φᵤ) conditioned  on z;
17.     Choose a temporal factor t associated with topic z   from
        multinomial (Ψₜ) conditioned on t;
```

4   Experimental Results

4.1   Data

The exchanged information on forums or social networking is a very important source of data for us to analyze and extract information. The data for testing of model can be gathered from various sources to analyze such as policy, social, health counseling, economics, education ... in addition, developing social networking sites today like Facebook, Twitter, Zing me ... with the available APIs provided so that we easily collect data sources. Within the scope of this research, we focus on analyzing and testing the model on data of student forum of our university forum (named FTSN) which is considered as a social network, where users can register as a member. Users can discuss and exchange messages about learning, life, culture, society, union, associations, training ... in university. The size of this data is 350 MB from 2008 to 2009, the number of users are 2,494 (users) with discussions, the number of messages exchanged between users are 13,208 (posts). The number of posts this selective form of semi-automatic and has removed many posts outside Vietnamese or special characters, the messages do not make sense.

4.2   Implementation

We suggested a general model for discovering topics and analyzing authors' and recipients' interests with the temporal factor based on topic model (see Fig 4). There are 4 steps, including:

Step 1. The system collects and integrates data from the student forum which is transformed into social networks form. We need to pre-process data because the exchange information forums and social network always has garbage data such as special symbols, shortcut words, the local language, mistaken vocabulary, incorrect grammar, stop words, etc. Therefore, the data filter is very important. The system will filter out grimy information or garbage data out of the text. Then we will separate the words by some methods such as vnTokenizer [13] - the method used to determine which is the single word and which is the compound word in Vietnamese. Lastly, we will use vnTagger tool [13] to determine the POS tagging, which is noun, adjective or verb. When the system determines the POS tagging, it will help us summarize the content and get only the word that is meaningful to use for the following step.

Step 2. After the system has cleaned the data, this step will separate the topics into groups when we have exchange information from users on the social network. The similar contents will be coupled to a group that has the same topic. In this step, we use the LDA model [1][[9][10][19] to detect the latent topics based on the number of exchanged information on the social network. However, these topics don't have label yet. It has only general name such as topic 1, topic 2, topic 3 and so on.

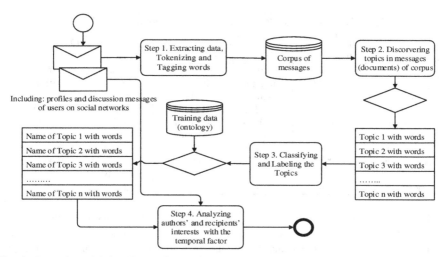

Fig. 4. General model for discovering topics and analyzing authors' and recipients' interests with the temporal factor based on topic model

Step 3. The above result is a collection of topics without labeling. In this step, we will know the label for each topic. We have built an ontology manually about the topics and the perfect training data for each topic on the ontology. Secondly, we use SVM algorithm [8], the above training data and the test data to separate the latent topics and set title for them (see Table 1).

Step 4. When B3 is finished, we have lists of topics. This step allows us to detect the interesting topic by real time. At this step, TART model is proposed to discover topically related authors and recipients for different time periods and show how authors and recipients interests are changed over time of the topics.

After finishing Step 4 of general model, the results are lists of topics with the relationships between the topics, authors and recipients with discussing probability interests and temporal factor (see Table 1, 2 and 3).

4.3   Results and Discussion

We apply TART model for FTSN data. In the test processing, we have the results as shown in Table 1. In Table 1, with 2000 posts firstly, 3 topics are discovered and labeled which are "social activity" - topic 4, "management learning" - topic 12, "recruitment jobs" - topic 17. Besides, there are some topics but they didn't depend on the topics that we surveyed. However, we could build any training data when we needed. Therefore, if we wanted to create a new survey for a new topic, we only created the training data for that topic and started to survey. Obviously, in labeling stage, the results of model depend on the training data, so the training data could be seen as the gold data. We must really notice when creating training data. If we have clear and perfect training data, the result will be very good.

Table 1 shows 3 topics which was discovered from data exchange of student forums our university. In which, the each labeled topic is described with 10 highest probability words.

Table 1. List of labeled topics and words distribution with probability

Topic 4 "social activity"		Topic 12 "management training"		Topic 17 "recruitment jobs"	
union	0.03189	Office	0.03925	company	0.03245
associate	0.01957	management	0.02322	job	0.01767
Uncle Ho	0.01769	credits	0.01933	fresher	0.01485
activity	0.01689	school fee	0.01402	staff	0.01093
union member	0.01433	teaching	0.01021	visit	0.00952
content	0.01277	test	0.00967	Salary	0.00711
mission	0.00677	leaning	0.00833	people	0.00429
school year	0.00597	schedule	0.00790	environment	0.00641
regulation	0.00535	classroom	0.00718	activity	0.00500
movement	0.00499	projector	0.00554	learning	0.00359

In table 2 and table 3 show details of 2 topics. In which, the first column describes list of authors in numbers and authors' probability interests on topics; the second column shows list of recipients and recipients' probability interests on topic; the last column shows each time period on topics that users discuss.

Table 2 shows "social activity" topic which was discovered from data exchange student forums and the authors' and recipients' probability interests of discussing with each time period on that topic. This topic, the author with code 24 has highest probability interests in Dec-2008 and the author with code 71 has lowest probability interests in Aug-2008.

Table 2. Results of analyzing "social activity" topic

Topic 4 "social activity"				
List of authors		List of recipients		Temporal factor
Authors	Probability	Recipients	Probability	
24	0.77380	49	0.72362	Dec-08
14	0.71324	191	0.70275	Jan-09
49	0.71000	69	0.67893	Jan-09
1515	0.65859	70	0.62439	Apr-09
69	0.59090	82	0.58021	Feb-09
236	0.53703	14	0.56231	Jan-09
2665	0.52321	16	0.54689	Apr-09
14	0.51121	71	0.50216	Dec-08
24	0.50000	15	0.48982	Dec-08
71	0.47652	989	0.46327	Aug-08

Table 3 shows "management learning" topic which was discovered from data exchange on the student forum and the authors' and recipients' probability interests of discussing with each time period on that topic.

Observing in table 2 and table 3, we recognize that each topic have a lot of users (authors and recipients) discussing together. Therefore, it is necessary to represent 10 authors and 10 recipients related with these topics for each time period (month) and probability interests of each user. Such as, user 49 is one of users discussed all three of topics and probability interests of user 49 are various for time periods. In "social activity" topic,

Table 3. Results of analyzing "management learning" topic

Topic 12 "management learning"				
List of authors		List of recipients		Temporal fac-tor
Authors	Probability	Recipients	Probability	
161	0.95443	49	0.87716	Dec-08
168	0.94324	87	0.88351	Jan-09
87	0.92909	1	0.85461	Jan-09
1	0.88618	42	0.80201	Apr-09
14	0.85300	789	0.85332	Feb-09
3	0.85294	213	0.82645	Jan-09
4	0.82263	12	0.78342	Apr-09
49	0.74398	168	0.73860	Dec-08
78	0.60227	132	0.70231	Dec-08
12	0.58333	49	0.69030	Aug-08

user 49 is a author for sending that topic with probability interests at 0.71 in Jan-09 and is a recipient for receiving the topic with probability interests at 0.72362 in Dec-08.

In addition, observing in table 2 and table 3 above, we realize that each user has different probability interests for time periods on each topic. For example, user 14 has probability interests at 0.71324 in Jan-09 but user 14 has probability interests at 0.51121 in Dec-08. These things show that each user can depend on some factors, such as: environments, friends, time or communities that users have relative together. Moreover, topics are discussed in time periods are important basics so that we can analyze the trend and the change of users' interests for applications are mentioned in section I.

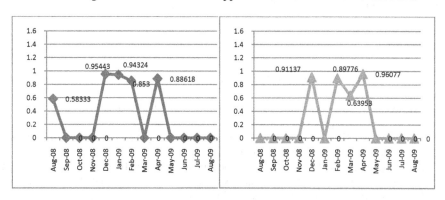

Fig. 5. Users' interests on "social activity" topic

Fig. 6. Users' interests on "management learning" topic

Fig 5 and Fig 6 show "social activity" and "management learning" topics. Fig 5, in Aug-08, the probability interests of users' interests on "social activity" topic is 0.58333 and in Jan-09, the probability interest is 0.88618, so on.

However, Fig 6, in Aug-08, on "management learning" topic, the probability interests of users' interests is 0.0 and in Jan-09, the probability is 0.0. The things show that each time period, users' interests on each topic are often different.

5   Conclusions and Future Works

In this work, we proposed a Temporal-Author-Recipient-Topic (TART) model based on topic modeling and the ideas of ART model, LDA model for social network analysis. Our proposed model combined temporal factor with authors' and recipients' interests based on topic model. Labeling and classifying topics, we used ontology and SVM method to give the results more exactly. Finally, TART model is applied to analyze users' interests over time.

Moreover, the model can be applied to discover topics conditioned on message sending and receiving, clustering and labeling topic. Our model can discover relationships of users (authors and recipients), and analyzing huge message data on social network. All parts of model are integrated on a social network analysis system based on topic model. The experiment of the model on the collection of Vietnamese texts has proved the efficiency of our model.

The proposed model will create a useful component in systems, such as recommendation, expert-finding systems. Our model can analyze to understand the interactions between users in an organization in order to make recommendations about improving organizational efficiency, especially in university systems [16][17]. In the future, we will analyze relationships of the trend and the change of users' interests based on discussion messages and structure in community network.

Acknowledgement. This research is funded by Viet Nam National University Ho Chi Minh City (VNU-HCMC) under Grant number B2013-26-02. We would like to thank Mr. Duy Doan who supported to us when we researched this paper.

References

1. Blei, D.M., Ng, A.Y., Jordan, M.I.: Latent Dirichlet Allocation, Journal of Machine Learning Research, pp. 993–1022 (2003)
2. Griffiths, T.: Gibbs Sampling in the generative model of Latent Dirichlet Allocation (2004). Gruffydd@psych.stanford.edu
3. Rosen-Zvi, M., Griffths, T., et al.: Probabilistic Author-Topic Models for Information Discovery. In: 10th ACM SigKDD, Seattle (2004)
4. McCallum, A., Corrada, A., Wang, X.: The Author-Recipient-Topic Model for Topic and Role Discovery in Social Networks: Experiments with Enron and Academic Email, Department of Computer Science, University of MA (2004)
5. Darling, W.M.: A Theoretical and Practical Implementation Tutorial on Topic Modeling and Gibbs Sampling - School of Computer Science, University of Guelph (2011)
6. Lau, J.H,: Automatic Labelling of Topic Models, Dept of Computer Science and Software Engineering, University of Melbourne (2011)
7. Phan, X.-H., Nguyen, C.-T., Le, D.-T., Nguyen, L.-M., Horiguchi, S., Ha, Q.-T.: A hidden topic-based framework towards building applications with short Web documents. IEEE Transactions on Knowledge and Data Engineering (IEEE TKDE) 23(7), 961–976 (2011)
8. Joachims, T.: Transductive Inference for Text Classification using Support Vector Machines. In: International Conference on Machine Learning (ICML) (1999)

9. Bíró, I., Szabó, J.: Latent Dirichlet Allocation for Automatic Document Categorization, Research Institute of the Hungarian Academy of Sciences Budapest (2008)
10. Bíró, I.: Document Classification with Latent Dirichlet Allocation, Ph.D. Thesis Summary, Faculty of Informatics, Eötvös Loránd University (2009)
11. Blei, D.M.: Introduction to Probabilistic Topic Models. Comm. ACM **55**(4), 77–84 (2012)
12. ui Wang, X., McCallum, A.: Topics over Time: A Non-Markov Continuous-Time Model of Topical Trends. In: ACM SIGKDD, Philadelphia, Pennsylvania, USA (2006)
13. http://vlsp.vietlp.org:8080/demo/, vnTokenizer and vnTagger tools
14. Aggarwal, C.C.: Social Network Data Analytics, IBM Thomas J. Watson Research Center (2011)
15. Mora-Soto, A., Sanchez-Segura, M., Medina-Dominguez, F., Amescua, A.: Collaborative Learning Experiences Using Social Networks, International Conference on Education and New Learning Technologies (EDULEARN09), Barcelona, Spain, pp. 4260–4270 (2009)
16. Calvó-Armengol, A., Patacchini, E., Zenou, Y.: Peer Effects and Social Networks in Education. Rev. Econ. Stud. **76**(4), 1239–2167 (2009)
17. Yang, H.-L., Tang, J.H.: Effects of social network on students' performance: A web-based forum study in Taiwan. Journal of Asynchronous Learning Networks **7**(3), 93–197 (2003)
18. Nguyen, M., Ho, T., Do, P.: Social Networks Analysis Based on Topic Modeling. In: The 10th IEEE RIVF International Conference on Computing and Communication Technologies, RIVF 11/2013, pp. 119–122, IEEE, Vietnam (2013)
19. Ho, T., Duy, D., Do, P.: Discovering Hot Topics On Social Network Based On Improving The Aging Theory. Advances in Computer Science: an International Journal **3**(3), 4/2014 (2014)

Classifying Continuous Classes with Reinforcement Learning RULES

Hebah ElGibreen[1](✉) and Mehmet Sabih Aksoy[2]

[1] Information Technology Department, College of Computer and Information Sciences,
King Saud University, Riyadh 11415, Saudi Arabia
hjibreen@ksu.edu.sa
[2] Information System Department, College of Computer and Information Sciences,
King Saud University, Riyadh 11543, Saudi Arabia
msaksoy@ksu.edu.sa

Abstract. Autonomous machines are interesting for both researchers and regular people. Everyone wants to have a self control machine that do the work by itself and deal with all types of problems. Thus, supervised learning and classification became important for high-dimensional and complex problems. However, classification algorithms only deals with discrete classes while practical and real-life applications contain continuous labels. Although several statistical techniques in machine learning were applied to solve this problem but they act as a black box and their actions are difficult to justify. Covering algorithms (CA), however, is one type of inductive learning that can be used to build a simple and powerful repository. Nevertheless, current CA approaches that deal with continuous classes are bias, non-updatable, overspecialized and sensitive to noise, or time consuming. Consequently, this paper proposes a novel non-discretization algorithm that deal with numeric classes while predicting discrete actions. It is a new version of RULES family called RULES-3C that learns interactively and transfer experience through exploiting the properties of reinforcement learning. This paper will investigate and assess the performance of RULES-3C with different practical cases and algorithms. Friedman test is also applied to rank RULES-3C performance and measure its significance.

Keywords: Continuous classes · Classification · Covering algorithm · RULES family · Reinforcement learning

1 Introduction

Due to the recent flood of new technologies, researchers and developers are striving for adaptability and sustainability of information systems over different needs and requirements. Supervised learning and classification became important for high-dimensional and complex problems [1]. Although classification only predicts discrete classes, in practical and real-life applications class labels contain continuous labels, and predictions need to be based on these numeric values. For example, in CPU performance, it is possible that the desired actions is to know if the performance is good, intermediate, or bad; while the given sample measure the performance in numbers.

© Springer International Publishing Switzerland 2015
N.T. Nguyen et al. (Eds.): ACIIDS 2015, Part II, LNAI 9012, pp. 116–127, 2015.
DOI: 10.1007/978-3-319-15705-4_12

Different statistical methods and techniques have been developed in ML. However, these methods act as a black box and their model is difficult to understand or predict. An intelligent agent needs to learn through its discovered knowledge to reuse and apply it to new situations. Such discovery can be done when the agent makes a suitable generalization, which Asgharbeygi et.al [2] identified as inductive learning. Inductive learning (IL) is one field of ML and has several characteristics that draw the attention [3, 4]. Its model simplicity and transparency makes it more appealing, especially in domains that needs to understand the systems. IL learns a set of rules from given instances and create a classifier that can generalize the result over new instances. One type of IL, called covering algorithm (CA), became attractive due to its appealing properties; summarized in [5]. Therefore, CA is a promising area of research, especially when the users need to understand how the systems work. Nevertheless, CA methods are designed to predict nominal actions from discrete classes. It is difficult to handle continuous labels because they have an infinite space.

The problem of continuous classes can be defined as the ability to induce new knowledge about numeric conclusions. Depending on the desired action, researchers dealt with the problem differently. Depending on the desired action, researchers dealt with the problem differently. First, when the desired action is nominal and classification is desired, pre-discretization is usually applied. However, even though pre-discretization reduces the time of rule induction, but it can seriously affect the rules' quality resulting from the CA [6]. There is a great tradeoff between the number of intervals and the consistency of the rule. It can also cause major problems, on the long run, because it has fixed intervals. It is difficult to update the interval and it reduces the future accuracy of the algorithm. Second, when the desired action to predict is numeric, the problem was recognized as a regression problem. In this type of prediction, IL has mainly two approaches: regression and regression via classification (RvC). Based on the study conducted by Elgibreen and Aksoy [7], both approaches are still lacking and create more complexity and inflexibility. Inducing rules from regression trees can become NP-complete problem, its resulting model is difficult to understand, and its models are unable to provide good point estimates with noisy data. Regression trees are sensitive to data changes and require a high amount of data. RvC, on the other hand, can take the advantages of both classification and regression models to predict continuous classes. Its methods, however, are similar to the pre-discretization approach used when the predicted action is nominal. It fixes the intervals and cause high computation and time complexity when applied online. When fuzzy set or genetic programming approaches are used, RvC performance is highly dependent on a function that is defined by the user, and becomes complex and difficult to understand. Clearly then, classification based on continuous classes in CA remains a fertile ground for new research. Due to the need for nominal prediction while it is possible to get numeric labels, a new approach must be proposed to offer better performance while addressing the current shortcomings.

Therefore, this paper proposes a new algorithm called RULES with Continuous Classes for Classification (RULES-3C). It is inspired by how human learn their behaviors through interacting with their environment. This property was introduced through adopting reinforcement learning (RL) [8]. The use of RL offers several appealing properties, in addition to solving the problems of continuous classes, but it has never been done in CA. The proposed algorithm generalizes classification over datasets with numeric labels, where discrete actions are directly predicted based on continuous classes without discretization. RULES-3C contributes to CA through exploiting the properties of RL to introduce aspects of the human mind aspects and improves the current literature of CA by solving its current shortcomings. Finally, its model is helpful for both decision makers and expert systems due to the simple 'if-then' representation of its model. At the end of this paper, to show how RULES-3C can improve the performance of classification, it is compared with seven classification algorithms over 30 well-known datasets and the result is validated with 10-fold cross-validation. The performance is analyzed based on Demsar suggestion [9], where Friedman test with Nemenyi post hoc is applied to examine the significant between the algorithms.

The paper is organized as follows. First, RL is explained and RULES-3C is presented with its technical details. The experiment results are discussed afterwards and the paper is concluded at the end.

2   Reinforcement Learning

In order to understand the proposed algorithm Reinforcement Learning (RL) needs to be explained first. This field of advance ML was inspired of how living being learns their actions through interacting with their environment. RL is *"the problem faced by an agent that must learn behavior through trial-and-error interactions with a dynamic environment"* [8]. It is a learning process that discovers the appropriate action at a time through examining its current state in order to obtain a new state in which the rewards are maximized. The main goal of RL is to reach a decision in a series of tasks throughout trial-and-error interactions with the environment [10]. Consequently, unlike most ML methods, learners in RL are not told what to do but they discover the best rewarding action by trying them. The basic components of RL model are the agent and environment, which are connected together through actions and perceptions [8]. At a time (t), the environment presents a state (s_t) and the agent performs an action (a_t) based on this state [11]. Rewards, on the other hand, are what the agent would get after applying an action; to know if the action is rewarding (towards the goal) or a penalty (bad movement). In order to apply RL, a control method must be chosen to decide what action should be selected in a certain state. One of the simplest and best-known control methods for RL is Q-Learning [12], which repeatedly applies actions given the current state. Q-learning collects action values to find the optimal policy without the need for an environment model [13]. In particular, the action value $Q(s, a)$ can be calculated using (1), where s is the current state, s' is the new state, a is the action applied to move from s to s', r is the reward for applying a over s, and $(0 \leq \alpha \leq 1)$ and $(0 \leq \gamma \leq 1)$ are two constants that are used to normalize the reward values.

$$Q(s, a) + \alpha [r + \gamma \max_{a' \in A(s_{t+1})} Q(s', a') - Q(s, a)] \tag{1}$$

In Q-Learning, when the agent visits a state, it selects an action based on a policy, such as greedy policy. The reward of the selected action is then collected and the action value is calculated to select the optimal action. These steps are repeated until the stop condition is reached. RL has been used as rule classifier for the classification task [14-18], and it was used to cluster data [19-21], in addition to its use in intelligent systems and robotics. Thus, the properties of RL can be taken advantage of in different learning paradigms. However, in our knowledge no one yet used it in CA to predict discrete actions based on continuous classes.

3   RULES-3C Algorithm

Due to the significance of continuous classes in classification, and the problems of current approaches in CA, a new version of RULES family is proposed in this section. Known as RULES with Continuous Classes for Classification (RULES-3C), this algorithm is developed to perform classification even if the data contain continuous classes. The main idea of RULES-3C is to produce classification model using continuous classes without discretization. The intervals are not fixed and approximation function is not needed. It discovers the best threshold for a label depending on its relationship with other examples and classes. It handles all type of attribute and does not require them to be numeric, as in most regression and RvC methods. It reduces rule overlapping during the learning process by using two thresholds label. In RULES-3C, the properties of RL have been exploited. It is designed to be easily updated to be directly integrated with any incremental rule induction. Like RULES-TL [22], RULES-3C transfers previous knowledge to the agent in the form of rules. RL knowledge transfer is used to reduce the gap between the slow pace of ML and the speed of human learning [23, 24]. Therefore, transferring knowledge reduces the time, improve performance, and increase agent intelligence. As illustrated in Fig. 1 (step 1), RULES-3C starts by reading the training set and storing its parameters and examples. In step 2, past knowledge is transferred (if found) and mapped to the target task representation, similarly to RULES-TL. If there are mapped rules, they are used to mark covered examples and then added to the rule set repository. Step 3 initializes the RL state space by storing all possible class values that are available in the training set as states. In the initialization step, all the parameters are initialized to zero, including its reward and Q-values, and two signs are assigned for every state cell (including <= and >).

```
RULES-3C (Training set (Tset), minimum negative (mN), minimum posi-
tive (mP), beam width (w))
1.  Read training set
2.  If past knowledge exist:
    a. TL = Read previously discovered rules, if exist
    b. ∀ r ∈ TL if at least one example E is covered
    i.  Mark E as covered
    ii. RuleSet = RuleSet+ r
3.  RL = Initialize RL.stateSpace
4.  For each E ∈ Tset & E not covered yet
    a. Th = RL.Execute(E, Tset);
    b. If (Th.contains("=")) //if the state have <= sign
    i.  If (Th>E.class) Then Threshold = "[E.class, Th]"
    ii. Else Threshold = "[Th, E.class]"
    c. Else  //if the state have > sign
    i.  If (Th>E.class) Then Threshold = "[E.class, Th)"
    ii. Else Threshold = "(Th, E.class]"
    d. InduceR = Induce_One_Rule (Threshold, E, Nset, mP, mN, w)
    e. Mark all examples covered by InduceR
    f. RuleSet = RuleSet+ InduceR
5.  Merge similar rules
6.  Order RuleSet by the rules score
7.  Return RuleSet
```

Fig. 1. RULES-3C main procedure

In addition to the signs, the number of examples that belong to that range is also stored in the state metadata. However, the matching examples themselves are not stored in the future. This is because it might cause "out of memory" problem. Hence, only the needed information is stored when needed. After initializing and constructing the state space in Fig. 1, the algorithm starts the rule discovery. In step 4, the algorithm takes every uncovered example as seed and discovers the best possible value for its class using RL agent (step 4.a). Then, the alternative threshold is created using the resulting label and the seed label (step 4.b). This way, the rule overlapping is reduced and unknown areas are not included in the coverage. To obtain the alternative labels using RL agent called in Fig. 1(step 4.a), RL method is executed as in Fig. 2.(a). This method is based on Q-learning and starts from the state that covers the seed example (step 1). The agent tries to choose the next state based on their Q values before choosing the subsequent action based on the reward value. The agent chooses to go up or down depending on the actions' reward values. Then, in step 4, a loop starts to search for better actions and stops if the reward is no longer improving or it reaches its stop condition. The stop condition is chosen based on various trials, where it was found that 25% of the state space is enough to know whether the reward is improving or not.

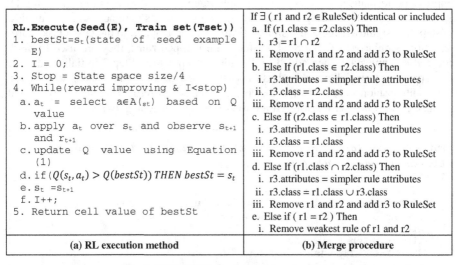

| (a) RL execution method | (b) Merge procedure |

Fig. 2. RULES-2C supplementary procedures

In this loop, the RL procedure selects several actions based on a greedy policy, where the best actions are chosen (step 4.a). In step 4.b and c, the new state and its reward are observed and the Q-value of the previous action is computed, using (1). Based on the computed Q-value, the algorithm observes its history and store the state that has the highest Q-value (step 4.d). After optimizing the search over the state space, the algorithm returns the best state value observed as the alternative label. This label is used to create the threshold that can be used to induce a new rule. The actual reward of each action is computed using the distance between the seed and examples that belong to current state; in addition to the average number of negative examples (comparing to the seed) in the state; as shown in Equation (2), where the average distance is computed using (3), average negative computed using (4), and SE is the seed example. Note that less distance and number of negatives increases the reward. However, if the state includes only positive or only matching examples then the reward will be zero to reduce overspecialization. Cells that do not match the seed example have a penalty (-1) not a reward to prevent choosing unrelated cells.

$$
r_t = \begin{cases} \dfrac{1}{AvgD*AvgNeg} & if \ AvgD > 0 \ and \ AvgNeg > 0 \\[3mm] 0 & if \ AvgD = 0 \ or \ AvgNeg = 0 \\[3mm] -1 & if \ SE \notin current \ state \end{cases} \tag{2}
$$

$$
AvgD = \frac{\Sigma_{I \in example \ index \ of \ current \ state} \ dis(SE,I)}{number \ of \ examples \ in \ current \ state * number \ of \ attributes} \tag{3}
$$

$$
AvgNeg = \frac{number \ of \ negative \ examples \ in \ current \ state \ (comparing \ to \ SE)}{number \ of \ examples \ in \ current \ state} \tag{4}
$$

Going back to Fig. 1, after discovering the alternative threshold, it is used to induce the best rule by executing Induce_One_Rule procedure. This procedure is similar to the one executed in the proceeding versions (RULES-TL and RULES-6) thus is will not be explained in details. What is important is to know that the discovered threshold is used as class label instead of the original numeric label. This threshold is not fixed and changes from seed to another; thus, it is not discretized. Moreover, as in its proceeding versions, m-estimate [25] is applied as the quality metric; while continuous attributes are managed by PKID discretization technique [26]. After inducting the best rule, this rule is used to mark all covered examples (step 4.e). When all examples are marked, redundant rules are merged in step 5 and the merge procedure summarized in Fig. 2.(b) is applied. In this procedure, rules are merged to reduce the rule space and produce simpler model that can generalize to noise in the future. Merge is applied instead of removal because the class label produced contains two threshold values, so if only identical rules are removed then the result will not be optimized. Basically, the merge procedure can be considered as post pruning technique applied after rule set discovery to simplify the model and avoid overspecialization.

From Fig. 2.(b), for every identical or sub rules, if the class threshold is identical then the shared conditions along with the class label is stored while the parents' rules are removed (step 1.a). If classes are inside each other (step 1.b and 1.c), simpler rule is stored with larger threshold. If the classes intersect (step 1.d), the simpler rules are stored along with the thresholds union. Finally, when the thresholds are not equal and do not intersect or includes (step 1.e), if the rules are identical the stronger one is stored and the other is removed, otherwise both are stored. Note that if the rules conditions conflict or gives different values, their classes will not be compared because it means that these rules are not matching. Finally, after producing the final rules set this set is ordered by its rules scores and stored for future discovery (Fig. 1 step 6). This way, RULES-3C can directly deal with conflict based on the stored decision list. Nevertheless, after discovering the knowledge prediction is required. In contrast to numeric action prediction, when the action predicted is discrete the class label threshold is chosen and used for classification. When a certain rule cover the example its class label is used to predict the unseen example class. If the example label is inside the chosen threshold, this means that the algorithm made a correct classification; otherwise it incorrectly classified the example. Note that if none of the discovered rules can cover the unseen example, then the average of the training set class values is chosen. Moreover, if conflict occurs then the strongest rule is chosen for prediction.

4   Experiment

In order to measure the performance of the proposed algorithm and to compare it with the existing methods, KEEL tool [27] is used. The experiments are conducted on a PC with Intel®Core™ i7 CPU, 2.40 GHz processes, and 16GB RAM. In order to show how reliable the proposed algorithm is, 30 dataset with continuous classes are used to test it. These datasets are taken from KEEL dataset repository [28] and gathered as real-life benchmark while validated using 10-fold cross-validation. Note that numeric attributes are discretized using PKID to focus on the problem of continuous classes. Moreover, classification algorithms that cannot handle continuous labels, excluding RULES-3C, are using unsupervised pre-discretization (PKID) to produce discrete actions from the numeric input. RL parameters, including the Alpha (α) and Gamma (γ) have been initialized based on the standard discovered by Sutton and Barto [29]; both are equal to 0.5 to give equal weight for past and current knowledge.

In order to show that performance is improved by using RULES-3C, seven well known classification algorithms are compared. These algorithms represent the mostly known families in CA and DT that are usually used to benchmark new developments. One of the latest versions of RULES algorithm, called RULES-6, is compared to analyze the effect of RL in RULES family. Another three CA are also compared to differentiate RULES-3C with CA, including Slipper, Ripper, and DataSqueezer. Finally, three DT algorithms are also compared, including C45, PUBLIC, and C45RulesSA, to differentiate between RULES-3C and DT algorithms. In order to assess an algorithm accuracy, the error rate and variance is measured for every algorithm. The error rate is used to measure how well an algorithm predict its test set, in order to know if the model underfit its training set or not. The variance, on the other hand, is used to measure the difference between the training and test prediction in order to know if the model overfit its training set or not. Underfit measure the current accuracy level of an algorithm and, hence, can be measured by computing the percentage of correct prediction. Overfit, however, is measured by the variance between the training set and test set, as suggested by Zahálka and Železný [30].

Starting from the error rate, in Fig. 3, DataSqueezer has the worst error rate in most of the datasets. RULES-3C, on the other hand, shows an obvious improvement in many datasets. In Abalone, Delta_ail, Delta_elv, and Quake datasets, which include a medium number of examples and small number of attributes, RULES-3C shows a significant reduction in error. In Plastic dataset, which has small number of examples and attributes, and in ailerons dataset, which has large number of examples and attributes, RULES-3C also significantly reduced the error. Finally, Compative and Flare datasets have a medium number of examples and attributes, and RULES-3C significantly reduced the error in these datasets. After RULES-3C, the error rates of rest of the algorithms have similar behavior. C45, in particular, comes in second place after RULES-3C. Nevertheless, there is one dataset where RULES-3C shows a significant increase in error rate. As shown in Fig. 3, RULES-3C is the worst in Mortgage dataset, which has medium number of examples and attributes. Even though Compative and Flare datasets have the same properties as Mortgage but RULES-3C behaves differently in each dataset. This is because of the continuous classes' effect. Depending on the results of RL agent, RULES-3C behaves differently in similar datasets. However, regardless of the effect of reasoning under uncertainty that is introduced due to the use of RL, RULES-3C has better performance in most datasets, and similar performance in the other datasets.

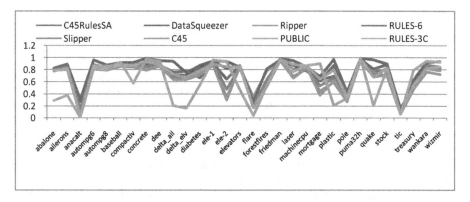

Fig. 3. Error rate using 10-fold cross validation

In addition to the error rate, the variance is also recorded in Fig. 4. From the graph, it can be noticed that even though C45 error rate was not far away from RULES-3C and better than most CA algorithms but it excessively increase its variance. C45 variance is very high, where it comes in last place after Ripper. This indicates that C45 traded its underfit on the expenses of its overfit. This conclusion also applies over Ripper and Slipper. In contrast, RULES-6 and Datasqueezer traded their error rate on the expenses of the variance. In these two algorithms the error rate was relatively high indicating that they underfit their datasets. However, when it comes to the variance they both have relatively low variance to reduce the overfit effect. From Fig. 4, the variance of RULES-6 and Datasqueezer come in second place after RULES-3C and PUBLIC. In C45RulesSA, both variance and error rate are relatively high. Although it is not the worst but in both measures the performance is not one of the best. In PUBLIC, however, even though its error rate is similar to C45RulesSA the variance is one of the best. PUBLIC presented in Fig. 4 shows how low its variance is. Finally, when it comes to RULES-3C, there is no trading between the underfit and overfit measures. It has relatively low error rate (similarly to C45 and Ripper) while introducing a model with low variance (similarly to PUBLIC and Datasqueezer). Therefore, it can be concluded that adopting RL into RULES family introduce a model that has an accuracy as good as the DT algorithms while improving the variance of its family to surpass other DT and CA.

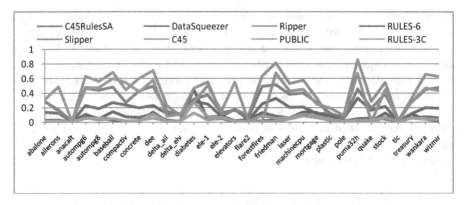

Fig. 4. Variance using 10-fold cross validation

Nevertheless, comparing the performance using simple figures is not enough and, thus, the performance needs to statistically study. In particular, to asses RULES-3C's predictive value and determine whether any improvement is significant, the statistical methods that Demsar [9] proposed are applied using XLSTAT tool [31]. Using 10-fold cross validation, the average error rate and variance of every dataset over all its folds is tested using Friedman test along with Nemenyi post hoc test. The result is illustrated in the CD diagram, as shown in Fig. 5. In this graph, the CD line is the critical bar; algorithms that that do not significantly differ by the Nemenyi test are connected with a horizontal bar; better algorithms are on the right hand side; and the numbers on the horizontal line indicate the rank mean of every algorithm. The graph shows that RULES-3C belongs to the first group in both error rate and variance. In the error rate (Fig. 5.a), RULES-3C rank in third after C45 and Ripper, but the difference is insignificant. RULES-3C also belongs to the same groups of the other algorithms, except for Datasqueezer. Hence, introducing RL into classification can produce similar to better result than conventional CA and DT algorithms. In the variance (Fig. 5.b), RULES-3C also comes in third place after PUBLIC and Datasqueezer but, again, the difference is insignificant. Actually, such reduction in variance is introduced to insure the balance between the underfit and overfit measures. In contrast to the other algorithm, RULES-3C managed to preserve both error rate and variance and belong to the best group in both measures. Its variance is significantly better than other CA and DT algorithms, except for PUBLIC, RULES-6, and Datasqueezer; which confirm that RULES-3C can manage noise in the future and induce a model that is not overspecialized.

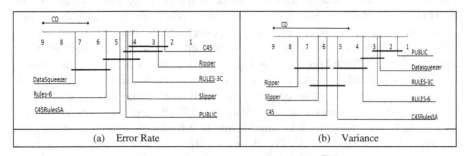

Fig. 5. Error rate and variance CD diagram using Nemenyi post hoc test

5   Discussion

From all the empirical and technical details of RULES-3C, it was concluded that adopting RL into RULES family introduce a model that is better than other CA and as good as the DT algorithms; while improving the variance of its family to surpass other DT and CA. From the properties of RULES-3C, the gaps of current approaches exposed in CA literature are covered. The complexity of regression tree and fuzzy-based algorithms is eliminated because the complex regression functions and approximation are discarded. The result of RULES-3C is understandable and comprehensible, in contrast to regression trees, due to the use of simple IF...THEN rules. The result is generalized and rule overlapping and coverage of unknown area problems are managed due to the use of two threshold label. Like RvC, RULES-3C takes advantages of both classification and regression models to deal with continuous classes, but intervals are not fixed and discrete actions are predicted. In contrast to RvC, continuous learning is introduced due to the use of RL, so integrating the algorithm into incremental learning is straightforward. When applying RULES-3C for classification, the experiment demonstrated how this algorithm outperforms other CA and DT algorithms. RULES-3C produces a model that is as accurate as DT algorithms with better variance. It improves the accuracy of CA and manages data with continuous classes without discretization. As a result, introducing RL into RULES family produced a simple model without the need for discretization and solved the current issues of continuous classes' problem while predicting discrete actions.

6   Conclusion

Due to the importance of understanding how decisions are reached and to help decision-makers, CA is the focus of this paper. Although it produces simple and powerful results, its classification model is still lacking when dealing with continuous classes and further improvement is still needed. Thus, this paper proposed an extended version of RULES family called RULES-3C. This algorithm solved the problem of continuous classes in classification, in addition to the good properties obtained from the use of RL. Due to the use of RL, learning starts from scratch and gathered knowledge can be used to improve the threshold selection. It tries to discover the best discrete value for a class depending on its relationship with other examples and classes. Moreover, RULES-3C cumulatively learns throughout the lifetime of an agent, so incremental learning can be directly applied without additional procedure. From the empirical results, it was found that RULES-3C surpasses the other DT and CA discussed in the literature. In the future, numeric attributes should also be handled directly, as is the case with labels, instead of invoking discretization.

Acknowledgements. This research project is supported by a grant from King AbdulAzaiz City for Science & Technology.

References

1. Escalante-B, A.N., Wiskott, L.: How to Solve Classification and Regression Problems on High-Dimensional Data with a Supervised Extension of Slow Feature Analysis. Journal of Machine Learning Research **14**, 3683–3719 (2013)
2. Asgharbeygi, N., Nejati, N., Langley, P., Arai, S.: Guiding Inference Through Relational Reinforcement Learning. In: Kramer, S., Pfahringer, B. (eds.) ILP 2005. LNCS (LNAI), vol. 3625, pp. 20–37. Springer, Heidelberg (2005)
3. Aksoy, M.S., Mathkour, H., Alasoos, B.A.: Performance Evaluation of RULES-3 Induction System for Data Mining. International Journal of Innovative Computing, Information and Control **6**, 3339–3346 (2010)
4. Kotsiantis, S.B.: Supervised Machine Learning: A Review of Classification Techniques. Informatica **31**, 249–268 (2007)
5. Kurgan, L.A., Cios, K.J., Dick, S.: Highly Scalable and Robust Rule Learner: Performance Evaluation and Comparison. IEEE Systems, Man, and Cybernetics—Part B Cybernetics **36**, 32–53 (2006)
6. Pham, D., Bigot, S., Dimov, S.: RULES-5: a rule induction algorithm for classification problems involving continuous attributes. In: Institution of Mechanical Engineers, pp. 1273–1286 (2003)
7. ElGibreen, H., Aksoy, M.S.: Inductive Learning for Continuous Classes and the Effect of RULES Family. International Journal of Information and Education Technology **5**, 564–570 (2014)
8. Kaelbling, L.P., Littman, M.L., Moore, A.W.: Reinforcement Learning: A Survey. Journal of Artificial Intelligence Research **4**, 237–285 (1996)
9. Demšar, J.: Statistical Comparisons of Classifiers over Multiple Data Sets. Journal of Machine Learning Research **7**, 1–30 (2006)
10. Moriarty, D.E., Schultz, A., Grefenstette, J.: Evolutionary Algorithms for Reinforcement Learning. Journal of Artificial Intelligence Research **11**, 241–276 (1999)
11. Nissen, S.: Large Scale Reinforcement Learning using Q-SARSA(λ) and Cascading Neural Networks. Master Thesis, Department of Computer Science, University of Copenhagen, Denmark (2007)
12. Watkins, C.: Learning from Delayed Rewards. PhD Thesis, Cambridge University, Cambridge, England (1989)
13. Ertel, W.: Reinforcement Learning. In: Introduction to Artificial Intelligence, pp. 257–277. Springer (2011)
14. Garcia, F., Martin-Clouaire, R., Nguyen, G.L.: Generating Decision Rules By Reinforcement Learning For A Class Of Crop Management Problems. In: 3rd European Conference of the European Federation for Information Technology in Agriculture, Food and the Environment (EFITA'01). Montpellier (2001)
15. Lagoudakis, M.G., Parr, R.: Reinforcement Learning as Classification: Leveraging Modern Classifiers. In: Twentieth International Conference on Macine Learning (ICML-2003). Washington DC (2003)
16. Lanzi, P.L.: Learning classifier systems from a reinforcement learning perspective. Soft Computing - A Fusion of Foundations, Methodologies and Applications **6**, 162–170 (2002)
17. Wiering, M.A., van Hasselt, H., Pietersma, A.D., Schomaker, L.: Reinforcement Learning Algorithms for solving Classification Problems. In: IEEE International Symposium on Approximate Dynamic Programming and Reinforcement Learning (ADPRL), pp. 91–96. Paris, France (2011)

18. Hwang, K.-S., Chen, Y.-J., Jiang, W.-C., Yang, T.-W.: Induced states in a decision tree constructed by Q-learning. Information Sciences **213**, 39–49 (2012)
19. Tamee, K., Bull, L., Pinngern, O.: A Learning Classifier Systems Approach to Clustering (2006)
20. Likas, A.: A Reinforcement Learning approach to on-line clustering. Neural Computation **11**, 1915–1932 (1999)
21. Barbakh, W., Fyfe, C.: Clustering with Reinforcement Learning. In: Yin, H., Tino, P., Corchado, E., Byrne, W., Yao, X. (eds.) IDEAL 2007. LNCS, vol. 4881, pp. 507–516. Springer, Heidelberg (2007)
22. ElGibreen, H., Aksoy, M.S.: Multi Model Transfer Learning with RULES Family. In: Perner, P. (ed.) MLDM 2013. LNCS, vol. 7988, pp. 42–56. Springer, Heidelberg (2013)
23. Wilson, A., Fern, A., Tadepalli, P.: Transfer Learning in Sequential Decision Problems: A Hierarchical Bayesian Approach. In: Workshop on Unsupervised and Transfer Learning, pp. 217–227 (2012)
24. Konidaris, G., Scheidwasser, I., Barto, A.G.: Transfer in reinforcement learning via shared features. Journal of Machine Learning Research **13**, 1333–1371 (2012)
25. Dzeroski, S., Cestnik, B., Petrovski, I.: Using the m-estimate in rule induction. Journal of Computing and Information Technology **1**, 37–46 (1993)
26. Yang, Y., Webb, G.: Proportional k-Interval Discretization for Naive-Bayes Classifiers. In: Raedt, L., Flach, P. (eds.) ECML 2001. LNCS (LNAI), vol. 2167, pp. 564–575. Springer, Heidelberg (2001)
27. Webmaster.Team. KEEL: A software tool to assess evolutionary algorithms for Data Mining problems including regression, classification, clustering, pattern mining and so on (2012). http://keel.es/
28. Alcalá-Fdez, J., Fernandez, A., Luengo, J., Derrac, J., García, S., Sánchez, L., Herrera, F.: KEEL Data-Mining Software Tool: Data Set Repository, Integration of Algorithms and Experimental Analysis Framework. Journal of Multiple-Valued Logic and Soft Computing **17**, 255–287 (2011)
29. Sutton, R., Barto, A.: Reinforcement Learning: An Introduction, 1st edn. MIT Press, Cambridge, MA (1998)
30. Zahálka, J., Železný, F.: An experimental test of Occam's razor in classification. Machine Learning, vol. 82, pp. 475–481, 2011/03/01(2011)
31. Addinsoft. XLSTAT (2014). http://www.xlstat.com/en/

Fuzzy Association Rule Mining
with Type-2 Membership Functions

Chun-Hao Chen[1], Tzung-Pei Hong[2,3(✉)], and Yu Li[2]

[1] Department of Computer Science and Information Engineering,
Tamkang University, Taipei 251, Taiwan
chchen@mail.tku.edu.tw
[2] Department of Computer Science and Engineering,
National Sun Yat-sen University, Kaohsiung, Taiwan
[3] Department of Computer Science and Information Engineering,
National University of Kaohsiung, Kaohsiung, Taiwan
tphong@nuk.edu.tw, m023040059@student.nsysu.edu.tw

Abstract. In this paper, a fuzzy association rule mining approach with type-2 membership functions is proposed for dealing with data uncertainty. It first transfers quantitative values in transactions into type-2 fuzzy values. Then, according to a predefined split number of points, they are reduced to type-1 fuzzy values. At last, the fuzzy association rules are derived by using these fuzzy values. Experiments on a simulated dataset were made to show the effectiveness of the proposed approach.

Keywords: Data mining · Fuzzy association rule · Membership functions · Type-2 fuzzy set

1 Introduction

Association rule analysis is most commonly used in attempts to derive useful information and extract useful relationship of items from large data sets or database for solving specific issue, which is an expression $X \rightarrow Y$, where X and Y are a set of items [1]. Since transactions always have quantitative values, how to handle the quantitative values becomes an interesting issue. Because the fuzzy theory is usually utilized to handle continuous values, hence lots of mining algorithms have been proposed to induce fuzzy rules from quantitative transaction database. They can be divided into two kinds according to the type of minimum support thresholds, namely *single-minimum-support fuzzy-mining (SSFM)* [2, 3, 5, 6, 11, 13, 16, 19, 20, 21, 22] and *multiple-minimum-support fuzzy-mining (MSFM)* [9, 10, 12, 14, 15] approaches. The SSFM uses only one minimum support for all items, and each item has its own minimum support in MSFM.

However, according to [17], they described that the data used to mine fuzzy rules may be uncertain. In other words, how to handle data uncertainty has an important influence on the quality of derived rules. And, they also indicated type-2 fuzzy sets are able to model uncertainties well. From the mentioned approaches, only type-1 membership func-

© Springer International Publishing Switzerland 2015
N.T. Nguyen et al. (Eds.): ACIIDS 2015, Part II, LNAI 9012, pp. 128–134, 2015.
DOI: 10.1007/978-3-319-15705-4_13

tions are utilized for handling quantitative values for deriving fuzzy association rules. Although many type-2 fuzzy rule learning approaches have been proposed, they are used for control problems [4, 7], financial analysis [8], spatial analysis [18], etc.

This study thus proposes a fuzzy association rule mining algorithm with type-2 membership functions for dealing with the uncertainties of data. The proposed approach first transforms quantitative transactions into upper and lower fuzzy values by the given type-2 membership functions. Then, by using the predefined split number of points, the upper and lower fuzzy values of a linguistic term are aggregated to a fuzzy value. At last, the derived fuzzy values are utilized to mine large itemsets for mining fuzzy association rules. Experimental results on a simulation dataset are also made to show the comparison of the proposed approach and existing ones in terms of number of derived rules.

2   Framework of the Proposed Approach

In this section, the framework of the proposed approach is described, and is shown in Figure 1.

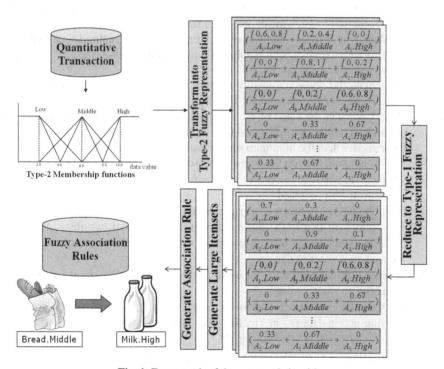

Fig. 1. Framework of the proposed algorithm

Fig. 1 shows that the proposed algorithm first transforms the quantitative transactions into fuzzy representation by the given type-2 membership functions. Then, according to a predefined split number of point, the derived type-2 fuzzy values of items are then reduced to type-1 fuzzy values. At last, the large itemsets and fuzzy association rules are generated from the derived type-1 fuzzy values.

3   The Proposed Mining Algorithm

In this section, the proposed approach is stated. As described in previous section, it first transforms quantitative values into fuzzy values according to the given type-2 membership functions. Then, the large itemsets and fuzzy rules are generated by the transformed fuzzy value with the minimum support and minimum confidence. The detail of the proposed approach is described as follows:

INPUT: A body of n quantitative transaction data, a given set of h type-2 membership functions for data values, a split number of point sn, a predefined minimum support α, and a predefined minimum confidence λ.

OUTPUT: A set of fuzzy association rules.

STEP 1: Transform the quantitative value v_{ij} of each transaction datum D_i ($i=1$ to n) for each item I_j into fuzzy values f_{ijl}^{lower} and f_{ijl}^{upper} using the given type-2 membership functions for each R_{jl}, where R_{jl} is the l-th fuzzy region of item I_j, $1 \leq l \leq h$, f_{ijl}^{lower} and f_{ijl}^{upper} is v_{ij}'s lower and upper fuzzy membership values in region R_{jl}.

STEP 2: Reduce type-2 fuzzy values to type-1 fuzzy values by centroid type-reduction method with the given split number of point. In other words, the f_{ijl}^{lower} and f_{ijl}^{upper} are reduced to a fuzzy value f_{ijl}.

STEP 3: Calculate the scalar cardinality of each fuzzy region R_{jl} in the transactions. Thus $count_{jl}$ is the summation of f_{ijl} in transactions.

STEP 4: Check whether the value of each fuzzy region R_{jl} is larger than or equal to the predefined minimum support value α. If the value of a fuzzy region R_{jl} is equal to or greater than the minimum support value, put it in the large 1-itemsets (L_1). If L_1 is not null, then do the next step; otherwise, exit the algorithm.

STEP 5: Set $r = 1$, where r is used to represent the number of items in the current itemsets to be processed.

STEP 6: Join the large r-itemsets L_r to generate the candidate $(r+1)$-itemsets C_{r+1} in a way similar to that in the *apriori* algorithm except that two items generated from the same order of data points in subsequences cannot simultaneously exist in an itemset in C_{r+1}.

STEP 7: Do the following substeps for each newly formed $(r+1)$-itemset s with fuzzy items $(s_1, s_2, \ldots, s_{r+1})$ in C_{r+1}:

(a) Calculate the upper and lower fuzzy value $(f_{f_j}^{lower}$ and $f_{f_j}^{upper})$ of each transaction data $D^{(i)}$ in s by the given type-2 membership functions as:
$$[f_{s_1}^{lower\ (i)}, f_{s_1}^{upper\ (i)}] + [f_{s_2}^{lower\ (i)}, f_{s_2}^{upper\ (i)}] + \ldots + [f_{s_{r+1}}^{lower\ (i)}, f_{s_{r+1}}^{upper\ (i)}].$$

(b) Reduce each $f_{sj}^{lower(i)}$ and $f_{sj}^{upper(i)}$ of fuzzy item I_j into a fuzzy value $f_{sj}^{(i)}$ with the predefined split number of point sn.

(c) Calculate the fuzzy value of each transaction data $D^{(i)}$ in s as
$$f_s^{(i)} = f_{s_1}^{(i)} \wedge f_{s_2}^{(i)} \wedge \ldots \wedge f_{s_{r+1}}^{(i)},$$
where $f_{s_j}^{(i)}$ is the membership value of $D^{(i)}$ in region s_j. The minimum operator is used for the intersection.

(d) Calculate the count of s in the transactions. The $count_s$ is the summation of $f_s^{(i)}$, and If the support ($= count_s / n$) of s is larger than or equal to the predefined minimum support value α, put it in L_{r+1}. If L_{r+1} is null, then do the next step; otherwise, set $r = r + 1$ and repeat STEPs 6 to 7.

STEP 8: Construct the fuzzy association rules for each large q-itemset s with items $(s_1, s_2, \ldots, s_q)$, $q \geq 2$, using the following substeps:

(a) Form each possible fuzzy association rule as follows:
$$s_1 \wedge \ldots \wedge s_{k-1} \wedge s_{k+1} \wedge \ldots \wedge s_q, k=1 \text{ to } q.$$

(b) Calculate the confidence values of all association rules by the following formula:
$$\frac{\sum_{i=1}^{n} f_s^{(i)}}{\sum_{i=1}^{n} (f_{s_1}^{(i)} \wedge \ldots \wedge f_{s_{k-1}}^{(i)}, f_{s_{k+1}}^{(i)} \wedge \ldots \wedge f_{s_q}^{(i)})}.$$

STEP 9: Output fuzzy the association rules with confidence values larger than or equal to the predefined confidence threshold λ.

4   Experimental Results

In this section, experiments on a simulated dataset were made to show the performance of the proposed approach. The synthetic dataset had 64 items. The parameters of the dataset include the transaction length, the purchased items and their quantities. The transaction length in a transaction was randomly generated in a uniform distribution of the range [1, 19]. The purchased items in each transaction were then selected from the 64 items in an exponential distribution with the rate parameter of 16. Their quantities were then assigned from an exponential distribution with the rate parameter of 5. In the following experiments, comparison results of the proposed approach and the previous approach were given to show the merits of the proposed approach. The membership functions used in the previous approach are shown in Fig. 2.

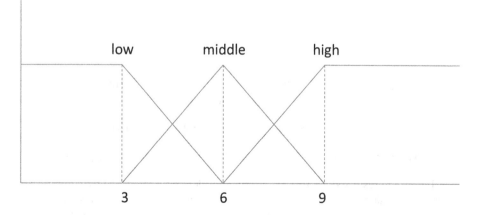

Fig. 2. The membership functions used in the previous approach

Fig. 2 shows that three linguistic terms, includes *Low, Middle, and High*, are used in the experiments. And, the type-2 membership functions used in the proposed approach are shown in Fig. 3.

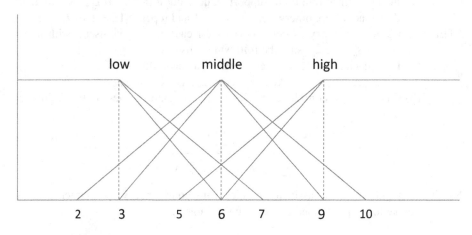

Fig. 3. The type-2 membership functions used in the proposed approach

Fig. 3 shows that there are three linguistic terms, includes *Low, Middle, and High*. The experiments were made to show the relationship between the number of rules and minimum supports of them. The results are shown in Fig. 4.

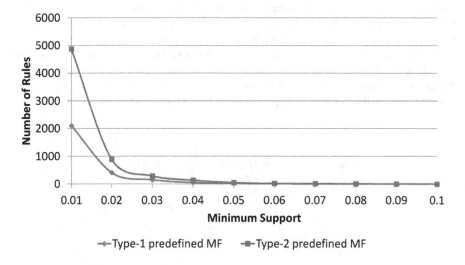

Fig. 4. The relationship between the number of rules and minimum supports

In Fig. 4, the results show that the numbers of derived rules of the two approaches are decreasing along with the increasing of minimum supports. And, it also show that the proposed approach is better than the previous approach in terms of number of derived rules.

5   Conclusion and Future Work

In this study, we have proposed an approach for mining fuzzy association rules by the given type-2 membership functions for dealing with data uncertainty in mining process. The main contributions of this paper are: (1) A fuzzy association rule mining approach with type-2 membership functions for dealing with data uncertainty is proposed; (2) By using type-2 membership functions, the experimental results show that the proposed approach is better than the previous approach in terms of number of rules. Of course, there are many directions to be improved, e.g., mining more actionable rules with some criteria. In the future, we will try to enhance the proposed approach and extend it to mine different types of rules.

Acknowledgment. This research was supported by the Ministry of Science and Technology of the Republic of China under grant MOST 103-2221-E-032 -029.

References

1. Agrawal, R., Srikant, R.: Fast algorithms for mining association rules.In: International Conference on Very Large Data Bases, pp. 487–499 (1994)
2. Au, W.H., Chan, K.C.C.: Mining fuzzy association rules in a bank-account database. IEEE Transactions on Fuzzy Systems **11**(2), 238–248 (2003)
3. Chan, K.C.C., Au, W.H.: An Effective Algorithm for Discovering Fuzzy Rules in Relational Databases. The IEEE International Conference on Fuzzy Systems **2**, 1314–1319 (1998)
4. Castillo, O., Melin, P., Pedrycz, W.: "Design of interval type-2 fuzzy models through optimal granularity allocation. Applied Soft Computing **11**(8), 5590–5601 (2011)
5. Chiu, H.P., Tang, Y.T., Hsieh, K.L.: Applying cluster-based fuzzy association rules mining framework into EC environment. Applied Soft Computing **12**(8), 2114–2122 (2012)
6. Dubois, D., Prade, H., Sudkamp, T.: On the representation, measurement, and discovery of fuzzy associations. IEEE Transactions on Fuzzy Systems **13**(2), 250–262 (2005)
7. El-Nagar, A.M., El-Bardini, M.: Derivation and stability analysis of the analytical structures of the interval type-2 fuzzy PID controller. Applied Soft Computing **24**, 704–716 (2014)
8. Fazel Zarandi, M.H., Rezaee, B., Turksen, I.B., Neshat, E.: A type-2 fuzzy rule-based expert system model for stock price analysis. Expert Systems with Applications **36**(1), 139–154 (2009)
9. Hu, Y.C.: Determining membership functions and minimum fuzzy support in finding fuzzy association rules for classification problems. Knowledge-Based Systems **19**(1), 57–66 (2006)
10. Huang, T.C.K.: Discovery of fuzzy quantitative sequential patterns with multiple minimum supports and adjustable membership functions. Information Sciences **222**(10), 126–146 (2013)
11. Hong, T.P., Kuo, C.S., Chi, S.C.: Mining association rules from quantitative data. Intelligent Data Analysis **3**(5), 363–376 (1999)
12. Hu, Y.H., Wu, F., Liao, Y.J.: An efficient tree-based algorithm for mining sequential patterns with multiple minimum supports. Journal of Systems and Software **86**(5), 1224–1238 (2013)

13. Kuok, C.M., Fu, A.W.C., Wong, M.H.: Mining fuzzy association rules in databases. ACM SIGMOD Record **27**(1), 41–46 (1998)
14. Lee, Y.-C., Hong, T.-P., Lin, W.-Y.: Mining Fuzzy Association Rules with Multiple Minimum Supports Using Maximum Constraints. In: Negoita, M.G., Howlett, R.J., Jain, L.C. (eds.) KES 2004. LNCS (LNAI), vol. 3214, pp. 1283–1290. Springer, Heidelberg (2004)
15. Lee, Y.C., Hong, T.P., Wang, T.C.: Multi-level fuzzy mining with multiple minimum supports. Expert Systems with Applications **34**(1), 459–468 (2008)
16. Mangalampalli, A., Pudi, V.: FPrep: fuzzy clustering driven efficient automated preprocessing for fuzzy association rule mining. In: 2010 IEEE International Conference on Fuzzy Systems, pp. 1–8 (2010)
17. Mendel, J.M., John, R.I.B.: Type-2 fuzzy sets made simple. IEEE Transactions on Fuzzy Systems, vol. 10, No. 2 (2002)
18. Martino, F.D., Sessa, S.: Type-2 interval fuzzy rule-based systems in spatial analysis. Information Sciences **279**(20), 199–212 (2014)
19. Pei, B., Zhao, S., Chen, H., Zhou, X., Chen, D.: FARP: Mining fuzzy association rules from a probabilistic quantitative database. Information Sciences **237**, 242–260 (2013)
20. Sowan, B., Dahal, K., Hossain, M.A., Zhang, L., Spencer, L.: Fuzzy association rule mining approaches for enhancing prediction performance. Expert Systems with Applications **40**(17), 6928–6937 (2013)
21. Tajbakhsh, A., Rahmati, M., Mirzaei, A.: Intrusion detection using fuzzy association rules. Applied Soft Computing **9**(2), 462–469 (2009)
22. Yue, S., Tsang, E., Yeung, D., Shi, D.: Mining fuzzy association rules with weighted items. In: The IEEE International Conference on Systems, Man and Cybernetics, pp. 1906–1911 (2000)

Dependence Factor for Association Rules

Marzena Kryszkiewicz[(⊠)]

Institute of Computer Science, Warsaw University of Technology, Nowowiejska
15/19, 00-665 Warsaw, Poland
mkr@ii.pw.edu.pl

Abstract. Certainty factor and lift are known evaluation measures of association rules. These measures, nevertheless, do not guarantee accurate evaluation of strength of dependence between rule's constituents. In particular, even if there is a strongest possible positive or negative dependence between rule's constituents X and Y, these measures may reach values quite close to the values characteristic for rule's constituents independence. In this paper, we first re-examine both certainty factor and lift. Then, in order to better evaluate dependence between rule's constituents, we offer and examine a new measure – a dependence factor. Unlike in the case of the certainty factor, when defining our measure, we take into account the fact that for a given rule $X \rightarrow Y$, the minimal conditional probability of the occurrence of Y given X may be greater than 0, while its maximal possible value may less than 1. In the paper, a number of properties and relations of all investigated measures are derived.

1 Introduction

Certainty factor and *lift* are known evaluation measures of association rules. The former measure was offered in the expert system Mycin [7], while the latter is widely implemented in both commercial and non-commercial data mining systems [2]. Nevertheless, they do not guarantee accurate evaluation of the strength of dependence between rule's constituents. In particular, even if there is a strongest possible positive or negative dependence between rule constituents X and Y, these measures may reach values quite close to the values indicating independence of X and Y. This might suggest that one deals with a weak dependence, while in fact the dependence is strong. In this paper, we first re-examine both certainty factor and lift. Then we offer and examine a new measure - a *dependence factor* - to evaluate dependence between rule's constituents more accurately than by means of the certainty factor or lift. Unlike in the case of the certainty factor, when defining our measure, we take into account the fact that that for a given rule $X \rightarrow Y$, the minimal conditional probability of the occurrence of Y given X may be greater than 0, while its maximal possible value may less than 1. We end the paper with the comparison of dependence evaluation capabilities of the dependence factor, certainty factor and lift.

As certainty factor [7] and lift [2] were introduced first as measures for rules, we will define them in the rule context, however our work is more general and can be applied whenever a strength of dependence (or independence) between any events is

© Springer International Publishing Switzerland 2015
N.T. Nguyen et al. (Eds.): ACIIDS 2015, Part II, LNAI 9012, pp. 135–145, 2015.
DOI: 10.1007/978-3-319-15705-4_14

to be determined. However, we would like to stress that in our paper, we focus on a particular aspect related to dependence of co-occurrence of X and Y rather than on rule interestingness aspects of $X \rightarrow Y$ in their entirety and complexity [2-10].

Our paper has the following layout. In Section 2, we briefly recall basic notions of association rules, their basic measures (support, confidence) as well as lift and certainty factor. In Section 3, we derive maximal and minimal values of these measures taking into account constraints imposed by marginal probabilities on a joint probability. In Section 4, we offer a new rule measure - dependence factor, and derive its properties as well as relations with the certainty factor. There, we also illustrate the differences among dependence evaluation capabilities of the dependence factor, certainty factor and lift by means of an example. Section 5 concludes our work.

2   Basic Notions and Properties

In this section, we recall the notion of association rules after [1].

Definition 1. Let $I = \{i_1, i_2, ..., i_m\}$ be a set of distinct literals, called *items* (e.g. products, features). Any $X \subseteq I$ is called an *itemset*. A *transaction database* is denoted by $\mathcal{D}$ and is defined as a set of itemsets. Each itemset T in $\mathcal{D}$ is a *transaction*. An *association rule* is an expression associating two itemsets:

$$X \rightarrow Y, \text{ where } \varnothing \neq Y \subseteq I \text{ and } X \subseteq I \setminus Y.$$

Itemsets and association rules are typically characterized by *support* and *confidence*, which are simple statistical parameters.

Definition 2. *Support of an itemset* X is denoted by $sup(X)$ and is defined as the number of transactions in $\mathcal{D}$ that contain X; that is,

$$sup(X) = |\{T \in \mathcal{D} \mid X \subseteq T\}|.$$

Support of a rule $X \rightarrow Y$ is denoted by $sup(X \rightarrow Y)$ and is defined as the support of $X \cup Y$; that is,

$$sup(X \rightarrow Y) = sup(X \cup Y).$$

Clearly, the probability of the event that itemset X occurs in a transaction equals $sup(X) / |\mathcal{D}|$, while the probability of the event that both X and Y occur in a transaction equals $sup(X \cup Y) / |\mathcal{D}|$. In the remainder, the former probability will be denoted by $P(X)$, while the latter by $P(XY)$.

Definition 3. The *confidence* of an association rule $X \rightarrow Y$ is denoted by $conf(X \rightarrow Y)$ and is defined as the conditional probability that Y occurs in a transaction provided X occurs in the transaction; that is:

$$conf(X \rightarrow Y) = sup(X \rightarrow Y) / sup(X) = P(XY) / P(X).$$

A large amount of research was devoted to *strong association rules* understood as those association rules the supports and confidences of which exceed user-defined support threshold and confidence threshold, respectively. However, it has been argued that these two measures are not sufficient to express different interestingness, usefulness or unexpectedness aspects of association rules [2-10]. In fact, a number of additional measures of association rules was proposed. Among them very popular measures are *lift* [2] and *certainty factor* [7].

Definition 4. The *lift* of an association rule $X \rightarrow Y$ is denoted by $lift(X \rightarrow Y)$ and is defined as the ratio of the conditional probability of the occurrence of Y in a transaction given X occurs there to the probability of the occurrence of Y; that is,

$$lift(X \rightarrow Y) = \frac{conf(X \rightarrow Y)}{P(Y)}.$$

Sometimes, lift is defined in an equivalent way in terms of probabilities only:

Property 1.

$$lift(X \rightarrow Y) = \frac{P(XY)}{P(X) \times P(Y)}.$$

Definition 5. The *certainty factor* of an association rule $X \rightarrow Y$ is denoted by $cf(X \rightarrow Y)$ and is defined as the degree to which the probability of the occurrence of Y in a transaction can change when X occurs there as follows:

$$cf(X \rightarrow Y) = \begin{cases} \dfrac{conf(X \rightarrow Y) - P(Y)}{1 - P(Y)} & \text{if } conf(X \rightarrow Y) > P(Y), \\ 0 & \text{if } conf(X \rightarrow Y) = P(Y), \\ -\dfrac{P(Y) - conf(X \rightarrow Y)}{P(Y) - 0} & \text{if } conf(X \rightarrow Y) < P(Y). \end{cases}$$

The definition of the certainty factor is based on the assumption that the probability of the occurrence of Y in a transaction given X occurs there ($conf(X \rightarrow Y)$) can be increased from $P(Y)$ up to 1 and decreased from $P(Y)$ down to 0. In Fig. 1, we visualize the meaning of the absolute value of the certainty factor as the ratio of the lengths of respective intervals.

Fig. 1. Calculating the absolute value of the certainty factor as the ratio of the lengths of respective intervals when $conf(X \rightarrow Y) > P(Y)$ (on the left-hand side) and when $conf(X \rightarrow Y) < P(Y)$ (on the right-hand side)

In fact, *cf* can be expressed equivalently in terms of unconditional probabilities (by multiplying the numerator and denominator of the formula in Definition 5 by $P(X)$) or lift (by dividing the numerator and denominator of the original *cf* formula by $P(Y)$).

Property 2.

a) $\quad cf(X \rightarrow Y) = \begin{cases} \dfrac{P(XY) - P(X) \times P(Y)}{P(X) - P(X) \times P(Y)} & \text{if } P(XY) > P(X) \times P(Y), \\ 0 & \text{if } P(XY) = P(X) \times P(Y), \\ -\dfrac{P(X) \times P(Y) - P(XY)}{P(X) \times P(Y) - 0} & \text{if } P(XY) < P(X) \times P(Y). \end{cases}$

b) $\quad cf(X \rightarrow Y) = \begin{cases} \dfrac{lift(X \rightarrow Y) - 1}{1/P(Y) - 1} & \text{if } lift(X \rightarrow Y) > 1, \\ 0 & \text{if } lift(X \rightarrow Y) = 1, \\ -\dfrac{1 - lift(X \rightarrow Y)}{1 - 0} & \text{if } lift(X \rightarrow Y) < 1. \end{cases}$

Both lift and certainty factor are related to the notion of (in)dependence of events, where two events are treated as independent if the product of probabilities of their occurrences equals the probability that the two events co-occur. Otherwise, they are regarded as dependent. Note that this notion of dependence does not indicate which event is a reason of the other. In Table 1, we provide equivalent conditions in terms of P, *conf*, *lift* and *cf* for independence, positive dependence and negative dependence, respectively, between two itemsets.

Table 1. Conditions for independence, positive dependence and negative dependence

(In)dependence	(In)dependence condition	Equivalent conditions in terms of measures for $X{\rightarrow}Y$	Equivalent conditions in terms of measures for $Y{\rightarrow}X$
Y and X are dependent positively	$P(XY) > P(X) \times P(Y)$	$conf(X{\rightarrow}Y) > P(Y)$ $lift(X{\rightarrow}Y) > 1$ $cf(X{\rightarrow}Y) > 0$	$conf(Y{\rightarrow}X) > P(X)$ $lift(Y{\rightarrow}X) > 1$ $cf(Y{\rightarrow}X) > 0$
Y and X are independent	$P(XY) = P(X) \times P(Y)$	$conf(X{\rightarrow}Y) = P(Y)$ $lift(X{\rightarrow}Y) = 1$ $cf(X{\rightarrow}Y) = 0$	$conf(Y{\rightarrow}X) = P(X)$ $lift(Y{\rightarrow}X) = 1$ $cf(Y{\rightarrow}X) = 0$
Y and X are dependent negatively	$P(XY) < P(X) \times P(Y)$	$conf(X{\rightarrow}Y) < P(Y)$ $lift(X{\rightarrow}Y) < 1$ $cf(X{\rightarrow}Y) < 0$	$conf(Y{\rightarrow}X) < P(X)$ $lift(Y{\rightarrow}X) < 1$ $cf(Y{\rightarrow}X) < 0$

In general, one may distinguish between *symmetric (two direction) measures* of association rules and *asymmetric (one direction)* ones.

Definition 6. A measure m is called *symmetric (two direction)* if for any X and Y $m(X{\rightarrow}Y) = m(Y{\rightarrow}X)$. Otherwise, it is called an *asymmetric (one direction) measure*.

Property 3.

a) $conf(X{\rightarrow}Y) = conf(Y{\rightarrow}X)$ is not guaranteed to hold.

b) $lift(X{\rightarrow}Y) = lift(Y{\rightarrow}X)$.

c) $cf(X \rightarrow Y) = cf(Y \rightarrow X)$ is not guaranteed to hold if $conf(X \rightarrow Y) > P(Y)$.

d) $cf(X \rightarrow Y) = cf(Y \rightarrow X)$ if $conf(X \rightarrow Y) \leq P(Y)$.

As follows from Property 3, *conf* is an asymmetric measure and *lift* is a symmetric measure. On the other hand, we observe that strangely *cf* has a mixed nature – asymmetric for positive dependences and symmetric for negative dependences and independences. This observation provoked us to revisit the definition of *cf* and to propose its modification. In our proposal, we take into account the fact that sometimes it may be infeasible to increase the probability of the occurrence of Y in a transaction given X occurs there ($conf(X \rightarrow Y)$) from $P(Y)$ up to 1 as well as it may be infeasible to decrease it from $P(Y)$ down to 0.

3   Maximal and Minimal Values of Rule Measures

In this section, we first recall global maximal and minimal values of rule measures (Table 2). Next we derive maximal and minimal values of rule measures for given values of $P(X)$ and $P(Y)$.

Table 2. Global maximal and minimal values of rule measures

measure	max	min
$P(XY)$	1	0
$conf(X \rightarrow Y)$	1	0
$lift(X \rightarrow Y)$	∞	0
$cf(X \rightarrow Y)$	1	−1
	if Y depends on X positively	if Y depends on X negatively

In the remainder of the paper, we denote *maximal probability* and *minimal probability* of the co-occurrence of X and Y given $P(X)$ and $P(Y)$ are fixed by $max\_P(XY|_{P(X), P(Y)})$ and $min\_P(XY|_{P(X), P(Y)})$, respectively. Analogously, *maximal confidence* and *minimal confidence* (*maximal lift*, *minimal lift*, *maximal certainty factor*, *minimal certainty factor*) *of X→Y* given $P(X)$ and $P(Y)$ are fixed are denoted by $max\_conf(X \rightarrow Y|_{P(X), P(Y)})$ and $min\_conf(X \rightarrow Y|_{P(X), P(Y)})$ ($max\_lift(X \rightarrow Y|_{P(X), P(Y)})$, $min\_lift(X \rightarrow Y|_{P(X), P(Y)})$, $max\_cf(X \rightarrow Y|_{P(X), P(Y)})$, $min\_cf(X \rightarrow Y|_{P(X), P(Y)})$), respectively.

Property 4.

a) $max\_conf(X \rightarrow Y|_{P(X), P(Y)}) = \dfrac{max\_P(XY|_{P(X),P(Y)})}{P(X)}$

b) $min\_conf(X \rightarrow Y|_{P(X), P(Y)}) = \dfrac{min\_P(XY|_{P(X),P(Y)})}{P(X)}$

c) $max\_lift(X \rightarrow Y|_{P(X), P(Y)}) = \dfrac{max\_conf(XY|_{P(X),P(Y)})}{P(Y)} = \dfrac{max\_P(XY|_{P(X),P(Y)})}{P(X) \times P(Y)}$

d) $min\_lift(X \rightarrow Y|_{P(X), P(Y)}) = \dfrac{min\_conf(XY|_{P(X),P(Y)})}{P(Y)} = \dfrac{min\_P(XY|_{P(X),P(Y)})}{P(X) \times P(Y)}$

e) $max\_cf(X \rightarrow Y|_{P(X), P(Y)}) = \dfrac{max\_conf(X \rightarrow Y|_{P(X),P(Y)}) - P(Y)}{1 - P(Y)}$

$= \dfrac{max\_P(XY|_{P(X),P(Y)}) - P(X) \times P(Y)}{P(X) - P(X) \times P(Y)} = \dfrac{max\_lift(XY|_{P(X),P(Y)}) - 1}{\dfrac{1}{P(Y)} - 1}$

f) $min\_cf(X \rightarrow Y|_{P(X), P(Y)}) = -\dfrac{P(Y) - min\_conf(X \rightarrow Y|_{P(X),P(Y)})}{P(Y) - 0}$

$= -\dfrac{P(X) \times P(Y) - min\_P(XY|_{P(X),P(Y)})}{P(X) \times P(Y) - 0} = -\dfrac{1 - min\_lift(XY|_{P(X),P(Y)})}{1 - 0}$

In Proposition 1, we show how to calculate $min\_P(XY|_{P(X), P(Y)})$ and $max\_P(XY|_{P(X), P(Y)})$. We note that neither $max\_P(XY|_{P(X), P(Y)})$ necessarily equals 1 nor $min\_P(XY|_{P(X), P(Y)})$ necessarily equals 0. Figure 2 illustrates this.

Proposition 1.

a) $max\_P(XY|_{P(X), P(Y)}) = \min\{P(X), P(Y)\}$

b) $min\_P(XY|_{P(X), P(Y)}) = \begin{cases} 0 & \text{if } P(X) + P(Y) \le 1 \\ P(X) + P(Y) - 1 & \text{if } P(X) + P(Y) > 1 \end{cases}$

$= \max\{0, \ P(X) + P(Y) - 1\}$

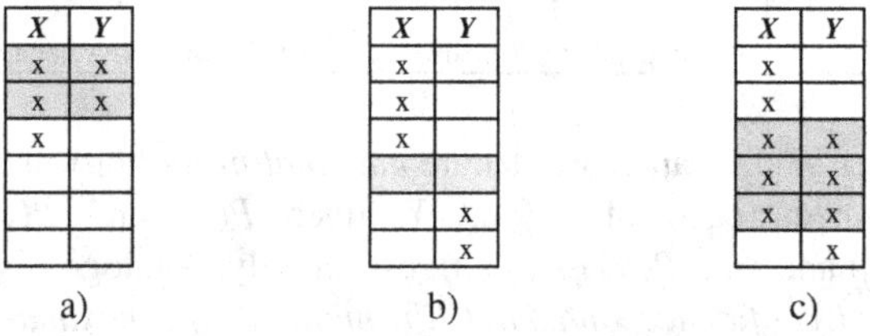

X	Y
x	x
x	x
x	

X	Y
x	
x	
x	
	x
	x

X	Y
x	
x	
x	x
x	x
x	x
	x

a)   b)   c)

Fig. 2. a) $max\_P(XY|_{P(X), P(Y)}) = \min\{P(X), P(Y)\} = \min\left\{\frac{3}{6}, \frac{2}{6}\right\} = \frac{2}{6}$; b) $min\_P(XY|_{P(X), P(Y)}) = 0$ if $P(X) + P(Y) \le 1$; c) $min\_P(XY|_{P(X), P(Y)}) = P(X) + P(Y) - 1 = \frac{5}{6} + \frac{4}{6} - 1 = \frac{3}{6}$ if $P(X) + P(Y) > 1$

The next proposition follows from Property 4 and Proposition 1.

Proposition 2.

a) $max\_conf(X \rightarrow Y|_{P(X), P(Y)}) = \dfrac{\min\{P(X), P(Y)\}}{P(X)} = \begin{cases} 1 & \text{if } P(X) \le P(Y), \\ \dfrac{P(Y)}{P(X)} & \text{if } P(Y) < P(X). \end{cases}$

b) $min\_conf(X{\to}Y|_{P(X),\,P(Y)}) = \dfrac{\max\{0,P(X)+P(Y)-1\}}{P(X)}$

$$= \begin{cases} 0 & \text{if } P(X)+P(Y) \le 1, \\ \dfrac{P(X)+P(Y)-1}{P(X)} & \text{if } P(X)+P(Y) > 1. \end{cases}$$

c) $max\_lift(X{\to}Y|_{P(X),\,P(Y)}) = \dfrac{\min\{P(X),P(Y)\}}{P(X){\times}P(Y)} = \dfrac{1}{max\{P(X),P(Y)\}}.$

d) $min\_lift(X{\to}Y|_{P(X),\,P(Y)}) = \dfrac{\max\{0,P(X)+P(Y)-1\}}{P(X){\times}P(Y)}$

$$= \begin{cases} 0 & \text{if } P(X)+P(Y) \le 1, \\ \dfrac{P(X)+P(Y)-1}{P(X){\times}P(Y)} & \text{if } P(X)+P(Y) > 1. \end{cases}$$

e) $max\_cf(X{\to}Y|_{P(X),\,P(Y)}) = \dfrac{\min\{P(X),P(Y)\}-P(X){\times}P(Y)}{P(X)-P(X){\times}P(Y)}$

$$= \dfrac{\dfrac{1}{max\{P(X),P(Y)\}}-1}{\dfrac{1}{P(Y)}-1} = \begin{cases} 1 & \text{if } P(X) \le P(Y), \\ \dfrac{\dfrac{1}{P(X)}-1}{\dfrac{1}{P(Y)}-1} & \text{if } P(X) > P(Y). \end{cases}$$

f) $min\_cf(X{\to}Y|_{P(X),\,P(Y)}) = -\dfrac{P(X){\times}P(Y)-\max\{0,P(X)+P(Y)-1\}}{P(X){\times}P(Y)-0}$

$$= \dfrac{\max\{0,P(X)+P(Y)-1\}}{P(X){\times}P(Y)}-1 = \begin{cases} -1 & \text{if } P(X)+P(Y) \le 1 \\ \dfrac{P(X)+P(Y)-1}{P(X){\times}P(Y)}-1 & \text{if } P(X)+P(Y) > 1. \end{cases}$$

In Table 3, we summarize real achievable maximal and minimal values of $P(XY)$, $conf(X{\to}Y)$, $lift(X{\to}Y)$ and $cf(X{\to}Y)$ for given values of $P(X)$ and $P(Y)$.

Table 3. Real achievable maximal and minimal values of $P(XY)$, $conf(X{\to}Y)$, $lift(X{\to}Y)$ and $cf(X{\to}Y)$ for given values of $P(X)$ and $P(Y)$

measure	max for given values of $P(X)$ and $P(Y)$	min for given values of $P(X)$ and $P(Y)$
$P(XY)$	$\min\{P(X), P(Y)\}$	$\max\{0,\ P(X)+P(Y)-1\}$
$conf(X{\to}Y)$	$\dfrac{\min\{P(X),P(Y)\}}{P(X)}$	$\dfrac{\max\{0,P(X)+P(Y)-1\}}{P(X)}$
$lift(X{\to}Y)$	$\dfrac{\min\{P(X),P(Y)\}}{P(X){\times}P(Y)}$	$\dfrac{\max\{0,P(X)+P(Y)-1\}}{P(X){\times}P(Y)}$
$cf(X{\to}Y)$	$\dfrac{\min\{P(X),P(Y)\}-P(X){\times}P(Y)}{P(X)-P(X){\times}P(Y)}$ if Y depends on X positively	$-\dfrac{P(X){\times}P(Y)-\max\{0,P(X)+P(Y)-1\}}{P(X){\times}P(Y)-0}$ if Y depends on X negatively

4 Dependence Factor

In this section, we propose a *dependence factor* of a rule $X \to Y$ as a modification of the certainty factor, which, unlike the certainty factor, is based on real maximal and minimal values of $conf(X \to Y)$ for given values of $P(X)$ and $P(Y)$.

Definition 7. The *dependence factor* of $X \to Y$ is denoted by $df(X \to Y)$ and is defined as the ratio of the actual change of the probability of the occurrence of Y in a transaction given X occurs there to its maximal feasible change as follows:

$$df(X \to Y) = \begin{cases} \dfrac{conf(X \to Y) - P(Y)}{max\_conf(X \to Y \mid_{P(X),P(Y)}) - P(Y)} & \text{if } conf(X \to Y) > P(Y), \\ 0 & \text{if } conf(X \to Y) = P(Y), \\ -\dfrac{P(Y) - conf(X \to Y)}{P(Y) - min\_conf(X \to Y \mid_{P(X),P(Y)})} & \text{if } conf(X \to Y) < P(Y). \end{cases}$$

The dependence factor not only determines by how much the probability of the occurrence of Y in a transaction changes given X occurs there with respect to by how much it could have changed, but also it determines by how much the probability of the occurrence of X and Y in a transaction differs from the probability of their common occurrence under independence assumption with respect to by how much it could have been different (see Proposition 3a). In addition, df determines by how much the value of the lift of a rule $X \to Y$ differs from the value 1 (that is, from the value indicating independence of rule's constituents in terms of the lift measure) with respect to by how much it could have been different (see Proposition 3b).

Proposition 3.

a) $$df(X \to Y) = \begin{cases} \dfrac{P(XY) - P(X) \times P(Y)}{max\_P(XY \mid_{P(X),P(Y)}) - P(X) \times P(Y)} & \text{if } P(XY) > P(X) \times P(Y), \\ 0 & \text{if } P(XY) = P(X) \times P(Y), \\ -\dfrac{P(X) \times P(Y) - P(XY)}{P(X) \times P(Y) - min\_P(XY \mid_{P(X),P(Y)})} & \text{if } P(XY) < P(X) \times P(Y). \end{cases}$$

b) $$df(X \to Y) = \begin{cases} \dfrac{lift(X \to Y) - 1}{max\_lift(X \to Y \mid_{P(X),P(Y)}) - 1} & \text{if } lift(X \to Y) > 1, \\ 0 & \text{if } lift(X \to Y) = 1, \\ -\dfrac{1 - lift(X \to Y)}{1 - min\_lift(X \to Y \mid_{P(X),P(Y)})} & \text{if } lift(X \to Y) < 1. \end{cases}$$

Theorem 1.

a) If $P(XY) > P(X) \times P(Y)$, then $df(X \to Y) \in (0, 1]$.
b) If $P(XY) = P(X) \times P(Y)$, then $df(X \to Y) = 0$.
c) If $P(XY) < P(X) \times P(Y)$, then $df(X \to Y) \in [-1, 0)$.

Proof. Follows from Proposition 3a.

As follows from Proposition 3a, the dependence factor is a symmetric measure.

Theorem 2. $df(X{\rightarrow}Y) = df(Y{\rightarrow}X)$.

Based on Proposition 1 and Proposition 3a, we will express $df(X{\rightarrow}Y)$ in terms of $P(XY)$, $P(X)$ and $P(Y)$, which will be useful for examining properties of this measure.

Theorem 3.

$$df(X \rightarrow Y) = \begin{cases} \dfrac{P(XY) - P(X) \times P(Y)}{\min\{P(X), P(Y)\} - P(X) \times P(Y)} & \text{if } P(XY) > P(X) \times P(Y), \\ 0 & \text{if } P(XY) = P(X) \times P(Y), \\ -\dfrac{P(X) \times P(Y) - P(XY)}{P(X) \times P(Y) - \max\{0, P(X) + P(Y) - 1\}} & \text{if } P(XY) < P(X) \times P(Y). \end{cases}$$

One may easily note that $df(X{\rightarrow}Y)$ reaches 1 when $P(XY)$ is maximal for given values of $P(X)$ and $P(Y)$; that is, when $P(XY) = \min\{P(X), P(Y)\}$ or, in other words, when the dependence between X and Y is strongest possible positive for given values of $P(X)$ and $P(Y)$. Analogously, $df(X{\rightarrow}Y)$ reaches -1 when $P(XY)$ is minimal for given values of $P(X)$ and $P(Y)$; that is, when $P(XY) = \max\{0, P(X) + P(Y) - 1\}$ or, in other words, when the dependence between X and Y is strongest possible negative for these probability values.

Table 4. Maximal and minimal values of $df(X{\rightarrow}Y)$ for any given values of $P(X)$ and $P(Y)$

measure	max for any given values of $P(X)$ and $P(Y)$	min for any given values of $P(X)$ and $P(Y)$
$df(X{\rightarrow}Y)$	1 (if X and Y are dependent positively)	-1 (if X and Y are dependent negatively)

Based on Theorem 3 and Property 2a, we derive relations between the dependence factor and the certainty factor as follows:

Theorem 4.

a) $df(X{\rightarrow}Y) \geq cf(X{\rightarrow}Y)$ if $P(XY) > P(X) \times P(Y)$,
b) $df(X{\rightarrow}Y) = cf(X{\rightarrow}Y) = 0$ if $P(XY) = P(X) \times P(Y)$,
c) $df(X{\rightarrow}Y) \leq cf(X{\rightarrow}Y)$ if $P(XY) < P(X) \times P(Y)$,
d) $df(X{\rightarrow}Y) = \max\{cf(X{\rightarrow}Y), cf(Y{\rightarrow}X)\}$ if $P(XY) > P(X) \times P(Y)$,
e) $df(X{\rightarrow}Y) = cf(X{\rightarrow}Y)$ if $P(XY) < P(X) \times P(Y)$ and $P(X) + P(Y) \leq 1$,
f) $df(X{\rightarrow}Y) < cf(X{\rightarrow}Y)$ if $P(XY) < P(X) \times P(Y)$ and $P(X) + P(Y) > 1$.

Tables 5-6 illustrate our findings expressed as Theorem 4. In particular, Table 5 shows values of $lift(X{\rightarrow}Y)$, $cf(X{\rightarrow}Y)$ and $df(X{\rightarrow}Y)$ for $P(X) = 0.6$ and $P(Y) = 0.3$; that is, in the case when $P(X) + P(Y) \leq 1$. For these values of $P(X)$ and $P(Y)$, the maximal possible value for $P(XY)$ equals $\min\{P(X), P(Y)\} = 0.3$. The fact of reaching the maximal possible value by $P(XY)$ for the given values of $P(X)$ and $P(Y)$ is reflected by the value of $df(X{\rightarrow}Y) = 1$, which means that the dependence between X and Y is strongest possible positive. On the other hand, $cf(X{\rightarrow}Y) = 0.29$ does not reflect this fact. In general, the real dependence of Y on X may be underestimated when expressed in terms of $cf(X{\rightarrow}Y)$. Also the value 1.67 of $lift(X{\rightarrow}Y)$ itself does not reflect the strong positive dependence between X and Y in the considered case in the view that the lift may reach very large values (close to infinity) in general.

Table 5. Comparison of values of $lift(X{\rightarrow}Y)$, $cf(X{\rightarrow}Y)$ and $df(X{\rightarrow}Y)$ when $P(X) + P(Y) \le 1$

P(X)	P(Y)	P(XY)	P(X)×P(Y)	lift(X→Y)	cf(X→Y)	cf(Y→X)	df(X →Y)=df(Y→X)
0.60	0.30	0.30	0.18	1.67	0.29	1.00	1.00
0.60	0.30	0.25	0.18	1.39	0.17	0.58	0.58
0.60	0.30	0.20	0.18	1.11	0.05	0.17	0.17
0.60	0.30	0.18	0.18	1.00	0.00	0.00	0.00
0.60	0.30	0.15	0.18	0.83	-0.17	-0.17	-0.17
0.60	0.30	0.10	0.18	0.56	-0.44	-0.44	-0.44
0.60	0.30	0.00	0.18	0.00	-1.00	-1.00	-1.00

Table 6. Comparison of values of $lift(X{\rightarrow}Y)$, $cf(X{\rightarrow}Y)$ and $df(X{\rightarrow}Y)$ when $P(X) + P(Y) > 1$

P(X)	P(Y)	P(XY)	P(X)×P(Y)	lift(X→Y)	cf(X→Y)	cf(Y→X)	df(X →Y)=df(Y→X)
0.80	0.60	0.60	0.48	1.25	0.38	1.00	1.00
0.80	0.60	0.55	0.48	1.15	0.22	0.58	0.58
0.80	0.60	0.50	0.48	1.04	0.06	0.17	0.17
0.80	0.60	0.48	0.48	1.00	0.00	0.00	0.00
0.80	0.60	0.45	0.48	0.94	-0.06	-0.06	-0.37
0.80	0.60	0.40	0.48	0.83	-0.17	-0.17	-1.00

Table 6 shows values of $lift(X{\rightarrow}Y)$, $cf(X{\rightarrow}Y)$ and $df(X{\rightarrow}Y)$ for $P(X) = 0.8$ and $P(Y) = 0.6$; that is, in the case when $P(X) + P(Y) > 1$. For these values of $P(X)$ and $P(Y)$, the minimal possible value of $P(XY)$ equals $P(X) + P(Y) - 1 = 0.4$. Then the dependence between X and Y is strongest possible negative. This is reflected by the value of $df(X{\rightarrow}Y) = -1$. On the other hand, $cf(X{\rightarrow}Y) = -0.17$ does not reflect this fact by itself. Also the value 0.83 of $lift(X{\rightarrow}Y)$ itself does not reflect the strong negative dependence between X and Y as it is positioned closer to the value 1 characteristic for independence rather than to the value 0.

5   Conclusions

In this paper, we have offered a new measure called dependence factor (df) to evaluate the dependence between rule's constituents. $df(X{\rightarrow}Y)$ always reaches 1 when the dependence between X and Y is strongest possible positive and -1 when the dependence between X and Y is strongest possible negative for any given values of $P(X)$ and $P(Y)$. Unlike the dependence factor, the certainty factor itself as well as lift are misleading in expressing the strength of the dependence. In particular, if there is strongest possible positive dependence of Y on X, $cf(X{\rightarrow}Y)$ is not guaranteed to reach its global maximum value 1 (in fact, its value can be quite close to 0 that suggests independence). On the other hand, if there is strongest possible negative dependence between X and Y, $cf(X{\rightarrow}Y)$ is not guaranteed to reach its global minimum value -1 (in fact, its value can be quite close to 0). Similarly, lift may reach values close to the value 1 (which means independence in terms of this measure) even in the cases when the dependence between X and Y is strongest possible positive or strongest possible negative.

Thus, we find the dependence factor more accurate measure of a rule constituents' dependence than the certainty factor or lift. In the paper, we have also derived a number of properties of the investigated measures. In particular, we have found the relations between the dependence factor and the certainty factor.

References

1. Agrawal, R., Imielinski, T., Swami, A.N.: Mining association rules between sets of items in large databases. In: ACM SIGMOD Int. Conf. on Management of Data, pp. 207–216 (1993)
2. Brin, S., Motwani, R., Ullman, J.D., Tsur, S.: Dynamic itemset counting and implication rules for market basket data. In: ACM SIGMOD 1997 Int. Conf. on Management of Data, pp. 255–264 (1997)
3. Hilderman, R.J., Hamilton, H.J.: Evaluation of interestingness measures for ranking discovered knowledge. In: Cheung, D., Williams, G.J., Li, Q. (eds.) PAKDD 2001. LNCS (LNAI), vol. 2035, pp. 247–259. Springer, Heidelberg (2001)
4. Lallich, S., Teytaud, O., Prudhomme, E.: Association rule interestingness: measure and statistical validation. Quality Measures in Data Mining 2007, 251–275 (2006)
5. Lenca, P., Meyer, P., Vaillant, B., Lallich, S.: On selecting interestingness measures for association rules: User oriented description and multiple criteria decision aid. European Journal of Operational Research 184, 610–626 (2008)
6. Piatetsky-Shapiro, G.: Discovery, analysis, and presentation of strong rules. In: Knowledge Discovery in Databases, pp. 229–248. AAAI/MIT Press (1991)
7. Shortliffe, E., Buchanan, B.: A model of inexact reasoning in medicine. Mathematical Biosciences 23, 351–379 (1975)
8. Silberschatz, A., Tuzhilin, A.: on subjective measures of interestingness in knowledge discovery. In Proc. of KDD 1995, pp. 275–281 (1995)
9. Suzuki, E.: Pitfalls for categorizations of objective interestingness measures for rule discover. In: Gras, R., Suzuki, E., Guillet, F., Spagnolo, F. (eds.) Statistical Implicative Analysis. SCI, vol. 127, pp. 383–395. Springer, Heidelberg (2008)
10. Suzuki, E.: Interestingness measures - limits, desiderata, and recent results. In: QIMIE/PAKDD (2009)

Collective Intelligent Systems
for E-market Trading, Technology
Opportunity Discovery
and Collaborative Learning

A Novel Framework of Consumer Co-creation for New Service Development

Chao-Fu Hong[1(✉)], Mu-Hua Lin[2], and Hsiao-Fang Yang[2]

[1] Department of Information Management, Aletheia University, Taipei, Taiwan
au4076@au.edu.tw
[2] Management Information Systems,
National Chengchi University, Taipei, Taiwan
95356503@nccu.edu.tw, hfyang.wang@gmail.com

Abstract. Consumer co-creation can be seen as an attractive approach for companies for a variety of reasons. In particular, ideas generated through co-creation will more closely mirror consumers' needs. Additionally, in Web 2.0, the consumers can be easily post their consuming article on the Internet. In the present study, we develop a new consumer co-creation framework: obtaining consuming data from the Internet, and using Grounded Theory (Strauss and Corbin, 1998) based Human-Centered Computing System (Hong, 2009) to investigate consumers' needs or creations and to aid the company designing new products or services.

Keywords: Co-creation · Consuming tribe · Lead user · Innovative idea

1 Introduction

The present study attempts to develop a new consumer co-creation framework for exploring innovative ways of designing new products: using Google blog search to obtain relevant consuming textual data and applying Grounded Theory along with the social identity theory, self-categorization theory (Oakes 1987, Ashforth and Mael 1989), and social presence (Shen and Khalifa, 2007) to do self-categorization to dis-cover the innovative value of a product by using co-occurrence analysis method, such as using social identity to identify the underlying knowledge and innovation or weak tie.

As discussed earlier, lead users are important partners when a company attempts to create innovative products. But the problem is that before companies attempt to design an innovative product, they hardly know who the lead user is (Etgar 2008; Franke, Keinz, and Steger 2009). Fortunately, especially in the era of web 2.0, it is quite easy for consumers to upload what they have written down about their consuming value to the Internet. In the present study, the researchers believe that very much consuming information is posted on the Internet. Although the virtual world might not be similar to the physical society, it may be easier, more effective and less costly to identify a trend or a framework created by innovative lead users on the Internet than the traditional method may.

© Springer International Publishing Switzerland 2015
N.T. Nguyen et al. (Eds.): ACIIDS 2015, Part II, LNAI 9012, pp. 149–158, 2015.
DOI: 10.1007/978-3-319-15705-4_15

Finally, the researchers used a case study, the Tsmsui travel plan, to demonstrate that the innovative values discovered by using the new consumer co-creation framework, which is similar to lead users in the physical society, to help or test authors how to get good results to evidence our method is useful.

2   Literature Review

2.1   Informational Social Influence, Grounded Theory (GT) and Qualitative Chance Discovery

The weak tie (bridge) is generally not so closely connected with other clusters, and it gives a different piece of information. Then depending on the organization of different clusters, researchers may find short cuts (bridge, weak tie or friends of other groups) between different clusters and the short cuts, in turn, can bring potential innovation. Therefore, rare association analysis can be used to identify purchased goods with low support and high confidence and is also used to solve the problem of low frequency, it still fails to link with future trend analysis until Ohsawa et al. (1998) proposed the KeyGraph algorithm. KeyGraph algorithm is used to calculate the frequency of the node in the shopping cart and that of the co-occurrence of two nodes, in which strong clusters are expected to emerge. Next, the key-value derived from all the nodes connecting to each other is calculated. Among the strong clusters, we may find nodes with high key values and low frequency, which are called chance nodes. Then the researcher may integrate the hints given by chance nodes and strong clusters to build a chance scenario, which is called chance discovery.

The above discussion describes an uncertain scenario: before the researcher who follows the steps can identify relevant clusters, organize the clusters, and choose the weak tie or strong tie to reveal chances, he or she has to read and comprehend a mass of relevant theories, such as the GT. Therefore, the researcher not only has to do qualitative analysis to define the objective words but also has to decide on the constraints. Then these objective words and constrains are entered into the information retrieval system to extract small but enough precise and meaningful data, like Grounded-theory-based chance discovery (Hong, 2009).

2.2   Consumers Co-creation in New Product Development (NPD)

Involving consumers in the NPD process can improve product quality, reduce risk, and increase market acceptance (Business Wire 2001). Therefore, ideas generated through co-creation with customers will more closely mirror consumers' needs. It has been clearly recognized that successful NPD depends on a deep understanding of consumers' needs and product development efforts that meet those needs (Hauser, Tellis, and Griffin 2006). However, by involving consumers more actively in the NPD process, new ideas for a product can be generated, which are more likely to be valued by consumers, thereby increasing the likelihood of success of a new product.

Then, in the consumer co-creation process, the firms need to identify the consumers or consumer tribes who have the highest potential for co-creation (Ernst, Hoyer,

Krafft, and Soll 2010; Franke, Keinz, and Steger 2009), and, then, they can recruit those valuable customers to form a larger group of consumers in the co-creation process. This is to explore whether a larger group of people can lead to successfully customizing the product or focusing on a smaller particular group of customers, such as lead users can be more effective in developing and marketing a product. Co-creators need to be those who perceive themselves highly involved and knowledgeable consumers who often differ significantly from the majority of consumers and the majority of consumers may eventually purchase the product. Again, we need a better understanding of the needs, wants, preferences, and motivation of different groups of co-creating consumers. Therefore, one of the underlying issues of consumers' co-creation is that finding early adopters or lead users from the market is very difficult and the cost to do the search can be very high.

Therefore, to explore innovative ways of designing new products, we integrated GT, informational social influence, social presence (Shen & Khalifa, 2007), and text mining to define the objectives and constraints to extract relevant data and did co-occurrence analysis to obtain innovative ideas. The new consumer co-creation framework will be described in the next section.

3   Methodology

To achieve the goal of discovering innovative ideal, the researchers of the present study developed a new consumer co-creation framework for exploring innovative ways of designing new products: using Google blog search to obtain relevant consuming textual data and applying GT along with the social identity theory, self-categorization theory (Oakes 1987, Ashforth and Mael 1989), and social presence (Shen and Khalifa, 2007) to do self-categorization to discover the innovative value of a product by using co-occurrence analysis method, such as using social identity to identify the underlying knowledge and innovation or Deutsch and Gerard's (1955) theory of informational social influences (SI) that may lead humans to conform to the expectations of others. The research flowchart is shown as follows.

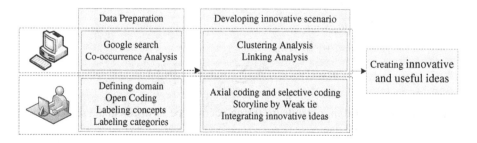

Fig. 1. Research flowchart

3.1   Human-Centered Computing System (HCCS)

The researchers realize that only a limited number of lead users existing in whole society, and they always use their informative social influence; for example, they may show their creativity by using innovative products in new ways and share them with the majority voluntarily to earn their social status. That is, either in the physical society or in the virtual community only little innovative consuming information can be found from lead users, and the little information is mixed with most of the consuming information. In the present study we used GT and employed a text-mining method to develop a HCCS for creating an alternative innovation. We followed the steps listed below in our research process.

Phase 1: Preparation for Data and Initial Analysis
To create a new innovative scenario, the researchers first collected data relevant to innovative uses of a product to extract useful data and then combine GT with text mining to process the data. The detailed process is listed as follows: H labels indicate the steps to be done by human beings, and C labels indicate the steps to be done by computer systems.

Step 1: Processing the raw data.
1-1-1H) The researcher defines the domain and relevant key words he/she intends to study.
1-1-1C) The researcher selects the data corresponding to keywords from the Internet.
1-1-2H) Based on his/her domain knowledge, the researcher interprets the texts, and at the same time, segments texts into words, removes irrelevant words, and marks meaningful words with conceptual labels.

Step 2: Word co-occurrence analysis (open coding).
1-2-1C) Use equation (1) to calculate the association values of all relevant words as shown below:

$$N \ is \ all \ words$$
$$i = 1 \ to \ N-1$$
$$j = i+1 \ to \ N \qquad\qquad (1)$$
$$assoc(w_i, w_j) = \sum_{s \in allD} \min(|w_i|_s, |w_j|_s)$$

In the above algorithm, s stands for the co-occurrence of words in the sentence, and D stands for all textual data.
1-2-2C) To visualize the analysis result, the computer system can reveal an association diagram showing the association among the co-occurrence words.
1-2-1H) The researcher identifies keywords as concepts and the clusters as categories derived from the co-occurrence association diagram, which helps the researcher preliminarily figures out the various theme values presented in the data.

Phase 2: Developing the Weak tie Storyline to Process Innovative Scenario
Based on the analysis done in phase 1, in which the researcher discovers various innovative ways of using ideas, clusters can be developed from data provided by lead users (axial coding), and the researcher draws a storyline based on links, using clusters developed by the innovative ideas to design an advertising scenario to improve the information social influence (selective coding).

Step 1: To generate uses of innovative product clusters (Axial coding)
Based on the focus of "rare but meaningful" uses of innovative ideas, the researcher extracts the sentence data to create a scenario of how the products are used innovatively. The process is illustrated as follows.

2-1-1H) The researcher needs to define what the uses of innovative products $(w_{innovation})$ are.

2-1-2C) Innovative product use $(w_{innovation})$ extracts out some rare but meaningful sentences. This variable is used to confirm the theme and to remove irrelevant sentences with a view to narrowing down the data range and to sifting out valid sentences. That is, valid sentences must include words related to $w_{innovation}$, which is shown as equation (2).

$$i = first\ sentence\ to\ the\ last\ sentence$$
$$valid\ sentence\ set = if\left(\{w_1, w_2, ...\}_i \cap \{w_{innovation}\}\right) \neq \phi \qquad (2)$$

2-1-3C) Use equation (1) to calculate the association value of all words in the set of valid sentences, and then create an association diagram.

Step 2: Link innovative ideas (weak tie recognition) to extract uses of these products from the lead users' (selective coding).

The process is listed as follows.

2-2-1H) The researcher needs to decide what the various uses of the innovative ideas are: $(w_{innovation_1}, w_{innovation_2}, ...)$.

2-2-2C) Various innovative ways of using ideas $(w_{innovation_1}, w_{innovation_2}, ...)$ extract the rare but meaningful sentences.

$$i = first\ valid\ sentence\ to\ the\ last\ valid\ sentence$$
$$use\ of\ innovative\ product\ set \qquad (3)$$
$$= if\left(\{w_1, w_2, ...\}_i \cap \{w_{innovation_1}, w_{innovation_2}, ...\}\right) \neq \phi$$

2-2-3C) Use equation (1) to calculate the association value of all words in terms of innovative uses of ideas, and then create an association diagram.

2-2-2H) Based on his/her domain knowledge, the researcher identifies the lead users' innovative uses of ideas to create the advertising scenario.

4   Case Study

Phase 1: Preparation for Data and Initial Analysis

Because the information is introduced for Taiwanese, only weblog written by Taiwanese is valid. So, the authors collected data posted on blogs relevant to Tamsui travel and Tamsui holiday by Taiwanese. These data ranged from January 1, 2011 to December 31. Using Google blogs (http://blogsearch.google.com/blogsearch) and the keywords are "Tamsui travel" or "Tamsui holiday", to search for the data, the authors obtained 218 related data from blog articles. Here, we try to prove that innovative idea discovery based on the phase 1 of HCCS is useful for creating innovative service.

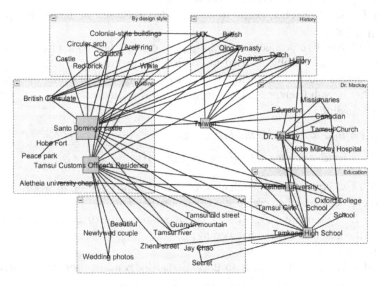

Fig. 2. Visualizing the travel characteristics of Tamsui

Phase 2: Developing the Weak Tie Storyline to Process Innovative Scenario

Step 1: To generate uses of innovative product clusters (Axial coding).

In this step, each place is not only presents the characteristics of landscape and building style, but also helps authors to integrate the related clusters (graphs) to clearly sketch the whole innovative scenario. First, after analyzed the results of phase 1, the authors can identify some important places, such as Santo Doming castle, little White House, Oxford college, Tamkang high school. Then, the characteristics of each graph are presented in following:

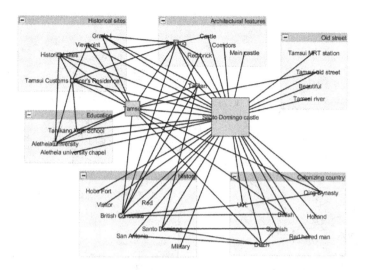

Fig. 3. The characteristics of Santo Doming castle

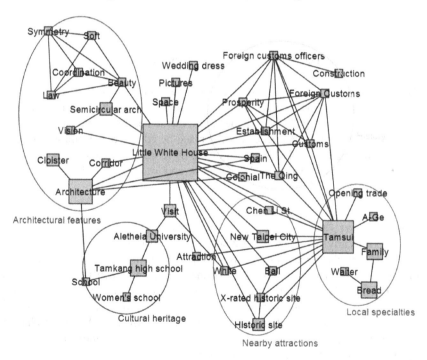

Fig. 4. The characteristics of Little White House

In Fig.3 and Fig.4, authors could recognize that the garden, carved statue, spire, corridors, arch et al., are presented in everywhere of Santo Doming castle and little White House. The European culture is very rich in there. Therefore, the visitors always take a photo in there. Especially, in spring the flours of cherry bloom.

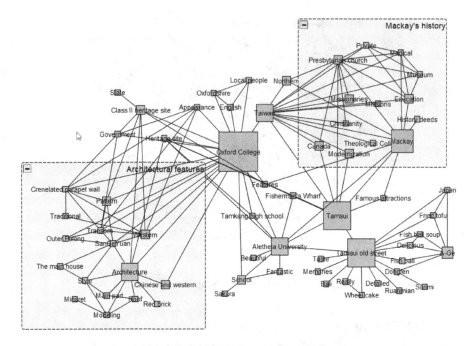

Fig. 5. The characteristics of Oxford College

The word of Oxford College is presents two parts in Fig.5: one is that about 130 years ago, Dr. Mackey who came from Canada and he was a pastor of Christianity, too. At that time, he brought the west education, medicine, and the belief of Christianity to Taiwan. The Oxford College was the place he educated Taiwanese. The other part is the building was included by European style and Taiwan style, which let us understand Dr. Mackey was very like Taiwan and wanted to dedicate himself for Taiwan and his belief.

Step 2: Link innovative ideas (weak tie recognition) to extract uses of these products from the lead users' (selective coding).
Now, authors clearly recognize each places characteristics, and have to co-creation with the relevant characteristics for discovering innovative ideas. First, the authors identify each European colonial building as one concept, and then sifted out concept's data is very clean. Second, Tamsui Oxford College, little white house et al. were used to integrate the relevant data, and word frequency and word-word co-occurrence frequency can be set in very low level to externalize all information. Third, the characteristics of British manor, such as garden, carved statue, spire, corridors, arch et al., are emerged to bridge the (sub tribes), as shown in Fig.6.

The beauty of British manor and Taiwanese are recognized from Fig.6. Authors are not surprised that the characteristics of Taiwan building style and living style are emerged on Fig.6. But some interesting characteristics, such as the beauty of British manor could attract many Taiwanese to travel Tamsui. Previously describing scenario as a hint can help authors to emphasize the beauty of European colonial building and garden, European culture, is an important concept of storyline for travel design. The other is the Dr. Mackey who was a pioneer of medicine and education in Taiwan, brought a great medical technique to Taiwan and helped Taiwanese to leave the diseases. So, Taiwanese visit Tamsui is not only enjoy in European culture, but also can memory Dr. Mackey.

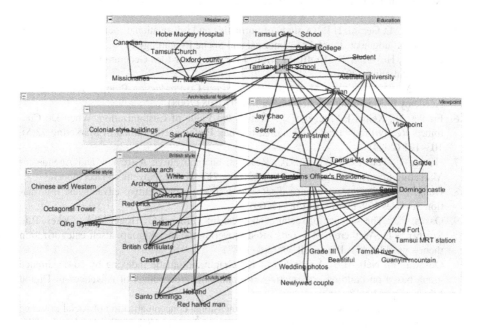

Fig. 6. The characteristics of Tamsui, spring, and Oxford College

5   Conclusion

Consumer co-creation model is one of important methods for company to develop new product. But how to discover and invite useful consumers into brainstorm meeting is a big problem of this method. Therefore, this study proposes an innovative co-creation method; collecting consuming data from internet, the HCCS is used to analyze the consuming data for discovery innovative ideas, and help experts to design a new product. At last, the Tsmsui travel plan is used to help or test authors how to get good results to evidence our method is useful.

Reference

1. Ashforth, B.E., Mael, F.: Social identity theory and the organization. Academy of Management Review **14**(1), 20–39 (1989)
2. Business Wire: New Study Identifies Customer Involvement as Primary Success Factor in New Product Development. Business/ Technology Editors, Business Wire, New York (2001)
3. Deutsch, M., Gerard, H.B.: A study of normative and informational social influences upon individual judgment. Journal of Abnormal and Social Psychology **51**, 629–636 (1955)
4. Ernst, H., Hoyer, W.D., Krafft, M., Soll, J.H.: Consumer Idea Generation. working paper, WHU, Vallendar (2010)
5. Etgar, M.: A Descriptive Model of the Consumer Co-Production Process. Journal of the Academy of Marketing Science **36**(Spring), 97–108 (2008)
6. Franke, N., Keinz, P., Steger, C.J.: Testing the Value of Customization: When Do Customers Really Prefer Products Tailored to Their Preferences? Journal of Marketing **73**(5), 103–121 (2009)
7. Hauser, J., Tellis, G.J., Griffin, A.: Research on Innovation: A Review and Agenda for Marketing Science. Marketing Science **25**, 686–717 (2006). (November-December)
8. Hong, C.-F.: Qualitative chance discovery: Extracting competitive advantages. Information Sciences **179**(11), 1570–1583 (2009)
9. Oakes, P.J.: The salience of social categories. In: Turner, J.C., Hogg, M.A., Oakes, P.J., Reicher, S.D., Wetherell, M.S. (eds.) Rediscovering the social group: a self-categorization theory, pp. 117–141. Blackwell, Oxford (1987)
10. Ohsawa, Y., Nels, E.B., Yachida, M.: KeyGraph: Automatic indexing by co-occurrence graph based on building construction metaphor. In: Proceedings of Advances in Digital Libraries, pp. 12–18 (1998)
11. Shen, K.N., Khalifa, M.: Exploring multi-dimensional conceptualization of social presence in the context of online communities. In: Jacko, J.A. (ed.) HCI 2007. LNCS, vol. 4553, pp. 999–1008. Springer, Heidelberg (2007)
12. Strauss, A.C., Corbin, J.M.: Basics of qualitative research: Techniques and procedures for developing grounded theory (2nd edn.). Sage Publications, Inc. (1998)

Recognizing and Evaluating the Technology Opportunities via Clustering Method and Google Scholar

Tzu-Fu Chiu[1](✉) and Chao-Fu Hong[2]

[1] Department of Industrial Management and Enterprise Information,
Aletheia University, New Taipei City, Taiwan, R.O.C
chiu@mail.au.edu.tw
[2] Department of Information Management, Aletheia University,
New Taipei City, Taiwan, R.O.C
cfhong@mail.au.edu.tw

Abstract. In order to recognize and evaluate the technology opportunities in an industry, a research framework has been formed via combining several data-mining methods, where clustering method is employed to generate the clusters, similarity measure is adopted to identify the variant patents, and association analysis is used to find out the rare topics. Patent data contains plentiful technological information from which it is worthwhile to extract further knowledge. Consequently, the variant patents were identified, the rare topics were found, and the technology opportunities for companies were recognized. Finally, the Google Scholar has been utilized to evaluate the recognized technology opportunities.

Keywords: Technology opportunity · IPC-based clustering · Similarity measure · Association analysis · Google Scholar · Patent data

1   Introduction

For attempting to recognize and evaluate the technology opportunities, a clustering method is employed to generate the clusters; the similarity measure is adopted to identify the variant patents; and the association analysis is used to figure out the rare topics. In addition, as up to 80% of the disclosures in patents are never published in any other form [1], it would be worthwhile for researchers and practitioners to recognize the technology opportunities upon the patent datasets. Therefore, a research framework will be formed to identify the variant patents, to find out the rare topics, and to recognize the technology opportunities for companies and the industry.

2   Related Work

As this study is aimed to recognize and evaluate the technology opportunities, a research framework needs to be constructed via a consideration of clustering method, similarity measure, association analysis, and Google Scholar [2] search.

© Springer International Publishing Switzerland 2015
N.T. Nguyen et al. (Eds.): ACIIDS 2015, Part II, LNAI 9012, pp. 159–169, 2015.
DOI: 10.1007/978-3-319-15705-4_16

Therefore, the related areas of this study would be technology opportunity analysis, IPC-based clustering [3], similarity measure, association analysis, and Google Scholar.

2.1   Technology Opportunity Analysis

Technology opportunity analysis (TOA) draws on bibliometric methods, augmented by expert opinion, to provide insight into specific emerging technologies [4]. TOA performs value-added data analysis, collecting bibliographic and/or patent information and digesting it to a form useful to the research or technology managers, strategic planners, or market analysts. TOA can identify the following topics: component technologies and how they relate to each other; who (companies, universities, individuals) is active in developing those technologies; where the active developers are located nationally and internationally; how technological emphases are shifting over time; and institutional strengths and weaknesses as identified by research profiles. TOA had been applied in the personal digital assistant (PDA) area [5] and the blue light-emitting diode (LED) area [6].

2.2   IPC-Based Clustering

Cluster analysis is the process of grouping a set of physical or abstract objects into classes of similar objects, which is an unsupervised classification and also called as clustering, or data segmentation [7]. An IPC (International Patent Classification) is a classification derived from the International Patent Classification System (supported by WIPO) which provides for a hierarchical system of language independent symbols for the classification of patents and utility models according to the different areas of technology to which they pertain [8].

IPC-based clustering, proposed by one of the authors, is a modified clustering method which utilizes the professional knowledge of the patent office examiners (implied in the IPC field) to tune the clustering mechanism and to classify the patents into a number of clusters effectively [3]. It mainly includes the following five steps: (1) IPC code group generation, (2) centroid of IPC code group generation, (3) producing initial clustering alternative, (4) producing refined clustering alternative, and (5) optimal alternative selection. IPC-based clustering will be applied in this study for partitioning the patent data.

2.3   Similarity Measure

Similarity measure is a way to measure the likeness between two objects (e.g., documents, events, behaviors, concepts, images, and so on). The methods for measuring similarity vary from distance-based measures, feature-based measures, to probabilistic measures [9]. In distance-based measures, there are Minkowski family, intersection family, inner product family, Shannon's entropy family and so on [10]. Here, the shorter the distance is, the bigger the similarity will be. Among distance-based measures, the Euclidean Distance is one the most popular methods, which can be defined as in Equation (1), where x_i and x_j are vectors with l features (i.e., x_{ik} and x_{jk}) [11].

$$dis(\mathbf{x}_i, \mathbf{x}_j) = \sqrt{\sum_{k=1}^{l}(x_{ik} - x_{jk})^2} \quad (1); \qquad Ja(e_i, e_j) = \frac{Freq(e_i \cap e_j)}{Freq(e_i \cup e_j)} \qquad (2)$$

2.4   Association Analysis

Association analysis is a useful method for discovering interesting relationships hidden in large data sets. The uncovered relationships can be represented in the form of association rules or co-occurrence graphs [7]. An event map, a sort of co-occurrence graphs, is a two-dimension undirected graph, which consists of event clusters, visible events, and chances [12]. The co-occurrence between two events is measured by the Jaccard coefficient as in Equation (2), where e_i is the ith event in a data record (of the data set D). The event map is also called as an association diagram in this study.

2.5   Google Scholar

Google Scholar [2] is a freely accessible web search engine that indexes the full text of scholarly literature across an array of publishing formats and disciplines. The Google Scholar index includes most peer-reviewed online journals of Europe and America's largest scholarly publishers, plus scholarly books and other non-peer reviewed journals. It is similar in function to the freely available CiteSeerX and getCITED, and also similar to the subscription-based tools, Elsevier's Scopus and Thomson ISI's Web of Science [13]. In this study, the Google Scholar will be utilized to evaluate the recognized technology opportunities.

3   A Research Framework for Recognizing and Evaluating the Technology Opportunity

As this study is attempted to recognize and evaluate the technology opportunity in the thin-film solar cell industry, a research framework has been developed based on the related work and shown in Fig. 1. It consists of six phases: data preprocessing, cluster generation, variant patent identification, rare topic recognition, new findings, and evaluations; and will be described in the following subsections.

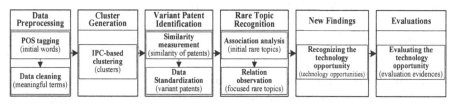

Fig. 1. A research framework for recognizing the technology opportunity

3.1   Data Preprocessing

In first phase, the patent data of thin-film solar cell (during a certain period of time) will be downloaded from the USPTO [14]. For considering an essential part to represent a complex patent data, the Abstract, Assignee, and Issue Date fields are selected as the objects for this study. Afterward, two processes, POS tagging and data cleaning, will be executed to clean up the source textual data.

(1) **POS Tagging:** An English POS tagger (i.e., a Part-Of-Speech tagger for English) from the Stanford Natural Language Processing Group [15] will be employed to perform word segmenting and labeling on the patents (i.e., the abstract field). Then, a list of proper morphological features of words needs to be decided for sifting out the initial words.

(2) **Data Cleaning:** Upon these initial words, files of n-grams, stop words, and synonyms will be built so as to combine relevant words into compound terms, to eliminate less meaningful words, and to aggregate synonymous words. Consequently, the meaningful terms will be obtained from this process.

3.2   Cluster Generation

Second phase is designed to conduct the cluster generation via IPC-based clustering so as to obtain the different clusters of thin-film solar cell.

IPC-Based Clustering: In order to carry out the homogeneity analysis, an IPC-based clustering is adopted for separating patents into clusters. The clusters will be named using the IPC and its description.

3.3   Variant Patent Identification

Third phase, including similarity measure and variant patent sifting, is used to calculate the cluster centroid, to measure the similarity of each patent with its cluster centroid, and to find out the variant patents.

(1) **Similarity Measure:** A cluster centroid is calculated by averaging all patents of a cluster as in Equation (3), where p_{ij} is the jth patent in the ith cluster and cen_i is the centroid of the ith cluster. The similarity (x_{ij}) between each patent (p_{ij}) and its centroid (cen_i) is measured by the Euclidian Distance as in Equation (4).

$$cen_i = (1/m) \cdot \left(\sum\nolimits_{j=1}^{m} p_{ij} \right) \quad (3); \qquad x_{ij} = dis\left(p_{ij}, cen_i \right) = \sqrt{\sum\nolimits_{j=1}^{m} \left(p_{ij} - cen_i \right)^2} \quad (4)$$

(2) **Variant Patent Sifting:** The sample mean $(\bar{x}_i)$ of the ith cluster is calculated as in Equation (5). The sample standard deviation (S_i) of the ith cluster is computed by Equation (6). The z-score (z_{ij}) of each patent is calculated as in Equation (7); and a variant patent will be determined if the abstract z-score $(|z_{ij}|)$ is no less than 1.65 as in Equation (8).

$$\overline{X}_i = (1/m)\cdot\left(\sum_{j=1}^{m} x_{ij}\right) \quad (5); \qquad S_i = (1/(m-1))\cdot\left(\sqrt{\sum_{j=1}^{m}\left(x_{ij} - \overline{X}_i\right)^2}\right) \qquad (6)$$

$$z_{ij} = \left(x_{ij} - \overline{X}_i\right)/S_i \quad (7); \qquad \left|z_{ij}\right| \geq 1.65 \qquad (8)$$

3.4   Rare Topic Recognition

Fourth phase, containing association analysis and relation observation, is used to draw the association diagram and to find out the rare topics.

(1) Association Analysis: An association diagram will be drawn via the variant patents, so that a number of rare topics can be generated. These rare topics will be named using the domain knowledge.

(2) Relation Observation: According to the relations between initial rare topics and companies as well as issue years, the focused rare topics will be recognized. As the variant patents in recent years are more likely to be the clues of technology opportunity, the time frame is divided into three periods of time: earlier (1999 to 2002), middle (2003 to 2006), and later (2007 to 2010). Therefore, the initial rare topics which possess strong links with companies and in the middle and later periods will be identified as the focused rare topics.

3.5   New Findings

Fifth phase, technology opportunity recognition will try to figure out the technology opportunities based on the variant patents and focused rare topics.

Technology Opportunity Recognition: According to the variant patents and focused rare topics, the technology opportunities will be recognized. Both the focused rare topics and technology opportunities will be provided to facilitate the decision-making of managers and stakeholders.

3.6   Evaluations

In last phase, the technology opportunities found in the above Subsection will be evaluated by observing the growing status of the opportunities using the related articles searched from the Google Scholar.

Evaluating the Technology Opportunity: In order to generate the growing status of the technology opportunities, the number of related articles (published) of each focused rare topic were collected from the Google Scholar during 1990-2012 (settings: search English pages; articles without patents; topic title). The growing status (Q) was measured by dividing the information of the recent period (2008-2012) by the early period (1990-2007), which denoted the increasing degree of developing speed of a rare topic while turning from the early period to recent period. If the growing status (Q) is greater enough, for example up to 2 times ahead, it would be plausible to state that there is an evidence for supporting "this topic is growing more rapidly in recent years". And it would also be appropriate for this topic to be regarded as a promising technology opportunity.

4   Experimental Results and Explanations

The experiment has been implemented according to the research framework. The experimental results would be explained in the following five subsections: result of data preprocessing, result of cluster generation, result of variant patent identification, result of rare topic recognition, and result of technology opportunity recognition.

4.1   Result of Data Preprocessing

As the aim of this study was to recognize the technology opportunities via patent data, the patents of thin-film solar cell were the target data for the experiment. Mainly, the Abstract, Assignee, and Issue Date fields were used in this study. 213 issued patents of thin-film solar cell during year 1999 to 2010 were collected from USPTO. The POS tagger was then triggered to do data preprocessing. Consequently, the patents were cleaned up and the meaningful terms were obtained.

4.2   Result of Cluster Generation

After executing the programs of IPC-based clustering, eleven clusters of thin-film solar cell were generated. The No. of cluster, IPC, and Num. of patents (number of comprising patents), were listed in Table 1. For example, the cluster No. 1 with IPC: H01L031/18 was consisting of 78 patents.

Table 1. 11 clusters of thin-film solar cell via IPC-based clustering

No.	IPC	Num. of patents	No.	IPC	Num. of patents	No.	IPC	Num. of patents
1	H01L031/18	78	5	H01L027/142	14	9	H01L031/032	7
2	H01L031/06	7	6	H01L031/042	7	10	H01L031/075	12
3	H01L031/00	18	7	H01L031/052	15	11	H01L021/762	10
4	H01L021/00	27	8	H01L031/0336	18			

4.3   Result of Variant Patent Identification

Using similarity measure, the variant patents were identified via comparing the distance of each patent with the mean of distance of its cluster. A patent with the comparing difference no less than 1.65 standard deviation was regarded as a variant patent. For examples, two variant patents in Cluster H01L027/142 were identified as Patent 06013870 ($dis = 1.2122$; $z = 2.3604$) and 06274804 ($dis = 1.2122$; $z = 2.3604$), while S was 0.3668 and $\bar{X}$ was 0.3463. There was no variant patent identified in Cluster H01L031/0336 as the differences were not greater than $1.65S$, while S was

0.0297 and $\bar{x}$ was 0.9586. Graphs of Cluster H01L027/142 and Cluster H01L031/0336 of z-score were shown in Fig. 2. The result of all variant patents (17 patents) in eleven clusters was summarized in Table 2.

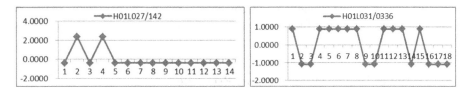

Fig. 2. Variant patents of Cluster H01L027/142 and Cluster H01L031/0336

Table 2. The result of all variant patents in eleven clusters

No.	Cluster ID.	Variant Patent	No.	Cluster ID.	Variant Patent
1	H01L031/18	07641937; 07811633	7	H01L031/052	--
2	H01L031/06	--	8	H01L031/0336	--
3	H01L031/00	06239352; 06414235; 07507903; 07851700	9	H01L031/032	--
4	H01L021/00	07252781; 07288617; 07795452	10	H01L031/075	06124545; 06133061; 07122736
5	H01L027/142	06013870; 06274804	11	H01L021/762	06133112; 06534383
6	H01L031/042	06537845			

4.4   Result of Rare Topic Recognition

Through association analysis, an association diagram of 17 variant patents was drawn using the meaningful terms of Abstract field of the variant patents as in Fig. 3. In the diagram, 10 initial rare topics were identified and named according to the domain knowledge, including roll-to-roll-processing, SOI-substrate, hot-wire-method, and so on. By inserting the Assignee and Issue Year fields into the association diagram, eight companies were linked to the rare topics, such as Luch-Daniel (US), Canon Kabushiki Kaisha (JP), Merck Patent GmbH (DE), and so on, also shown in Fig. 3.

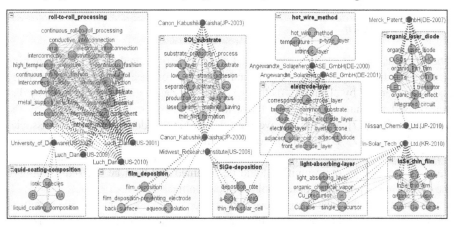

Fig. 3. An association diagram with initial rare topics and linked companies

In order to recognize the focused rare topics, the relations between initial rare topics and companies as well as issue years were observed. The relation judging criteria were: the topic with one strong link (i.e., greater than or equal to 3 linking lines) to a company in the later period of issue years (2007~2010) was the most focused rare topic; the topic with one strong link to a company in the middle period of issue years (2003~2006) was the more focused rare topic; and the other options were the less focused rare topics

4.5   Result of Technology Opportunity Recognition

According to the association diagram with initial rare topics, links of companies and issue years, as well as relation criteria, the focused rare topics were recognized: the most focused rare topics were roll-to-roll-processing, organic-laser-diode, light-absorbing-layer, and InSe-thin-film; the more focused rare topics were SOI-substrate, liquid-coating-composition, and SiGe-deposition. The technology opportunities were formed reasonably by the most and more focused rare topics as follows.

(1) Roll-to-roll-processing topic: It linked to Luch-Daniel (US) and University-of-Delaware (US); and to year 2010, 2009, 2001, and 2003.

(2) Organic-laser-diode topic: It linked to Merck Patent GmbH (DE) and Nissan Chemical Ltd (JP); and to year 2007 and 2010.

(3) Light-absorbing-layer topic: It linked to In-Solar Tech Co., Ltd (KR); and to year 2010.

(4) InSe-thin-film topic: It linked to In-Solar Tech Co., Ltd (KR); and to year 2010.

(5) SOI-substrate topic: It linked to Canon-Kabushiki-Kaisha (JP); and to year 2003 and 2000.

(6) Liquid-coating-composition topic: It linked to University-of-Delaware (US); and to year 2003.

(7) SiGe-deposition topic: It linked to Midwest-Research-Institute (US) and Ange-wandte-Solarenergie-ASE-GmbH (DE); and to year 2006 and 2000.

5   Evaluation

The methods of this study were derived from statistical analysis and data mining which have been regarded as proper methodologies for measuring the homogeneity and heterogeneity by clustering methods [7, 16], for calculating the similarity by distance-based measures [9, 10, 11], for sifting the variant data from datasets by the standardization [17], and for discovering relationships hidden in datasets by association analysis [7, 12]. By means of collecting the information of related articles (published) of each focused rare topic from Google Scholar (1990-2012), the growing status of technology opportunities could be obtained and then be utilized to observe the development potentials of technology opportunities.

According to the searching process described in Subsection 3.6, the result of growing status of technology opportunities from Google Scholar was summarized in Table 3. The column definitions are as follows: "A" is the number of articles in the early period (1990-2007); "B" is the number of articles in the recent period (2008-2012); "a" is the yearly rate of the early period via "A" dividing by total amount and number of years (a = A/(A+B)/18); "b" is the yearly rate of the recent period via "B" dividing by total amount and number of years (b = B/(A+B)/5). "Q" is the growing status, as a quotient of "b" (yearly rate of recent period) dividing by "a" (yearly rate of early period). An example of No. 1 topic (Roll-to-roll-processing) is: A = 461; B = 1460; a = A/(A+B)/18 = 461/1921/18 = 1.33%; b = B/(A+B)/5 = 1460/1921/5 = 15.2%; Q = b/a = (15.2%)/(1.33%) = 11.4.

Table 3. A summary of the growing status of technology opportunities

No.	Topic title	A	B	a	b	Q
1	Roll-to-roll-processing	461	1460	1.33%	15.2%	11.4
2	Organic-laser-diode	59	29	3.72%	6.59%	1.7
3	Light-absorbing-layer	408	391	2.84%	9.79%	3.4
4	InSe-thin-film	42	20	3.76%	6.45%	1.7
5	Liquid-coating-composition	16	8	3.70%	6.67%	1.8
6	SOI-substrate	3400	3330	2.81%	9.90%	3.5
7	SiGe-deposition	304	149	3.73%	6.58%	1.7

(1) Evidence of roll-to-roll-processing topic: Since the growing status of this topic (Q = (15.2%)/(1.33%)) is 11.4 times ahead, it is regarded as an evidence to strongly support: "this topic is a promising technology opportunity".

(2) Evidence of organic-laser-diode topic: Since the growing status of this topic (Q) is 1.7 times ahead, it is regarded as an evidence to evenly support this topic.

(3) Evidence of light-absorbing-layer topic: Since the growing status of this topic (Q) is 3.4 times ahead, it is regarded as an evidence to strongly support this topic.

(4) Evidence of InSe-thin-film topic: Since the growing status of this topic (Q) is 1.7 times ahead, it is regarded as an evidence to evenly support this topic.

(5) Evidence of liquid-coating-composition topic: Since the growing status of this topic (Q) is 1.8 times ahead, it is regarded as an evidence to evenly support this topic.

(6) Evidence of SOI-substrate topic: Since the growing status of this topic (Q) is 3.5 times ahead, it is regarded as an evidence to strongly support this topic.

(7) Evidence of SiGe-deposition topic: Since the growing status of this topic (Q) is 1.7 times ahead, it is regarded as an evidence to evenly support this topic.

Consequently, the strongly supported technology opportunities were roll-to-roll-processing, light-absorbing-layer, and SOI-substrate topics; and the evenly supported technology opportunities were organic-laser-diode, InSe-thin-film, liquid-coating-composition, and SiGe-deposition topics.

6   Conclusions

The research framework for recognizing and evaluating the technology opportunity has been formed and applied to thin-film solar cell using patent data. The experiment was performed and the experimental results were obtained. Eleven clusters of thin-film solar cell during 1999 to 2010 were generated via IPC-based clustering. Seventeen variant patents were found through similarity measure and standardization. Ten rare topics were identified using association analysis. The technology opportunities were recognized and evaluated as: roll-to-roll-processing, light-absorbing-layer, SOI-substrate, organic-laser-diode, InSe-thin-film, liquid-coating-composition, and SiGe-deposition topics. The variant patents and technology opportunities on thin-film solar cell would be helpful for managers and stakeholders to facilitate their decision-making.

In the future work, the threshold of z-score for sifting out variant patents can be set to different levels (i.e., different from 1.65) to obtain more or less variant patents. In addition, the data source can be expanded from USPTO to WIPO or TIPO in order to recognize the technology opportunity on thin-film solar cell widely.

Acknowledgements. This research was supported by the Ministry of Science and Technology of the Republic of China under the Grants MOST 103-2410-H-156-010.

References

1. Blackman, M.: Provision of patent information: a national patent office perspective. World Patent Inf. **17**(2), 115–123 (1995)
2. Google, Google Scholar (2014/4/25). http://scholar.google.com.tw/schhp?hl=en&as_sdt=0
3. Chiu, T.F.: A proposed ipc-based clustering method for exploiting expert knowledge and its application to strategic planning. Journal of Information Science **40**(1), 50–66 (2014)
4. Porter, A.L., Detampel, M.J.: Technology opportunities analysis. Technol. Forecast. Soc. Chang. **49**(3), 237–255 (1995)
5. Lee, S., Yoon, B., Park, Y.: An approach to discovering new technology opportunities: Keyword-based patent map approach. Technovation **29**(6), 481–497 (2009)
6. Li, X., Wang, J., Huang, L., Li, J., Li, J.: Empirical research on the technology opportunities analysis based on morphology analysis and conjoint analysis. Foresight **12**(2), 66–76 (2010)
7. Han, J., Kamber, M.: Data Mining: Concepts and Techniques, 2nd edn. Morgan Kaufmann Publishers, CA (2006)
8. WIPO: Preface to the International Patent Classification (IPC) (2013/10/20). http://www.wipo.int/classifications/ipc/en/general/preface.html
9. Scholarpedia, Similarity measures (2013/10/20). http://www.scholarpedia.org/article/Similarity_measures
10. Cha, S.H.: Comprehensive survey on distance/similarity measures between probability density functions. International Journal of Mathematical Models and Methods in Applied Sciences **1**(4), 300–307 (2007)
11. Feldman, R. & Sanger, J.: The Text Mining Handbook: Advanced Approaches in Analyzing Unstructured Data. Cambridge University Press (2007)

12. Ohsawa, Y., Benson, N.E., Yachida, M.: KeyGraph: automatic indexing by co-occurrence graph based on building construction metaphor. In: Proceedings of the Advanced Digital Library Conference (IEEE ADL 1998), pp. 12–18 (1998)
13. Wikipedia, Google Scholar (2014/4/25). http://en.wikipedia.org/wiki/Google_Scholar
14. USPTO. 'the United States Patent and Trademark Office' (2013/10/20). http://www.uspto.gov/
15. Stanford Natural Language Processing Group, Stanford Log-linear Part-Of-Speech Tagger (2009/8/15). http://nlp.stanford.edu/software/tagger.shtml
16. Tseng, Y.H., Lin, C.J., Lin, Y.I.: Text Mining Techniques for Patent Analysis. Inf. Process. Manage. **43**, 1216–1247 (2007)
17. Anderson, D.R., Sweeney, D.J., Williams, T.A.: Statistics for business and economics. CengageBrain (2008)

Raising EFL College Learners' Awareness
of Collocations via COCA

Yuh-Chang Lin[(✉)]

Center for General Education, Aletheia University, No. 32, Chen-Li Street,
Tamsui, New Taipei City 25103, Taiwan, R.O.C.
au1258@au.edu.tw

Abstract. This study aimed to raise EFL learners' awareness of collocations via
COCA. Three learners participated in this experimental study. They were college students aged from 21 to 25. Four instruments and several exercises were
employed in this study. The instruments included a standardized test, a questionnaire, two surveys, and a corpus called Corpus of Contemporary American
English (COCA). Exercises involve translations and collocational grids.
Through one-month experiment, learners' awareness of collocations was raised.
Although learners' awareness of collocations was raised, their application of
collocational knowledge to real-life settings was still far from satisfaction. It is
suggested that attention should be paid to the advancement of collocational
knowledge, if learners intend to convey their messages in English clearly and
economically.

Keywords: Awareness of collocations · COCA · Collocational knowledge ·
Low-achieving learners

1 Introduction

EFL (English as a foreign language) college students in Taiwan have learned English
for at least six years (three years in their junior high, and another three years in their
senior high). Through this six-year exposure to English, a great majority of them
spend time developing their reading ability, a receptive language skill, for two main
entrance exams in Taiwan, senior high schools and colleges, for these two exams
assess mainly test-takers' reading ability. Under the influence of the entrance exams,
much attention has been paid to the advancement of reading ability.

Too much focus on meanings but little attention paid to form leads to EFL learners' inability to write clearly. They tend to make errors when they write in the target
language (here English). Of the errors made by Taiwanese learners, misuse of collocations is one of them, which poses difficulties for native speakers in reading texts
written by EFL writers. For example, coming across such an expression as *I would
like to have a cup of red tea*, native speakers can be confused about the meaning in
that they cannot retrieve from their mental lexicon for the exact meaning of *red tea*.
Actually, the expression should be *black tea*.

Quite a number of Taiwanese learners of English tend to literally translate Chinese
into English when they express in a productive way. This may result from their lack

© Springer International Publishing Switzerland 2015
N.T. Nguyen et al. (Eds.): ACIIDS 2015, Part II, LNAI 9012, pp. 170–179, 2015.
DOI: 10.1007/978-3-319-15705-4_17

of collocational knowledge. In order to raise EFL learners' awareness of collocations and improve their accuracy in composing lexical items, the researcher intends to adopt an online resource, Corpus of Contemporary American English (COCA), to explore the effect of employing it on raising EFL college students' awareness of collocations.

2   Related Literature

2.1   Theoretical Aspect

Collocations had been explored since 300 B.C. (Robins [1], cited in Hsu [2]), when Greek Stoic scholars had included collocations in their study of lexical semantics. These scholars did not support the notion of "one word, one meaning". To them, collocations were an important consideration while determining the meaning of a lexical item. They claimed that word meanings do not emerge in isolation, and its meanings may differ depending on the collocations in which they appear.

The approach to the investigation into collocations can be divided into two categories: lexis, and the integration of lexis and grammar. Researchers who apply lexical approach to investigating collocations include Palmer, Firth, Halliday, McIntosh, and Lewis. They hold that lexis is a crucial criterion in determining the collocability of words. For example, Palmer [3] views collocations as lexical items whose meanings are not inferable from their component parts. Firth [4] regards collocations as words collocating with other words in collocational or conventional order and examines collocations at syntagmatic level. In order to describe the collocability of words, Halliday [5] invents the notion of *set* to illustrate the distinct combination of words. McIntosh's [6] addition of the notions of *tolerance of compatibility*, *potential of collocability*, *range* and *range-extension* to the field of collocations improve learners' vision of collocations. Lewis [7] considers lexis as an alternative to grammar-based approach to the study of collocations. All these researchers support lexical approach to collocations.

Contrary to the lexis-oriented convention to the study of collocations, some linguists and researchers proposed that for a clearer explanation of collocations, lexis as well as grammar had better be seen as interpenetrating factors in determining the collocability of words. Among the researchers advocating this belief are Sinclair; Greenbaum; Mitchell; Bolinger; Kjellmer; Benson, Benson, and Ilson; Nation.

The researchers stated above have shared an approach, an integration of lexis and grammar, to the study of collocations. Sinclair [8] proposes *node* and *span* to describe the collocational features of words. "Open-choice" and "idiom" principles are Sinclair's another approach to examining collocability of words. Greenbaum [9] notes that examining a collocation simply from lexical considerations may have its shortcomings for it might relax syntactic restrictions on collocations. Mitchell [10] describes collocations as lexico-grammatical combinations and further proposes a distinction between collocations and colligations. Bolinger [11] considers selectional restrictions of generative grammars as one of the restrictions on the collocability of words. Kjellmer [12] defines collocations as "lexically determined and grammatically restricted sequences of words". Benson et al. [13] divides collocations into two groups: lexical collocations and grammatical collocations. Nation [14] adopts ten scales to identify collocations. All these researchers have reached a consensus on the study of collocations – an integration of lexis and grammar.

2.2   Empirical Aspect

Empirical studies on collocations have primarily focused on the following three perspectives: measuring collocational knowledge, development of collocational knowledge, and the teaching of collocations.

Measuring Collocational Knowledge. A significant amount of studies on collocations are related to measuring learners' collocational knowledge. Some researchers (e.g., Zhang [15]; Al-Zahrani [16]) explore the correlation between learners' collocational knowledge and their writing performance. Others (e.g., Aghbar [17]; Bahns and Eldaw [18]; Bonk [19]) investigate the relationship between learners with different language proficiency and their collocational knowledge. They employ different instruments, tests and questionnaires for example, to conduct their experiments. The results show that learners' collocational knowledge closely correlates with their language proficiency. However, there is still a gap between L2 learners' collocational knowledge and their writing performance.

Development of Collocational Knowledge. For the purpose of having a general idea on the development of learners' collocational knowledge, Cowie and Howarth [20] employed a lexical collocation test (a verb-noun type), and Gitsaki [21] adopted an essay writing task, a translation task and a blank-filling test as instruments to conduct their experiments to see the results. Cowie and Howarth's finding shows that there is an overlap on the development of collocational knowledge between less proficient native speakers and more proficient non-native speakers. Gitsaki's findings show that beginning learners' acquisition of lexical collocations is better than that of grammatical ones. However, their collocational knowledge applied in a productive way remains limited. Intermediate learners' knowledge of grammatical collocations seems to increase, but their proper use of collocations is still unsatisfied. Post-intermediate learners are found to have a better performance in the use of their collocational knowledge. On the whole, L2 learners still have difficulty in applying their collocational knowledge to the real-world settings.

The Teaching of Collocations. Being aware of the importance of collocational knowledge, some researchers have focused their studies on the teaching of collocations with a view to providing teachers with a clearer concept of how to teach collocations. Cowie [22] proposes that the instruction of ready-made units at a basic level of discourse is as important as that of lexical innovation, which many researchers suggest too early to do it. Bahns [23] subsequently argues that EFL learners should be taught how to use collocation dictionaries. He further claims that not all collocations should be taught by teachers. For example, collocations with grammatical similarity between two languages need not to be taught. He points out that teaching materials composed of frequently-used collocational items can be useful in EFL classrooms.

Farghal and Obiedate [24], conducting a similar study to Bahns', proposes that collocations are an important yet neglected knowledge in EFL classrooms. Their finding supports Bahns' studies. They employed two tests to assess two groups of learners and found that learners adopt four different strategies to take the tests—strategies such as lexical simplification, namely synonym, transfer, avoidance and paraphrasing.

What these researchers suggest is that learners, ESL or EFL learners in particular, should be equipped with collocational knowledge so that they may use the language accurately and economically.

In order to help those learners who are still in a puzzle at the notion of collocations, this study is thus conducted with the following two research questions:

1. Can learners' awareness of collocations be raised via COCA?
2. Can low-achieving learners' knowledge of collocations be improved by explicit teaching?

2.3   COCA

COCA is a free online corpus. Anyone can access it. New users can gain access to it and use it directly. After 10-15 queries, users are asked to register for continued use of the corpus. By then, users are asked to provide their account numbers (email addresses) and passwords to gain access to the system. Afterwards, when users intend to exploit COCA, their account numbers and passwords are required. There are four functions listed under the word *display*. They are *list, chart, kwic*, and *compare*. Of them, *kwic* is the only function employed by this study. *Kwic* stands for *key word in context*. For example, learners may be asked to check if *white hair* or *gray hair* has more tokens (the exact number of certain lexical items) appearing in COCA. They may key in either *white hair* or *gray hair* in a box with *word(s)* to the left of the box and click the button *search*. The number of the tokens for either *white hair* or *gray hair* tends to show on the screen. The number of the tokens for *white hair* and *gray hair* are 996 and 1083 respectively in COCA. With significant number of the tokens for each expression, one may conclude that these two expressions are widely used by Americans. Learners are expected to learn the skill of examining whether a term is used by most Americans or not via COCA. If a term is widely used in COCA, learners may employ it without much doubt. If a term cannot be found in COCA, learners are suggested to find another way to express it.

3   Methodology

3.1   Participants

Participants are three college students who took a summer session English course called Freshman English. One of them is male, and the other two are female. Their years of studying English vary from 7 to 12 years. This is an intensive course spanning one month. Class meets two times a week and 4 hours each time. Total teaching hours are 32 hours.

3.2   Instruments

Four instruments and several exercises are employed in this study. The instruments include a standardized test, a questionnaire, two surveys, and a corpus called Corpus of Contemporary American English (COCA). The standardized test is an English proficiency test developed by Language Testing Center (LTC) in Taiwan. It contains three parts: vocabulary and structure, fill-in-the-blanks, and reading comprehension. Another instrument is a questionnaire. It includes three parts: learners' personal data, their English learning experience, and their English vocabulary learning strategies. There are two surveys conducted during the study. One was carried out before the class began, and the other was administered at the end of the class. The survey questions are primarily about learners' knowledge of collocations before and after taking this course. Still another instrument is COCA, a freely searchable 450-million-word Corpus, the largest corpus of American English currently available.

Exercises involve five translations and five collocational grids. Translations are primarily about phrase translation from Chinese to English. For example, in a Chinese sentence,麥當勞及肯德基之類的食物被視為<u>垃圾食物</u>, there is a phrase underlined, which is 垃圾食物. Following the Chinese, there is an English sentence made based on the meaning of the Chinese sentence. Its translation goes like this: Food such as MacDonald's and Kentucky's is generally taken as ___ ___. There are two blanks provided for learners to fill in with proper words.

Collocational grid exercises are to test learners' knowledge of collocations. One example of the collocational grid exercise is shown in figure 1.

Fig. 1. A Collocational Grid Exercise for *ride*, *drive*, and *fly*

	a car	a bicycle	an airline	a bus
ride		+		+
drive	+			+
fly			+	

3.3   Procedures

First of all, participants are asked to take a standardized English proficiency test called General English Proficiency Test (GEPT), which is a test developed by Language Testing Center (LTC) in Taiwan. The level of the test is intermediate. It is a test designed as a benchmark for evaluating high school students' English ability in Taiwan. Learners are allowed to take the test for 45 minutes. The length of time is stipulated by LTC. The purpose of asking learners to take the test is to know their language levels of proficiency.

Participants are also required to complete a questionnaire on English vocabulary learning strategies before the class starts. It is made of 39 questions with an expert validity from two English professors who have taught English for 3 and 17 years respectively. The questionnaire is administered for the purpose of collecting learners' personal data, their English learning experience, and their vocabulary learning strategies before and after taking this course. Moreover, results of the questionnaire completed before the class is taken as a benchmark to explore whether learners' awareness of collocations has been raised.

Additionally, four translation exercises were practiced during class. There are ten questions in each exercise. Learners were asked to do an exercise for three times. For the first time, they did the exercise without consulting any reference books. When they completed the questions, learners' answers were checked by the researcher. Those correct answers were marked with a red circle. Then, the researcher gave back the original test questions to learners. Without erasing any answers written in the first stage, learners may refer to any resources they can access for the second-time exercise. Still, the answers written by learners for the second-time exercise were checked with a red triangle, and learners were asked to keep their answers intact. After this, learners were asked to do the third-time exercise on the same questions. This time, learners were asked to check their correct or incorrect answers based on the number of the tokens for a specific term they examined via COCA. If they fail to discover acceptable answers to any questions by every effort they've made, answers will be provided accordingly. Still, learners were asked to double check the answers via COCA.

Five collocational grid exercises were administered in this study. They were carried out by the learners in a way similar to the steps in undertaking the translation exercises.

3.4   Data Collected

Data are collected from the following instruments: a standardized test, a questionnaire, two surveys, five translation tests, and five collocational grid exercises.

4   Results

4.1   Before the Class

On the standardized test, three learners' scores are 60, 37, and 31 respectively. The total score for the test is 100. One learner passed the test, and the other two failed the test. Their scores are significantly below the average; therefore, I call them low-achieving learners. While the one who passed the test, I call her an intermediate learner.

A survey was conducted before class began. There are four open-ended questions in the survey. They are: (1) What is your way of learning words? (2) Have you ever had experience in writing in English? If yes, have you ever encountered any difficulty in determining whether a word or a phrase is correct or not? (3) Following the previous question, how do you find a suitable phrase while you are writing in English (e.g., consulting a dictionary, asking classmates for answers, asking teachers for answers, using an electronic dictionary, or depending on other resources)? (4) Have you ever heard of COCA on the Internet? If yes, who introduced it to you? After being introduced to this Internet resource, have you ever tried to use it? If yes, how many times did you access it?

In answering the first question, one learner said that he tended to read the word out loud first and then try to memorize it. If he failed to pronounce the word, he tended to commit it to memory by rotten practice. The other two shared a method: they read out loud the word and at the same time write it down on a piece of paper for several times. As to the answers to the second question, one said she would compose it in other way; that is, she paraphrased the expression which she failed to come up with proper terms directly. Another one said that she would try her best to use the terms which she was familiar with. Still another one said that he had no experience writing in English. In response to the third question, one said when she ran into difficulty in finding proper terms in writing in English, she tended to find them through consulting Internet, asking her elder sister, or using an electronic dictionary. Another one said that he would use an electronic dictionary or refer to Internet resources. Still another one said that she would check what she had in mind by keying in the terms into Google to see whether it is proper or not. If her terms were not proper, she would key in Chinese into Google and looked for proper English translation. Answering the fourth question, one said that she heard of COCA via teachers; however, she has not used it yet. The other two said that they had never heard of it.

4.2   During the Class

As to the translation exercises, three learners' performance was shown in the following tables. In order to distinguish the intermediate learner from the other two, I labeled the intermediate learner as learner A, and the other two as learner B and learner C.

Table 1. Learner A's Translation Score

	Exercise 1	Exercise 2	Exercise 3	Exercise 4	Exercise 5
First-time	30	20	30	20	30
Sec.-time	50	90	70	90	70

Note: First-time exercise was done by learners without referring to any resources. For sec.-time (second-time) exercise, learners might freely consult any reference materials. The total score is 100.

Table 2. Learner B's Translation Score

	Exercise 1	Exercise 2	Exercise 3	Exercise 4	Exercise 5
First-time	0	0	0	0	0
Sec.-time	45	50	30	50	30

Table 3. Learner C's Translation Score

	Exercise 1	Exercise 2	Exercise 3	Exercise 4	Exercise 5
First-time	0	10	0	0	0
Sec.-time	50	50	20	55	30

Concerning collocational grid exercises, learners' performance for the second-time exercise was recorded in the following tables from 4 to 6.

Table 4. Learner A's Performance

	Exercise 1	Exercise 2	Exercise 3	Exercise 4	Exercise 5
Total No.	9	4	6	8	8
First-time	6	2	4	5	2
Sec.-time	8	4	6	8	8

Note: Total number means the sum of possible combinations based on Oxford Collocations Dictionary for students of English [20] and The BBI Combinatory Dictionary of English [6]. First-time means that learners are not allowed to consult any reference materials. Second-time means that learners do the exercise for the second time, when they are allowed to consult any reference materials available to them.

Table 5. Learner B's Performance

	Exercise 1	Exercise 2	Exercise 3	Exercise 4	Exercise 5
Total No.	9	4	6	8	8
First-time	6	3	0	3	1
Sec.-time	8	4	6	8	6

Table 6. Learner C's Performance

	Exercise 1	Exercise 2	Exercise 3	Exercise 4	Exercise 5
Total No.	9	4	6	8	8
First-time	5	2	0	3	2
Sec.-time	6	3	5	6	5

There are three questions for the second survey conducted at the end of the course. They are: (1) What is a collocation? Please give examples to illustrate it. (2) How do you translate a Chinese phrase into English when you fail to find an English equivalent for any part of the phrase? For example, how do you find an English equivalent of a Chinese phrase "化濃妝"? (3) What do you learn from this summer session course? What impresses you the most?

Learner A's answers are as follows: (1) She just gave two examples: One is *drive the bus* (√), and the other is *ride the bus* (×). (2) She will google for the answer. She will key in the Chinese phrase '化濃妝' and then leave a space. After this measure, she will key in '中文' and press 'Enter' button to look for possible answers. (3) She learned collocations from this course, and COCA impressed her the most.

Learner B's responses are in the following: (1) He has no response to this question. (2) He will access the Internet and use COCA to look for English translations. (3) He learned how to find an acceptable phrase in English, and COCA impressed him the most.

Learner C's answers are listed as follows: (1) Two different words combine together and the meaning of the sum is a little different from the meaning of its component parts. For example, 'return trip' means the journey to go back home. For an EFL learner, it is sometimes hard to understand its meaning from the meanings of its component parts. (2) First of all, she will look for the English equivalent of the Chinese word '畫' and then look for the English expressions for the Chinese phrase '濃妝'. When she has two English lexical items, she will key in those two items into COCA and see whether its tokens meet certain requirements as an acceptable expression. (3) What she learned from this course are figuring out the way to employ COCA, paying attention to a word's environment, knowing the importance of word classes in the combination of words, and realizing that phrases should be learned as a single item.

5   Discussion and Suggestions

The researcher intends to begin this section with response to the two research questions: (1) Can learners' awareness of collocations be raised via COCA? (2) Can learners' knowledge of collocations be improved by explicit teaching?

From the translation exercises, collocational grid exercises, and learners' response to the surveys before the class began and after the class was over, the researcher would like to say that learners' awareness of collocation has been raised. For the intermediate learner, her awareness of collocation has improved greatly. While for the low achievers, even though their awareness of collocations seemed limited after they took the course, they felt that COCA benefited them a lot in terms of finding an acceptable lexical expression in their language learning process.

Can learners' knowledge of collocations be improved by explicit teaching? The answer is yes. Although learners' different levels of language proficiency might lead to different learning effects, explicit teaching does good to the learning of collocations. From learners' performance on translation exercises and collocational grid exercises, the intermediate learner did better than those two low-achieving learners. Learners' language proficiency has effect on their performance of collocational knowledge.

It seems that this study has achieved its goal of raising EFL college learners' awareness of collocations. However, it is not without its limitation. Limited number of participants and brief duration of the experiment might contribute to a result far from being general. One may take this study as a pilot study. That is, more participants and longer duration are needed for a study to make a generalized conclusion.

6   Conclusions

Collocation is one of the important language abilities ESL or EFL learners, even native low achievers, should develop. It may help facilitate learners' proper employment of English in a productive way. In this study, learners' awareness of collocations is raised to different degrees based on their language proficiency. Low achievers' knowledge of collocations are developed too although the progress is limited. By explicit teaching plus proper use of COCA, learners' knowledge of collocations can be advanced.

References

1. Robins, R.H.: A short history of linguistics. Longman, London (1967)
2. Hsu, J.Y.: Development of collocational proficiency in a workshop on English for general business purposes for Taiwanese college students (Doctoral dissertation). Available from ProQuest Dissertation and Theses database. (UMI No. 3040680) (2002)
3. Palmer, H.E.: Second interim report on English collocations. Kaitahusha, Tokyo (1933)
4. Firth, J.R.: Modes of meaning. In: Firth, J.R. (ed.) Papers in linguistics, 1934–1951, pp. 190–215. Oxford University Press, Oxford (1957)
5. Halliday, M.A.K.: Lexis as a linguistic level. In: Bazell, C.E., Catford, J.C., Halliday, M.A.K., Robins, R.H. (eds.) In memory of J. R. Firth, pp. 148–162. Longman, London (1996)
6. McIntosh, A.: Patterns and ranges. Language **37**, 325–337 (1961)
7. Lewis, M.: The lexical approach: the state of ELT and a way forward. Language Teaching Publications, London (1993)
8. Sinclair, J.M.: Beginning the study of lexis. In: Bazell, C.E., Catford, J.C., Halliday, M.A.K., Robins, R.H. (eds.) In memory of J. R. Firth, pp. 410–430. Longman, London (1996)
9. Greembaum, S.: Verb-intensifier collocations in English. The Hague: Mouton. Mitchell, T.F.: Linguistics "going-on": Collocations and other lexical matters on the syntagmatic record. Archivum Linguisticum 2, 35–69 (1970, 1971)
10. McIntosh, C., Francis, B., Poole, R. (eds.): Oxford collocations dictionary for students of English. Oxford University Press, New York (2009)
11. Bolinger, D.: Meaning and memory. Forum Linguisticum **1**(1), 1–14 (1976)

12. Kjellmer, G.: Some thoughts on collocational distinctiveness. In: Aarts, J., Meijs, W. (eds.) Corpus linguistics: Recent development in the use of computer corpora in English language research, pp. 163–171. Rodopi, Amsterdam & Atlanta (1984)
13. Benson, M., Benson, E., Ilson, B.: The BBI combinatory dictionary of English, 3rd edn. John Benjamins Publishing Company, Amsterdam/Philadelphia (2010)
14. Nation, I.S.P.: Learning vocabulary in another language. Cambridge University Press, Cambridge (2010)
15. Zhang, X.: English collocations and their effect on the writing of native and non-native college freshmen. Unpublished doctoral dissertation, Indiana University of Pennsylvania, Pennsylvania (1993)
16. Al-Zahrani, M.S.: Knowledge of English lexical collocations among male Saudi college students majoring in English at a Saudi university. Unpublished doctoral dissertation, Indiana University of Pennsylvania, Pennsylvania (1998)
17. Aghbar, A.A.: Fixed expressions in written tests: Implications for assessing writing sophistication. East Lansing, National Center for Research on Teacher Learning, MI (ERIC Document Reproduction Service No. ED 352808) (1990)
18. Bahns, J., Eldaw, M.: Should we teach EFL students collocations? System **21**(1), 101–114 (1993)
19. Bonk, W.J.: Testing ESL learners' knowledge of collocations. East Lansing, National Center for Research on Teacher Learning, MI (ERIC Document Reproduction Service No. ED442309) (2000)
20. Cowie, A., Howarth, P.: Phraseological competence and written proficiency. In: Blue, G., Mitchell, R. (eds.) Language and Education. Multilingual Matters, Clevedon (1996)
21. Gitsaki, C.: Second language lexical acquisition: A study lexical acquisition: A Study of the development of collocational knowledge. International Scholars Publications, Maryland (1999)
22. Cowie, A.P.: Multiword lexical units and communicative language teaching. In: Arnaud, P., Bejoint, H. (eds.) Vocabulary and applied linguistics, pp. 1–12. Macmillan, Basingstoke (1992)
23. Bahns, J.: Lexical collocations: a contrastive view. ELT Journal **47**(1), 56–63 (1993)
24. Farghal, M., Obiedat, H.: Collocations: A neglected variable in EFL. IRAL **33**(4), 315–333 (1995)

Effects of Question Prompts and Self-explanation on Database Problem Solving in a Peer Tutoring Context

Min-Huei Lin[1(✉)], Ming-Puu Chen[2], and Ching-Fan Chen[3]

[1] Department of Information Management, Aletheia University,
New Taipei City, Taiwan
au4052@au.edu.tw
[2] Graduate Institute of Information & Computer Education,
National Taiwan Normal University, Taipei, Taiwan
mpchen@ice.ntnu.edu.tw
[3] Department of Educational Technology, Tamkang University,
New Taipei City, Taiwan
cfchen@mail.tku.edu.tw

Abstract. In the prior research, using the computing system (HCCS) indeed identified the gaps between doing and knowing when students learned concepts and performed problem solving tasks of database management. To bridge the emerged gaps by interacting with well-designed question prompts and self-explanations treatments, the purpose of this study is to examine the effects of question prompts, self-explanation, and the interaction of question prompts and self-explanation on undergraduate students' database conceptual knowledge and problem solving performance in a peer tutoring context. One hundred and fifty-one undergraduates from three classes in a private university participated in the 8-week experimental instruction. The results reveals that (a) the interaction of question prompts and self-explanation was not significant, and (b) for the problem solving performances, the integral question prompt group outperformed the gradual question prompt group and the scenario-based self-explanation group outperformed the elaboration-based self-explanation group.

Keywords: Question prompt · Self-explanation · Problem solving · Peer tutoring

1 Introduction

Some core topics organized in the knowledge area, information management, of the ACM/IEEE CS2013 Body of Knowledge, are primarily concerned with the capture, digitization, representation, organization, transformation, and presentation of information; data modeling and abstraction. The student needs to develop conceptual and physical data models, determine which IM methods and techniques are appropriate for a given problem, and be able to select and implement an appropriate IM solution that address relevant design concerns including scalability, accessibility and usability. The topics include database systems, data modeling, indexing, relational databases, query languages.

Database management contains data modeling, development, applications and implementations. To learn the higher level concepts of database, students need to have

© Springer International Publishing Switzerland 2015
N.T. Nguyen et al. (Eds.): ACIIDS 2015, Part II, LNAI 9012, pp. 180–189, 2015.
DOI: 10.1007/978-3-319-15705-4_18

the abstract concepts. Instructional strategies such as problem-based learning have been suggested to reduce misconceptions and enhance meaningful learning [2]. The DB instructional students showed that novices are lack of DB-specific knowledge, so the doing and knowing about actions of problem solving are usually inconsistent. When teachers teach and strive to achieve successful students learning outcomes, most of them have experienced difficulty in communicating with students. There are only a few students can learn well, early and timely, lots of students learn helplessness or give up halfway under successive frustration instead.

Question prompts as scaffoldings have been found effective in helping students focus attention to specific aspects of their learning process and monitor and evaluate their learning through elaboration on the questions asked [6]. Researchers have examined the use of question prompts in scaffolding student knowledge construction, integration, and problem-solving processes in various content domains. The overall findings have consistently pointed to the advantages of the use of question prompts in directing students' attention to important aspects of the problem, activating their schema, eliciting their explanations, and prompting them for self-monitoring and self-reflection. Ge and Land (2005) conducted a critical analysis of question prompts in terms of their cognitive and metacognitive functions in supporting ill-structured problem-solving processes, which led to the theoretical assumption that question prompts could also be effective in supporting ill-structured problem solving and a couple of empirical studies have also examined the use of question prompts in scaffolding ill-structured problem solving in the classroom setting [7].

Self-explanation is an act which the learners perform when they explains how they solve a problem. Self-explanation process is not mere communication of information to other people by conversation. Rather it is one of the mental representation activities. Self-explanation refers to a reflective activity explaining to oneself a learning material in order to understand meaning from the material or to repair misunderstanding during studying worked-out examples or reading exploratory texts [4]. It seems obvious that a student performs better at problem-solving tasks, generates inferences which facilitate conceptual understanding, and repairs flawed mental models as well when being encouraged to use the self-explanation strategy during learning [3].

In the study of Ge and Land (2004), the quantitative outcomes revealed that question prompts had significantly positive effects on student problem-solving performance but peer interactions did not show significant effects. The qualitative findings, however, did indicate some positive effects of peer interactions in facilitating cognitive thinking and metacognitive skills. The study suggested that the peer interaction process itself must be guided and monitored with various strategies, including question prompts, in order to maximize its benefits. Peer tutoring refers to an instructional method that uses pairings of high-performing students to tutor lower-performing students in a class-wide setting or in a common venue outside of school under the supervision of a teacher. Currently, there is sufficient research that documents the benefits of peer tutoring as a supplement to traditional instruction. Peer tutoring has been used across academic subjects, and has been found to result in improvement in academic achievement for a diversity of learners within a wide range of content areas [8]. Wu and Looi (2011) designed and implemented a reflective tutoring framework by incorporating question prompts within an inquisitive simulated tutee environment that seeks to scaffold learner's reflection in pursuing tutoring activities [12].

Lin, Chen and Chen (2013) have used the intelligent human-computer computing system (HCCS) to identify the gaps between doing and knowing when students learned concepts and performed problem solving tasks of database management [10]. In this study, the strategies of question prompts and self-explanations are used to bridge the gaps between doing and knowing in learning database management. Hence, this study examined the effects of question prompts and self-explanation in scaffolding undergraduate students' database problem solving processes through peer tutoring.

2   Literature Review

2.1   Problem Solving

Although most educators regard problem solving as a critical skill for life and there are different perspectives on problem solving as an educational goal, method, or as a skill. As an educational goal, problem solving refers to the use of weak methods to solve unfamiliar, new domain-general problems, the use of strong methods to solve domain-specific well-structured problems, the use of knowledge-based methods to find an acceptable solution for ill-structured problems. In modern education, authentic learning tasks that are based on real-life tasks are increasingly seen as the driving force for teaching and learning because they are instrumental in helping learners to integrate their knowledge, skills and attitudes. The use of problem solving as an educational method completely ignore human working memory limitations. Several alternatives to conventional problem solving have been devised to teach problem solving in more efficient and effective way, including the use of goal-free problems, worked examples, and completion problems. Within the skills perspective, problem solving is seen as something that develops over time as a function of practice. The phase models typically link problem solving to one or more phases in a process of expertise development or skill acquisition. The 4C/ID-model uses the four components, learning tasks, procedural information, supportive information and part-task practice to build an environment for teaching problem solving [13].

When solving database problems, students have no clear path to a solution. During the problem solving learning, students need to learn to integrate various database concepts and applications into solving various phases of problems. The links between learning and problem solving suggest that students can learn heuristics and strategies and become better problem solvers, and it is best to integrate problem solving with academic content. Through problem solving learning, content, skills, and attitudes could be integrated in the learning context in order to promote learning by problem solving. Therefore, problem solving could be employed as a prospective means for learning database management and applications. This study aims to investigate the conceptual and operational framework for scaffolding the problem solving learning process in database management and applications.

2.2   Question Prompts

Question prompts include procedural prompts, elaboration prompts, reflection prompts, knowledge integration prompts, justification prompts, self-monitoring prompts, each of which serves different cognitive and metacognitive purposes. Procedural prompts are designed to help students to complete specific tasks, and they have been successfully used to help students learn cognitive strategies in specific content areas. Elaboration prompts are designed to prompt learners to articulate thoughts and elicit explanations. Reflection prompts encourage reflection on a meta-level that students do not generally consider [5]. Lin and Lehman found that justification prompts facilitated transfer to a contextually dissimilar problem. Davis and Linn found that self-monitoring prompts embedded in the Web-knowledge integration environment encouraged students to think carefully about their activities and facilitated planning and reflection. The gradual question prompts, in the present study, are to scaffold students with attending to important aspects of a problem at different phases and assist them to plan, monitor, and evaluate the solution during the processes of problem solving, and the integral question prompts are to scaffold students' problem solving with the top-down style.

2.3   Self-explanation

The self-explanation effect can be explained with two fundamental reflection mechanisms: inference generation and conceptual revision. It seems obvious that a student performs better at problem-solving tasks, generates inferences which facilitate conceptual understanding, and repairs flawed mental models as well when being encouraged to use the self-explanation strategy during learning. According to cognitive load theory, generating self-explanation requires high cognitive load by requiring that learners monitor their understanding and represent incoming information at the same time. Kwon, Kumalasari and Howland (2011) examined the effects of self-explanation on conceptual understanding and problem-solving performance. The main results of the study were that generating an explanation to oneself was more effective than completing an explanation provided by an expert on problem-solving performance, and when students generated more correct explanations on problems, they solved the problems more accurately. The results also revealed the superiority of "open self-explanation" over completing an expert's partial explanation. The "open self-explanation" might encourage students to generate more inferences to make sense of their explanations and the inferences could be elements of new knowledge. However, the "completing other-explanation" might force students to make sense of a given explanation and thus hinder inference generation [9]. This study examines the effect of self-explanation in reducing the gaps between doing and knowing.

2.4   Peer Tutoring Context

Higher Education has been facing many challenges. The student body is itself becoming more diverse in age, experience level, motivation, and learning need. Evidence shows that students can learn by teaching their peers, yet the magnitude of learning gains is often small. To account for such findings, Roscoe and Chi (2007) considered peer tutors' behaviors in terms of knowledge-building versus knowledge-telling. Learning by teaching arises from knowledge-building opportunities inherent to the teaching process that may depend on tutors' self-monitoring, such as explaining and questioning. To explain well, peer tutors may need to evaluate their own knowledge gaps or confusion, and recover from these problems using logical reasoning or other knowledge-building strategies. Tutors also need to organize their domain knowledge to offer explanations that are well-structured. Roscoe (2014) considered the hypothesis that peer tutors' knowledge-telling bias results from inadequate self-monitoring. This hypothesis is tested by analyzing the dialog of untrained, novice tutors and tutees to identify tutors' overt knowledge-building and self-monitoring activities, and assessing how these activities relate to each other and learning. The results showed that tutors' comprehension-monitoring and domain knowledge, along with pupils' questions, were significant predictors of knowledge-building, which was in turn predictive of deeper understanding of the material. Moreover, tutorial interactions and questions appeared to naturally promote tutors' self-monitoring [11]. Arco-Tirado (2011) examined the impact of a peer tutoring program on preventing academic failure and dropouts among first-year students. The result showed differences in favor of the treatment group on grade point average, performance rate, success rate and learning strategies.

This study examines the effects of the question prompts that guide the peer tutors to interact with tutees, and scaffold novice tutors in promoting knowledge-building behavior [1].

3   Purpose and Research Questions

The purpose of this study is to examine the effects of question prompts, self-explanation, and the interaction of question prompts and self-explanation on undergraduate students' database conceptual knowledge and problem solving performance in a peer tutoring context by comparing gradual prompts with integral prompts and conditional event-driven self-explanation with factual knowledge-elaborated self-explanation in a private university in New Taipei City, Taiwan.

The study tried to answer the following research questions:

1. Do gradual prompts differ from integral prompts in terms of database conceptual knowledge and problem solving performance for undergraduate students in a peer tutoring context?
2. Does conditional event-driven self-explanation differ from factual knowledge-elaborated self-explanation in terms of database conceptual knowledge and problem solving performance for undergraduate students in a peer tutoring context?
3. Does the combination of question prompts and self-explanation affect students' database conceptual knowledge and problem solving performance in a peer tutoring context?

4   Method

4.1   Research Design

A 2 x 2 factorial qusi-experimental design was implemented to address the research questions. Two treatment variables used - question prompts and self-explanation. Dependent variables included conceptual knowledge and problem solving.

Treatment variable 1. The treatment levels of question prompt were gradual question and integral question prompts. The gradual question prompts prompted students to attend to important aspects of a problem at different phases and assisted them to plan, monitor, and evaluate the solution during the process of problem solving. In terms of modeling the problem as an issue tree, the integral question prompts referred to a set of questions with specific procedures for problem solving that prompted students to initiate the problem solving at the top level, then solve the elements in the next level, and repeat the process until the tree was exhausted.

Treatment variable 2. The treatment levels of self-explanation were scenario-based self-explanation and elaboration-based self-explanation. The scenario-based self-explanation asked students to reason their actions from a view of problem scenario in the environment of web application systems. The later asked students to reason and elaborate their action according to the factual knowledge.

4.2   Participants

151 undergraduate students from three database management and application classes at a private university in northern Taiwan, participated in the 8-week experimental instruction. 140 participants were sophomores, all of them possessed basic computer and programming skills, and were randomly assigned to one of the four groups to receive the four-hour weekly treatment.

4.3   Materials and Tasks

Students learn database concepts and applications by designing and implementing web applications. The problem solving situations were provided to the students. With the help of the problem solving scenarios, real-life events which closely related to the target concepts were introduced to serve as learning contexts for problem solving. The lectures and practices of theoretical concepts and practical skills for data modeling and database applications were synchronized with the phases of problem solving and covered in the four tasks depicted in the figure 1 that described the topics, goals and learning activities. To minimize the intrusion of web programming, the sample programs and a sample web application were used to help students in acquiring the knowledge and skills of web applications development and applying them in their own web application systems.

4.4   Instruments

The instruments utilized in the present study were the pre-test, post-test, two versions of worksheets for the gradual-prompt group and integral-prompt group, and two versions of worksheets for the two self-explanation groups in each task, the grading rubrics for the students' responses to question prompts, the self-explanation, and the performances for problem solving.

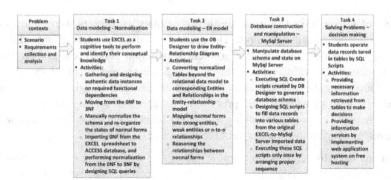

Fig. 1. The procedure of performing tasks of problem solving

4.5   Treatment and Data Collection Procedure

The database management and application is a required course for one academic year, the participants across all the four groups were instructed to perform the tasks about web applications problem solving during the second term. The experiment was carried out during the regular class schedule for database management and application, and consisted of four main phases as follows: assessing prerequisite knowledge, exploring the problem contexts, performing the four tasks, and assessing learning outcome. In the first week of instruction, students explored the problem contexts and identified the problem. In the following seven weeks, the task1 took students two weeks to finish the database normalization, the task2 took students two weeks to perform the transformation between the ER-model and Relational data model, the task3 took students one week to construct the database, and students spent two weeks on implementing web application systems. Whenever students performed tasks, the instructor provided the question prompts, and the peer tutor of each group used the question prompts to assist less-knowledgeable classmates. Tutors were not trained to use any specific question prompting, explaining, or monitoring strategies. After the peer tutoring session, all students performed the specific self-explanation and complete the worksheets. The data of students' responses to question prompts and self-explanations in the problem solving performances were gathered by means of the worksheets. After finishing the four tasks, students' performances of problem solving were scored.

4.6   Data Analysis

The effects of question prompts and self-explanation on database conceptual knowledge acquisition and problem solving performance in problem representation, developing solutions, and monitoring and evaluating a plan of action were examined by means of a 2 × 2 Multivariate Analysis of Variance (MANOVA). The significance level was set to 0.05 for the study.

5   Results and Discussion

Box's test of equality of covariance matrices revealed that the observed covariance matrices of the dependent variables were not equal across groups (Box's M = 33.440, F =3.624, p = 0.000).In the present study, two-way MANCOVA was employed to examine the effects of question prompt and self-explanation on learners' conceptual knowledge and database problem-solving performances with pre-test as the covariate.

5.1   Analysis of Question Prompt and Self-explanation on Database Conceptual Knowledge

The mean scores on database conceptual knowledge posttest for the question prompt and self-explanation groups are shown in Table 1. The gradual group scored higher than the integral group in conceptual knowledge learning. The scenario-based group scored higher than the elaboration-based group in conceptual knowledge learning. The MANCOVA summary of question prompt and self-explanation are shown in Table 2. The non-significant/significant Pillai' Trace indicated that there was not a significant interactive effect between question prompts and self-explanation in conceptual knowledge learning (Pillai' trace = .078, p = .925, η^2 = .001). This means that the effect of question prompts in conceptual knowledge was not significantly different for the scenario-based self-explanation than it was for the elaboration-based self-explanation counterparts. The both main effects of question prompt and self-explanation were not significant on database conceptual knowledge. As the conceptual knowledge mean scores shown in Table 1, it indicated that using the gradual question prompt was better than the integral question prompt, using the scenario-based self-explanation was better than the elaboration-based self-explanation for provoking students' conceptual knowledge.

Table 1. Summary of adjusted group means of conceptual knowledge learning and problem solving performance for question prompt and self-explanation

Dependent measure	Source	Aspect	Mean	SD	n
Conceptual	Question	Gradual	55.77	2.552	70
knowledge learning	prompt	Integral	53.69	2.371	81
	Self-	Scenario-based	56.03	2.443	76
	explanation	Elaboration-based	53.41	2.457	75
Database problem	Question	Gradual	193.97	1.735	70
solving perfor-	prompt	Integral	224.07	1.612	81
mance					
	Self-	Scenario-based	224.87	1.661	76
	explanation	Elaboration-based	193.17	1.670	75

Possible interpretations of these results were that the knowledge and comprehension as the first two levels of Bloom's Taxonomy of cognitive skills, in spite of more general prompts offered in the integral prompt group, students receiving gradual prompts could engage in connecting, planning, monitoring, and reflecting during approaching solutions as well as those receiving integral prompts, and students making self-explanations in the authentic problem context could elicit more relative connection between conceptual knowledge and real lives and perform better than those in the elaboration-based self-explanation groups.

5.2 Analysis of Question Prompt and Self-explanation on Database Problem Solving Performance

The experimental data has been collected from students since February, 2013, who have experiences in learning database management for one semester in September, 2012. They have the basis for programming languages, algorithms, flowchart, entity-relationship model and diagrams, mysql and php homepage design.

The mean scores on database problem solving performances for the question prompt and self-explanation groups are also shown in Table 1. The integral question prompt group scored higher than the gradual question prompt group in database problem solving performance. The scenario-based group scored higher than the elaboration-based group in database problem solving performance. Moreover, the two-way interaction was not significant, and both the main effects of question prompt and self-explanation were significant on database problem solving performance ($F_{(1,151)}$ = 79.906, p = 0.000, η^2 = 0.524; $F_{(1,151)}$ = 89.962, p = 0.000, η^2 = 0.554). As shown by the mean scores in Table 1, the integral question prompt group outperformed the integral question prompt group, and the scenario-based self-explanation group outperformed the elaboration-based self-explanation group in database problem solving performance. The problem solving in the application level of Bloom's Taxonomy of cognitive skills, the results confirmed that using the integral question prompt strategy with the whole view of top-down procedure to remind and apply students' conceptual knowledge did enhance their problem solving performance. Moreover, the purposeful arrangement of self-explanation with events or activities in the authentic context did enhance students' database problem solving performance.

Table 2. Summary of MANCOVA of question prompt and self-explanation on conceptual knowledge learning and problem solving performance

Source	Dependent Variable	Type III Sum of Squares	df	Mean Square	F	Sig.
Question prompt	Conceptual knowledge	161.683	1	161.683	.359	.550
	Problem Solving	33269.766	1	33269.766	159.744	.000
Self-explanation	Conceptual knowledge	257.369	1	257.369	.571	.451
	Problem Solving	37718.512	1	37718.512	181.105	.000
Question prompt *Conceptual knowledge		67.112	1	67.112	.149	.700
Self-explanation	Problem Solving	2.081	1	2.081	.010	.921
Total	Conceptual knowledge	535763.000	151			
	Problem Solving	6775344.000	151			

a. R Squared = .223 (Adjusted R Squared = .202)

b. Computed using alpha = .05

c. R Squared = .702 (Adjusted R Squared = .694)

6   Conclusions

This study found that there was no difference between gradual question prompt and integral question prompt, and there was no difference between scenario-based self-explanation and elaboration-based self-explanation in student' conceptual knowledge. Students received integral question prompts outperformed those receiving gradual question prompt in performing the problem solving tasks. Students adopted the scenario-based self-explanation outperformed those adopting the elaboration-based self-explanation. Students learn a procedure only when they need to use the procedure to gather information. The integral question prompts guided students to identify a problem and develop a solution strategy from the whole view. Students adopted the scenario-based self-explanation could be able to concentrate on the problems at hand and call on the schemas when needed.

This study contributes to the research on database problem solving in the peer tutoring context by suggesting integral question prompt and scenario-based self-explanation.

References

1. Arco-Tirado, J.L., Fernández-Martín, F.D., Fernández-Balboa, J.-M.: The impact of a peer-tutoring program on quality standards in higher education. Higher Education **62**, 773–788 (2011)
2. Chen, C.: Teaching problem solving and database skills that transfer. Journal of Business Research **63**, 175–181 (2010)
3. Chi, M.T.H., de Leeuw, N., Chiu, M.-H., La Vancher, C.: Eliciting self-explanations improves understanding. Cognitive Science **18**, 439–477 (1994)
4. Chi, M.T.H., Bassok, M., Lewis, M.W., Reimann, P., Glaser, R.: Self-explanations: How students study and use examples in learning to solve problems. Cognitive Science **13**, 145–182 (1989)
5. Davis, E.A., Linn, M.: Scaffolding students' knowledge integration: Prompts for reflection in KIE. International Journal of Science Education **22**, 819–837 (2000)
6. Ge, X., Land, S.M.: Scaffolding Students' problem-solving processes in an ill-structured task using question prompts and peer interactions. ETR&D **51**, 21–38 (2003)
7. Ge, X., Chen, C.H., Davis, K.A.: Scaffolding novice instructional designers' problem-solving processes using question prompt in a web-based learning environment. J. Educational Computing Research **33**, 219–248 (2005)
8. Greenwood, C.R., Delquadri, J.: Classwide peer tutoring and the prevention of school failure, Preventing School Failure. Alternative Education for Children and Youth **39**, 21–25 (1995)
9. Kwon, K., Kumalasari, C.D., Howland, J.L.: Self-Explanation Prompts on Problem-Solving Performance in an Interactive Learning Environment. Journal of Interactive Online Learning **10**, 96–112 (2011)
10. Lin, M.-H., Chen, M.-P., Chen, C.-F.: Exploring peer scaffolding opportunities on experiential problem solving learning. In: Badica, C., Nguyen, N.T., Brezovan, M. (eds.) ICCCI 2013. LNCS (LNAI), vol. 8083, pp. 572–581. Springer, Heidelberg (2013)
11. Roscoe, R.D.: Self-monitoring and knowledge-building in learning by teaching. Instructional Science **42**, 327–351 (2014)
12. Wu, L., Looi, C.-K.: A Reflective tutoring framework using question prompts for scaffolding of reflection. In: Biswas, G., Bull, S., Kay, J., Mitrovic, A. (eds.) AIED 2011. LNCS (LNAI), vol. 6738, pp. 403–410. Springer, Heidelberg (2011)
13. van Merriënboer, J.J.G.: Perspectives on problem solving and instruction. Computers & Education **64**, 153–160 (2013)

Intelligent Information Systems
in Security & Defense

FP-tree and SVM for Malicious Web Campaign Detection

Michał Kruczkowski[1,2],
Ewa Niewiadomska-Szynkiewicz[1,3]([✉]), and Adam Kozakiewicz[1,3]

[1] Research and Academic Computer Network (NASK), Warsaw, Poland
{michal.kruczkowski,adam.kozakiewicz}@nask.pl
[2] Institute of Computer Science, Polish Academy of Science, Warsaw, Poland
[3] Institute of Control and Computation Engineering, Warsaw University of
Technology, Warsaw, Poland
ens@ia.pw.edu.pl

Abstract. The classification of the massive amount of malicious software variants into families is a challenging problem faced by the network community. In this paper (The work was supported by the EU FP7 grant No. 608533 (NECOMA) and "Information technologies: Research and their interdisciplinary applications", POKL.04.01.01-00-051/10-00.) we introduce a hybrid technique combining a frequent pattern mining and a classification technique to detect malicious campaigns. A novel approach to prepare malicious datasets containing URLs for training the supervised learning classification method is provided. We have investigated the performance of our system employing frequent pattern tree and Support Vector Machine on the real database consisting of malicious data taken from numerous devices located in many organizations and serviced by CERT Polska. The results of extensive experiments show the effectiveness and efficiency of our approach in detecting malicious web campaigns.

Keywords: Malware campaign · URL · FP-tree · FP-growth · SVM

1 Introduction to Campaign Identification

Recently, numerous attacks have threatened the operation of the Internet [3,8, 10]. One of the major threats on the Internet is malicious software, often referred to as malware. Malware is a software designed to perform unwanted actions on computer systems such as gather sensitive information, or disrupt and even damage computer systems. In case of finding a server vulnerability (similar to the disease [7]) affecting multiple web pages, the attacker can automate the exploitation and launch an infection campaign in order to maximize the number of potential victims. Hence, the campaign determines a group of incidents that have the same objective and employ the same dissemination strategy [4]. Detecting malicious web campaigns has become a major challenge in Internet today [3,9,12]. Therefore, strategies and mechanisms for effective and fast detection of malware are crucial components of most network security systems.

© Springer International Publishing Switzerland 2015
N.T. Nguyen et al. (Eds.): ACIIDS 2015, Part II, LNAI 9012, pp. 193–201, 2015.
DOI: 10.1007/978-3-319-15705-4_19

In this paper we propose and investigate the use of frequent pattern mining [1,5] and supervised learning classification [6,13] for malicious campaign identification. The detection process is based on the analysis of URLs (Uniform Resource Locators). Each URL expressed by (1)

$$scheme : //domain : port/path?query\_string\#fragment\_id \qquad (1)$$

consists of the following components, i.e., scheme, domain name or IP address, port, path, query string and a fragment identifier (some schemes also allow a username and password in front of the domain). The query string consists of pairs of keys and values (also called attributes). By malicious URL we denote such URL that is created with malicious purposes. It can serve malicious goals, for example download any type of malware, which can be contained in phishing or spam messages, etc. It is obvious that the ultimate goal of obfuscation is to make each URL unique. Due to the fact that malicious URLs are generated by tools which employ the same obfuscation mechanisms for a given campaign we assume that usually intruders keep some parts of the URL static, while other parts are changed systematically and in an automated fashion. Moreover, since the duration of a given campaign is relatively short, we assume that the obfuscation mechanisms employed for such campaign do not change significantly during this time. We have designed and developed the system for malware campaigns identification. Our system utilizes FP-tree data structure of tokenized malicious URLs to form a training dataset for learning Support Vector Machine classifier. Finally, the trained SVM is used for on-line classification of new malware URLs into related or unrelated with malicious campaigns.

2   FP-Tree Structure and FP-Growth Algorithm

Frequent pattern mining is a widely used technique for discovering interesting relations between data items in large databases. It is employed today in many application areas including malware detection, Web usage mining, etc. In our research we use the frequent pattern tree structure (FP-tree) – a prefix structure for storing quantitative information about frequent patterns in a database – and the FP-growth algorithm for frequent pattern discovery using a divide-and-conquer strategy, both developed by J. Han et al. [5]. The FP-growth algorithm operates in two steps: 1) the FP-tree compact structure is constructed, 2) frequent patterns are extracted from the FP-tree. In the first step the occurrence of patterns in the input transaction database is counted. Next, infrequent patterns are discarded, frequent patterns are sorted by descending order of their frequency in the database, and the FP-tree structure is built. Common, usually most frequent patterns are shared. Therefore, FP-tree provides high compression close to tree root and can be processed quickly. Recursive growth is applied to extract the frequent patterns. FP-growth starts from the bottom of the tree structure (longest branches), by finding all patterns matching given condition. New tree is created, etc. Recursive growth ends when no patterns meet the condition, and processing continues on the remaining main branches of the original FP-tree.

3   Support Vector Machine Classifier

The Support Vector Machine (SVM) [13] is a supervised learning classification method widely used in data mining research. The concept of SVM is to classify each data sample into one of two categories: positive class denoted by "+1" and negative class denoted by "-1". Thus, the goal is to determine a decision boundary, which divides data into two sets (one for each class), a plane for $n \leq 3$ or hyperplane for $n > 3$, where n denotes the size of a dataset. Next, all the measurements on one side of this boundary are classified as belonging to "+1" class and all those on the other side as belonging to "-1" class. The problem is that many such hyperplanes can be determined, and the best one has to be selected. Hence, SVM tries to learn the decision boundary which gives the best generalization. A good separation is achieved by the hyperplane that has the largest distance to the nearest data sample of any class – a wider margin implies the lower generalization error of the classifier. To select the maximum margin hyperplane an optimization problem is formulated and solved for a given training dataset. The original SVM model – linear classifier – was developed by V.N. Vapnik, [13]. The hyperplane is calculated as a solution of a quadratic optimization problem. However, in many practical applications the datasets are not linearly separable in the data space. To determine the hyperplane the original space is mapped into a much higher-dimensional space using nonlinear kernel functions. The kernel function $K(x_i, x_j)$ defines the similarity between a given pair of objects. A large value of $K(x_i, x_j)$ indicates that x_i and x_j are similar and a small value indicates that they are dissimilar. Various kernel functions and described in literature [6].

Finally, the trained SVM can be used to classify the samples from a new dataset based only on the knowledge about their attributes.

4   FP-SVM System

The architecture of our system for malicious campaigns identification (FP-SVM) is presented in Fig. 1. It consists of three main modules: a database collecting malicious data, a frequent pattern mining module that implements the FP-growth algorithm for frequent patterns discovery and a data classification module that uses the SVM method to classify malicious datasets containing URLs as related or not to a campaign. To produce the SVM classifier a training dataset collecting malicious data (precisely URLs) that have already been classified into campaigns is required. Unfortunately, access to databases collecting malicious data divided into campaigns (if they exist) is restricted. Therefore, the only sensible solution is to define some patterns of URLs related with campaigns based on analysis of existing databases collecting malicious data, and generate a set of samples containing these patterns. It is assumed that all these samples are related with campaigns, and can be used to train the SVM classifier. In our FP-SVM system the FP-growth algorithm is applied to produce the training dataset. A fixed number N of URLs are selected from the database, and the

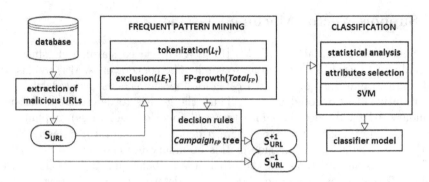

Fig. 1. The architecture of the FP-SVM system

input dataset $S_{URL} = \{URL_1, URL_2, \ldots, URL_N\}$ for frequent pattern mining is created. Each sample from S_{URL} is tokenized. The simple heuristic rules to break up a given URL (stream of text) into shorter strings described in literature [2,4] are adopted. Each sample is cut by a specific characters that are typical for URLs (1), i.e., "/", ".", "?", "#", etc. As a final result of this operation we obtain a set of tokens L_T. This set of tokens becomes input for further processing such as parsing and typical subsequence mining. The goal is to reduce the size of the dataset consisting of extracted tokens and finally speed up the FP-tree generation. The typical URL's attributes which do not carry valuable information, such as the schemes: "http", "https", domain name parts "www", "org", "com", "waw", etc. and extensions: "exe", "php", "html", "xhtml" are excluded from the set L_T. Once the final set of tokens LE_T is built, the FP-growth algorithm is employed to discover frequent tokens and the FP-tree structure $Total_{FP}$ storing quantitative information about frequent tokens from LE_T is constructed. In this tree each node (besides root) represents an extracted token that is shared by all subtrees consisting of itself and all the nodes beneath it. Each path in the tree shows a set of tokens that co-occur in URLs. Thus two URLs that contain several identical frequent tokens and differ in several infrequent tokens share a common path. The root is the node that has no superior and separates all disjoint subtrees. The $Total_{FP}$ tree structure is analysed. Simple decision rules are used for data processing. These rules define the characteristics of each URL that is suspected to belong to any campaign. The final FP-tree structure $Campaign_{FP}$ formed by URLs with these characteristics is created. $S_{URL}^{+1} and S_{URL}^{-1}$. Both these datasets form a training set of samples that is used to produce the SVM classifier model. The relevant attributes used for URL classification are selected. The commonly used attributes assigned to URLs are: date, time, address (IP, ASN), length of address, domain name, length of domain name, number of subdomains, path name, length of path name, number of subpaths, length of query, number of queries, country code, confidence of code. The selection of adequate attributes is a key feature that guarantees the effective and efficient classification. The

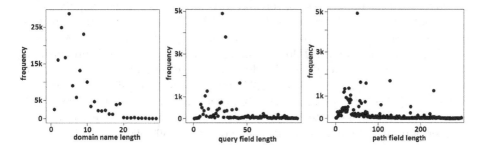

Fig. 2. Distribution of URLs; various attributes

preliminary statistical analysis on the malicious dataset is often performed to select the set of these attributes. The results of such analysis performed on the example set of malicious data is presented in Fig. 2. This figure shows the frequency of occurrence of respectively, lengths of domain name, query and path. It can be seen that in case of two attributes, i.e., length of domain name and length of path the distribution is similar After preliminary analysis of the available data the following attributes have been selected in the FP-SVM system: date, time, address (IP, ASN), length of address, domain name, number of subdomains, path name, number of queries, country code, confidence of code. Next step of the SVM algorithm is to learn the decision boundary. Four variants of the SVM classifier with linear and nonlinear kernels are implemented in our FP-SVM system. The following nonlinear kernels are provided: polynomial function: $K(x_i, x_j) = (\gamma x_i^T x_j + r)^d$, radial basis function: $K(x_i, x_j) = exp(\gamma \parallel x_i x_j \parallel^2)$, sigmoid function: $K(x_i, x_j) = tanh(\gamma x_i^T x_j + r)$, where $\gamma > 0$, r, and d denote kernel parameters. Finally, the trained SVM classifier can be employed by malware detection systems to classify malicious URLs taken directly from the Internet as related or unrelated to known malicious campaigns. Note that suspicious, unverified URLs can also be analyzed – if they are found likely to be part of a campaign, their malicious status is confirmed. It is obvious that the classification model has to be continuously updated. The data about new Internet security threats should be used in training process.

5   Performance Evaluation

5.1   The N6 Platform

The FP-SVM system was used to classify heterogeneous data from a real malware URL database of the n6 platform to identify malicious campaigns. The n6 platform [11] developed at NASK (Research and Academic Computer Network) is used to monitor computer networks, store and analyse data about incidents, threats, etc. The n6 database collects data taken from various sources, including security organizations, software providers, independent experts, etc., and monitoring systems serviced by CERT Polska. The datasets contain URLs of malicious websites, addresses of infected machines, open DNS resolvers, etc. Most of the data is updated daily. Information about malicious sources is provided by the platform as URL's, domain, IP addresses, names of malware, etc. We have performed a preliminary analysis of 27 700 560 URLs collected in year 2013 and stored in the n6 database. Figure 3 depicts the amount of malicious URLs observed per day (in a selected month) and per month. As we can see, the average number of observed URLs is about 80 000 daily or 2 300 000 monthly.

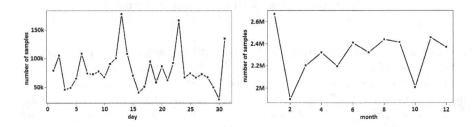

Fig. 3. Malicious URLs detected during a day and during a month

5.2 Case Study Results

The aim of the experiments was to validate the FP-SVM system on the n6 dataset. First, N malicious URLs were selected from the n6 malware database. They formed the S_{URL} training set. Next, frequent pattern analysis was applied and the $Total_{FP}$ tree with nodes representing tokens extracted from URLs from the S_{URL} dataset was constructed. The following rule was used to extract the subtree $Campaign_{FP}$ from the original $Total_{FP}$ tree. We assumed that all URLs containing m common tokens in a sequence were suspected to be related with the same campaign. Hence, short branches with less than m nodes were excluded from the $Total_{FP}$ tree. The $Total_{FP}$ tree and the approach to the $Campaign_{FP}$ tree generation are presented in Fig. 4. Table 1 presents the results of the application of the FP-growth algorithm and our decision rule with $m = 4$ to the S_{URL} dataset. It contains the number of detected malware campaigns and average number of URL's tokens related with one campaign.

$S_{URL}^{+1} and S_{URL}^{-1}$ produce the SVM classifier model. Various sizes of S_{URL} were considered, i.e., $N = n \cdot 100000$, $n = 1, 2, 3, 4, 5$. The decision borders for classification were calculated and validated for each size of S_{URL}. Finally, the trained SVM classifier was applied to malware campaign detection. It was used to classify the dataset consisting of 80 000 malicious URLs (unrelated with URLs from training dataset) into two categories: +1 – URLs related with any campaign, -1 – URLs unrelated with any campaign. Then, the quality of the classification was assessed. The following commonly used criteria were considered: classification accuracy (CA) – ratio of number of correctly identified URLs to the size of the

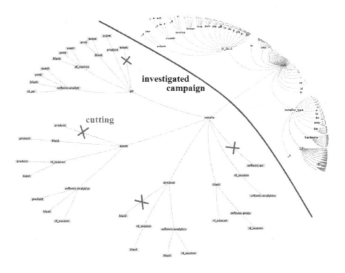

Fig. 4. Frequent patterns tree and a campaign extraction

Table 1. Identification of malicious campaigns; the S_{URL} dataset

Number of URLs (number of IP)	500 000 (108 745)
Number of unique URLs (number of IP)	69 906 (5 321)
Number of detected campaigns	72
Average number of related attributes	84 per campaign

dataset, sensitivity – the proportion of positives that are correctly identified as such, specificity – the proportion of negatives that are correctly identified as such, accuracy of a test. The accuracy of the test is measured by AUC (Area Under ROC Curve – the measure that shows how well the test separates the URLs being tested) and another popular measure – F-measure. The value of precision shows how close the separated URLs are to each other. The values of all mentioned criteria obtained for various variants of SVM classifier (providing linear and nonlinear kernel functions) and a training dataset S_{URL} consisting of 250 000 malware URLs are collected in Table 2. Figure 5 shows the accuracy of the classification for various sizes of the training dataset.   In general, the

Table 2. Evaluation of SVM classification; $N = 250000$ URLs

	Radial	Sigmoid	Polynomial	Linear
CA	0.7035	0.5571	0.5000	0.3495
Sensitivity	0.8750	0.7083	0.4375	0.6042
Specificity	0.8202	0.6685	0.8258	0.6011
AUC	0.9250	0.8653	0.8367	0.7823
F-measure	0.6885	0.4823	0.4200	0.3919
Precision	0.5676	0.3656	0.4038	0.2900

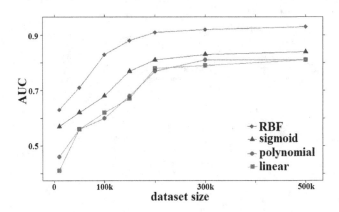

Fig. 5. Area Under ROC Curve values for various sizes of the training dataset S_{URL}

results presented in Fig. 5 and Table 2 confirm that the accuracy of classification strongly depends on the size and quality of a training dataset. Moreover, it is very important to choose the adequate kernel function. The achieved classification accuracy ranged from about 35% to 70%, accuracy of the test from 78% to 93%, sensitivity from 44% to 88%, and specificity from 60% to 82%. Explicitly, the worst results were obtained for the SVM variant implementing a linear kernel function. The best results – the best values of all criteria – were obtained when employing the radial basis kernel function. It is worth to mention that due to our assumption that the same obfuscation mechanisms is used to the malicious URLs generation, and usually intruders keep some parts of the URL static we are able to identify the URLs related to a given campaign. In our approach we use the most frequent tokens (the static parts of malicious URLs) in the classification process.

6   Summary and Conclusion

Data mining methods hold a great potential toward detection of malicious software. To identify malicious web campaigns we have adopted a technique incorporating a combination of frequent pattern mining and supervised learning methods. In general, the presented results of experiments confirm that our classification system employing the FP-growth algorithm and the SVM method gives satisfactory results. This technique can be successfully used to analyse a huge amount of dynamic, heterogenous, unstructured and imbalanced network data. However, accuracy of this approach depends heavily on the selection and calibration of parameters, especially the choice of a kernel function and of the rules for selection of URLs for the training dataset. As a final conclusion we can say that it can be expected that our FP-SVM system can be successfully implemented in intrusion detection systems as a malicious campaign sensor. In our future work we plan to extend our system with the tool mapping a detected malicious software to a given campaign.

References

1. Argawal, C., Li, Y., Wang, J.: Frequent pattern mining with uncertain data. In: Proc. of 15th Inter. Conf. on Knowledge Discovery and Data Mining (ACM SIGKDD), pp. 29–38 (2009)
2. Calais, P., Pires, D., Neto, D., Meira, W., Hoepers, C., Steding-Jessen, K.: A campaign-based characterization of spamming strategies. In: CEAS 2008, pp. 1–6 (2008)
3. Gandotra, E.: Malware analysis and classification: A survey. Journal of Information Security **5**, 56–64 (2014)
4. Gao, H., Hu, J., Wilson, C., Li, Z., Chen, Y., Zhao, B.: Detecting and characterizing social spam campaigns. In: Proc. of the 10th ACM SIGCOMM Conference on Internet Measurement, pp. 35–47 (2010)
5. Han, Y., Pei, Y., Yin, Y.: Mining frequent patterns without candidate generation. In: Proc. of SIGMOD, pp. 1–12 (2000)
6. Jebara, T.: Multi-task feature and kernel selection for svms. In: Proc. of Inter. Conf. on Machine Learning, pp. 55–63 (2004)
7. Radu, V.: Application. In: Radu, V. (ed.) Stochastic Modeling of Thermal Fatigue Crack Growth. ACM, vol. 1, pp. 63–70. Springer, Heidelberg (2015)
8. Kozakiewicz, A., Felkner, A., Kijewski, P., Kruk, T.: Application of bioinformatics methods to recognition of network threats. JTIT, pp. 23–27 (2007)
9. Kruczkowski, M., Niewiadomska-Szynkiewicz, E.: Support vector machine for malware analysis and classification. In: Proc. of IEEE/WIC/ACM Inter. Conf. on Web Intelligence, pp. 1–6 (2014)
10. Lasota, K., Kozakiewicz, A.: Analysis of the similarities in malicious dns domain names. In: Lee, C., Seigneur, J.-M., Park, J.J., Wagner, R.R. (eds.) STA 2011 Workshops. CCIS, vol. 187, pp. 1–6. Springer, Heidelberg (2011)
11. NASK: n6 platform (2014). http://www.cert.pl/news/tag/n6
12. de Oliveira, I.L., Grégio, A.R.A., Cansian, A.M.: A malware detection system inspired on the human immune system. In: Murgante, B., Gervasi, O., Misra, S., Nedjah, N., Rocha, A.M.A.C., Taniar, D., Apduhan, B.O. (eds.) ICCSA 2012, Part IV. LNCS, vol. 7336, pp. 286–301. Springer, Heidelberg (2012)
13. Vapnik, V.: The Nature of Statistical Learning Theory. Springer, New York (1995)

Simulation of Human Behavior in Different Densities as a Part of Crowd Control Systems

Michał Kapałka[✉]

Military University of Technology, Warsaw, Poland
mkapalka@wat.edu.pl

Abstract. This paper presents a novel approach to the crowd behavior modeling and simulation in different densities used as a decision support tool in crowd control systems. Non-invasive examination of movement and behavior of people in different densities and situations may lead to the detection and prevention of crisis situations involving crowds especially at high densities. In order to get approximated model of human acting we use microscopic approach for pedestrian representation. In model each person in crowd is described as an agent with individual attributes, behavior rules and own proxemics space. Representation of environment uses a combination of two approaches: coarse network model and fine network model simultaneously. The model and simulator can be used as a part of systems for supporting and controlling pedestrian movement, evacuations, demonstrations or situations where high congestion of people can be critical.

Keywords: Crowd modeling · Crowd control system · Crowd simulation · Pedestrian movement model

1 Introduction

Public spaces should be safe for moving pedestrians. Typically, security is provided by the physical objects in accordance with accepted safety standards. As shown by numerous disasters involving crowds [1], sometimes it is not enough to prevent injury or even loss of life. Awareness of the threat posed by a large gathering of people in a small area makes it necessary to look for better solutions to this problem. One of them is the use of computer simulation of pedestrian movement in a virtual environment to analyze possible threats. On the basis of the work of many crowd researchers [2,3,4,5] of the crowd there are many known phenomena of movement of the crowd such as: lane formation, stop-and-go waves or oscillations at bottlenecks. Many of the characteristics associated with the movement of the crowd has been well described and have become a tool for testing the adequacy of the new tools. In the field of crowd modeling there are many sophisticated models [6] that can reproduce real human behavior in many scenarios. The next step is to use these models in real crowd control systems to have a tool for non-invasive testing physical or non-physical methods for making pedestrian movement safer. In this paper model for decision support in area of managing crowd movement is presented.

© Springer International Publishing Switzerland 2015
N.T. Nguyen et al. (Eds.): ACIIDS 2015, Part II, LNAI 9012, pp. 202–211, 2015.
DOI: 10.1007/978-3-319-15705-4_20

2   Crowd Behavior Model

The main purpose of the model is to reproduce real human behavior during movement in different densities. Pedestrians act different in different congestion of people around them. With that information it is necessary to take into account the fact that pedestrians change behavior with density changes. In our model the space is based on hybrid representation using fine and coarse network models simultaneously. Each pedestrian is represented by an agent with its individual characteristics. For different behavior simulation rule based system was used. To obtain behavior changes based on actual density around pedestrians we take into consideration concept of personal distances described by E. Hall [7].

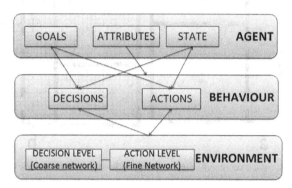

Fig. 1. Base elements of crowd behavior model

2.1   Space Representation

We use hybrid space representation to reproduce natural human perception of surroundings. People, when they want to go somewhere they do not see all the way to the target as a single trajectory. At a higher level, they see important points: front door, staircase, bus stop, pedestrian crossing. That level in our model is represented by a coarse network E^C and will be named as **decision level**,

$$E^C = \langle G, f^C \rangle \tag{1}$$

where: $G = \langle W, U \rangle$ is a graph, W is a set of nodes where each represents a logical space such as single room (e.g. corridor, staircase), U is a set of edges where each represents physical possibility to move between nodes and f^C is a set of functions related to graph G. In our model we use some modification of a classical graph representation of a space. We split the space near each exit from a rooms space and set it as a single node. We do that to have a way to describe each exit with function of pass cost. In real situations people often chose less crowded exits to walk.

Second level in our space representation is a **action level**. On that level we describe space as a fine network E^F,

$$E^F = \langle X, Y, C, C^S, C^D, f^F \rangle \tag{2}$$

Where: $X, Y \in N_+$ is space size as number of cells horizontal and vertical, $C = \{\langle x, y \rangle \in N_+^2 : x \leq X, y \leq Y\}$ is a grid of square cells, $C^S \in 2^N$ is a set of possible cell states (in basic approach we use three states: unavailable space, free space for pedestrian, space occupied by pedestrian), $C^D = \{\langle x, y \rangle \in \{-1,0,1\}^2 \setminus \langle 0,0 \rangle\}$ is a set of possible moving directions for pedestrians and f^F is a set of functions related to elements of E^F.

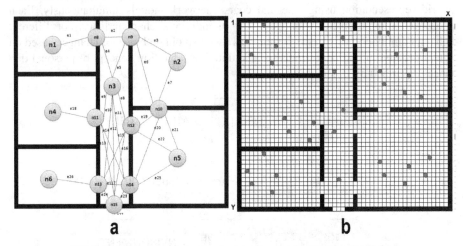

Fig. 2. Space representation on decision level (a) and action level (b)

Hybrid space representation E in our model use described 2 levels: decision and action simultaneously and is described as:

$$E = \langle E^C, E^F, E^O, f^E \rangle \qquad (3)$$

where: E^O is a set of phenomena and interactive objects and f^E is a set of functions related to elements E^C, E^F, E^O.

2.2   Pedestrian Model

In this model each pedestrian is an individual and heterogeneous agent. In simulation the model from the viewpoint of pedestrians can be treated as an multi-agent system where agents (pedestrians) interact with each other while moving to desired locations. Model of single pedestrian can be described as

$$P = \langle P^A, P^S, P^G, B \rangle \qquad (4)$$

where: P^A is a list of agent attributes, P^S is a state vector of agent, P^G is ordered list of agent goals and B is model of agent behavior. In basic approach the main attributes of agent are: energy, preferred velocity, preferred personal distances, interaction time, knowledge of the environment – these attributes have impact on how, when and where the agent will move in the space. In each moment P^S describes current state of an agent: position, velocity and energy. The goals of each agent can be different, the ordered list P^G can be changed during simulation many times according to situation around agent.

Using approach with multi-agent systems to simulate crowd movement is well known. One of the first who proposed this conception was Raynolds[9] and later this method became one of the fundaments of crowd models [11,13,16,17]. In our model we use this approach with extended pedestrian model to reproduce individual and different behavior and role in crowd. One of the our assumptions that can be achieved using agent system is to reproduce sophisticated pedestrian phenomena (e. g. supervised children and family evacuation, leadership in crowd or riot movement).

2.3   Pedestrian Behavior Model

Agents acts individually, but their actions depends where other agents are and what they do. The basic algorithm of agent action is shown in Fig. 3

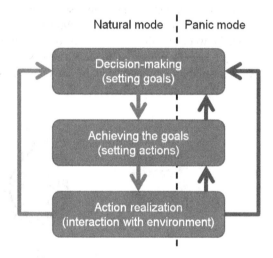

Fig. 3. Basic algorithm of agent behavior

Agent can act in two modes. The normal mode is chosen when agent is safe. That means there is no risk for him to lose extra energy, health or life. In this mode his behavior is based on basic goals such as: go to specific place or follow a leader. In panic mode agent can act differently. Panic is a situation when agent are in stress caused by a specific threat (e.g. fire, smoke, running agents, crowded place). In this mode the basic goal of agent can be changed to respect "instinct" of escape from threat.

Agent behavior in model is rule based. For each simulation we can change the rule sets to achieve desired pedestrian acting. The basic rules are focused on reproduction of real human movement including: obstacle avoiding, choosing "best" routes, avoiding other agents, forming lanes, changing velocity in specific situations. Some of the rules are based on social forces described by D. Helbing [8]. The open character of creating rules for agents gives the possibility for creating agents with specific roles in crowd movement (e.g. during simulation of evacuation from building in fire there is sometime a need to simulate and chose best routes for movement of firefighters who usually goes in the opposite direction then crowd). Another advantage of model is possibility to defining group behaviors. Using simple rule sets described by Reynolds [9] we can reproduce realistic group movement.

Behavior rules are dependent on congestion of agents. This is author interpretation of proxemics, a phenomena related with personal distances between pedestrians. The idea of personal distances are known to crowd researchers where one of the firsts who described it in crowd modeling were Fruin[3], Helbing[8] and Raynolds[9]. As shown in some later works [14,15] still this idea can be used with new interpretation. In our model we use proxemic spaces to reproduce phenomena of unequal distribution of pedestrian density (e.g. on mass events we can notice empty and overcrowded places) by possibility of defining a sets of different personal spaces to each pedestrians and change them in different scenarios. We use cellular automata to check and change defined in model proxemic spaces. Changing proxemic space alters behavior rules of agent. The personal distances are strictly related with fundamental diagrams which represents the relationship between density and velocity of pedestrians. These relationship is one of the major rule in presented model. The basic proxemics spaces (with corresponding cells available for pedestrian in set C) defined by author is showed on Fig. 4.

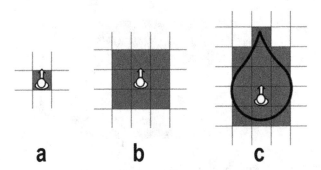

a   **b**   **c**

Fig. 4. Basic proxemics spaces for pedestrian: intimate distance (a), social distance (b), public distance (c)

Pedestrian behavior model can be described as

$$B = \langle B^G, B^A, B^P, B^D, B^S, B^R, B^{GR}, B^{IR}, f^B \rangle \tag{5}$$

where: B^G is a set of defined goals, B^A is a set of defined actions, B^P is a set of defined personal distances, B^D is a set of defined density levels, B^S is a set of stress levels, B^R is a set of defined behavior rules, B^{GR} is a set of global behavior rules, B^{IR} is a set of individual behavior rules and f^B is a set of functions related to elements $B^G, B^A, B^P, B^D, B^S, B^R, B^{GR}, B^{IR}$.

2.4   Methods and Algorithms

In presented solution we use large set of methods and algorithms to be able to reproduce human behavior. Pedestrian behavior and movement on decision and action level are achieved using additionally:

- graph algorithms [10] for finding shortest paths with time-dependent cost function for nodes
- potential fields method for finding best trajectories between nodes
- attraction masks for computing actual moving direction of agents
- calibration phase to reproduce dependencies from fundamental diagrams
- discreet simulation method in two modes: with event and with steps

One of the most frequently used goals by each agent is reaching specific nodes. In real situations people knows the best route for them. Typically it will be the shortest one, but not always. Exception of this rule can be unfamiliarity of the environment. In that situation sometimes people takes a longer known to them way. In model each agent has own knowledge of network and nodes on decision level. Choosing a way to desired node is based on shortest path algorithm with functions: $f_{UC}^E : U \to N$ determining static cost moving between nodes, $f_{NMC}^E : W, t \to N$ determining dynamic, changing in time cost of crossing single node (this function is strictly related with pedestrian congestion in node).

On action level pedestrian move from cell to cell (between two adjacent nodes) is determined by using potential fields method and attraction masks which represent dynamic changes in agent surrounding. In potential fields method each cell has computed a potential which reflects real distance from target. The trajectory is determined in two steps: first a desired direction probability mask is computed by seeking free cell with smaller potential, the second step is to calculate real move direction.

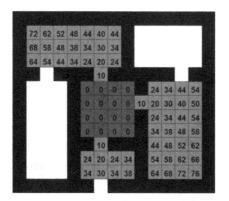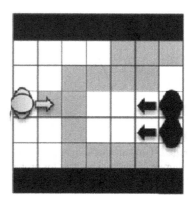

Fig. 5. Conception of potential fields method and changing cells attraction

The second step is needed to avoid collision between pedestrians. In real situations if pedestrians are walking against themselves one of them or both will change direction before collision. In model we applied a attraction masks for each agent. These masks change probability of choosing cells surrounding other agent.

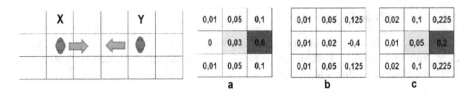

Fig. 6. Probability mask for determining best direction: individual probability for agent X (a), changing cell attraction mask for agent X from agent Y (b), result direction probability mask for agent X (c)

3   Experiments and Possible Applications

Stage preceding the execution of simulation experiments and collecting the desired output characteristics is a calibration step. This step is necessary to obtain the desired adequacy of the model. The calibration process involves appropriate selection of input variables such as .: average speed of movement of persons, the average space occupied by a person, the strength of interpersonal interaction, response times, or the probability of interaction between individuals competing for space. The output characteristics from one of calibration processes are presented on Fig. 7.

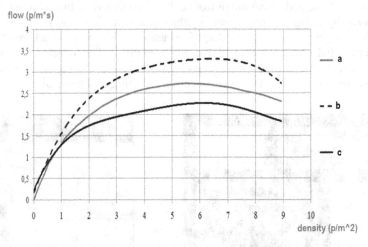

Fig. 7. Fundamental diagram and output characteristics from calibration process: fundamental diagram [4] (a), output data for probability of pedestrians interaction set to 0.3 (b), output data for probability of pedestrians interaction set to 0.4 (c)

Presented model can be successfully used to simulate the evacuation process. To simulate evacuation process one base rule is create: *reach nearest safe area*. For experiment basic input data included: average velocity 1,2 m/s, space occupied by each pedestrian 0,25m x 0,25m, reaction time 10s. In experiment we use university building with 3 lecture rooms (areas 1-3), one corridor (intersection area) and one exit to safe area. The simulation results of evacuation experiment are shown in Fig. 8.

Successful experiments of evacuation process was a first step to start using model as a support tool in developing and testing conception of dynamic signage system for evacuation process in complex buildings[12, 13].

In addition to providing numerical output data from experiments, implemented simulator is capable to give real time preview of simulation process. This feature is important for training workshops where experts can see moving crowd on a screen and interactively react in simulated emergencies. 2D preview of a simulation evacuation process from 2 floors building is presented on Fig. 9.

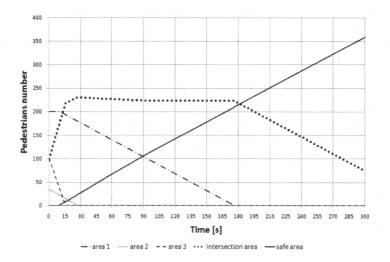

Fig. 8. Pedestrians in different areas during evacuation process, experiment results

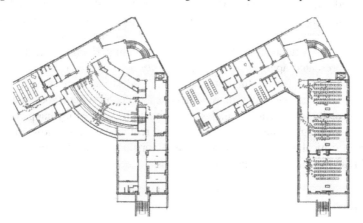

Fig. 9. 2D visualization of simulation process, 2 floors of a building

4   Crowd Control Systems

Systems for managing pedestrian movement should be well prepared. Today all around the world we have directives and rules how make public and gathering places for large number of peoples to be safe. The main problem in most of this situations is that all these methods are static and not prepared for unusual scenario. We can find elements of today's crowd control system at mass events, in every buildings with evacuation plans with evacuation route signs. But signs and crowd control barriers are not enough. Especially when situation changes to different than expected. For example a good static evacuation plans can be deadly when part of a route are destroyed or dangerous to walk in.

The next step in making pedestrian movement safer is to support static methods with analytical systems which are able in reasonable time find a threat and help to prevent it. In real situations when in one area gathers a large number of pedestrians the dynamic crowd control is the only way to prevent deaths (e.g. crowd control system at Jamarat Bridge [11]).

Proper system for crowd control should be prepared to work in several areas:

- monitoring and detection – this is very important thing to know positions and numbers of pedestrians in each areas,
- adequate model of the environment and pedestrian behavior
- cooperation with other supporting models
- effective and dynamic information system
- trained staff to support crowd controlling

Presented in this paper crowd behavior model can be an important part of crowd control system. Microscopic representation of pedestrians, open and rule based behavior modeling and high resolution of space representation are advantages that allow to use model to simulate pedestrian movement in different densities and scenarios (e.g. evacuations, supervised evacuations in schools, tunneling movement during mass events, controlling demonstrations and even preventing riots). Using hybrid space model with presented approaches gives possibility to simulate crowd situations from beginning when density is still low and single pedestrians can cause a dangerous event. This model can be a real improvement as a decision support tool in crowd controlling especially in detection of undesirable situations before they happen (e.g. monitoring and making ahead simulations during real events can show that without intervention critical situation may occur) and verification of efficacy and safety of potential reacting methods during real situations.

References

1. List of human stampedes. http://en.wikipedia.org/wiki/List_of_human_stampedes
2. Le Bon, G.: The Crowd, A study of the popular mind. Kitchener (2001)
3. Fruin, J.J.: Pedestrian Planning and Design, Metropolitan Association of Urban Designers and Environmental Planners, New York (1971)

4. Predtetschenski, W.M., Milinski, A.I.: Personenströme in Gebäuden - Berechnungsmethoden für die Projektierung, Verlaggesellschaft Rudolf Müller Köln-Braunsfeld (1971)
5. Zhang, J., Seyfried, A.: Empricial characteristics of different types of pedestrian streams, Procedia Engineering (2012)
6. Kuligowski, E., Peacock, R.: A review of building evacuation models, Washington (2005)
7. Hall, E.T: The hidden dimension, New York (1966)
8. Helbing, D.: Social force model for pedestrian dynamics, Physical Review E, May (1995)
9. Reynolds, C.: Flocks, Herds, and Schools: A distributed behavioral model. Proceedings of SIGGRAPH 1987 in Computer Graphics **21**(4), July 1987
10. Dijkstra's algorithm. http://en.wikipedia.org/wiki/Dijkstra%27s_algorithm
11. Modelling the Jamarat bridge. http://www.gkstill.com/CV/Projects/Jamarat.html
12. Cisek, M., Kapałka, M.: The use of fine – coarse network model for simulating building evacuation with information system. In: Fifth International Conference on Pedestrian and Evacuation Dynamics, Gaithersburg MD (2010)
13. Cisek, M., Kapałka, M.: The Evaluation Indicator Problem in Determining Optimal Evacuation Schedule for Selected Adverse Event Scenarios, International Scientific and Technical Conference "Emergency Evacuation of People from Buildings" EMEVAC, Warszawa (2011)
14. Wąs, J., Gudowski, B., Matuszyk, P.J.: Social Distances Model of Pedestrian Dynamics. In: El Yacoubi, S., Chopard, B., Bandini, S. (eds.) ACRI 2006. LNCS, vol. 4173, pp. 492–501. Springer, Heidelberg (2006)
15. Ezaki, T., Yanagisawa, D., Ohtsuka, K., Nishinari, K.: Simulation of space acquisition process of pedestrians using Proxemic Floor Field Model. Physica A (2012)
16. Heliovaara, S., Korhonen, T., Hostikka, S., Ehtamo, H.: Counterflow model for agent-based simulation of crowd dynamics Build. Environ. **48**(0) (2012)
17. Wąs, J.: Robert Lubaś, Towards realistic and effective Agent-based models of crowd dynamics. Neurocomputing **146** (2014)

The Qualitative and Quantitative Support Method for Capability Based Planning of Armed Forces Development

Andrzej Najgebauer, Ryszard Antkiewicz, Mariusz Chmielewski, Michał Dyk,
Rafał Kasprzyk, Dariusz Pierzchała, Jarosław Rulka, and Zbigniew Tarapata[✉]

Cybernetics Faculty, Military University of Technology,
Gen. Sylwestra Kaliskiego Str. 2, 00-908 Warsaw, Poland
{anajgebauer,rantkiewicz,mchmielewski,mdyk,rkasprzyk,
dpierzchala,jrulka,ztarapata}@wat.edu.pl

Abstract. The paper introduces and presents the model and method of Capability Based Planning in the area of the Armed Forces Development. The model of development contains: the mathematical description of different capabilities' assessment , the problem's formulation of assessment of required, existing and lacking capabilities. The method of allocation of the capabilities to response on the threats scenarios is the next step explained in the paper. The verification method of allocated capabilities and their synergy - as simulations of possible conflicts of the fixed sides equipped with the capabilities - is proposed. Finally, the set of analytical and simulation tools for CBP is discussed.

Keywords: Capability based planning · Allocation of capability · Conflict simulation

1 Introduction

In terms of systemic change in the country, including changes in constructing the State budget on the Armed Forces (AF), the Ministry of Defence (MOD) faces new challenges. Both NATO's transformation as well as AF cause the need for a new approach to the operation of AF [6, 7]. One of the important directions of the transformation is the issue concerning development of the capability of the Armed Forces and the identification of operational needs [2, 4, 7]. The process is known as Capability Based Planning (CBP). As a key step of CBP, a division of AF into functional systems might be performed: direct fighting system, command support system, system of identification, system of troops' protection, logistics system. The paper covers the selected elements of the quantitative methods designed for: the evaluation of capabilities required, the assessment of the existing and the identification of capabilities' gaps for the threats scenarios highlighted for Country. Moreover, the most important issue considered in the paper is measuring the capabilities. The two phases of CBP - the planning and programming processes of development of AF - can be supported by both analytical methods of operations research and simulation methods. The paper

© Springer International Publishing Switzerland 2015
N.T. Nguyen et al. (Eds.): ACIIDS 2015, Part II, LNAI 9012, pp. 212–223, 2015.
DOI: 10.1007/978-3-319-15705-4_21

introduces the originally implemented methods combining process modelling, simulation process controlling, task' scenarios planning and capabilities (firing, manoeuvre and movement) assessment. The main contribution of our approach, in comparison to the others [9, 10], there are original models and methods of AF development. We propose the new model for the optimization (allocation) of CBP problem (during a planning phase). Moreover a discrete event simulator is used for the identification of operational capabilities needs (during a programming phase).

2 Qualitative and Quantitative Estimation of Capabilities for AF

2.1 The General Approach

An operational capability is an ability to perform the certain task by AF to achieve the expected military and non-military effects. Let us consider several groups of capabilities: to command -and reconnaissance, to direct firing (destruction), survivability and protection of troops, to logistic support of troops' operations. The process of defense capabilities assessment requires several steps, typically conducted on the strategic level of command. The process starts from inferring and reviewing of both defense priorities and possible threats (based on the current geo-political and military situation in the region) as well as possible allied responsibilities. The prospective threats arise not only from kinetic forms of operations but they involve also the threats from a cyberspace. Such distinction comes not only from meaning of threat factors but also from the operation area. Analyzing the threats to a country and possible operations of AF in the context, we have considered the following types of scenarios: intensive conflict (kinetic), asymmetric conflict during external mission, cyber war, natural disaster in the country and terrorist attack However, the list of possible scenarios can be widened and must not be limited to the above. The identified threats are a source for events, goals and mission list, which form reaction scenarios for the Armed Forces. The most crucial part of such analysis is a mission description containing a sketch of developed solution to overcome that particular threat. Each mission should be described by a network of tasks in order to elaborate and specify the concrete solution. The network is, in fact, a form of a schedule which formulates activities and a timeline of a particular mission. Each stage of this process involves utilization of operational concepts and capability partitions. This process can be also perceived as a form of requirements analysis for problem involving country security while coping with identified threats.

The main steps to define capability requirements are as follows [9]:

1. Description of planning scenario in aspects of: common characteristics of threats; threats being identified – a detailed description of a selected threat (e.g. conflict of high intensity);
2. Description of planning situation: operation area; Reaction Scenario for the Armed Forces: threat estimation, mission; mission to task decomposition;
3. Definition of required capabilities;
4. Comparison of required and possessed capabilities and then assessment of lacking capabilities.

The computer system we proposed can support all the mentioned steps of the process of defining capabilities requirements. At first, we decompose a mission into a network of tasks. In the next step, each task is described by dedicated card. These cards are fulfilled by analysts while the parameters ought to be determined in order to planned use the reference modules (military units) to complete the mission. In the phase we assume that the other capabilities of the reference modules are unrestricted. The assumption allows determining the required capabilities in a way of resolving the reference modules' allocation to military tasks problem (section 2.3). On the basis of currently possessed capabilities of the real units we can determine the deficiencies. The next phase – programming - there is determining of the needs for different capabilities achievement by using the simulation modelling and experimentation. The computer simulation method can be realized in a sense of assessment of options for obtaining lacking capabilities:

- for a fixed option in order to obtain operational capabilities (with respect to all defined planning situations) the values of the given criteria are calculated: loss of capability, cost of losing a capability, level of completeness of a mission, probability of a mission success;
- by calculating the values of the given criteria for different options in order to obtain operational capabilities it is possible to choose the best one.

2.2   Capabilities Allocation Problem

In order to formulate the problem of capabilities allocation the mathematical description of the network of tasks is proposed. Military tasks network is the 2 - tuple:

$$S_p = \langle G, TT \rangle \tag{1}$$

where:

$$G = (NTT, U) \tag{2}$$

G - directed graph without cycles and loops;

- $NTT = \{1, .., N_{TT}\}$ - set of military task numbers;

- $U \subset NTT \times NTT$ - set of arcs which represent precedence relations among military tasks - a task must not start before all its predecessors are finished;

- $TT = \{type_{tt}(\bullet), t(\bullet), \tau(\bullet), Env_{tt}(\bullet), OPFOR_{tt}(\bullet), EF_{tt}(\bullet)\}$ - set of functions defined on the nodes of graph G. For $k \in NTT$ we have:

  o $type_{tt}(k)$ - type of task k; $type_{tt}(k) \in \{attack, counterattack, delay, ...\}$;

  o $t(k)$ - start time of task k, $t(k) = null$ means start time is not determined for k-th task;

  o $\tau(k)$ - duration of k-th task, $\tau(k) = null$ means duration is not determined for k-th task;

  o $Env_{tt}(k) = (Terrain(k), Weather(k), Season(k), Climate(k))$ - functions describing the environmental conditions of k-th task's realization:

    ▪ $Terrain(k)$ - type of terrain;

    ▪ $Weather(k)$ - type of weather during task's realization;

- ▪ *Season(k)* - season of year;
- ▪ *Climate(k)* - type of climate;
- ○ $OPFOR_{tt}(k)$ - description of opposing forces which would fight against own forces during *k-th* task;
- ○ $EF_{tt}(k) = \left(effect_{tt}(k, j) \right)_{j=1,...,N_{EF(k)}}$, $effect_{tt}(k, j)$ - expected value of *j-th* effect of *k-th* task's realization;

The own and opposite forces are defined using, so called, reference modules. The reference modules are prepared by military analytics taking into account prospective conditions of a battlefield. In essence, a reference module is related to a battalion. The mathematical model of a reference module is described as follows:

- $RMd_{A(B)}(i) = \left(Zd_{RMd}^{A(B)}(i) \right)_{i=1,...,N_{RMd}^{A(B)}}$, where $N_{RMd}^{A(B)}$ - number of types of reference modules for side A – the own forces (for side B – the opposite forces);

- $C_{RMd}^{A(B)}(i) = \left(id_{CRMd}(i, j), v(i, j) \right)_{j=1,...,N_{CRMd}^{A(B)}(i)}$, where:

  - ○ $id_{CRMd}^{A(B)}(i, j)$ - id of *j-th* capability of *i-th* reference module;

  - ○ $v^{A(B)}(i, j)$ - quantity of *j-th* capability of *i-th* reference module.

Letus consider the following decision variables on CBP:

$$\left(\overline{X}, \overline{T} \right) \tag{3}$$

where:

$$\overline{X} = \left[x_{ik} \right]_{\substack{i=1,...,N_{RMd}^{A} \\ k=1,...,N_{TT}}}, \ \overline{T} = \left[T_k \right]_{k=1,...,N_{TT}} \tag{4}$$

and:

- x_{ik} - number of *i-th* type modules of side A assigned to *k-th* task's realization;
- T_k - the start time of *k-th* task.

We define the following computable functions in order to formulate the optimization problem:

- $g_k(\overline{X}, \overline{T})$ - duration of *k-th* task. taking into account modules allocation given by $\overline{X}$. This function is important for tasks which duration is not determined in a scenario;

- $\overline{ef}_k(\overline{X}, \tau) = \left(ef(k, j, \overline{X}, \tau) \right)_{j=1,...,N_{EF}(k)}$, $ef(k, j, \overline{X}, \tau)$ - value of *j-th* effect of *k-th* task's realization. taking into account modules allocation given by $\overline{X}$ and duration of *k-th* task equals to τ ;

- $F_1(\overline{X}, \overline{T}) = \sum_{l=1}^{N_C} Loss_l(\overline{X}, \overline{T}) \cdot w_l$ - percent of capabilities lost during all operations,

where:

- ○ $Loss_l(\overline{X}, \overline{T})$ - percent of *l-th* capability loss;

- ○ w_l - weight of *l-th* capability and $\sum_{l=1}^{N_C} w_l = 1$;

- $F_2(\overline{X},\overline{T}) = \sum_{i=1}^{N_{RMd}^A} \sum_{k=1}^{N_{TT}} x_{ik}$ - number of allocated modules;

- $\overline{F_3}(\overline{X},\overline{T}) = \left(F_3(\overline{X},\overline{T},l)\right)_{l=1,...,N_C}$, where:

  o $F_3(\overline{X},\overline{T},l) = \sum_{k=1}^{N_{TT}} \sum_{i=1}^{N_{RMd}^A} x_{ik} \cdot \sum_{j=1}^{N_{CRMd}^A(i)} v(i,l) \cdot I\{l = id_{CRMd}(i,j)\}$ - quantity of l-th

    capability in all allocated modules;

- $F_4(\overline{X},\overline{T}) = \sum_{i=1}^{N_{RMd}^A} K_i \sum_{k=1}^{N_{TT}} x_{ik}$ - a cost of all allocated reference modules, where K_i is a

cost to acquire and maintain reference module of i-th type.

The problem of capability allocation is formulated as the following multi-objective optimization problem:

$$\min_{(X,T)} \left(F_1(\overline{X},\overline{T}), F_2(\overline{X},\overline{T}), \overline{F_3}(\overline{X},\overline{T}), F_4(\overline{X},\overline{T}) \right) \tag{5}$$

subject to the constraints:

1) $\forall k \in NTT_d \quad g_k(\overline{X},\overline{T}) = \tau(k)$;

2) $\forall k \in NTT_s \quad T_k = t(k)$;

3) $\overline{ef}_k\left(\overline{X}, g_k(\overline{X},\overline{T})\right) \ge EF_{tt}(k)$ for $k = 1,..,N_{TT}$;

4) $\forall r \in \Gamma^{-1}(k) \quad T_k \ge T_r + g_r(\overline{X},\overline{T})$ for $k = 1,..,N_{TT}$;

5) $\forall k \; \Gamma^{-1}(k) = \varnothing \quad T_k = t_0$;

6) $\forall k \; \Gamma(k) = \varnothing \quad T_k \le t_{end}$.

The problem (5) can be solved by a scalarization of multi-objective optimization problem or using method of compromise solutions [5].

Let $\overline{M} = (M_r)_{r=1,...,N_M}$ be the vector of missions defined for all scenarios and $C^R(M_r,l)$ means quantity of l-th capability required for r-th mission. Model (1) of tasks' network is defined for each mission. Let $(\overline{X}_r^*,\overline{T}_r^*)$ be a Pareto optimal solution of the problem (5) formulated for mission M_r. Therefore vector $\overline{F_3}(\overline{X}_r^*,\overline{T}_r^*) = \left(F_3(\overline{X}_r^*,\overline{T}_r^*,l)\right)_{l=1,...,N_c}$ gives us quantities of all capabilities required for realization of operation described by task's networks (1) for M_r. Then we can consider that $C^R(M_r,l) = F_3(\overline{X}^*,\overline{T}^*,l)$ for $r = 1,..,N_M$. In order to estimate capabilities of our armed forces required for realization of all type of missions we take the following formula:

$$C_{AF}^R(l) = \max_{r=1,...,N_M} C^R(M_r,l), \text{ for } l = 1,..N_c. \tag{6}$$

Let $C_{AF}^{P}(l)$ be quantity of *l-th* capability possessed now by our armed forces for $l = 1, .., N_C$. We can evaluate the degree of satisfaction of needs of capabilities using the formula:

$$Satis_{AF}(l) = \frac{C_{AF}^{P}(l)}{C_{AF}^{R}(l)}, \text{ for } l = 1, .., N_C .$$ (7)

The lack of capabilities can be evaluated using the formula:

$$Lac_{AF}(l) = \max\left\{0, 1 - Satis_{AF}(l)\right\} = \max\left\{0, 1 - \frac{C_{AF}^{P}(l)}{C_{AF}^{R}(l)}\right\}, \text{ for } l = 1, .., N_C .$$ (8)

2.3 The Simulation Support of Needs Identification for Operational Capabilities

A constructive computer simulation approach is relatively fast and sufficiently accurate both to (A) supporting an identification of needs and (B) an assessment of options for acquiring the operational capabilities. The procedure for applying the simulation method is shown on the following example. For each considered variant of obtaining operational capabilities relating to all defined planning situations, and using the simulation we can determine the values of the following criteria: loss of ability, the cost of loss of capability, the degree of realisation of the mission (task network), the probability of the mission.

In general, both tasks (A and B) might be resolved based upon of simulation models with different resolution (accuracy) and properties suitable to both continuous (eg. Lanchester's equations) and discrete models. This is a direct reason for the use of both the techniques of simulation – the simulation community uses for such a fusion the term "hybrid simulation". The combination of both techniques in a single model allows *de facto* the integration of physical, behavioural, doctrinal and procedural aspects of a complex combat system.

Let's define the computer simulation as a quantitative and qualitative method for a representation in a computer program both structural and behavioural characteristics of systems. It enables experimentation with the model (instead of running) and observing the results of running models on the background of a simulation time flow.

Typically, simulation time is a non-negative and non-decreasing real variable $t \in R^{+} \cup \{0\}$. For the defined simulation moments t_i, t_j, t_k, for T as a collection of the moments when a system state changes, and for the two simulation techniques mentioned earlier there are applicable the following conditions:

- for a continuous passage of time: T is a subset of points in a certain range of non-negative real numbers such that $(\forall t_i, t_k)(\exists t_j)(t_i < t_j < t_k)$ – for any two moments there is a moment between them;
- for a discrete passage of time: T is a countable subset such that $(\exists t_i, t_k)(\neg \exists t_j)(t_i < t_j < t_k)$ – there are two such moments that there is no time between them.

In our work we adopt the discrete approach along with the assumptions that for each object there are defined: the state vector with attributes and either the events or periodically executed equations that change the values of attributes (and finally a state of the system). The term event e from a finite set of events E is understood as a scheduled for a specific simulation moment $t \in T$ algorithmic change in object's state: $f_e^s : T \times S \to S$ and $S(t) = S_i$ for $t \in [t_i, t_{i+1})$. On the other side we have the equations (mostly difference) that require a simulation-based numerical solution based upon the assumption *"Future state = Present state + Step change"*. A sequence of chronologically (in a sense of simulation time) ordered change in state is the process of simulation. In general, it may be perceived as a multi-dimensional stochastic process where the individual elements of a state vector describe the various parameters of the system at the time t. The current simulation time is calculated either with the time stamp of the first event from the event calendar: $t^* = \min\left(t : e_i = \left\langle t, f_e^s \right\rangle, i = 1..2^{E \times T}\right)$ or incrementally by a constant time step: $t^* = t^{i+1} = t^i + \Delta t$. The practical computer implementation of the hybrid model has been realised using the *DisSim* package. The main classes (presented on the class diagram - Fig.1) are responsible for [3, 8]:

- the package *simcore* – the whole experiment, i.e. the passage of simulation time, events ordering and changes of state;
- the package *broker* – sending messages between simulation objects;
- the package *random* – (pseudo) random number generation;
- the package *monitors* – monitoring, collecting and statistical analysis.

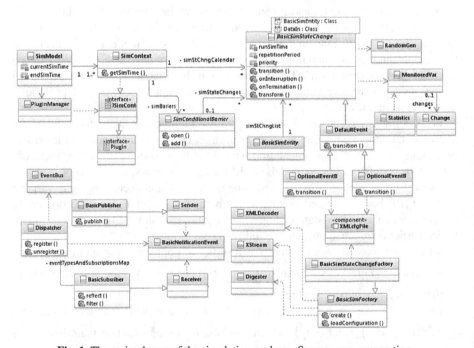

Fig. 1. The main classes of the simulation package. Source: own preparation

Each modelled object is the specialization of the abstract *BasicSimEntity*. The state changes that are assigned to simulation objects are subclasses of the generic *Basic-SimStateChange*. They correspond to either the events or the equations. The main method of *BasicSimStateChange* – *transition()* – is to change a state. The essential attributes are: *runSimTime* – the scheduled simulation time for a "one shot" change of state (an event); *priority* – a priority of an event; and *repetitionPeriod* – a parameter defining a constant time step for simulation-based numerical solution. The managing class *SimManager* is responsible for controlling e.g.: start/stop/pause an experiment as well as the passage of time (as soon as possible, astronomical).

3   Computer System for Support of Capability Based Planning for the Armed Forces Development

The military mission specification used in Planning Scenario has an abstract and general description form. The idea is to present a mission in a way which is appropriate to formulate quantitative conclusions about the required capabilities for a future form of the Armed Forces.

In this particular context an activity diagram concept was applied. The activity diagram is a formalized graphical representation of any process. This graphical language uses simple symbols to define relationships between activities (tasks), like: decision point, sequence, parallelism and others. In our case the activities are military tasks to be performed to accomplish a mission of the Armed Forces. We proposed a specialized Military Tasks Realization Network Editor to build the models of any military missions its Graphical User Interface is presented on the Fig. 2.

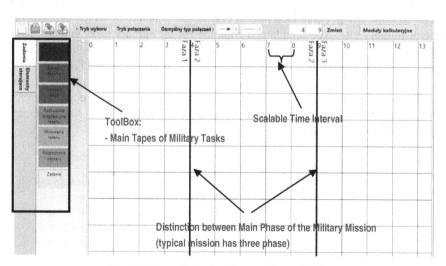

Fig. 2. Military Tasks Realization Network Editor

Using Military Tasks Realization Network Editor the military experts are able to define a mission of the Armed Forces in a very detailed and formal way. Military Tasks Networks for missions "conflict of high intensity" is presented on the Fig. 3.

It is noteworthy that one particular mission described in the Planning Scenario can be decomposed to number of Military Tasks Networks depending on the experts' experience. This fact is very important and has to be taken into account while determining the quantities of capabilities required by the Armed Forces.

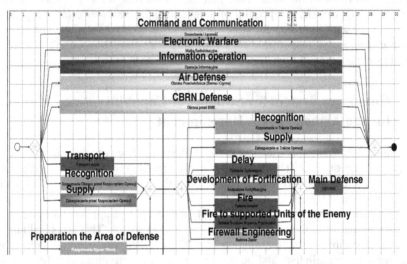

Fig. 3. Military Tasks Realization Network for a particular threat – "conflict of high intensity"

To assess the amount of capabilities needed to accomplish each one of task in Military Tasks Realization Network it is necessary to describe each task in a specific way. It can be done using Tasks Description Forms, the contents of which depends on a type of military task. The template is common for all military tasks and consists of five classes of detail: *Basic information deal with Task, Condition of realization, Task's realization effect, Realization time* and *Visual parameters* (see Fig. 4).

Task's realization effect and *Conditions of task's realization* are the classes of details describing military tasks, which is crucial to assess the work to be done. Therefore, these classes are highly specialized for every type of military task. As an example, logistics tasks will be described.

Właściwości		
(Task realization Effect) Efekt realizacji	(Realization Time) Czas realizacji	(Visual parameters) Parametry wizualne
Informacje podstawowe (Basic Information deal with Task)		Warunki realizacji (Conditions of realization)
Identyfikator zadania: (Task Identifier)	SP1/SytuacjaPlanistyczna1/Sieć1/Transport	
Nazwa zadania: (Task Name)	Transport wojsk	
Treść zadania: (Short Task Description)	Transport wojsk uczestniczących w realizacji zadań w fazie 2 i 3	

✔ Ok

Fig. 4. Military Task Description Form

The capabilities connected with logistics deal with: transport (material, equipment and people), material supply and auxiliary capabilities like: loading, unloading and storing.

Task's realization effect of transport capability is described by the three elements: type of material/personnel/platform being transported, amount and destination - for example: type=NAS 3610 - 463L (NATO standard palette), count=100, destination=Warsaw.

Task's realization effect of material supply is described by the four elements: type of material/personnel, an amount, destination and supply frequency (e.g. twice per day), for example: type=NAS 3610 - 463L (NATO standard palette), count=100, destination= Warsaw, frequency=twice per day.

We define the following types of platforms being transported: transport palettes (open), transport containers (closed), personnel with personal equipment, personnel without personal equipment and calculation platform. A calculation platform is unified platform for transportation of tanks, vehicles, etc.

Conditions of task's realizations for both transport and material supply capabilities are as follows: kind and type of transportation means (e.g. air-plane, air-helicopter, sea-surface, sea-underwater, etc.), distance, weather conditions, percent of day/night.

A reference transport module is a unit which contains transport carriers (trucks, transport planes, container ships, etc.) described by the dedicated parameters (dimensions, own weight, range, capacity, etc.).

The measure for both transport and material supply capabilities of a reference transport module are: weight of material being transported in a time interval (e.g. 1000t in 2 h), count of personnel being transported in a time interval (e.g. 2000 people in 1h), count of palettes/containers/calculation platforms being transported in a time interval (e.g. 100 palettes in 1 h).

In the Table 1 we present the measure of transport capabilities for an exemplified reference transport module consisting of three transport planes 'Hercules C-103E'.

The presented approach produces consistent, verifiable formal model for mission description. The model, in further steps, is evaluated in order to calculate the required capability needs by using model and methods presented in section 2.2.

One of the important features of the method and software itself is the computer discrete combat simulator [1] - it was used for a verification of the capability allocation

Table 1. The measure of transport capabilities for an exemplified reference transport module consisting of three transport planes 'Hercules C-103E'

Conditions of task's realizations			*Capability measure for a reference transport module*					
Distance [km]	Weather conditions	Percent of day/night	Total weight [kg]	Transp. time [h]	Count of person.	Transp. time [h]	Count of platforms Rosomak	Transp. time [h]
1200	Very good	50/50	39400	2,40	248	2,40	3	2,40
7000	Very good	50/50	39400	18,6	248	18,6	3	18,6
8000[1]	Very good	50/50	39400	20,7	248	20,7	3	20,7

[1] *under possibility to tank fuel in the air.*

(see Fig.5). It is worth mentioning that the method assumes that a constructed planning scenario, in conjunction with operation guidelines, will be processed in order to produce a simulation scenario. In the described method the simulation has been used mainly for the estimation purposes. It means that in order to evaluate a capability dependency matrix we apply the computer simulation methods that - through experimentation on a specific scenario - determine the relationships between the particular capabilities.

The simulator consists of the set of tools for a preparation, an execution and an evaluation of a game scenario: *Scenario Editor, Military equipment dictionary and editor, Simulation Manager. Scenario Editor* enables fast preparation of a game scenario (military units (location, equipment), initial tasks, command chain, others), calibration of simulation models and evaluation of simulation results. *Simulation Manager* enables management of simulation. It calculates and shows states of military potential for own and opposite forces, reports from all units during the simulation, controls a simulation experiment. *Military equipment dictionary and editor* enables set parameters of weapons (strength (potential), max speed, typical speed on ground, typical speed on water, caliber, shooting parameters, tracked or wheeled vehicle, kind of fuel, crew, other parameters).

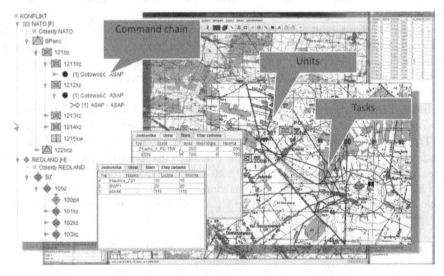

Fig. 5. The set of tools for a preparation and an evaluation of a scenario in the simulator [1]

4   Conclusions

The mathematical models and methods supporting of long-term capability planning process for the Armed Forces are presented in the paper. These models and methods allow assessing quantitatively required, possessed and lacking capabilities of AF in the context of defined threats. Particularly, it is presented the application of the computer discrete simulation for an estimation of lacking capabilities and evaluation the variants of gaining them.

All the introduced mathematical and simulation methods were implemented inside the Computer System for Support of Capability Based Planning of the Armed Forces Development. Some elements of this system have been verified and tested using real data and military experts (combat simulator, Military Tasks Realization Network Editor). The system has been developed as a result of a task co-financed by Ministry of National Defence Republic of Poland.

Acknowledgments. This work was partially supported by the research co-financed by the National Centre for Research and Development and realized by Cybernetics Faculty at MUT: No DOBR/0069/R/ID1/2012/03, titled "System of Computer Based Support of Capability Development and Operational Needs Identification of the Polish Armed Forces".

References

1. Antkiewicz, R., Najgebauer, A., Rulka, J., Tarapata, Z., Wantoch-Rekowski, R.: Knowledge-based pattern recognition method and tool to support mission planning and simulation. In: Jędrzejowicz, P., Nguyen, N.T., Hoang, K. (eds.) ICCCI 2011, Part I. LNCS, vol. 6922, pp. 478–487. Springer, Heidelberg (2011)
2. Davis, P.K.: Analytic Architecture for Capabilities-Based Planning. Mission-System Analysis, and Transformation, RAND MR-1513-OSD (2002)
3. Pierzchała, D., Dyk, M., Szydłowski, A.: Distributed military simulation augmented by computational collective intelligence. In: Jędrzejowicz, P., Nguyen, N.T., Hoang, K. (eds.) ICCCI 2011, Part I. LNCS, vol. 6922, pp. 399–408. Springer, Heidelberg (2011)
4. Fauske, M.F., Vestli, M., Glærum, S.: Optimization Model for Robust Acquisition Decisions in the Norwegian Armed Forces. Interfaces **43**(4), 352–359 (2013)
5. Marler, R.T., Arora, J.S.: Survey of multi-objective optimization methods for engineering. Structural and Multidisciplinary Optimization (26), 369–395 (2004)
6. NATO, The Use of Scenarios in Long Term Defence Planning. http://www.plausiblefutures.com/55074
7. NATO Research and Technology Board: Panel on Studies, Analysis and Simulation (SAS), Handbook in Long Term Defence Planning (2001)
8. Pierzchała, D.: Application of Ontology and Rough Set Theory to Information Sharing in Multi-resolution Combat M&S. Studies in Computational Intelligence **551**, 193–203 (2014)
9. Tagarev, T., Tsachev, T., Zhivkov, N.: Formalizing the Optimization Problem in Long-term Capability Planning. Information & Security **23**(1), 99–114 (2009)
10. Xiong, J., Yang, K., Liu, J., Zhao, Q., Chen, Y.: A two-stage preference-based evolutionary multi-objective approach for capability planning problems. Knowledge-Based Systems (31), 128–139 (2012)

Agent-Based M&S of Smart Sensors for Knowledge Acquisition Inside the Internet of Things and Sensor Networks

Michał Dyk[✉], Andrzej Najgebauer, and Dariusz Pierzchała

Faculty of Cybernetics, Military University of Technology,
Kaliskiego Str. 2, 00-908 Warsaw, Poland
{michal.dyk,andrzej.najgebauer,dariusz.pierzchala}@wat.edu.pl
http://www.wcy.wat.edu.pl

Abstract. The paper presents a formal model of heterogeneous sensor network and crucial parts of its simulation framework. The main goal of this work is to prepare a virtual environment in which the methods for knowledge acquisition from ubiquitous smart devices can be evaluated. The proposed model allows to freely define sensors and description of observed phenomena. Moreover, the simulation framework *SenseSim* allows to apply changes in devices' behaviour by using macroprogramming. A core part of the simulator is an agent-based event-driven simulation library called *DisSim*.

Keywords: Internet of things · Sensor networks · Distributed multi-resolution simulation

1 Introduction

In general, a simulation-based knowledge acquisition technique is widely acclaimed and applied throughout the whole life cycle of IT systems [1] [9] [19] [20] [21]. It reduces the effort requested for gathering knowledge from several sources. Furthermore, it allows for the collection of data from situations that in the real world are a rarity. Referring the issue to our work, it relates to sensor networks and observers - their correct functioning requires a lot of tests in all possible situations. In this paper we propose a formal model and a simulation framework which allow us to explore the process of knowledge acquisition from ubiquitous, smart sensors. We define a sensor as a device which is not only a passive observer but also its behaviour can be adjusted to a improve process of gathering data.

In recent years, sensor networks evolve dynamically into new concept called Internet of Things. Deficiency of formal definitions causes sometimes those terms are used interchangeably and seems not easy to distinguish them. In the paper we understand sensor network as a self-organized and self-sufficient structure composed of smart devices with their own network configuration, routing protocols, resources management etc. The Internet of Things (in short IoT) concept

© Springer International Publishing Switzerland 2015
N.T. Nguyen et al. (Eds.): ACIIDS 2015, Part II, LNAI 9012, pp. 224–234, 2015.
DOI: 10.1007/978-3-319-15705-4_22

was developed by Kevin Ashton in 1999 [1]. He described this phenomenon as a network of uniquely identifiable devices (or things) connected with each other in the Internet-like structure. He even claimed that The Internet of Things would be a kind of interface between the real world and the Web. In short we are talking about IoT when devices are directly connected into the Internet, so all network management, routing etc. are done by global network. IoT and sensor networks can interact together (some devices from a sensor network can be connected to the Internet) which is a natural way. Nowadays there are billions of devices connected is such a manner.

Despite the fact that sensors become cheaper and cheaper, simulation is still the most common and convenient way to perform research. There are several simulators that can also be used to research over sensor networks and IoT. They can be divided into the two groups:

- Simulators which model the low level aspects of network communication. Examples are: *NS2* [2], *QualNet* [4], *Omnet++* [3];
- Simulators which are developed specially for wireless sensor networks and which can be also used for high-level design. The examples are: *CupCarbon* [5], *WSNet* [6], *Ptolemy II* [7] and *Atarraya* [8].

The proposed solution, the simulator *SenseSim*, can be classified as the second category. It is an agent-based discrete-event simulator which allows to model various phenomena which may occur in an environment and various sensors which can be simple devices or more complex ones with multiple perceptual capabilities. It is perceived as compliant with Augmented Perception concept described in the work [9].

2   Observer Model

The main purpose of sensors connected into Internet of Things or other sensor networks is to collect data from an environment. The crucial question is then: how do those sensors observe the world? We propose a model in which sensors can have multiple perceptual capabilities (it also means that perception is not reserved only for living beings). With each capability instance of an so called observer is associated. We define an observer just like Bennet et. al. [10]. It is defined as six-tuple:

$$O = \langle (X, \chi), (Y, \nu), E, S, \pi, \eta \rangle \tag{1}$$

where X and Y are measurable spaces (with χ and ν sigma-algebras respectively), E and S are subsets of X and Y respectively, π is a surjective function with domain in X and values in Y, η is so called *conclusion kernel*. The example of an observer can be camera in a forest fire observeration system [11].

Figure 1 shows how an observer works. When O observes it does not interact with the object of perception itself. Space X is a mathematical construct and is called *configuration space*. It represents all properties of relevance to O. Space Y is a formal representation of premises about events which occur in X. Based on those premises the observer can conclude what happen in the external world. Set E is called a *distinguished configuration* and represents events of interest of

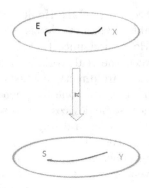

Fig. 1. Visualization of an observer. Source: on the basis of [10].

an observer. Set S is called *distinguished premises* and holds the premises about event E. Transformation between spaces X and Y is realised by function π, called *perspective*. Let us suppose that some point $x \in X$ represents the property of relevance to O. Then O, in consequence of interaction with the outside world, does not see x but its representation $y = \pi(x)$, where $y \in Y$. If x is in E then y is in S. However all that O receives is y, not x. In other words, the observer must decide whether event E really occurred, basing on premises S. Function π is surjection, so O does not really know which point $x \in E$ corresponds to given point $y \in S$. That is why with observer's definition comes conclusion kernel η. It provides, for each point in S, the probability distribution supported on E. η gives the final result of the observer - the probability that for given premises S event E occurred in the real world. Proposed definition of the observer O is for now just an abstract construct which represents some perceptual capability. To become a perceiver it has to be embedded in an environment. It is done by defining *scenario* for the observer, as follows:

$$(C, R, \{Z_t\}_{t \in R}, \Xi) \tag{2}$$

where:

- C - measurable space, which elements are *states of affairs*;
- R - countable, totally ordered set called *active time*;
- $\{Z_t\}_{t \in R}$ - sequence of mesurable functions taking values in $C \times Y$ [1].

Scenario is then a stochastic process with state space $C \times Y$ and indexed by R. It is also called *perception trajectory*. Function Z_t is an observation at time t and takes the value (c_t, y_t) with $c_t \in C$ and $y_t \in Y$; c_t is a state of affairs at time t and y_t is corresponding premise, also called observation, at time t. States of affairs should be understood as parts of the external world and are

---

[1] It can be also understood as a set of random variables.

subjects of observation. Naturally there is a link between states of affairs and a configuration space X of the observer which is a function $\Xi : C \to X.^2$

3 Sensor Network Model

In this section we introduce our model of (wireless) sensor network, which also can be applied to devices connected into IoT. In the description we apply the following designations:

- $\Lambda \in \mathbb{N}^{\mathbb{N}}$ - set of sensors' identifiers;
- $L(s,t)$, $L_s(t) \in \mathbb{R}_{>0} \times \mathbb{R}_{>0}$ - spatial location of sensor $s \in \Lambda$ at time t described by coordinates (x_t, y_t) in Cartesian or geographic (latitude - longitude) coordinate system;
- $d(L(i,t), L(j,t)) \in \mathbb{R}$ - Euclidean distance between sensors i and j at time t;
- $r_s \in \mathbb{R}_{>0}$ - radio range of sensor $s \in \Lambda$;
- $V_s \in \mathbb{N}$ - velocity of sensor $s \in \Lambda$;
- $t \in T$ - current time^3.

Notice: For the paper we assume that terms: *device*, *sensor* and *node* are synonymous.

There are many approaches in wireless sensor networks modelling. Most of them base on graph theory [12]. Similarly, in our model a wireless sensor network is an undirected graph:

$$WSN(t) = \langle \Lambda, E(t), b \rangle . \tag{3}$$

Function b is defined on edges and determines bandwidth, in bits per seconds, of the links between nodes (graph vertexes): $b : \Lambda \times \Lambda \to \mathbb{N}$. $E(t)$ is a family of subsets of Λ

$$E(t) = \left\{ \{s_x, s_y\} : s_x, s_y \in \Lambda, s_x \neq s_y, s_x \in N_{s_y}(t), s_y \in N_{s_x}(t) \right\} \tag{4}$$

which defines graph edges. As N_{s_x} and N_{s_y} we define the neighbours of sensor s_x and s_y respectively. Neighbourhood is defined as follows:

$$N_i(t) = \left\{ s \in \Lambda : d(L_s(t), L_i(t)) \leq min\{r_i, r_s\} \right\} . \tag{5}$$

It means that neighbours are those devices which together are within radio range. If it is true that $\underset{i,j\in\Lambda}{\forall}\ r_i = r_j$ then a network graph becomes Unit Disk Graph (UDG) [13], which is a widely used concept for sensor network modelling [12][14][16]. Of course UDG model is idealistic. In many cases even small obstacles can change connectivity. For our purposes, it is good enough approximation because we do not focus on low level wireless connectivity issues. However

---

^{2} Authors are aware that this description can be a bit vague. For detailed description of scenarios and states of affairs we propose reading [10] p. 64-78. At this point we should define state of affairs as some phenomenon (like fire) which properties (like temperature) can be extracted into space X by function Ξ.

^{3} Can be understood as simulation time, $T \in \mathbb{R}_{\geq 0}$.

our simulator *SenseSim* has a modular design so more adequate models, like COHERENT_NET [15], for network management can be used.

Neighbourhood may change in time because of devices movement so sets $N(t)_i$ and $E(t)$ are time-depended.

We model sensor as the eight-tuple which state can also change in time:

$$s(t) = \langle L(t), V(t), r, \{O_i\}_{i=1..n}, \{A_t^i\}_{i=1..n}, P(t), R(t), SNT(t) \rangle \qquad (6)$$

where:

- $\{O_i\}_{i=1..n}$ - a set of observers associated with perceptual capabilities of a sensor s;
- $\{A_t^i\}_{i=1..n}$ - a set of observations from perceptual capabilities at time t;
- $P(t)$ set of macroprograms executed by device during its work. They can adjust its behaviour which can help in data acquisition;
- $R(t) = \langle r_1(t), r_2(t), \ldots, r_i(t) \rangle$ set of resources (battery energy, CPU usage, memory usage) available at time t;
- $SNT(t)$ - devices' subnet on which a sensor s has complete knowledge.

In our model a sensor is a device with multiple perceptual capabilities. With each one an observer defined by the equation (1) is associated. It means that a single device can observe the world in various ways. Set $\{A_t^i\}_{i=1..n}$ holds observations from each capability. Naturally sensor can perceive only when some phenomenon occurs. Phenomenon is defined as the four-tuple:

$$e(t) = \langle \psi(t), \{\lambda_i(t)\}_{i=1..n}, t_b, t_e \rangle \qquad (7)$$

where $\psi(t)$ is a set of points which represent spatial area where phenomenon occurs; $\lambda_i(t)$ is a set of functions which describes observation for each perceptual capability i at time t; t_b and t_e describe the moments of time when phenomenon begins and ends respectively. Obviously, phenomenon $e(t)$ requires the existence of the scenario defined as (2). Function $\lambda_i(t)$ is then built using a projection pr of $C \times Y$ onto second coordinate. Having that, we can write $\lambda(t) = pr Z_t$. Observation of the sensor at time t and capability i is then: $A_t^i = \lambda_i(t)$.

A communication between sensors is a dynamic process with the three possible states: *in-progress* - value 0; *success* - value 1; *failure* - value -1. Assuming t as current time, τ as time when communication started and tm_{s_i,s_j} as time needed to transfer a message between sensors s_i and s_j, conditions for transition between states are as follows:

$$\begin{aligned} W_{0,0} &: t < \tau + tm_{s_i,s_j} \wedge s_j \in N_i(t), \\ W_{0,1} &: t > \tau + tm_{s_i,s_j} \wedge s_j \in N_i(t), \\ W_{0,-1} &: t < \tau + tm_{s_i,s_j} \wedge s_j \notin N_i(t), \end{aligned} \qquad (8)$$

Value tm_{s_i,s_j} is calculated by the function $Tm : M \times \mathbb{N} \to T$ which arguments are: a message $m \in M$ (where M is a set of messages) and value calculated by function b on graph's edge between s_i and s_j. In short, communication is successful when both a sender and a receiver stays within radio range.

In cases when $s_j \notin N_{s_i}$ (receiver is not a neighbour of sender) the message is sent using hop by hop method with route defined as:

$$R_{s_i,s_j}(t) = \langle s_i, s_1, ..., s_k, s_j \rangle, s_1 \in N_{s_i}(t), s_k \in N_{s_j}(t),$$
$$\forall_{s_l} \left(s_{l-1} \in N_{s_l}(t) \wedge s_{l+1} \in N_{s_l}(t) \right), l = 2, .., k-1 \tag{9}$$

Each node decides to which neighbour the message should be sent in next hop which is a decision variable $NX = \{s \in N_{s_s}\}$, where s_s is a sensor from which the current node received a message. Decision is undertaken basing on node's knowledge of the other devices that is a part of the sensor's definition (6) $D = SNTs_s(t)$. The criteria for decision are denoted by $W(D, NX)$ and they are defined by a particular routing algorithm. Each node calculates the next hop using the function rt:

$$rt : \Lambda \times \Lambda \times W(D, NX) \rightarrow \Lambda^k, k \in \mathbb{N}, \tag{10}$$

which, basing on information about sender, receiver and criteria, returns a set of sensors for the next hop. The form of this function depends on a routing algorithm. For example, for *flooding* it should return all the neighbours; for SPIN [17] [18] it should return the set of all neighbours with similar perceptual capabilities (similar observers); for MEDR [16] it should be the set of neighbours with minimal MED statistic and shortest path.

4   Agent-Based Simulation Framework

For a simulation purposes we have built and still develop *SenseSim* which is an agent-based discrete-event simulator written in Java. It has modular structure, which allows to easily expand its functionality. For example, a module responsible for wireless communication between devices can be replaced with more complex one if the communication issues should be more precisely simulated. The simulation core is *DisSim* library.

Let us define the computer simulation as a quantitative and qualitative method of both modelling in formal language and representation in a computer program of structural and behavioural characteristics of systems, which enables experimentation with the model (instead of running) and observing its behaviour during a simulation time flow. A typical simulation model represents both static and dynamic parts of a system. A static part is a (semi-)formal description of structure of modelled objects along with their attributes and relationships. In a dynamic simulation (with a passage of simulation time) the complementary models and algorithms for determining the states of objects (and thus the whole system) have to be considered. At any simulation moment t each position of the system state (and respective feature of the system) might be seen as: $< object\_attribute, value, time >$. Thus, the state of the modelled system $S(t)$ is formed by all the attributes of objects existing at the simulation time t. In a dynamic simulation the values of the attributes can be determined by: set of defined transformations, state change functions, formal rules, etc. As a change in a state may be also allowed changes in the structure of system's objects which result from their creation or removal. Typically, simulation time is assumed to be a real, non-negative, non-decreasing variable $t \in \mathbb{R}_{\geq 0}$. For the entire simulation model the exact one unit of time should be applied. If we assume the simulation moments t_i, t_j, t_k (time with a length of zero) and T as a collection of the moments when a system state changes occur, we have two conceptually different simulation methods:

- when T is a subset of points in a certain range of non-negative real numbers such that $(\forall t_i, t_k) \, (\exists t_j) \, (t_i < t_j < t_k)$ that is: for any two moments there is a moment between them it means a continuous passage of time;
- or when T is a countable subset such that $(\exists t_i, t_k) \, (\neg \exists t_j) \, (t_i < t_j < t_k)$ that is: there are two such moments that there is no time between them it means a discrete passage of time.

For the purpose of our consideration we adopt the discrete approach along with the assumption that for each modelled object are defined: the state vector with attributes and the formal descriptions of events. The term event e from a finite set of events ES is understood as a scheduled for a specific simulation moment $t \in T$ algorithmic change in object's state. Therefore, a sequence of chronologically (according to simulation time) ordered events is the process of simulation. Depending on the circumstances it may be perceived as a deterministic or stochastic process with a realisation of a time series in a domain of simulation time. In general, it can form a multi-dimensional stochastic process where the individual elements of a state vector describe the various parameters of the system at the time t. The current simulation time is equal to the time stamp of the first event from the event calendar:

$$t^* = min(t : e_i = < t, f_e^{OS} >, i = 1..2^{ES \times T}). \tag{11}$$

The presented discrete-event approach has been chosen as a basis for a dynamic perspective of the agent-based model of software agents.

We propose to describe an agent using the object-oriented paradigm. Then, each significant measurable feature (physical or abstract) of an agent should be assigned to an attribute of the selected object class from the modelled reality and:

- $OA = \{a = < id, c >\}, c \in OC^{OA}, id \in \mathbb{N}$ is a set of simulated agents of the class c; id is a unique identification key;
- OC^{OA} is a non-empty set of classes of modelled agents;
- CX_c is a non-empty set of attributes defined for the agent class $c \in OC^{OA}$;
- AV_{x^c} is a set of proper values for the attribute $x \in CX_c$ from the object class $c \in OC^{OA}$.

At any simulation time t a system state is defined by the four-tuple $\{< a, x, v, t >\}$. The state transition function:

$$f_e^{OS} : T \times OS \to OS \tag{12}$$

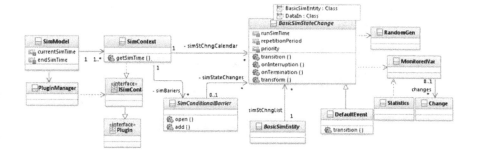

Fig. 2. The main classes of the *DisSim* package. Source: own preparation.

determines the state of the system at the simulation time t after the occurrence of the event e. In the discrete-event simulation the following simplification of the model is obvious: the state of the system is not changed until the occurrence of the next event - therefore $S(t) = S_i$ for $t \in [t_i, t_{i+1})$. Events must occur in an irreversible order: from the past, through the present to the future.

The practical Java implementation of the discrete-event agent-based model is the *DisSim* package. The main classes (presented on a simplified class diagram - Figure 2) are grouped into the following packages:

- *simcore* - the package groups classes responsible for the experiment management, i.e. the passage of simulation time, events ordering and changes of state;
- *broker* - the classes responsible for sending messages between simulation objects (agents, environment);
- *random* - the package with classes for (pseudo) random number generation;
- *monitors* – the classes dedicated to monitor, collect and statistical analysis of time series with simulation results and finally supported knowledge acquisition.

Each software observer (as a simulation agent) must inherit from the abstract class *BasicSimEntity*. The events assigned to simulation objects and associated with state changes are subclasses of a generic class *BasicSimStateChange*. The main methods of that class are as follows: *transition()* changes a state, *onTermination()* - removes an event before the occurrence, *reschedule()* reschedules an event for another time, *simTime()* reads the current simulation time. The attributes in *BasicSimStateChange* are control parameters: *runSimTime* - the scheduled simulation time for the change of state; priority a priority of the event; and *repetitionPeriod* - an optional parameter defining a constant time step for deterministic step-driven simulation. The most important class is *SimManager* - the singleton pattern based object responsible for controlling each: start, stop and pause during an experiment, as well as the passage of time (as soon as possible, astronomical), network communication, user's interactions, etc. However,

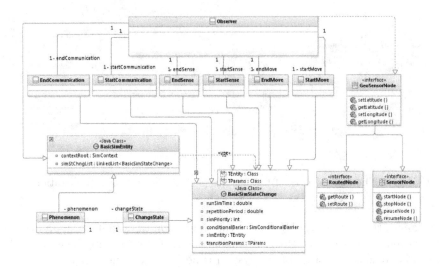

Fig. 3. The main classes of the observer and phenomenon. Source: own preparation.

its crucial responsibility is to indicate the only one event to perform at the next simulation time and to determine the current value of the simulation time on the basis of events' calendar. Figure 3 shows the main classes of an observer and phenomenon inside *SenseSim* simulator with respect to *DisSim* framework. It is software implementation of formal definitions (6) and (7). As described above, an observer inherits from *BasicSimEntity* thus its state is controlled by *DisSim* and can be modified during simulation. It has three main activities described by pairs of events: sensing (*StartSense*, *EndSense*), moving (*StartMove*, *EndMove*), communicating (*StartCommunication*, *EndCommunication*). All of them inherit from *BasicSimStateChange* class which ensures executing events in proper simulational order. Construction of phenomenon is quite similar. An simulation entity *Phenomenon* inherits from *BasicSimEntity*, however it has only one event *ChangeState*. Its purpose is to change state of the phenomenon accordingly to defined functions $\{\lambda_i(t)\}_{i=1..n}$ which describe state changes in phenomenon.

One of the unique aspects of *SenseSim* is ability to macroprogram simulated network. Macroprogramming is a way to program network of devices as a whole. There is one, high level, program which is distributed between sensors. Their responsibility is to interpret the program and adjust their behaviour to it. Our goal is to build environment in which the same macroprogramming language can be used during simulation and in a real network. Currently we evaluate Abstract Task Graph (in short ATaG) [22], however other solutions like Regiment [23] are also within range of our interest.

5   Conclusions

The simulation approach we proposed allows running the whole smart sensors based environment without building it. Through an agent-based simulation we can precisely design and iteratively test and then in the loop redesign and retest all the components (i.e. sensors and observers) until they obtain the appropriate level of functionality and quality.

The second practical advantage are the diverse levels of information that can be gathered during simulation experiments. The simulator is able to produce the results which would not be experimentally obtainable or measurable with current level of technology.

SenseSim is a tool able to build an environment for knowledge acquisition which can be used to fulfil user's information needs as proposed in [9]. Macroprogramming can change devices' behaviour so they can gather requested data and when its required fuse it. It is different approach than in many data fusion systems based on JDL process, which firstly gather observations from sensors and then proceed to explore information from them. Such solution has a broad range of applications. For example in military systems for building situation awareness or in augmented reality on battlefield solutions. Our model and simulator can also be apply for research over sensor networks and Internet of Things. Especially, possibility of changing behaviour of sensor opens wide range of experiments that can be performed to explore data and information flow in such networks.

Acknowledgments. This work was partially supported by the research co-financed by the National Centre for Research and Development and realized by Cybernetics Faculty at MUT: No DOBR/0069/R/ID1/2012/03, titled System of Computer Based Support of Capability Development and Operational Needs Identification of Polish Armed Forces.

References

1. Ashton, K.: That 'Internet of Things' Thing. RFID Journal, 22 June 2009 (2009). http://www.rfidjournal.com/articles/view?4986
2. Issariyakul, T., Hossain., E.: Introduction to network simulator ns2. Springer (2011). ISBN: 978-0-387-71759-3
3. Varga, A.: Omnet++ user manual, version 4.3, June 2011
4. Doerffel, T.: Simulation of Wireless Ad-hoc Networks with QualNet. Advanced Seminar Embedded Systems 2008/2009. Chemnitz (2009)
5. Mehdi, K., Lounis, M., Bounceur, A., Kechadi., T.: Cupcarbon: a multiagent and discrete event wireless sensor network design and simulation tool. In: 7th International Conference on Simulation Tools and Techniques (SIMUTools 2014), Lisbon, Portugal (2014)
6. Chelius, G., Fraboulet, A., Fleury, E.: Demonstration of worldsens: a fast prototyping and performance evaluation tool for wireless Sensor network applications & protocols. In: Second International Workshop on Multi-hop Ad Hoc Networks: from Theory to Reality, Italia. ACM (2006)
7. Baldwin, P., Kohli, S., Lee., E.A.: Modeling of sensor nets in ptolemy. In: The 3rd International Symposium Processing in Sensor Network, pp. 359–368 (2004)
8. Wightman, P.M., Labrador, M.A.: Atarraya: a simulation tool to teach and research topology control algorithms for wireless sensor networks. Simutools 2009. ACM (2009). ISBN 978-963-9799-45-5

9. Dyk, M., Najgebauer, A., Pierzchala, D.: Augmented perception using internet of things. In: Information Systems Architecture and Technology. Selected Aspects of Communication and Computational Systems, pp. 109–118. Oficyna Wydawnicza Politechniki Wrocawskiej, Wroclaw (2014). ISBN 978-83-7493-856-3

10. Bennet, B.M., Hoffman, D.D., Prakash, C.: Observer Mechanics. A Formal Theory of Perception. Academic Press (1989). ISBN 0-12-088635-9

11. Seric, L., Stipanicev, D., Stula, M.: Agent based sensor and data fusion in forest fire observer. In: Nada Milisavljevic, I. (ed.) Sensor and Data Fusion. InTech (2009). ISBN 978-3-902613-52-3

12. Schmid, S., Wattenhofer, R.: Modeling sensor networks. In: Boukerche, A. (ed.) Algorithms and Protocols for Wireless Sensor Networks. John Wiley & Sons (2008)

13. Clark, B.N., Colbourn, C.J., Johnson, D.S.: Unit disk graphs. Discrete Mathematics **86**(13), 165177 (1990)

14. Moscibroda, T., O'Dell, R., Wattenhofer, M., Wattenhofer, R.: Virtual coordinates for ad hoc and sensor networks. In: Proceedings of the 2004 Joint Workshop on Foundations of Mobile Computing (DIALM-POMC 2004), pp. 8–16. ACM, New York (2004)

15. Niewiadomska-Szynkiewicz, E., Sikora, A., Koodziej, J.: Modeling Mobility in Cooperative Ad Hoc Networks. Mobile Networks and Applications **18**(5), 610–621 (2013)

16. Feng, Y., Liu, M., Wang, X., Gong, H.: Minimum Expected Delay-Based Routing Protocol (MEDR) for Delay Tolerant Mobile Sensor Networks. Sensors 2010 **10**, 8348–8362 (2010). doi:10.3390/s100908348

17. Al-Karaki, J.N., Kamal, A.E.: Routing techniques in wireless sensor networks: a survey. IEEE Wireless Communications **11**(6), 6–28 (2004). doi:10.1109/MWC.2004.1368893

18. Singh, S.K., Singh, M.P., Singh, D.K.: Routing Protocols in Wireless Sensor Networks A Survey. International Journal of Computer Science & Engineering Survey (IJCSES) **1**(2) (2010)

19. Pierzchała, D., Dyk, M., Szydłowski, A.: Distributed military simulation augmented by computational collective intelligence. In: Jędrzejowicz, P., Nguyen, N.T., Hoang, K. (eds.) ICCCI 2011, Part I. LNCS, vol. 6922, pp. 399–408. Springer, Heidelberg (2011)

20. Pierzchala, D., Najgebauer, A., Antkiewicz, R., Chmielewski, M., Rulka, J., Wantoch-Rekowski, R., Tarapata Z., Drozdowski, T.: Knowledge-based approach for military mission planning and simulation. In: Gutirrez, C.R. (ed.) Advances in Knowledge Representation, pp. 251–272. InTech (2012). ISBN 978-953-51-0597-8

21. Rissino, S., Lambert-Torres, G.: Rough set theory fundamental concepts, principals, data extraction, and applications. In: Ponce, J., Karahoca, A. (eds.) Data Mining and Knowledge Discovery in Real Life Applications. InTech (2009). ISBN 978-3-902613-53-0

22. Bakshi, A., Prasanna, V., Reich, J., Larner, D.: The Abstract Task Graph: A Methodology for Architecture-Independent Programming of Networked Sensor Systems. In: Workshop on End-to-End, Sense-and-Respond Systems, Applications, and Services (EESR 2005), 5 June 2005

23. Newton, R., Morrisett, G., Welsh, M.: The regiment macroprogramming system. In: Proceedings of the 6th International Conference on Information Processing in Sensor Networks (IPSN 2007), pp. 489–498. ACM, New York (2007)

Analysis of Image, Video and Motion Data in Life Sciences

Camera Calibration and Navigation in Networks of Rotating Cameras

Adam Gudyś[1,2]([✉]), Kamil Wereszczyński[1,2], Jakub Segen[1],
Marek Kulbacki[1], and Aldona Drabik[1]

[1] Polish-Japanese Academy of Information Technology, Koszykowa 86, 02-008
Warsaw, Poland
{agudys,kw,js,mk}@pjatk.edu.pl
[2] Institute of Informatics, Silesian University of Technology, Akademicka 16,
Gliwice 44-100, Poland

Abstract. Camera calibration is one of the basic problems concerning intelligent video analysis in networks of multiple cameras with changeable pan and tilt (PT). Traditional calibration methods give satisfactory results, but are human labour intensive. In this paper we introduce a method of camera calibration and navigation based on continuous tracking, which requires minimal human involvement. After the initial precalibration, it allows the camera pose to be calculated recursively in real time on the basis of the current and previous camera images and the previous pose. The method is suitable if multiple coplanar points are shared between views from neighbouring cameras, which is often the case in the video surveillance systems.

1 Introduction

Current video surveillance systems are based on multiple cameras which can rotate and zoom (PTZ). Performing video analysis on data from multiple cameras usually requires the cameras to be calibrated with respect to a common world reference frame (WRF). Traditional camera calibration methods [2,10] require placing 3D reference points or markers in the view of the cameras, which is labour intensive, especially in the case of exterior cameras. Therefore, there is a need for methods able to calibrate such cameras with a minimum human effort. Calibration is understood as a computation of a camera model M consisting of two components, a pinhole camera model P and a distortion model D. These two elements together make it possible to project any point from the world reference frame to the image plane as well as to reconstruct 3D objects on the basis of their images. An important limitation of many video systems is lack of the information about height of objects being observed–only planar two-dimensional map is available. This must be taken into account by the calibration algorithm.

A calibration system suitable for a network of rotating (PT) exterior cameras, that requires a minimum of human involvement, without placing markers in camera views, is introduced in this paper. Its extension to a network of PTZ

© Springer International Publishing Switzerland 2015
N.T. Nguyen et al. (Eds.): ACIIDS 2015, Part II, LNAI 9012, pp. 237–247, 2015.
DOI: 10.1007/978-3-319-15705-4_23

cameras will be the subject of future work. The calibration procedure in the system is divided into three stages: image preparation, i.e., undistortion and background separation (I), computation of intrinsic and extrinsic camera parameters (II), and navigation (III). The problem of distortion introduced by imperfect lens geometry is known to affect negatively accuracy of 3D reconstruction, thus different techniques for dealing with it like straight-lines [3] or radial trifocal tensors [9] were presented. The system employs the first approach which, in spite of its simplicity, turned out to render satisfactory results. The presence of background separation is caused by the fact that only fixed background points are utilised by the following stages. Therefore, moving objects (foreground) should be filtered out to prevent from disturbing the calibration. This is done with a method based on Gaussian mixture models [1,11]. The next stage is the computation of intrinsic camera parameters with a use of a method suited for rotating cameras [6]. This is followed by the estimation of extrinsic camera parameters for a number of selected pan-tilt positions. This process is often performed with a use of correspondences between WRF and image points [5,7], thus it requires human assistance. As only two-dimensional information about the observed scene is available, the algorithm suited for coplanar points was employed [8]. Finally the navigation stage begins which, after some preliminary steps, allows camera pose to be calculated in the real time on the basis of the current camera image.

The system has been implemented in C++ with a use of OpenCV library.

2   Image Preparation

The camera model used most widely in a computer vision area is called a pinhole camera model. However, real-life cameras do not perfectly agree with this model, namely they introduce non-linear distortion. The distortion can be divided into two components: radial (along the direction from the center of the distortion to the considered point) and tangential (along the perpendicular direction), and can be written as an infinite series $x_u = x_d(1 + k_1 r_d{}^2 + k_2 r_d{}^4 + \ldots)$, where (x_u, y_u) and (x_d, y_d) are undistorted and distorted image points, respectively, and r_d is a distorted radius.

It has been proven, however, that considering only first order distortion parameter k_1 gives sufficient accuracy [10]. Let us denote (c_x, c_y) as a center of distortion and s_x as a distortion aspect ratio, which may differ from the image aspect ratio. The undistorted coordinates are expressed as follows:

$$\begin{cases} x_u = x_d + (x_d - c_x)k_1 r_d{}^2 \\ y_u = y_d + (y_d - c_y)k_1 r_d{}^2 \end{cases}, \text{ where } r_d = \sqrt{\left(\frac{x_d - c_x}{s_x}\right)^2 + (y_d - c_y)^2}. \quad (1)$$

The algorithm used in the library [3] is based on a fact that in the perfect pinhole camera, projections of straight lines are also straight lines. Consequently, when the image is distorted, straight lines are seen as curves. The idea is to select on the image curves, which are projections of straight lines, and estimate distortion parameters in the way that after applying undistortion transformation

Fig. 1. Synthetic undistortion experiment. The black square on the left is used as a reference. The red square in the middle was obtained by distorting the reference with the following model: $k = 0.67$, $c_x = -0.38$, $c_y = -0.25$. The blue square on the right is a result of the undistortion procedure.

these curves become straight. For this aim a standard non-linear optimisation method can be applied [4].

The algorithm works in two stages. At first only k parameter is being estimated while c_x, c_y, s_x are fixed (c_x, c_y are set to the centre of the image while s_x is set to 1). After algorithm converges to some value of k, the minimisation procedure is executed once again with all distortion parameters being optimised. As s_x parameter models tangential distortion which has a moderate influence on the final image, it is fixed to 1 in both optimisation steps which may positively influence the algorithm convergence. As suggested by the results of synthetic (Fig. 1) and real-life (Fig. 2) experiments, the undistortion procedure renders satisfactory results.

The next step of the image preparation is background separation. The original method based on Gaussian mixture model [1] analyses consecutive frames and estimates background colour of a pixel on the basis of how frequently different colours are observed. In the time perspective, pixels form a cloud of points, each described by a time stamp and a pixel colour. These points can be gathered into K clusters, each representing a Gaussian component. If a colour not belonging to any cluster is observed, a new Gaussian component is created with a mean set to this colour, large covariance matrix and with a small weight.

To formalise this idea, let q_n be a specific pixel on frame n; $p_q(c_n)$—the probability of pixel q having colour c in frame n; w_k—counted weight of k-th distribution component; K—current number of Gaussian distribution components; T, α—algorithm parameters. In that case, $p_q(c_n) = \sum_{k=1}^{K} w_j \eta(c_n, \theta_k)$, where $\eta(c, \theta_k)$ is the normal distribution of k-th Gaussian component. It is defined as $\eta(c, \mu_k, \Sigma_k)$, with μ_k being the mean and $\Sigma_k = \sigma_k^2$ being the covariance of k-th component.

The distributions are sorted decreasingly according to their fitness defined as w_k/σ_k, and first B components are selected to be updated (B is computed on the basis of T). The update of distribution w_k is done according to the equations:

$$
\begin{cases}
w_k^{n+1} = (1 - \alpha)w_k^n + \alpha\, p(\omega_k|c_N) \\
\mu_k^{n+1} = (1 - \alpha)\mu_k^n + \alpha\, \eta(c_{n+1}, \mu_k^n, \Sigma_k^n)\, c_{n+1} \\
\Sigma_k^{n+1} = (1 - \alpha)\Sigma_k^n + \alpha\, \eta(c_{n+1}, \mu_k^n, \Sigma_k^n)\left(c_{n+1} - \mu_k^{n+1}\right)^2
\end{cases}
\tag{2}
$$

where $p(\omega_k|c_n)$ equals 1 if ω_k is the first match component to c_n and 0 otherwise.

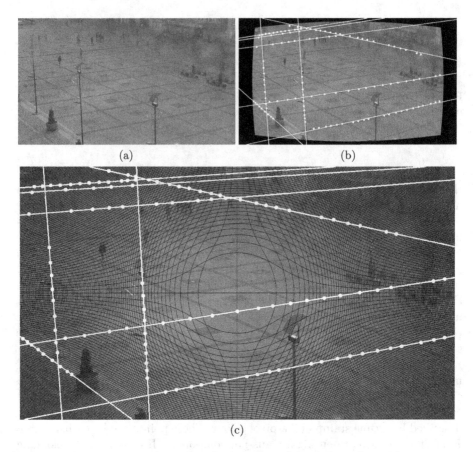

(a)   (b)

(c)

Fig. 2. Undistortion procedure on exemplary data. After taking the reference image (a), a synthetic distortion is applied on the image and several lines known to be straight in the real world are selected by the user (b). The undistortion procedure executed afterwards properly restores the original image (c). The radial grid is caused by some of output points not being covered by the undistortion transformation.

If none of the K distributions match pixel value c_n, the least probable component is replaced by a distribution mean $\mu_m^n = c_n$, an initially high variance, and a low weight parameter w_m^n. It was shown, that $\log_{1-\alpha} T$ frames are needed for satisfactory background separation, and the best values of parameters are $\alpha = 0.02$ and $T = 0.5$.

The problem concerning aforementioned approach is that the number of Gaussian components for all pixels is the same, which may render unsatisfactory results. In [11] a method of dealing with this issue known as adaptive Gaussian mixture model (AGMM) is presented. This variant adjusts the number of Gaussian components for each pixel independently and is used in the presented system. Results of performing background separation on the example sequence of frames are shown in Fig. 3.

$n = 1$ $n = 20$

$n = 50$ $n = 80$

Fig. 3. Results of AGMM background separation step based on n images

3   Intrinsic and Extrinsic Camera Parameters

Camera intrinsic parameters are: focal length f, skew s, coordinates of the principal point $P = (p_u, p_v)$, and scale factors in horizontal and vertical directions k_u, k_v. They are all gathered in K matrix.

$$K = \begin{bmatrix} fk_u & s & p_u \\ 0 & fk_v & p_v \\ 0 & 0 & 1 \end{bmatrix} \tag{3}$$

Extraction of these parameters is done according to [6]. The algorithm takes as an input a set of at least three overlapping images from rotating camera. No information about camera orientation during image acquisition is necessary. One of the images is selected as a reference one. The first step of the algorithm is to find transformations from the reference image to all the other ones. Let A, B and $P_{3\times3}$ be the reference image, the destination image, and the transformation from the reference to the destination, respectively. The computation of P is done as follows:

1. Find all key-points on the images A and B using feature detection algorithm, e.g. SIFT or SURF. Each key-point is described by a vector of descriptors.
2. Find correspondences between key-points on A and B using Euclidean distances in a descriptor space.
3. Pick M best pairs of corresponding key-points to calculate transformation from A to B using RANSAC [5]. Points should uniformly cover the overlap.

In order to check correctness of this step, a part of key-point pairs were used for testing. This allows pixel distances between points from A transformed by P and corresponding points from B to be compared. The intuitive way to verify whether P has been properly calculated is to transform A by P and draw it on B. As confirmed by the experiments both views fit to each other producing proper panorama. This operation is referred to as *image stitching*.

Having set of transformations from the reference image to the other images, the algorithm calculates matrix K of intrinsic camera parameters. For each transformation P, one can write the following equation based on the orthogonality of the rotation matrix:

$$(KK^T)P^{-T} = P(KK^T). \tag{4}$$

KK^T is a positive definite symmetric matrix with 6 degrees of freedom:

$$KK^T = \begin{bmatrix} a\ b\ c \\ b\ d\ e \\ c\ e\ f \end{bmatrix}. \tag{5}$$

Eq. 4 can be rephrased as a homogeneous system $AX = 0$ of nine linear equations with six unknowns $X = [a\ b\ c\ d\ e\ f]^T$. Elements of A can be computed straightforwardly. However, as equations in such system are not independent, at least two different P transformations are required to calculate X, which corresponds to three overlapping images. When KK^T elements are found (this can be done to the linear factor as KK^T appears on both sides of Eq. 4), matrix is normalised in the way that f is equal to 1. Then KK^T is decomposed.

The accuracy of the procedure was confirmed by simulated experiments. A set Q of four synthetic points in a world reference frame (WRF) was created. Extrinsic parameters were set in the way that the points from Q were visible by the camera. Additionally, non-trivial intrinsic camera parameters were assumed. The points from Q were projected into the image plane resulting in a reference view. Then the camera was rotated slightly in both, vertical and horizontal directions and images of Q were taken at each pose (resulting in 8 additional views overlapping with the reference one). For each pair of overlapping views a transformation was calculated (images of points from Q were used as key-points). Set of transformations was given as an input for Hartley et al.'s algorithm. Estimated intrinsic parameters were compared to the ones used for data generation proving that algorithm works properly. The whole procedure was successfully repeated for many matrices of intrinsic parameters.

The next step after calculating K matrix is to compute camera pose in a world reference frame for several selected pan-tilt positions. The pose consists of two components: translation vector $T_{3\times1}$ and rotation matrix $R_{3\times3}$. Calculations are based on the correspondence between coordinates of scene objects and their images. An important requirement to be met is that method must work correctly for coplanar scene objects. This is motivated by the fact that no height information is available.

The algorithm employed by the system is a POSIT strategy suited for coplanar points [8]. The method approximates perspective projection by a scaled

Fig. 4. Calibration results on exemplary dataset. Blue points were directly selected by the user. White points are world points (selected on the map) projected on the image plane using calculated intrinsic and extrinsic camera parameters.

orthographic projection and iteratively refines camera extrinsic parameters in order to minimise projection error. In the case of non-coplanar points, the algorithm returns a single pose. In the coplanar case, there are two poses returned, giving user the opportunity to choose the more appropriate one.

In order to test extrinsic camera calibration the same experimental scheme was used as previously. Several synthetic points were placed in WRF and images of these points were taken using assumed intrinsic and extrinsic camera parameters. Then the points from WRF as well as their images were given to Oberkampf et al.'s algorithm as an input. Estimated pose was compared with the one used for data generation. The procedure was repeated for both non-coplanar and coplanar WRF points under wide range of observation angles. In all cases the algorithm successfully estimated real camera pose.

The computation of both intrinsic and extrinsic camera parameters was further tested by experiments on real-life camera images. At the beginning, a number of overlapping images from Market Square in Bytom were acquired and undistorted with a use of straight-line approach. Then, they were put to Hartley's algorithm to extract intrinsic parameters. Next, a set of corresponding points was selected manually on the camera images and the 2D map of Market Square from GoogleMaps which was followed by the calculation of a camera pose. Finally, the map points were projected with a use of intrinsic and extrinsic camera parameters on the image plane and compared with the selected image points (Fig. 4).

4   Navigation: Continuous Tracking

Navigation is a process of real-time camera calibration which recalculates pose without human interference immediately after camera changes its pan-tilt orientation. In the presented study the navigation is performed with a use of a method referred to as a continuous tracking. It is assumed that the camera zoom during a single continuous navigation process is constant.

The navigation is preceded by the preparatory phase consisting in a generation of key-point clouds, each cloud corresponding to some camera orientation in a PT space. The key-point is defined as a pair of corresponding points in the image and in the WRF. The procedure of clouds generation is presented in Pseudocode 1.

The first key-point cloud is built upon an arbitrarily chosen reference image (Fig. 5-top-left). At the beginning a set of points in the image and corresponding points in the WRF (the map in the middle of Fig. 5) is selected manually by the user (step 2). These points are referred to as known points and are represented in Fig. 5 by green dots. Afterwards, point correspondences are used to calculate a homography between the world and the image space. This homography together with the intrinsic camera parameters allows initial pose to be computed (step 3) and is further used to increase the density of the current cloud (step 4). Namely, a set of additional feature points is found on the image with a use of SURF detector and is mapped to the WRF by the homography (yellow diamonds in Fig. 5).

The initial key-point cloud is employed for generation of consecutive clouds. The camera is rotated in a way that the significant part of a new image (Fig. 5-bottom-left) overlaps with the previous one (step 6). The SURF detector is run in order to find feature points on the new image (step 7). Since these points do not have WRF coordinates assigned, they are referred to as unknown and marked with blue squares. However, as the new image overlaps with the reference one, there are correspondences between known and unknown points. To find them, for each point the algorithm calculates its descriptor (a vector of 64 or 128 features). Then an attempt is made to match unknown points from the new image with known points from the reference cloud by calculating distances in a descriptor space (step 8). The points for which relevant correspondences were found are called adjustment points (yellow diamonds in Fig. 5-bottom-left). These points have known WRF positions, thus are further used to compute a new homography and estimate an updated camera pose (step 9). After that the algorithm goes back to step 6 changing the reference cloud if necessary.

The aim of the cloud generation procedure is to obtain full covering of the area observed by the camera keeping the number of acquired images at the minimal level to reduce computational and memory overhead. Fig. 6 shows the mechanism of altering camera orientation in a PT space during key-point clouds generation. The blue dot in the middle is a starting position. The camera pan and tilt are modified in the way that consecutive positions produce a circle in the PT space. This circle will be referred to as a generation. When the current generation is completed, the radius of the circle is increased by a given value,

Key-points cloud creation   2D map   Navigation

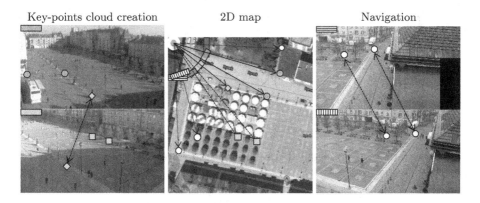

Fig. 5. Visualisation of the key-points cloud creation and navigation

---

Algorithm 1. Pseudo-code of the key-point clouds creation

1: Set initial orientation, acquire the reference image I.
2: Initialise cloud by manually selecting corresponding points from I and WRF.
3: Calculate homography H mapping I to WRF, estimate camera pose.
4: Increase density of the cloud using H and feature points from I detected by SURF.
5: **repeat**
6:     Alter camera orientation, acquire new image I, create empty cloud Ψ.
7:     Detect feature points on I using SURF.
8:     Find key-points matching to them in the closest cloud Ω.
9:     Compute homography H using points from I and matching key-points from Ω.
10:    Increase density of Ψ using points from I and H.
11: **until** entire area covered by clouds

---

and another generation of clouds is acquired. As described previously, for each processed PT position, a reference key-point cloud has to be selected in order to find matching feature points. In the first generation a starting cloud is used (the blue dot) for this purpose. For consecutive generations, the closest cloud from the previous generation is chosen to be used as a reference.

After all key-point clouds have been generated, the navigation procedure starts. It consists of continuous pose re-estimation on the basis of the current camera image and gathered key-point clouds. The navigation is done according to the following steps: (I) get an image from the camera; (II) select a reference key-point cloud C corresponding to the nearest PT to the current camera orientation; (III) using SURF find on the image a set Y of 70 feature points matching best to C; (IV) compute a homography H upon Y and corresponding WRF points from C; (V) estimate pose using H, Y, and intrinsic camera parameters. In Fig. 5 the current and the reference images are marked, respectively, with vertically and horizontally hatched rectangles.

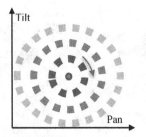

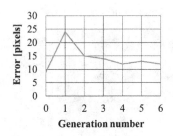

Fig. 6. Rotating camera method used in key-point creation

Fig. 7. Average error of navigation

5 Accuracy Evaluation

The aim of the experimental part was to assess the accuracy of the navigation procedure. For this purpose, approximately 30 different camera orientations in a PT space were tested. For each orientation, an image was acquired and ~ 10 characteristic points were selected manually. Those points were reprojected with a use of estimated camera parameters to the WRF (the map) and compared with their real positions. Reprojection error was measured in pixels. Note, that, according to the navigation method, for each camera orientation, the nearest key-point cloud is used for pose estimation. Thus, a generation containing this cloud was assigned for each orientation, and the reprojection errors were averaged within generations. The results can be seen in Fig. 7. As one can see, the error starts from 10 pixels at generation 0 and after the initial high peak to 25 for generation 1, it decreases and stabilises in the range 10-15 pixels.

6 Conclusions

In the paper we present a camera navigation system allowing rotating cameras to be calibrated in real time without extensive human effort. The preliminary steps include image undistortion, background separation, computation of intrinsic and extrinsic camera parameters for several pan-tilt positions, and generation of the key-points cloud. After that, the continuous tracking procedure starts which automatically recalculates camera pose immediately after changing pan or tilt.

As the method is based on key-points detection, it very sensitive to environment conditions like: weather, season, illumination and even time of a day (due to shadow changes). We plan to solve this issue basing on maximum likelihood methods leading to recognise environmental differences significant for continuous tracking process. The future work will also include addition of zoom as an adjustable parameter, i.e. the extension of the methods to PTZ cameras.

Acknowledgments. This work was supported by a project UOD-DEM-1-183/001 from the Polish National Centre for Research and Development.

References

1. Bowden, P., KaewTraKulPong, R.: An improved adaptive background mixture model for real-time tracking with shadow detection. In: Proc. of 2nd European Workshop on Advanced Video-Based Surveillance Systems, pp. 135–144 (2001)
2. Davis, J., Chen, X.: Calibrating pan-tilt cameras in wide-area surveillance networks. In: Proc. of ICCV 2003, pp. 144–149 (2003)
3. Devernay, F., Faugeras, O.: Automatic calibration and removal of distortion from scenes of structured environments. In: Proc. of SPIE 1995, pp. 62–72 (1995)
4. Devernay, F.: C/C++ Minpack (2007). http://devernay.free.fr/hacks/cminpack/
5. Fischler, M.A., Bolles, R.C.: Random sample consensus: a paradigm for model fitting with applications to image analysis and automated cartography. Commun. ACM **24**(6), 381–395 (1981)
6. Hartley, R.I.: Self-calibration from multiple views with a rotating camera. In: Eklundh, J.-O. (ed.) Computer Vision – ECCV 1994. LNCS, vol. 800, pp. 471–478. Springer, Heidelberg (1994)
7. Horaud, R., Conio, B., Leboulleux, O., Lacolle, B.: An analytic solution for the perspective 4-point problem. Comput. Vision Graph. **48**(2), 277–278 (1989)
8. Oberkampf, D., DeMenthon, D.F., Davis, L.S.: Iterative Pose Estimation Using Coplanar Feature Points. Comput. Vis. Image Und. **63**(3), 495–511 (1996)
9. Thirthala, S., Pollefeys, M.: The radial trifocal tensor: a tool for calibrating the radial distortion of wide-angle cameras. In: Proc. of CVPR 2005, pp. 321–328 (2005)
10. Tsai, R.Y.: A versatile camera calibration technique for high-accuracy 3D machine vision metrology using off-the-shelf TV cameras and lenses. IEEE Robot. Autom. Mag. **3**(4), 323–344 (1987)
11. Zivkovic, Z.: Improved adaptive gaussian mixture model for background subtraction. In: Proc. of ICPR 2004, pp. 28–31 (2004)

Expert Group Collaboration Tool
for Collective Diagnosis of Parkinson Disease

Marek Kulbacki[1(✉)], Jerzy Paweł Nowacki[1], Andrzej W. Przybyszewski[1]
Jakub Segen[1], Magdalena Lahor[1], Bartosz Jablonski[2], and Marzena Wojciechowska[1]

[1] Polish-Japanese Academy of Information Technology,
Koszykowa 86, 02-008 Warszawa, Poland
mk@pja.edu.pl
[2] Faculty of Electronics, Wroclaw University of Technology,
Wybrzeze Wyspiaskiego 27, 50-370 Wroclaw, Poland

Abstract. The paper presents the concept and implementation of distributed group collaboration tool intended for collective diagnosing of Parkinson Disease (PD). Collaborative decisions as a result of experts meetings and discussions on many clinical cases and many assessment methods, by geographically distributed domain experts seem to be more reliable option than assessment made by a single neurologist. Clinicians using our system are working on finding relationship between subjective Unified Parkinson's Disease Rating Scale (UPDRS) and completely new objective scale developed on the basis of selected parameters of patient's gait called by us Parkinson's Disease Gait Indexes (PDGI) [1].

Each expert using subjective UPDRS expresses classical assessment of the patient's stage or symptoms development. The obtained results are used as reference for other assessment methods in our database. An alternative objective PDGI scale is based on computations of patients' gait measurements from 4GAIT-Parkinson multimodal database [2]. The Motion Data Editor (MDE) [3] – a client part of distributed system, has been extended with modules computing and classifying gait indexes from motion data and collective scoring. For each PD data trial, MDE allows the simultaneous processing and visualization of four video data streams and more than five hundred independent kinematic and kinetic modalities, synchronized in time during measurement. Video recordings are the basis for the subjective UPDRS assessment by each of the experts. At the same time, the assessment result is a reference for comparison with the calculated PDGIs.

Keywords: Parkinson Disease · Expert Systems · Collective Collaboration · Motion Analysis

1 Introduction

The concept of finding solutions by subject-matter-experts has been known for a very long time. Knowledge acquisition and new knowledge development by multiple domain authorities supported by information technologies and multimodal human activity

© Springer International Publishing Switzerland 2015
N.T. Nguyen et al. (Eds.): ACIIDS 2015, Part II, LNAI 9012, pp. 248–257, 2015.
DOI: 10.1007/978-3-319-15705-4_24

measurements in the form of human data model give new opportunities and are novel practical concept. This paper presents the idea of searching by cooperating experts relationships between the gait indexes and the progress of Parkinson's disease as the result of experience and the difficulties encountered in previous studies in the field of rationalization and semi-automatic assessment of the PD patients. Agreed relationships between UPDRS and PDGI scales will help collectively decide about the quality and objectivity of the alternative scale. The main problem is the subjectivity of the assessment of the UPDRS. Consequently, determined the correlation coefficients with gait indexes assessments from PDGI are not always reliable. In addition, different gait indexes have different correlation coefficients, depending on the patient. Different clinicians may also interpret differently the meanings of the UPDRS. These problems are articulated in a popular statement "No two patients face Parkinson's in quite the same way". Patients vary substantially in their combination of symptoms, rate of progression, and reaction to treatment. If we would like to improve this analysis, we need to take into account an alternative scale based on a great variety of patients' measurements from different centers and also assessed by multiple neurologists.

In this work, we propose to apply an alternative objective PDGI scale with the quality assessed cumulatively by experts and the knowledge inducted from the database (intelligence knowledge of the database) to sense patient's symptoms severity changes. To build the knowledge we will use the expertise extracted from the unified database, dependent on cumulative knowledge of many neurologists instead the expertise of a single neurologist. Using a cloud based group collaboration system with PD data, it is relatively easy to gather expertise from multiple geographically distributed domain experts at the same time. Using the client application, Motion Data Editor, the experts will compose the Multimodal Motion Information View (MMIV) from the selected motion parameters and use it to load and observe selected measurements of patients with video recordings.

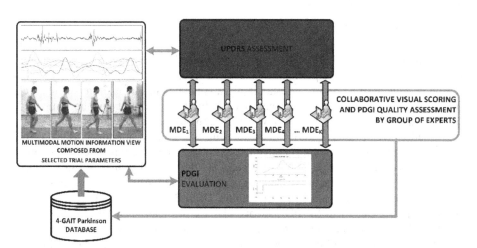

Fig. 1. Pipeline for collaborative assessment and scoring of PDGI

Video streams are essential for subjective UPDRS assessment of each of experts. The MMIV - visual correlation of gait parameters is necessary for scoring the quality of evaluated PDGI as shown in the Fig.1. Our reference 4GAIT-Parkinson dataset contains 1781 multimodal trials including 803 trials of gait grouped in 12 tasks, which were performed by 18 subjects [2]. We also developed and provide specification for multimodal data management - measurement protocol and a selection of spatiotemporal modalities required for data unification and calculation of PDGI's. By applying our measurement and data unification rules to larger databases such as National Parkinson Foundations (NPF) Database one can find additional significant information about specificity of the PD. As NPF database has information from different participating PD centers thanks to the Quality Improvement Initiative (QII) one can compare results of different treatments. But, due to a variety of cares (Thirty-nine leading medical centers worldwide deliver care to more than 50000 Parkinson's patients), some of results obtained from the most prominent expert centers might be inconsistent and require unification.

The advantage of our approach is that the expertise inducted from the unified database is related to many multimodal measurements and experiences of many neurologists. Motion Data Editor enables direct Internet based access to database and simultaneous presentation of many multimodal data of PD patients. Thanks to that by comparing similar symptoms in more advanced patients we may better predict disease near-future development. However, without actual objective measurements of symptoms performed by a neurologist one can argue that we do not have information about dynamical changes of the patient's actual state. We have only highly subjective patient's reports. Collective scoring and applying of PDGI approach is an alternative to UPDRS, that allows physicians to find spatiotemporal rules in motion, verify them collectively and will enhance subjective patient's reports making them more objective. These reports will help us to find and identify patient's dynamics in PD development.

2   Parkinson Disease

Parkinson's Disease (PD) - chronic progressive disease belongs to the group of motor and nervous system disorders. Movement dysfunctions in PD are caused by the loss of dopamine-producing brain cells in a region located in the midbrain (a substantia nigra). Currently only symptomatic methods of treatment are applied, because the reason of cells destruction in a substantia nigra is unknown. The main motor features of Parkinson's Disease are constrained but not limited to: rigidity of the limbs and trunk, tremor in hands, arms, legs, jaw and face, bradykinesia, impaired balance and coordination. The primary pharmacological drug is a specific aminoacid (L-DOPA), which after reaching the brain, is converted into dopamine. The fundamental pharmacological medication with L-DOPA in some PD after several year may cause symptoms fluctuations. Later the progression of neurodegeneration as well as dopaminergic treatment itself results in motor complications. Deep Brain Stimulation (DBS) [4][5] is an alternative treatment for PD patients who are medically resistant to pharmacotherapy. DBS of the subthalamic nucleus (STN) has become an established therapy for patient with PD. It is an effective and safe method of symptomatic treatment of PD patients as long as parameters of DBS are correctly tuned based on motion tasks evaluation. The evaluation carried out by experts is based mainly on the Unified Parkinson's Disease Rating Scale (UPDRS). The motor part of UPDRS consists of 14 points, which evaluate different motor skills based on discrete scale in range of 0-4, where 0 means normal ability to move. On the basis of the UPDRS classification we introduced PDGI and would like to make it more popular. Patients taking part in this research were operated on in the Department of Neurosurgery Medical University of Silesia in Katowice, Poland. As a result of research on gait recognition and identification of PD patients symptoms and stage classification of the disease progression we created 4GAIT-Parkinson database.

3   Motion Data Acquisition

All measurements were performed in multimodal Human Motion Laboratory (HML) of Polish–Japanese Academy of Information Technology R&D Center (PJAIT RDC) in Bytom, Poland. The HML allows to acquire motion data through simultaneous and synchronous measurement and recording of motion kinematics, muscle potentials by surface electromyography (sEMG), ground reaction forces (GRF) and four video streams in High Definition format. Data were collected from PD patients performing tasks under four experimental conditions called sessions, defined by pharmacological medication and subthalamic nucleus electrical stimulation: Session1: Stimulation OFF / Medication OFF, Session2: Stimulation ON / Medication OFF, Session3: Stimulation OFF / Medication ON and Session4: Stimulation ON / Medication ON. Medication ON/OFF means that patient was with/without drugs during the session. Stimulation ON/OFF means that patient was with stimulator turned on / turned off during the particular session.

The experimental scenario includes seven tasks and has been planned based on criteria taking into consideration in motor examination part of UPDRS scale. For kinematic measurement the configuration with 39 attached markers on a human body, tracked by 10 motion capture Vicon T40 cameras have been used. The assumed skeleton model, reconstructed on the basis of marker's 3D positions contains 24 segments with rotations in Euler Angles as well as Quaternion representations. There are also global rotation and translation with respect to the word frame, which results in 78 dimensional pose space. Muscle potentials of the lower body parts are measured by 16 electrodes of sEMG subsystem from Noraxon. Ground reaction forces (GRF) are measured by two Force Plates from Kistler and is base measure to estimate dozen kinematic parameters (powers, moments. The gait cycles corresponding to two adjacent steps can be detected on the basis of femur rotation analysis. Multimodal measurement is our basis for multi-featured analyses of neurological gait abnormalities from single trial.

4   Motion Data Analysis and Gait Indexes

In collaboration with major medical centers in Poland, PJAIT RDC developed new research methods and techniques with a focus on multimodal human motion measurements, analysis and synthesis needed by orthopedics and rehabilitation. In [12] the motion capture was used to assess gait abnormalities associated with coxarthrosis – a degenerative diseases of a hip joint. The Principal Component Analysis was applied as dimensionality reduction method to kinematic data. After that Fourier Coefficients of 3D gait trajectories have been calculated and gait descriptors were further classified by supervised machine learning.

The feature extraction and selection is being most often used for motion data classification in medical applications and uses the extracted features of motion time sequences. There is different kind of feature set proposed in the literature. They can directly reflect interpretable characteristics as described in previous section phase coordination index, decomposition or asymmetry indexes. In our previous works on human identification [13],[14],[15] and assessment of gait abnormalities [12] we proposed and applied statistical, timeline, histogram and Fourier transform features sets. In the statistical set there are mean values and variances of each pose attribute and in the histogram based one, we build separate histogram for each attribute. The Fourier feature set contains coefficients of a frequency domain into which motion data is transformed. The low pass filtering is carried out to reduce number of features. It takes into consideration only first twenty components with the lowest frequencies. In timeline feature extraction approach motion is divided into specified number of intervals which are described by their average values of every pose attribute.

The feature extraction usually gives great number of different features. For instance in case of Fourier extraction approach, for the pose description containing 75 attributes we obtained 3000 separate features. In most cases only small parts of the whole feature set are discriminative from the point of view of given classification or diagnostics task and the remaining parts contain mainly noise. Thus feature selection which determines most valuable subset of features could be carried out.

Over a dozen quantitative measurements describing way of walking were defined as a results of various studies related to disturbances of gait. Such parameters as speed, length of step, its variability and symmetry were used to diagnose the effect of bilateral posteroventral pallidotomy on the walking patterns of patients with PD [6]. Researchers have used some more complex measures as Decomposition Indices (DI), Phase Coordination Index (PCI), Arm Swing Asymmetry (ASA), Arm Swing Size Symmetry (ASSS), Freezing of Gait (FoG) [7]. The first measure is defined as percentage of stance phase when one joint is moving while the other is not, calculated for hip–knee, hip–ankle, and knee–ankle pairs while the second measure reflects temporal accuracy and consistency of step timing. Walking measures illustrating asymmetry of arm swing, trunk rotation and stride time [8] were used to improve the diagnosis of early stages of Parkinson disease [9]. Based on the literature review of the aforementioned publications, several indices describing gait have been implemented and calculated for data from PD Patients collected in 4GAIT-Parkinson dataset. These parameters were calculated for left and right sides separately (Fig 2). Based on the results, it was found that the left and right components of these indices have some additional diagnostic value, because they allow for detecting which body side is more affected by the disease [1]. We named them Parkinson's Disease Gait Indexes. The other research involved the analysis of resting tremor, which is the one of the most important motor symptoms of Parkinson disease for diagnosis and treatment. Based on data collected in PJAIT RDC a system for analysis of tremor in PD patients during standing have been developed. Trajectory of marker located on wrist has been used to calculate several parameters as mean and max amplitude Fourier spectra of 3D signals, for right and left side, in two frequency range 4-6 [Hz] (typical Parkinsonian termor [11]) and 3-7 [Hz]. The results show that, there are statistical differences between session with and without treatment [10].

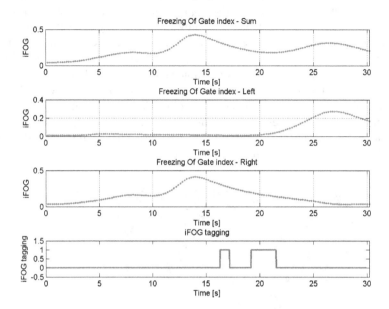

Fig. 2. An example of Gait Index. Summary of changes for Freezing of Gait factor in the subsequent time frames together with the observations made by a neurologist (iFOGtagging) [1]

5   Cloud Based Tool for Motion Data Processing

To fill in the gaps on tools and methods for motion processing, analysis and synthesis there have been undertaken research and development works at PJAIT RDC in Bytom, Poland. We developed decomposable and personal muscosceletal model applicable to medicine, algorithms for low level processing of motion data, the advanced motion database enabling to store processed and indexed data, together with the tools allowing the numerical analysis of chain of joints. The conducted research include development of the method, accepted by medical environment, for creation of low-dimensional representation of motion enabling perceptive comparison of movements, classification and motion visualization.

Results of the substantial studies provided practical achievement of 3 technological pillars: Human Motion Lab (HML), Cloud Based Human Motion Database (HMD) and Motion Data Editor (MDE) constituting the grounds for development of the existing and creation of the new advanced products and services in the area of motion analysis and synthesis.

Motion Data Editor is a software that originally was designed to support clinicians in viewing medical data for diagnosis of various human movement disorders. It has been reorganized and refactored to become a general purpose data processing software. MDE enables simultaneous communications with any number of Human Motion Databases with Internet access. It allows to process multimodal data from any number of measurements simultaneously and synchronously. It operates a few dozen industrial measured data formats. The application architecture uses data sources, processing and terminating elements and allows to create any number of information flow graphs and flexible system development by any elements composing information flow graph. It allows users to dynamically create new data flows without need of writing new fragments of code and compilation procedure, based on delivered nodes functionality (see Fig.3). They can now define and launch data flows without any knowledge about programming. Visual Data Flow (VDF) environment supports user in creating data flows by presenting graphically available nodes. It guides users, how particular nodes can be connected according to basic model rules and data types compatibility to infer new data from existing measurements.

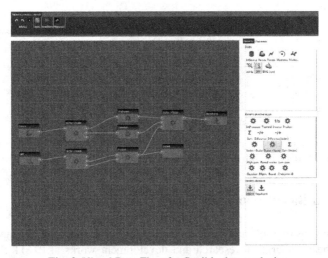

Fig. 3. Visual Data Flow for flexible data analysis

Visual Data Flow has been used to extend PD measurements by computing required Parkinson Disease Gait Indexes for all PD data trials. 4GAIT-Parkinson is one of the datasets created in Human Motion Lab and enabled in Human Motion Database. It was created from measurements of participants performing 12 tasks under four experimental conditions called sessions. Each session groups the following tasks: sway, gait/tandem gait, turnover, walk at normal/fast speed, heel to toe walk in a straight line, sensomotoric test related to tracking light spots, pulling back test, arising from chair and leg agility test.

MDE as a client application enables direct access to 4GAIT-Parkinson multimodal dataset. Various data types can be observed by clinicians in MDE from many perspectives using Visualizers. Their main task is to simplify visualization of multimodal data, manage presented data in a form of data series, where user can add and remove more data to/from scene of the Visualizer to compare them visually, depending on

Visualizer capabilities. The same data can be viewed in many Visualizers showing its various perspectives (i.e. 2D plot and 3D scene). With the help of Visualizers user can introduce new data perspectives named Custom Multimodal Motion Information Views for any kind of handled data types in MDE (Fig.4).

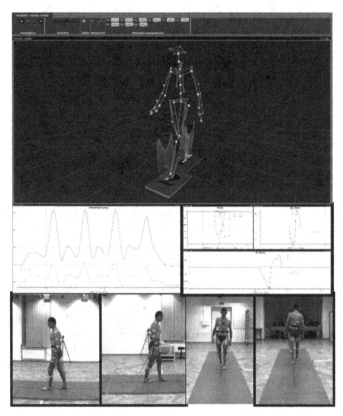

Fig. 4. Custom Multimodal Motion Information View in MDE (from up to down): 3D Visualizer with kinematic (joints, markers) and GRF data, , 2D Visualizers with motion, forces and moments trajectories, Video Visualizers with right, left, front and back camera view

Motion Data Editors gives physicians easy access to central repository with Parkinson multimodal data. MDE also enables very interesting collaborative functions: chat, scoring, easy integration with popular social networks like Twitter[TM], Facebook[TM], LinkedIn[TM].

Build in features for collaboration allow multiple physicians in different geographic locations to view the same data at the same time and score and discuss it. In particular they can access 4GAIT-Parkinson data individually, view selected parameters from measurements and collectively score the quality of proposed PDGI scale. Scoring criteria and function can be easily modified to specific requirements. That's why we decided to propose MDE as distributed group collaboration tool intended for collective diagnosis of Parkinson Disease cases and test it by two independent groups of physicians located in Boston, US and in Gliwice, Poland.

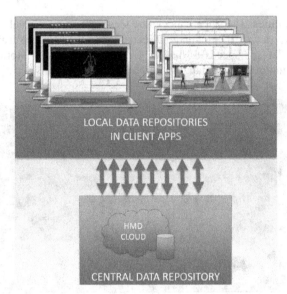

Fig. 5. MDE with HMD allows for data analysis and collective diagnosis on and off-line

6   Conclusions

In the paper the concept of the collaboration of experts group sharing, scoring and assessing of multimodal data was presented. Such an approach seems to be useful in cases, where there is a need for agreeing the expert evaluations and then finding appropriate measures, strongly correlated with the agreed assessment of the experts. Detection of described relation allows generating of the evaluation in the automatic mode only on the basis of the measured physical parameters. As an example of application we customized our tool in configuration with 4GAIT-Parkinson dataset and module for evaluation and collective quality assessment of PDGI.

We outlined our implemented cloud based group collaboration tool customized for collective diagnosis of Parkinson Disease data. We conducted early pilot tests on groups of experts from Poland and U.S. and our system received very positive feedback.

A popular statement that "No two patients face Parkinson's in quite the same way" may describe Parkinson's patient point of view on his/her disease. Patient's self-perception is subjective and depends on many factors but mostly on emotional states that are often related to the depressions and motor impairments. The PDGI supports objective assessment of neurological abnormalities. Collaborative expert's assessment of PDGI quality makes it more impartial.

An opinion of the neurologist is more objective as it is supported by objective but mostly not very precise interviews, tests and measurements of patient's symptoms such as UPDRS.

In the future after positive validation the quality of PDGI, applying multimodal tests should lead to automation of patients' disease stage assessment with partial doctor participation and replace subjective UPDRS measure by new objective measure.

Acknowledgement. This work was supported by projects NN 516475740 and NN 518289240 from the Polish National Science Centre.

Reference

1. Stawarz, M., Kwiek, S., Polański, A., Janik, Ł., Boczarska-Jedynak, M., Przybyszewski, A., Wojciechowski, K.: Algorithms for computing indexes of neurological gait abnormalities in patients after DBS surgery for Parkinson Disease based on motion capture data. Machine Graphics & Vision (2012)
2. Kulbacki, M., Segen, J., Nowacki, J.P.: 4GAIT: Synchronized MoCap, video, GRF and EMG datasets: acquisition, management and applications. In: Nguyen, N.T., Attachoo, B., Trawiński, B., Somboonviwat, K. (eds.) ACIIDS 2014, Part II. LNCS, vol. 8398, pp. 555–564. Springer, Heidelberg (2014)
3. Kulbacki, M., Janiak, M., Knieć, W.: Motion data editor software architecture oriented on efficient and general purpose data analysis. In: Nguyen, N.T., Attachoo, B., Trawiński, B., Somboonviwat, K. (eds.) ACIIDS 2014, Part II. LNCS, vol. 8398, pp. 545–554. Springer, Heidelberg (2014)
4. Rodriguez-Oroz, M.C., Obeso, J.A., Lang, A.E., et al.: Bilateral deep brain stimulation in Parkinson's disease: a multicentre study with 4 years follow-up. Brain **128**, 2240–2249 (2005)
5. Kenney, C., Simpson, R., Hunter, C., Ondo, W., Almaguer, M., Davidson, A., Jankovic, J.: Short-term and long-term safety of deep brain stimulation in the treatment of movement disorders. Neurosurg **106**, 621–625 (2007)
6. Siegel, K.L.; Metma, L.V.: Effects of Bilateral Posteroventral Pallidotomy on Gait of Subjects With Parkinson Disease. Neurology 57 (2000)
7. Mian, O.S., Schneider, S.A., Schwingenschuh, P., Bhatia, K.P., Day, D., Phil, B.L.: Gait in SWEDDs patients: Comparison with Parkinson's disease patients and healthy controls (2009)
8. Zifchock, R.A., Davis, I., Higginson, J., Royer, T.: The symmetry angle: a novel, robust method of quantifying asymmetry. Gait Posture **27**(4), 622–627 (2007)
9. Lewek, M.D., Poole, R., Johnson, J., Halawa, O., Huang, X.: Arm swing magnitude and asymmetry during gait in the early stages of Parkinson's disease. Gait Posture **31**, 256–260 (2010)
10. Stawarz, M., Polański, A., Kwiek, S., Boczarska-Jedynak, M., Janik, Ł., Przybyszewski, A., Wojciechowski, K.: A system for analysis of tremor in patients with parkinson's disease based on motion capture technique. In: Bolc, L., Tadeusiewicz, R., Chmielewski, L.J., Wojciechowski, K. (eds.) ICCVG 2012. LNCS, vol. 7594, pp. 618–625. Springer, Heidelberg (2012)
11. Findley, L.J., Gresty, M.A., Halmagyi, G.M.: Tremor, the cogwheel phenomenon and clonus in Parkinson's disease. J. Neurol Neurosurg Psychiatry **44**, 534–546 (1981)
12. Switonski, A., Mucha, R., Danowski, D., Mucha, M., Cieslar, G., Wojciechowski, K., Sieron, A.: Diagnosis of the motion pathologies based on a reduced kinematical data of a gait. Electrical Review 87(12B), 173–176 (2011)
13. Switonski, A., Mucha, R., Danowski, D., Mucha, M., Cieslar, G., Wojciechowski, K., Sieron, A.: Human identification based on a kinematical data of a gait. Electrical Review 87(12B), 169–172 (2011)
14. Świtoński, A., Polański, A., Wojciechowski, K.: Human Identification Based on Gait Paths. In: Blanc-Talon, J., Kleihorst, R., Philips, W., Popescu, D., Scheunders, P. (eds.) ACIVS 2011. LNCS, vol. 6915, pp. 531–542. Springer, Heidelberg (2011)
15. Switonski, A., Polanski, A., Wojciechowski, K.: Human identification based on the reduced kinematic data of the gait. In: International Symposium on Image and Signal Processing and Analisys, Dubrownik, pp. 650–655 (2011)

Dynamics Modeling of 3D Human Arm Using Switched Linear Systems

Artur Babiarz[✉], Adam Czornik, Jerzy Klamka, and Michał Niezabitowski

Faculty of Automatic Control, Electronics and Computer Science,
Institute of Automatic Control, Silesian University of Technology,
Akademicka 16 Street, 44-101 Gliwice, Poland
{artur.babiarz,adam.czornik,jerzy.klamka,michal.niezabitowski}@polsl.pl
http://www.ia.polsl.pl

Abstract. A novel approach to modeling human arm with hybrid systems theory is presented. The 3D human arm mathematical model is described. The arm model is built using three rotational link. Each of them is represented as a truncated cone prism. The shape of each link is changing during any motion. Based on the analysis of the arm motion a mathematical model of a switched linear system is proposed. The design process of state-dependent switching function and division of the state-space is shown. At the end, a few simulation results are presented.

Keywords: Dynamics of human arm · Switched system · Switching rule

1 Background and Significant

Over the past decade, greater and more specific attention undoubtedly focused on the modeling of mechanical systems such as exoskeletons. The exoskeletons are used primarily to support the movement of people with disabilities or to increase the strength of the human body [1], [2]. The models mentioned above objects mainly use knowledge about the structure of the human musculoskeletal system (also known as the locomotor system) [3]. The starting point for most of the work is the model described by the Euler-Lagrange formalism. In the literature, we can find simple models exoskeletons [4], [5], as well as models with kinematic chain which is the equivalent of the human locomotor system. Obviously, these objects are very complicated and simplification of dynamic models are acceptable [6], [7], [8]. A model of the exoskeleton with five degrees of freedom that was attached to human arm using an arm orthosis is described in [9], [10]. In addition, the articles dealt with the problem of measuring the orientation of the human arm by an orientation sensor and a rotary encoder. In [11], it is studied the trajectory planning of an anthropomorphic human arm and a concept of movement primitive. Besides, the authors present human arm triangle space as an intermediate space between joint space and task space.

The primary motivation for using switched linear systems comes partly from reaserch results published in [12], [13], [14], [15]. In this article, we focus on the

© Springer International Publishing Switzerland 2015
N.T. Nguyen et al. (Eds.): ACIIDS 2015, Part II, LNAI 9012, pp. 258–267, 2015.
DOI: 10.1007/978-3-319-15705-4_25

switched linear systems that can be described as follows:

$$\dot{x} = A_{\sigma(\cdot)}x + B_{\sigma(\cdot)}u,$$
$$y = C_{\sigma(\cdot)}x + D_{\sigma(\cdot)}u.$$

$$(1)$$

where: $x \in \mathcal{R}^n$ is the state, $u \in \mathcal{R}^m$ is control signal, $y \in \mathcal{R}^q$ is output, $\sigma(\cdot)$: $\mathcal{P} \to \{1, 2, \ldots, N\}$ is the switching rule, and $A_i, B_i, C_i, D_i, i = 1, 2, \ldots, N$ are constant matrices.

In the literature, there are no known cases of the above-mentioned systems for modeling the human arm. The initial work focused on a model of human upper limb, which the state space is divided into two regions only [16]. Subsequent research concentrated on modeling of lower or upper human limbs [17], [18]. In these studies, the authors show mathematical models of human arm or leg as switched linear systems. The state space of presented switched linear systems was divided for more regions than in [16].

1.1   Motivation

The research results presented in [14], [15] confirm that the shape of the human lower and upper limbs deformed during the execution of any movement. Moreover, the dimensions and shape of the human limbs do not depend on a single muscle, but group of muscles (a phenomenon so-called the muscle synergism).

Remark 1. Under the above conclusions, we can assume that the matrix of inertia and the distance from the center of gravity of each joint, are changed. In addition, changes of these parameters are dependent on the angular displacement of the arm. Besides, it is mentioned the muscles effect were omitted. They have influence on the shape of the each link only.

Furthermore, research results published in [19] justify the application of hybrid systems for modeling objects with complex biomechanical structure. On the other hand, results presented in the work [12], [13] indicate that the human arm is unstable in the considered range of motion. Due to the biomechanical limitations of motion of each link, a operating space of human limbs can be naturally divided. As a result, we can obtain a set of subsystems. The conclusion from the above analysis is the basis for modeling of the human arm using a switched system. Such a system has a switching function depending on the state vector. Consequently, we deduce the using theory of hybrid systems is a very novel approach to modeling and analysis of dynamic properties of human limbs.

The structure of this paper is as follows. At the beginning, the mathematical model of three-dimensional human arm is presented. The next section focuses on the description of the human arm dynamics using switched linear systems. In this section, the partition of the state-space and switching function are shown. The simulation results are presented in Section 3. Finally, we conclude our approach to mathematical modeling of human arm dynamics.

2 Mathematical Model of Human Arm

2.1 State-Space Model

The mathematical model of the arm can be obtained using the standard Euler-Lagrange description. The equation of motion is as follows:

$$M(q)\ddot{q} + C(q,\dot{q})\dot{q} + G(q) + B\dot{q} = u \tag{2}$$

where: $M(q) \in \mathcal{R}^{3\times3}$ - is a positive definite symmetric inertia matrix, $C(q,\dot{q}) \in \mathcal{R}^{3\times3}$ - is Coriolis and centrifugal forces matrix, $G(q) \in \mathcal{R}^3$ - is gravity forces vector, $B \in \mathcal{R}^{3\times3}$ - is the joint friction matrix, $u \in \mathcal{R}^3$ - is forces and moments acting on the system, $q \in \mathcal{R}^3$ - is angular displacement.

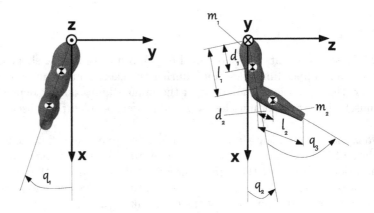

Fig. 1. Kinematics scheme of 3D human arm

The human arm model is presented in Fig. 1. The dynamic equation of a three-dimensional human arm is described by nonlinear state equation (2) where:

$$M(q) = \begin{bmatrix} m_{11} & m_{12} & m_{13} \\ m_{21} & m_{22} & m_{23} \\ m_{31} & m_{32} & m_{33} \end{bmatrix} \quad C(q,\dot{q}) = \begin{bmatrix} c_{11} & c_{12} & c_{13} \\ c_{21} & c_{22} & c_{23} \\ c_{31} & c_{32} & c_{33} \end{bmatrix}$$

$$G(q) = \begin{bmatrix} g_{11} \\ g_{21} \\ g_{31} \end{bmatrix} \quad B = \begin{bmatrix} b_{11} & b_{12} & b_{13} \\ b_{21} & b_{22} & b_{23} \\ b_{31} & b_{32} & b_{33} \end{bmatrix}$$

$$\begin{aligned}
m_{11} =\ & m_1 d_1^2 \cos^2(q_2) + I_{z1} + m_2 l_1^2 \cos^2(q_2) + m_2 l_1^2 \cos^2(q_2)\sin(q_3) \\
& + m_2 l_1^2 \sin(q_2)\cos(q_2)\cos(q_3) + m_2 l_1^2 \cos^2(q_2)\cos(q_3) \\
& + m_2 l_1^2 \sin(q_2)\cos(q_2)\sin(q_3) + m_2 d_2^2 \cos^2(q_2 + q_3),
\end{aligned}$$

$$m_{12} = 0, \quad m_{13} = 0, \quad m_{21} = 0,$$

$$m_{22} = 2m_2 l_1 d_2 \cos(q_3) + m_2 d_2^2 + m_2 l_1^2 + m_1 d_1^2 + I_{x1},$$

$$m_{23} = m_2 l_1 d_2 \cos(q_3) + m_2 d_2^2,$$

$$m_{31} = 0, \quad m_{32} = m_2 l_1 d_2 \cos(q_3) + m_2 d_2^2 \quad m_{33} = m_2 d_2^2 + I_{z2}$$

$$c_{11} = 0,$$

$$
\begin{aligned}
c_{12} = & - 2m_1 d_1^2 \cos(q_2) \sin(q_2) \dot{q}_1 - 2m_2 l_1^2 \cos(q_2) \sin(q_2) \dot{q}_1 \\
& - 2m_2 l_1^2 \sin(q_2) \cos(q_2) \sin(q_3) \dot{q}_1 + m_2 l_1^2 \cos^2(q_2) \cos(q_3) \dot{q}_1 \\
& - m_2 l_1^2 \sin^2(q_2) \cos(q_3) \dot{q}_1 - 2m_2 l_1^2 \sin(q_2) \cos(q_2) \cos(q_3) \dot{q}_1 \\
& + m_2 l_1^2 \cos^2(q_2) \sin(q_3) \dot{q}_1 - m_2 l_1^2 \sin^2(q_2) \sin(q_3) \dot{q}_1 \\
& - m_2 d_2^2 \sin(q_2) \cos(q_2) \dot{q}_1 - m_2 d_2^2 \cos(q_2) \sin(q_3) \dot{q}_1
\end{aligned}
$$

$$
\begin{aligned}
c_{13} = & \, m_2 l_1^2 \cos^2(q_2) \cos(q_3) \dot{q}_1 - m_2 l_1^2 \sin(q_2) \cos(q_2) \sin(q_3) \dot{q}_1 \\
& - m_2 l_1^2 \sin(q_2) \cos(q_2) \cos(q_3) \dot{q}_1 - m_2 d_2^2 \sin(q_2 + q_3) \dot{q}_1,
\end{aligned}
$$

$$
\begin{aligned}
c_{21} = & \, 0.5 m_2 l_1 d_2 \cos(q_2)(\cos(q_2 + q_3) - \sin(q_2 + q_3)) \dot{q}_1 \\
& - 0.5 m_2 l_1 d_2 (\sin(q_2 + q_3) + \cos(q_2 + q_3)) \dot{q}_1 \\
& - m_1 d_1^2 \cos(q_2) \sin(q_2) \dot{q}_1 - m_2 l_1^2 \cos(q_2) \sin(q_2) \dot{q}_1 \\
& - m_2 d_2^2 \cos(q_2 + q_3) \sin(q_2 + q_3),
\end{aligned}
$$

$$c_{22} = 0 \quad c_{23} = -m_2 l_1 d_2 \sin(q_3) \dot{q}_3$$

$$c_{31} = -(0.5 m_2 l_1 d_2 c_2 (\cos(q_2 + q_3) - \sin(q_2 + q_3)) + m_2 d_2^2 \cos(q_2 + q_3) \sin(q_2 + q_3)) \dot{q}_1$$

$$c_{32} = m_2 l_1 d_2 \sin(q_3) \dot{q}_2 \quad c_{33} = 0,$$

$$g_{11} = m_1 g d_1 \sin(q_1) \cos(q_2) - m_2 g \sin(q_1)(l_1 \cos(q_2) + d_2 \cos(q_2 + q_3))$$

$$g_{21} = m_1 g d_1 \cos(q_1) \sin(q_2) - m_2 g \cos(q_1)(l_1 \sin(q_2) + d_2 \sin(q_2 + q_3))$$

$$g_{31} = m_2 g d_2 \cos(q_1) \sin(q_2 + q_3)$$

$$b_{11} = b_{22} = b_{33} = 0.1, \quad b_{12} = b_{21} = b_{31} = 0.2, \quad b_{13} = b_{23} = b_{32} = 0.05.$$

where: m - is the mass, l_i - is the link length, d_i - is the distance from the joint to the center of mass, I_{zi}, I_{xi} - is the moment of inertia around the z and x axis, respectively.

The dynamics of model in terms of the state vector $\left[q^T, \dot{q}^T\right]^T$ can be expressed as:

$$\frac{d}{dt} \begin{bmatrix} q \\ \dot{q} \end{bmatrix} = \begin{bmatrix} \dot{q} \\ M(q)^{-1}[u - C(q, \dot{q})\dot{q} - G(q)] \end{bmatrix}, \tag{3}$$

where it is assumed that the inertia matrix M is invertible. Positive definiteness of M is seen directly by the fact that the kinetic energy is always nonnegative and is equal to zero if and only if all the joint velocities are zero. Thus, M is invertible and equation (3) is valid. Now, a new set of variables can be assigned to each of the derivatives. The new set of state variables and their equivalences can be expressed as follows:

$$
\begin{aligned}
x_1 &= q_1, \quad x_2 = q_2, \quad x_3 = q_3, \\
x_4 &= \dot{x}_1 = \dot{q}_1, \quad x_5 = \dot{x}_2 = \dot{q}_2, \quad x_6 = \dot{x}_3 = \dot{q}_3.
\end{aligned}
\tag{4}
$$

We can write the general output and state equations:

$$
\dot{x} = Ax + Bu, \tag{5}
$$
$$
y = Cx + Du, \tag{6}
$$

where:

$$
\begin{aligned}
\dot{x} &= \left[\dot{q}_1 \ \dot{q}_2 \ \dot{q}_3 \ \ddot{q}_1 \ \ddot{q}_2 \ \ddot{q}_3\right]^T, \ x = \left[q_1 \ q_2 \ q_3 \dot{q}_1 \dot{q}_2 \dot{q}_3\right]^T, \\
y &= \left[\ddot{q}_1 \ \ddot{q}_2 \ \ddot{q}_3\right]^T, \ u = \left[u_1 \ u_2 \ u_3\right]^T.
\end{aligned}
\tag{7}
$$

Matrices A, B, C and D are computed using the series expansion linearization method.

2.2 Switched Linear System

Model of the human arm has limitations of movement. The limitations arise from the physics of motion in the joint. We assume that the ranges of each joint movement are constrained:

$$
\begin{aligned}
-3.14\,[rad] &\leqslant x_1 \leqslant 0\,[rad], \\
-1.22\,[rad] &\leqslant x_2 \leqslant 3.14\,[rad], \\
0\,[rad] &\leqslant x_3 \leqslant 2.96\,[rad],
\end{aligned}
$$

According to section 1.1, we can design switched linear system which based on (3) and (7). It should be pointed out, that the switching function is state-dependent [17], [20]. Then, the mathematical model can be described by equations:

$$
\dot{x}(t) = A_{\sigma(x)}x(t) + B_{\sigma(x)}u(t), \tag{8}
$$
$$
y(t) = C_{\sigma(x)}x(t) + D_{\sigma(x)}u(t). \tag{9}
$$

Generally speaking, we consider the switched linear systems with state-dependent switching. The choice of active dynamics is determined strictly by the current state $x(t)$. So $\mathcal{R}^n$ is divided into a collection of disjoint regions $\Phi_1, ..., \Phi_i, ..., \Phi_N$ with $\Phi_1 \bigcup \cdots \bigcup \Phi_N = \mathcal{R}^n$, and then

$$
\dot{x} = \begin{cases}
A_1 x + B_1 u & \text{if} \quad x \in \Phi_1 \\
\vdots \\
A_i x + B_i u & \text{if} \quad x \in \Phi_i \\
\vdots \\
A_N x + B_N u & \text{if} \quad x \in \Phi_N
\end{cases}
$$

Furthermore, it is assumed that only the angular displacement (the first three elements of the state vector: x_1, x_2 and x_3) influences on the change of the shape parameters of the each link during the movement. Consequently, the angular velocity (the last three elements of the state vector: x_4, x_5 and x_6) may be any for each subsystem of model described by equations (10) and (11).

$$\dot{x} = \begin{cases} A_1 x + B_1 u & \text{if} & -1.57 \le x_1 < 0, x_2 = 0, x_3 \ge 0 \\ A_2 x + B_2 u & \text{if} & -1.57 \le x_1 < 0, x_2 > 0, x_3 > 0 \\ A_3 x + B_3 u & \text{if} & -1.57 \le x_1 < 0, x_2 < 0, x_3 = 0 \\ A_4 x + B_4 u & \text{if} & -1.57 \le x_1 < 0, x_2 < 0, x_3 > 0 \\ A_5 x + B_5 u & \text{if} & -3.14 \le x_1 \le -1.57, x_2 = 0, x_3 \ge 0 \\ A_6 x + B_6 u & \text{if} & -3.14 \le x_1 \le -1.57, x_2 > 0, x_3 > 0 \\ A_7 x + B_7 u & \text{if} & -3.14 \le x_1 \le -1.57, x_2 < 0, x_3 = 0 \\ A_8 x + B_8 u & \text{if} & -3.14 \le x_1 \le -1.57, x_2 < 0, x_3 > 0 \end{cases} \quad (10)$$

$$y = \begin{cases} C_1 x + D_1 u & \text{if} & -1.57 \le x_1 < 0, x_2 = 0, x_3 \ge 0 \\ C_2 x + D_2 u & \text{if} & -1.57 \le x_1 < 0, x_2 > 0, x_3 > 0 \\ C_3 x + D_3 u & \text{if} & -1.57 \le x_1 < 0, x_2 < 0, x_3 = 0 \\ C_4 x + D_4 u & \text{if} & -1.57 \le x_1 < 0, x_2 < 0, x_3 > 0 \\ C_5 x + D_5 u & \text{if} & -3.14 \le x_1 \le -1.57, x_2 = 0, x_3 \ge 0 \\ C_6 x + D_6 u & \text{if} & -3.14 \le x_1 \le -1.57, x_2 > 0, x_3 > 0 \\ C_7 x + D_7 u & \text{if} & -3.14 \le x_1 \le -1.57, x_2 < 0, x_3 = 0 \\ C_8 x + D_8 u & \text{if} & -3.14 \le x_1 \le -1.57, x_2 < 0, x_3 > 0 \end{cases} \quad (11)$$

The switching rule has been designed in accordance with mechanical movement limitations.

3   Experimental Results

In order to obtain the switched linear model, the operating points have been arbitrarily selected from each subspaces Ω_i, $i = 1, 2, \ldots, 8$ of a state space. Fixed parameters, used for linearization of the arm, are shown in Table 1. Whereas, Table 2 contains the variable model parameters: the distance from the joint to the center of mass and the moment of inertia for each Ω_i. The shape of each arm link is approximated a truncated cone, the dimensions of which depend on the configuration of the arm. Strictly speaking, the moment of inertia of each link depends on the value of the state vector elements x_1, x_2 and x_3.

Table 1. The parameters of human arm

	Link 1	Link 2
$m\ [kg]$	1.4	1.1
$l\ [m]$	0.3	0.33

Table 2. The parameters of switched model

	$d_1\ [m]$	$d_2\ [m]$	$I_{x1}\ [kgm^2]$	$I_{z1}\ [kgm^2]$	$I_{z2}\ [kgm^2]$
The case I	0.11	0.16	0.027	0.025	0.045
The case II	0.1	0.16	0.025	0.029	0.045
The case III	0.12	0.14	0.021	0.024	0.043
The case IV	0.1	0.15	0.024	0.023	0.046
The case V	0.14	0.16	0.024	0.022	0.043
The case VI	0.13	0.14	0.022	0.023	0.044
The case VII	0.13	0.15	0.024	0.023	0.044
The case VIII	0.15	0.14	0.024	0.023	0.042

The simulation results were obtained using Matlab Simulink package and S-Function [21]. The control signal is modeled as $u_1 = \sin(5t)$, $u_2 = \cos(2t)$, $u_3 = 0.1\sin(5t)$ and the initial condition is equal $x_0 = \left[-0.52\,[rad]; 0.52\,[rad]; 0.17\,[rad]; \right.$
$\left. 0\,[\frac{rad}{s}]; 0\,[\frac{rad}{s}]; 0\,[\frac{rad}{s}] \right]^T$. Figures 2-7 present time history of six elements of state vector.

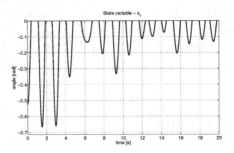

Fig. 2. A scope of x_1 state variable

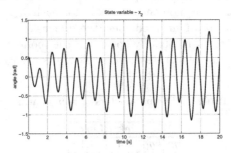

Fig. 3. A scope of x_2 state variable

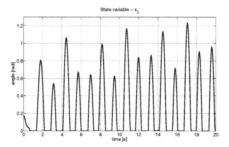

Fig. 4. A scope of x_3 state variable

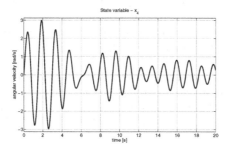

Fig. 5. A scope of x_4 state variable

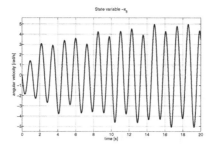

Fig. 6. A scope of x_5 state variable

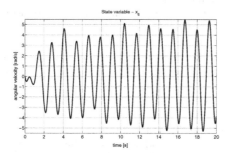

Fig. 7. A scope of x_6 state variable

4 Conclusions

In the article, we showed a way of obtaining a three degrees-of-freedom mathematical model of a human arm. Switched linear system model was used. Additionally, it was shown that the switching function should be state-dependent. The consideration on the use other elements of the state vector for the construction of a switching rule will be the next step in our research. In consequence, the research tasks should include also the analysis of switching dependent of the elements of state vector.

A fortiori, referring to the results of [19], [22] subsystems of the model of human arm might be unstable or on the stability boundary and may be unobservable. For this reason, we plan to perform an analysis of such properties as stability, controllability and observability [23], [24]. The subsequent research will address the application of the fractional order switched systems theory and discrete-time switched systems to modeling of human arm [25], [26].

Acknowledgments. The research presented here were funded by the Silesian University of Technology grant BK-265/RAu1/2014/2 (A.B.), the National Science Centre according to decision the National Science Centre granted according to decisions DEC-2012/05/B/ST7/00065 (A.C.), DEC-2012/07/N/ST7/03236 (M.N.) and DEC-2012/07/B/ST7/01404 (J.K.).

References

1. Kong, K., Tomizuka, M.: Control of exoskeletons inspired by fictitious gain in human model. IEEE/ASME Transactions on Mechatronics **14**, 689–698 (2009)
2. Pons, J.L., Moreno, J.C., Brunetti, F.J., Rocon, E.: Lower-Limb Wearable Exoskeleton In Rehabilitation Robotics, pp. 471–498. I-Tech Education and Publishing (2007)
3. Sekine, M., Sugimori, K., Gonzalez, J., Yu, W.: Optimization-Based Design of a Small Pneumatic-Actuator-Driven Parallel Mechanism for a Shoulder Prosthetic Arm with Statics and Spatial Accessibility. Evaluation. International Journal of Advanced Robotic Systems 286 (2013)
4. Csercsik, D.: Analysis and control of a simple nonlinear limb model. PhD Thesis, University of Technology (2005)
5. Choudhury, T.T., Rahman, M.M., Khorshidtalab, A., Khan, M.R.: Modeling of Human Arm Movement: A Study on Daily Movement. In: Fifth International Conference on Computational Intelligence, Modelling and Simulation (CIMSim), pp. 63–68 (2013)
6. Zawiski, R., Błachuta, M.: Model development and optimal control of quadrotor aerial robot. In: 17th International Conference on Methods and Models in Automation and Robotics (MMAR), pp. 475–480 (2012)
7. Błachuta, M., Czyba, R., Janusz, W., Szafrański, G.: Data Fusion Algorithm for the Altitude and Vertical Speed Estimation of the VTOL Platform. Journal of Intelligent and Robotic Systems **74**, 413–420 (2014)
8. Garrido, A.J., Garrido, I., Amundarain, M., Alberdi, M.: Sliding-mode control of wave power generation plants. IEEE Transactions on Industry Applications **48**, 2372–2381 (2012)

9. Mao, Y., Agrawal, S.K.: Transition from mechanical arm to human arm with CAREX: A cable driven ARm EXoskeleton (CAREX) for neural rehabilitation. In: 2012 IEEE International Conference on Robotics and Automation (ICRA), pp. 2457–2462 (2012)
10. Mao, Y., Agrawal, S.K.: Design of a Cable-Driven Arm Exoskeleton (CAREX) for Neural Rehabilitation. IEEE Transactions on Robotics 28, 922–931 (2012)
11. Ding, X., Fang, C.: A motion planning method for an anthropomorphic arm based on movement primitives of human arm triangle. In: International Conference on Mechatronics and Automation (ICMA), pp. 303–310 (2012)
12. Burdet, E., Tee, K.P., Mareels, I., Milner, T.E., Chew, C.-M., Franklin, D.W., Osu, R., Kawato, M.: Stability and motor adaptation in human arm movements. Biological Cybernetics 94, 20–32 (2006)
13. Chen, K.: Modeling of equilibrium point trajectory control in human arm movements. PhD Thesis, New Jersey Institute of Technology (2011)
14. Lee, D., Glueck, M., et al.: A survey of modeling and simulation of skeletal muscle. ACM Transactions on Graphics 28, 162 (2010)
15. Neumann, T., Varanasi, K., Hasler, N., Wacker, M., Magnor, M., Theobalt, C.: Capture and Statistical Modeling of Arm-Muscle Deformations. Computer Graphics Forum 32, 285–294 (2013)
16. Babiarz, A.: On mathematical modelling of the human arm using switched linear system. In: AIP Conference Proceedings, vol. 1637, pp. 47–54 (2014)
17. Babiarz, A., Czornik, A., Klamka, J., Niezabitowski, M., Zawiski, R.: The mathematical model of the human arm as a switched linear system. In: 19th International Conference on Methods and Models in Automation and Robotics (MMAR), pp. 508–513 (2014)
18. Babiarz, A., Czornik, A., Niezabitowski, M., Zawiski, R.: Mathematical model of a human leg - the switched linear system approach. In: International Conference on Pervasive and Embedded Computing and Communication Systems (PECCS) (accepted, 2015)
19. Babiarz, A., Bieda, R., Jaskot, K., Klamka, J.: The dynamics of the human arm with an observer for the capture of body motion parameters. Bulletin of the Polish Academy of Sciences: Technical Sciences 61, 955–971 (2013)
20. Liberzon, D.: Switching in systems and control. Springer (2003)
21. Goretti Sevillano, M., Garrido, I., Garrido, A.J.: Sliding-mode loop voltage control using ASTRA-matlab integration in tokamak reactors. International Journal ofInnovative Computing, Information and Control 8, 6473–6489 (2012)
22. Babiarz, A., Klamka, J., Zawiski, R., Niezabitowski, M.: An Approach to Observability Analysis and Estimation of Human Arm Model. In: 11th IEEE International Conference on Control and Automation, pp. 947–952 (2014)
23. Czornik, A., Niezabitowski, M.: Controllability and stability of switched systems. In: 18th International Conference on Methods and Models in Automation and Robotics (MMAR), pp. 16–21 (2013)
24. Klamka, J., Czornik, A., Niezabitowski, M.: Stability and controllability of switched systems. Bulletin of the Polish Academy of Sciences. Technical Sciences 61, 547–555 (2013)
25. Czornik, A., Świerniak, A.: Controllability of discrete time jump linear systems. Dynamics of Continuous Discrete and Impulsive Systems-Series B-Applications & Algorithms 12, 165–189 (2005)
26. Tejado, I., Valério, D., Pires, P., Martins, J.: Optimal Feedback Control for Predicting Dynamic Stiffness During Arm Movement. Mechatronics 23, 805–812 (2013)

Machine Learning on the Video Basis of Slow Pursuit Eye Movements Can Predict Symptom Development in Parkinson's Patients

Andrzej W. Przybyszewski[1,2(✉)], Stanisław Szlufik[3], Justyna Dutkiewicz[3],
Piotr Habela[2], and Dariusz M. Koziorowski[3]

[1] Department Neurology, University of Massachusetts Medical School,
Worcester, MA 01655, USA
Andrzej.Przybyszewski@umassmed.edu
[2] Polish-Japanese Institute of Information Technology, 02-008 Warszawa, Poland
{przy,piotr.habela}@pjwstk.edu.pl
[3] Department Neurology, Faculty of Health Science,
Medical University of Warsaw, Warsaw, Poland
stanislaw.szlufik@gmail.com, justyna_dutkiewicz@wp.pl,
dkoziorowski@esculap.pl

Abstract. We still do not know exactly how brain processes are affected by nerve cell deaths in neurodegenerative diseases such as Parkinson's (PD). Early diagnosis when symptom progressions are precisely monitored may result in improved therapies. In the case of PD, measurements of eye movements (EM) can be diagnostic. In order to better understand their relationship to the underlying disease process, we have performed measurements of slow (POM) eye movements in PD patients. We have compared our measurements and algorithmic diagnoses with doctor's diagnoses. We have used rough set theory and machine learning (ML), to classify how condition attributes predict the neurologist's diagnosis. We have measured pursuit ocular movements (POM) for three different frequencies and estimated patients' performance by gain and accuracy for each frequency. We have tested ten PD patients in four sessions related to combination of medication and DBS treatments. We have obtained a global accuracy in individual patients' UPRDS III predictions of about 80%, based on cross-validation. This demonstrates that POM may be a good biomarker helping to estimate PD symptoms in automatic, objective and doctor-independent way.

Keywords: Neurodegenerative disease · Rough set · Machine learning

1 Introduction

Our approach is to demonstrate an alternative to the mostly used statistical analysis of PD outcomes by using data mining and machine learning (ML) methods. We gave examples that our methods give a more precise description of individual patient's symptoms and development. We may propose an individual treatment adjusted to different

© Springer International Publishing Switzerland 2015
N.T. Nguyen et al. (Eds.): ACIIDS 2015, part II, LNCS 9012, pp. 268–276, 2015.
DOI: 10.1007/978-3-319-15705-4_26

patients that may lead more effectively than now to slowing of symptoms and improvements in quality of life. Our analysis is proposed on the basis of learning algorithms that intelligently process data of each patient in an individual and specific ways.

Our symptom classification method follows the principle of the complex object recognition such as those in visual systems. The ability of natural vision to recognize objects arises in the afferent, ascending pathways that classify properties of objects' parts from simple attributes in lower sensory areas, to more complex ones, in higher analytic areas. The resulting classifications are compared and adjusted by interaction with whole object ("holistic") properties (representing the visual knowledge) at all levels using interaction with descending pathways [1] that was confirmed in animal experiments [2]. These interactions at multiple levels between measurements and prior knowledge can help to differentiate individual patient's symptoms and response treatments variability in a way similar to a new, complex object inspection [3, 4]. Machine learning algorithms for analyzing subtle signal variations will hopefully lead to better analysis of individual patients' conditions. As it was demonstrated in [1, 3, 4] properties of the primates visual system can be well described by rough set theory, therefore we have applied the same concept to knowledge discovery from symptoms in PD.

Based on their experience, intuition and at least partly subjective measurements neurologists are giving diagnosis of individual patients. They use "Golden Standard" by estimation values of the Hoehn and Yahr scale and the UPDRS (Unified Parkinson's Disease Rating Scale). As different doctors are not always in the precise way perform exactly same procedure their diagnosis are partially subjective that may lead to different treatments. We propose to formalize the whole process and use the neurologist's diagnosis as decision attributes and their and our doctor-independent measurements as condition attributes.

2   Methods

Our experiments were performed on ten Parkinson Disease (PD) patients who had undergone the Deep Brain Stimulation (DBS) surgery mainly for treatment of their motor symptoms. They were qualified for the surgery and observed postoperatively in the Dept. of Neurology and got surgical DBS implementation in the Institute of Neurology and Psychiatry [5]. We conducted horizontal POM (pursuit ocular movement - as explained below) measurements in ten PD patients during four sessions designated as S1: MedOffDBSOff, S2: MedOffDBSOn, S3: MedOnDBSOff, S4: MedOnDBSOn. During the first session (S1) the patient was off medications (L-Dopa) and DBS stimulators was OFF; in the second session (S2) the patient was off medication, but the stimulator was ON; in the third session (S3) the patient was after his/her doses of L-Dopa and the stimulator was OFF, and in the fourth session (S4) the patient was on medication with the stimulator ON. Changes in motor performance, behavioral dysfunction, cognitive impairment and functional disability were evaluated in each session according to the UPDRS. The pursuit (POM) was recorded by head-mounted saccadometer (Ober Consulting, Poland). We have used an infrared eye track system coupled with a head tracking system (JAZZ-pursuit – Ober Consulting, Poland) in order to obtain high accuracy and precision in eye tracking and to compensate possible subjects' head movements relative to the monitor. Thus subjects did not need to be positioned in an unnatural chinrest.

A patient was sited at the distance of 60-70 cm from the monitor with head supported by a headrest in order to minimize head motion. We measured slow eye movements in response to a light spot with horizontal sinusoidal movements (three frequencies: 0.125, 0.25, 0.5Hz) from 10 deg to the left to 10 deg to the right (the exact rage of the spot amplitude (in degrees) depends on the patient's distance from the screen). At first the patient has to fixate eyes on the spot in the middle marker (0 deg) the spot was placed in 10 deg to the left and 10 deg to the right for the calibration. In the next step, patients had to look at the targets (small square) and follow its sinusoidal, horizontal movement.

In each test the subject had to perform 4 periods of POM with low and 10 with higher frequencies in Med-off (medication off) within two situations: with DBS off (S1) and DBS on (S2). In the next step the patient took medication and had a break for one half to one hour, and then the same experiments were performed, with DBS off (S3) and DBS on (S4). In this work we have analyzed POM data using the following population parameters averaged for both eyes: gain (eye movement amplitude/sinus amplitude) and accuracy (difference between sinusoid and eye positions) for three different frequencies.

2.1   Theoretical Basis

We represent our data in the form of information system that is also called the decision table. We define such an information system (after Pawlak [6]) as a pair $S = (U, A)$, where U, A are nonempty finite sets called the *universe of objects* and the *set of attributes*, respectively. If $a \in A$ and $u \in U$, the value $a(u)$ is a unique element of V (where V is a value set).

The *indiscernibility relation* of any subset B of A or $IND(B)$, is defined [6] as follows: $(x, y) \in IND(B)$ or $xI(B)y$ if and only if $a(x) = a(y)$ for every a $\in B$, where $a(x) \in V$. $IND(B)$ is an equivalence relation, and $[u]_B$ is the equivalence class of u, or a *B-elementary granule*. The family of all equivalence classes of $IND(B)$ will be denoted $U/I(B)$ or U/B. The block of the partition U/B containing u will be denoted by $B(u)$.

We define a *lower approximation* of symptoms set $X \subseteq U$ in relation to a symptom attribute B as $\underline{B}X = \{u \in U: [u]_B \subseteq X \}$, and the *upper approximation* of X as $\overline{B}X = \{u \in U: [u]_B \cap X \neq \phi\}$. It means that, symptoms are classified into two categories (sets). The lower approximation set X has the property that all symptoms with certain attributes are part of X, and the upper movement approximation set has property that only some symptoms with attributes in B are part of X (see [5]). The difference of $\overline{B}X$ and $\underline{B}X$ is defined as the boundary region of X i.e., $BN_B(X)$. If $BN_B(X)$ is empty set than X is *exact (crisp)* with respect to B; otherwise if $BN_B(X) \neq \phi$ and X is not *exact* (i.e., it is *rough*) with respect to B. We say that the B-lower approximation of a given set

X is union of all *B-granules* that are included in X, and the B-upper approximation of X is of the union of all *B-granules* that have nonempty intersection with X.

The system S will be called a decision table $S = (U, C, D)$ where C is the condition and D is the decision attribute [6]. In the table below (Table 2), as an example, the decision attribute D, based on the expert opinion, is placed in the last column, and condition attributes measured by the neurologist, are placed in other columns. One can interpret each row in the table as a rule. As the number of rules is same as the number of rows, and each row is related to different measurements, these rules can have many particular conditions. We would like to describe different symptoms in different patients by using such rules. On the basis of such rules, using the modus ponens principle we wish to find universal rules to relate symptoms and treatments in different patients [6].   As symptoms even for the same treatments are not always the same; our rules must have certain "flexibility", or granularity, which can be interpreted as the probability of finding certain symptoms in a group of patients under consideration. The granular computation simulates the way in which neurologists interact with patients. This way of thinking relies on the ability to perceive a patient's symptoms under various levels of granularity (i.e., abstraction) in order to abstract and consider only those symptoms that serve to determine a specific treatment and thus to switch among different granularities. By focusing on different levels of granularity, one can obtain different levels of knowledge, as well as a greater understanding of the inherent knowledge structure. As one of us has demonstrated [1, 2] that the visual system is using the granular computing in object recognition, we suggest that this approach is essential for human intelligent.

We define the **reduct** $B{\subset}A$. The set B is a reduct of the information system if $IND(B) = IND(A)$ and no proper subset of B has this property. In case of decision tables decision reduct is a set $B{\subset}A$ of attributes which cannot be further reduced and $IND(B) \subset IND(d)$. A decision rule is a formula of the form $(a_{i1} = v_1) \wedge ... \wedge (a_{ik} = v_k) \Rightarrow d = v_d$, where $1 \le i_1 < ... < i_k \le m$, $v_i \in Va_i$. Atomic subformulas $(a_{i1} = v_1)$ are called conditions. In this way, we can replace the original attribute a_i with new, binary attributes, which indicate whether actual attribute value for an object is greater or lower than c (see [7]), we define c as a cut. Thus a cut for an attribute $a_i \in A$, with V_{ai} will be a value $c \in V_{ai}$. A template of A is a propositional formula $v_i \in V_{ai}$. A generalized template is a formula of the form $\wedge(a_i \in T_i)$ where $T_i \subset V_{ai}$. An object satisfies (matches) a template if for every attribute a_i we have $a_i = v_i$ where $a_i \in A$. The template is a method to split the original information system into two distinct sub-tables. One of these sub-tables consists of the objects that satisfy the template, while the second contains all others. A *decomposition tree* is defined as a binary tree, whose every internal node is labeled by some template and external node (leaf) is associated with a set of objects matching all templates in a path from the root to a given leaf [8].

In a second test we have divided our data into two or more subsets. By training on all but one of these subsets (the training set) using machine learning (ML), we obtained classifiers that when applied to the remaining (test) set gave new numerical decision attributes, well correlated with neurologist decision attributes (based on a confusion matrix).

3   Results

The patients' mean age was 58.3±9.3(SD) years, mean disease duration was 10.9±1.6 years, mean UPDRS (related to all symptoms): S1: 59.4±16.2 S2: 29.9±13.3; S3: 51.2±14.4; S4: 18.2±11.4; mean UPDRS III (related only to motor symptoms): S1: 43.5±12.7 S2: 20.4±7.9; S3: 35.3±11.1; S4: 9.6±5.9.

The differences between UPDRS/UPDRS III: S1-S2, and S1-S4 were statistically significant (p< 0.001) and S1-S3 was not statistically significant. Our slow eye movement (POM) measurements did change significantly with the session number: gain - for slow/medium/fast sinusoids were: S1: 1.06±0.1/0.96±0.2/0.83±0.2 S2: 1.03±0.1/ 0.97±0.2/0.86±0.1 S3: 1.05±0.1/0.99±0.1/0.94±0.1 S4: 1.00±0.1/0.96±0.1/0.87±0.1. Accuracy – as sum of normalized differences between stimulus and eye position - for slow/medium/fast sinusoids were: S1: 0.70±0.13/0.62±0.16/0.54±0.18 S2: 0.67±0.18/ 0.68±0.16/0.61±0.20 S3: 0.73±0.13/0.70±0.18/0.63±0.18 S4: 0.78±0.13/0.76±0.15/ 0.66±0.18.

3.1   Rough Set and Machine Learning Approach

As described above we have used the RSES 2.2 (Rough System Exploration Program) [8] in order to find regularities in our data. At first our data was placed in the information table as originally proposed by Pawlak [6].

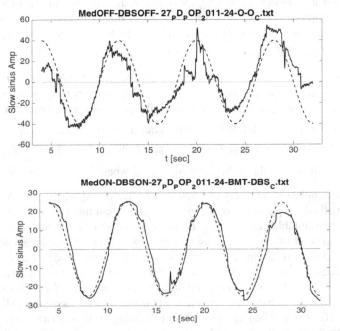

Fig. 1. Experimental recordings of POM from patient #27 before (upper part) and after (lower plot) medications and DBS treatments

Table 1. Extract from the information table

P#	age	sex	t_d	S#	UPDRS	HYsc	gxss	gxms	gxfs	accss	accms	accfs
28	54	1	8	1	58	2.0	0.94	1.04	0.97	0.71	0.86	0.81
28	54	1	8	2	40	1.0	1.04	0.98	0.93	0.91	0.93	0.86
28	54	1	8	2	40	1.0	1.17	1.07	0.91	0.82	0.90	0.67
28	54	1	8	4	16	1.0	1.08	1.00	0.90	0.86	0.89	0.69
38	56	0	11	1	49	2.5	0.90	0.94	0.94	0.73	0.76	0.66
38	56	0	11	2	22	1.5	1.04	1.03	0.93	0.89	0.85	0.76
38	56	0	11	3	37	2.5	0.99	1.01	1.03	0.83	0.81	0.69
38	56	0	11	4	12	1.5	1.08	1.11	1.03	0.81	0.77	0.76

The full table has 14 attributes and 40 objects (measurements). In the Table 1 are values of 11 attributes for two patient: P# - patient number, age – patient's age, sex – patient's sex: 0 - female, 1 – male, t_d – duration of the disease, S# - Session number, UPDRS – total UPDRS, HYsc – Hoehn and Yahr scale all measured by the neurologist and POM measurements: gxss - gain for slow sinus; gxms - gain for slow sinus; gxfs - gain for slow sinus; accss - accuracy for slow sinus; gxms - accuracy for medium sinus; gxfs - accuracy for fast sinus;

In the next step, we have performed reduction of attributes (see reduct in the Method section) to a minimum number of attributes describing our results. We have also created a discretization table: where single values of measurements were replaced by their range (as describe in the Method section on cut sets). As the result we have obtained the decision table (Table 2 –see below).

Table 2. Part of the decision discretized-table

Pat#	age	accfs	Ses#	HYsc	SchEng	gxms	gxfs	UPDRS III
28	*	"(0.75,Inf)"	1	*	"(-Inf, 85)"	"(1.04,Inf)"	"(0.845,Inf)"	"(28.0,Inf)"
28	*	"(0.75,Inf)"	2	*	"(-Inf, 85)"	"(0.97,1.04)"	"(0.845,Inf)"	"(16.5,28.0)"
28	*	"(0.39,0.75)"	3	*	"(-Inf, 85)"	"(1.04,Inf)"	"(0.845,Inf)"	"(16.5,28.0)"
28	*	"(0.39,0.75)"	4	*	"(85, Inf)"	"(0.97,1.04)"	"(0.845,Inf)"	"(-Inf,16.5)"
38	*	"(0.39,0.75)"	1	*	"(-Inf, 85)"	"(-Inf,0.97)"	"(0.845,Inf)"	"(28.0,Inf)"
38	*	"(0.75,Inf)"	2	*	"(-Inf, 85)"	"(0.97,1.04)"	"(0.845,Inf)"	"(-Inf,16.5)"
38	*	"(0.39,0.75)"	3	*	"(-Inf, 85)"	"(0.97,1.04)"	"(0.845,Inf)"	"(-Inf,16.5)"
38	*	"(0.39,0.75)"	4	*	"(85, Inf)"	"(1.04,Inf)"	"(0.845,Inf)"	"(-Inf,16.5)"

In the first column is the patient's number, in the second the patient's age not important (*); next was accfs – accuracy for fast sinus freq; Ses# -Session number,

Hoehn and Yahr scale were not considered important (stars); SchEng -Schwabe England scale; gxms – gain got medium sinus; gxfs – gain for fast freq. sinus and UPDRS III that was divided into different ranges: above 28, 16.5 to 28,, and below 16.5 (the last column). On the basis of this decision table we can write the following rule:

('Pat'=28)&('accfs'="(0.75,Inf)")&('Sess'=1)&('SchEng'="(-Inf,85)")&('gxms' ="(1.04,Inf)")& ('gxfs' ="(0.845,Inf)") => ('UPDRS III'="(28.0,Inf)")   (1)

We read this formula above (eq. 1), as stating that each row of the table (Table 1) can be written in form of this equation (eq. 1). It states that if we evaluate patient #28 *and* with accfs above 0.75 *and* in session #1 *and* with Schwabe England scale below 85. *and* gxms (gain fom medium freq. sinus) above 1.04 *and ... and* gxfs above 0.845 *then* patient's UPDRS is above 28.

These equations are parts of a data mining system bases on rough set theory [6]. We have tested our rule using the machine-learning concept. Randomly dividing our data into 4 groups, we took 3 groups as training set and tested the fourth. By changing groups belonging to the training and test sets, we have removed the effect of accidental group divisions. The results of each test were averaged – thus we have performed a 4-fold cross-validation. The results are gives as a confusion matrix (Table 3). As a machine-learning algorithm we have used the decomposition tree (see Methods).

Table 3. Confusion matrix for different session numbers (S1-S4)

		Predicted			
		28.0, Inf	**16.5, 28.0**	**-Inf, 16.5**	**ACC**
	28.0, Inf	0.5	0.5	0.0	**0.33**
Actual	**16.5, 28.0**	0.25	0.0	0.25	**0.0**
	-Inf, 16.5	0.0	0.25	2.25	**0.67**
	TPR	**0.5**	**0.0**	**0.67**	

TPR: True positive rates for decision classes, ACC: Accuracy for decision classes: the global coverage was 0.4, the **global accuracy was 0.774**, coverage for decision classes: 0.25, 0.3, 0.6.

Another question that result is, whether EM can help to estimate possible effects of different treatments in individual patients? In order to demonstrate an answer, we have removed EM measurements and added other typically measured attributes such as: the Schwab and England ADL Scale, and UPDRS III and UPDRS IV to the decision table and tried to predict the effects of different treatments as represented by sessions 1 to 4 (medication and stimulation effects- results are in Table 4) and compared them with predictions based on POM (results are in Table 5).

We have performed the same procedures once more to test results of patients' eye movement influence on our predictions.

Table 4. Confusion matrix for different session numbers (S1-S4)

		Predicted				
		1	2	3	4	ACC
	1	0.5	0.0	0.5	0.0	**0.3**
Actual	2	0.0	0.5	0.0	0.3	**0.4**
	3	0.8	0.0	0.2	0.0	**0.2**
	4	0.0	0.5	0.0	0.5	**0.4**
	TPR	**0.3**	**0.3**	**0.2**	**0.4**	

TPR: True positive rates for decision classes, ACC: Accuracy for decision classes: the global coverage was 0.64, the **global accuracy was 0.53**, coverage for decision classes: 0.5, 0.5, 0.75, 0.7.

Table 5. Confusion matrix for different session numbers (S1-S4)

		Predicted				
		1	2	3	4	ACC
	1	1.75	0.0	0.0	0.0	**0.75**
Actual	2	0.0	0.25	0.0	0.5	**0.25**
	3	0.0	0.0	0.5	0.0	**0.5**
	4	0.0	0.25	0.0	0.75	**0.33**
	TPR	**0.7**	**0.7**	**0.6**	**0.25**	

TPR: True positive rates for decision classes, ACC: Accuracy for decision classes, the global coverage was 0.45; the **global accuracy was 0.795**; coverage for decision classes: 0.58, 0.21, 0.38, 0.42.

In summary, two last results have demonstrated that adding eye movement (EM) results to classical measurements performed by the most neurologists, can result in improved predictions of disease progression measured, as measured by improvement in global accuracy from 0.5 to 0.8. The EM measurements may also partly replaces neurological measurements such as the UPDRS, as global accuracy of the total UPDRS predictions taken from EM data was 0.77 for the above 10 PD patients.

4   Discussion

In current therapeutic protocols, even with the large numbers of approaches and clinical trials, there have still been few conclusive results on therapeutic identification and measurement of PD symptoms. We have given an example comparing classical neurological diagnostic protocols with a new approach. The main difference between these types of measures is in their precision and objectivity. Our approach is doctor-independent and can be performed automatically. In the near future it may help in transforming some hospital-based to home-based treatments. In this scenario it will be possible to measure patient symptoms at home, and send these for consultation by neurologists.

5   Conclusions

We have presented a comparison of classical statistical averaging methods for PD diagnosis with rough set (RS) approaches. We used processed neurological data from PD patients in four different treatments and we have plotted averaged effects of the medication and brain stimulation in individual patients. As these effects are strongly patient dependent they could not give enough information to predict new patient's behavior. The RS and ML approaches are more universal giving general rules for predicting individual patient responses to treatments as demonstrated in UPDRS predictions.

Acknowledgements. This work was partly supported by projects Dec-2011/03/B/ST6/03816 and NN 518289240 from the Polish National Science Centre.

References

1. Przybyszewski, A.W.: The neurophysiological bases of cognitive computation using rough set theory. In: Peters, J.F., Skowron, A., Rybiński, H. (eds.) Transactions on Rough Sets IX. LNCS, vol. 5390, pp. 287–317. Springer, Heidelberg (2008)
2. Przybyszewski, A.W., Gaska, J.G., Foote, W., Pollen, D.A.: Striate cortex increases contrast gain of macaque LGN neurons. Visual Neuroscience **17**, 1–10 (2000)
3. Przybyszewski, A.W.: Logic in Visual Brain: Compute to Recognize Similarities: Formalized Anatomical and Neurophysiological Bases of Cognition. Review of Psychology Frontier **1**, 20–32 (2010) (open access)
4. Przybyszewski, A.W.: Logical rules of visual brain: From anatomy through neurophysiology to cognition. Cognitive Systems Research **11**, 53–66 (2012)
5. Pizzolato, T. Mandat, T.: Deep Brain Stimulation for Movement Disorders. Frontiers in Integrative Neuroscience **6**, 2 (2012) doi:10.3389/fnint.2012.00002 (Published online January 25, 2012)
6. Pawlak, Z.: Rough sets: Theoretical aspects of reasoning about data. Kluwer, Dordrecht (1991)
7. Bazan, J., Nguyen, H.S., Nguyen, T.T., Skowron A., Stepaniuk, J.: Desion rules synthesis for object classification. In: Orłowska, E. (ed.) Incomplete Information: Rough Set Analysis, pp. 23–57. Physica – Verlag, Heidelberg (1998)
8. Bazan, J., Szczuka, M.S.: RSES and RSESlib - a collection of tools for rough set computations. In: Ziarko, W.P., Yao, Y. (eds.) RSCTC 2000. LNCS (LNAI), vol. 2005, pp. 106–113. Springer, Heidelberg (2001)

Optimization of Joint Detector for Ultrasound Images Using Mixtures of Image Feature Descriptors

Kamil Wereszczyński[1,2]([✉]), Jakub Segen[1], Marek Kulbacki[1],
Konrad Wojciechowski[1], Paweł Mielnik[3], and Marcin Fojcik[4]

[1] Polish-Japanese Academy of Information Technology,
Koszykowa 86, 02-008 Warsaw, Poland
{kw,mk}@pj.edu.pl
[2] Institute of Informatics, Silesian University of Technology,
Akademicka 16, 44-100 Gliwice, Poland
[3] Revmatologisk Avdeling, Førde sentralsjukehus,
Svanehaugvegen 2, 6812 Førde, Norway
[4] Sogn og Fjordane University College, Vievegen 2, 6812 Førde, Norway

Abstract. Joint detector is an essential part of an approach towards automated assessment of synovitis activity, which is a subject of the current research work. A recent formulation of the joint detector, that integrates image processing, local image neighborhood descriptors, such as SURF, FAST, ORB, BRISK, FREAK, trainable classification (SVM, NN, CART) and clusterization, results in a large number of possible choices of classifiers, their modes, components of features vectors, and parameter values, and making such choices by experimentation is impractical. This article presents a novel approach, and an implemented environment for the parameter selection process for the joint detector, which automatically choses the best configuration of image processing operators, type of image neighborhood descriptors, the form of a classifier and the clustering method and their parameters. Its implementation uses new scripting tools and generic techniques, such as chain-of-responsibility design pattern and metafunction idiom. Also presented are novel results, comparing the effect of feature vectors composed from multiple SURF descriptors on the performance of the joint detector, which demonstrate the potential of mixture of descriptors for improving the classification results.

Keywords: Medical ultrasound images · Machine learning · Classifier · Image feature descriptor · Synovitis · Generic programming · Idiomatic programming

1 Introduction

Ultrasound images of joints, such is shown in Fig. 1 are examined by medical experts to detect and monitor the inflammation of synovial membrane covering

© Springer International Publishing Switzerland 2015
N.T. Nguyen et al. (Eds.): ACIIDS 2015, Part II, LNAI 9012, pp. 277–286, 2015.
DOI: 10.1007/978-3-319-15705-4_27

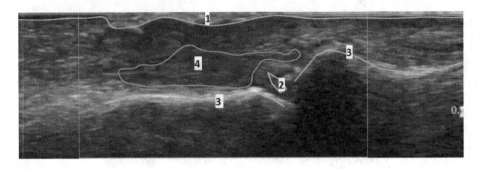

Fig. 1. Human fingers USG image with examplary biological structures marked: 1-skin, 2-joint, 3-bones, and 4-inflamation area

a joint, or synovitis, which is a condition usually caused by rheumatoid arthritis. There are two factors for disease progress determining: inflamation area and blood level, both specified by degree 0 (no inflamation/ blood), 1, 2 and 3 (the largest area / blood level). A research project MEDUSA [13], aiming automating the process of detection of synovitis and assessment of its degree is being conducted in Poland and Norway. The approach taken towards this goal assumes that, in the first step, the analyzed image will be registered with a structural model which represents a generic joint. The registration method is based on detecting a set of preselected features in an ultrasound image of a joint, which includes skin, bones and joint.

This theme of this paper is the detection of a joint, which is one of the key modules of the registration method. The complete registration method will be described in a separate article. In an earlier article [12], a learning approach to a joint detector has been described, which is based on using robust, invariant descriptors of a pixel neighborhood and constructs a detector in a form of a trainable classifier of pixel's neighborhood. Such a detector uses a number of control parameters that include discrete choices, such as selection of components, or the classifier form. However the subject of selection of the parameter values has not been addressed in depth. The current paper augments and expands the presentation in [12], describing an automated selection of the components and parameters of the joint detector, which is approached as an optimization problem. The basic learning approach to join detector is summarized in Section 2, the approach and implementation of parameter selection is described in Section 3, the measure used for comparison of parameters sets is shown in Section 4, and Section 5 presents new, improved results.

General joint detector output is a set of points where joint center is located. In ideal situation it should be one point. But due to inaccuracy of method it is considered to detect more joints: some of them are putted in joint area (positive detection) some outside it (false detections).

Joint area is a set of pixel. Thats why joint detector marks pixels on image that are laying in area of joint. It bases on trained classifier and local descriptors of pixel used as feature vectors. So joint detector task is a pixel classification task.

It consists of three modules: (1) image preprocessing, (2) feature extraction, (3) training or prediction. After this classification set of marked pixels are clustered; cluster centers are acknowledged as joint centers.

Finally one joint detector is distinguished from another by applied components. If two detectors have the same components and differ with theirs parameters only it means that they are two variants of the same detector.

The image set (over 3000 USG images) comes from Helse Førde HF (Norway) Hospital. It has different inflamation areas and blood levels covering all possible degrees.

2 Learning Approach to Joint Detector

In the first place damaged, irrelevant (comes from another parts of human body) and identical images were excluded from image set. Then images was divided into training and testing set randomly. Testing set has about 30% of all images.

2.1 Image Preprocessing

The image preprocessing phase is a series of image processing operations, such as filtering, gray scaling, etc. It leads to enhance image characteristic in a way specific and proper for trained classifier. One series is distinguished from another by used algorithms. It turned out that the order of using this operation has also influence for the results. Each operation has parameters that optimum could differ for different classifiers.

2.2 Feature Extraction

The basic element type used in the feature vector of a classifier that functions as a joint detector is the output of an application of a local descriptor to a pixel neighborhood. The local descriptors that were considered in this and the earlier work are robust, scale and rotation invariant descriptors, including SURF [1], ORB [7], BRISK [8] and FREAK [9]. The size of neighborhood used by these descriptors is one of the method parameters that are selected via the optimization process. The feature vectors of the detector are constructed using a weighted mixture of two, or more descriptors to ensure coverage small and large size characteristics. This mixture was chosen as feature vector for pixels.

In the preceding work [12] on joint detection problem SURF and its mixture gives best results in connection with SVM classifier. SURF and its mixtures has very small influence for AUC [6] analysis but noticeable changes false detection count. This paper presents more detailed research of exploring mixtures of SURF descriptors with different window size parameters and its influence for final classification results especially for false detection count.

2.3 Training and Prediction

For classification three methods were examined: Nearest Neighbour Classifier [4], Support Vector Machine [2] and Classification and regression trees [3]. As usual in supervised learning there are two modes of classifiers work: training mode and prediction mode. Once classifier is trained it could be used for prediction many times. In training phase classifier is fitted by pictures from a set called training set. Basing on this images knowledge base is created. Using it prediction is performed on images from outside of training set.

After labeling pixels belonging to joint using trained classifier clusterization is introduced for marking where the joint center is situated by one point. Joint center list is the output of detector. If given point lays in real joint area detection is recognized as positive detection else as false detection.

2.4 Evaluation

It is obvious, that amount of such created joint detectors is huge. So the problem is to find best of them. Evaluation consists in prediction for the set of pictures that wasn't used in training phase. The count of false and positive detections are counted and used for further analysis. All detectors that couldn't predict positive detection for even one image are acknowledge as not correct and excluded for analysis. For the rest of detector two measures are introduced: area under the ROC analysis (AUC) [6] and false detection count comparison (FDC). First all detectors are evaluated by AUC. Then on set of them with higher rank FDC is provided. So the best detector has comparatively high AUC value and the smallest count of false detection in training set with condition that in each image at least one positive detection is predicted.

The AUC analysis was conducted in the first phase of works on joint detection problem and detail described in [6]. It led to selection SVM as best classifier for joint detection problem. In current work part, described in this paper, several experiments with SVM classifier and different mixtures of SURF descriptor was made. Ranking was created using false detection count for each mixture of SURF.

3 Parameter Selection Method Overview

Scenario is description of one detector variant execution formulated in specific language. So scenario consists of description of all detector module containing operations, theirs parameters and order of execution. One scenario leads to one detector training and evaluating with one set of parameters. This means that one scenario creates one point in ROC. To create all curve there is necessity of execution numbers of such scenario with changing one parameter only. So there is a strong need of generation scripts automation, which was implemented actually.

Automation of script generation bases on so called **scenario template**. It distinct of scenario by indication of one or more values to be changed. Minimum, maximum and step value have to be defined by user in scenario template.

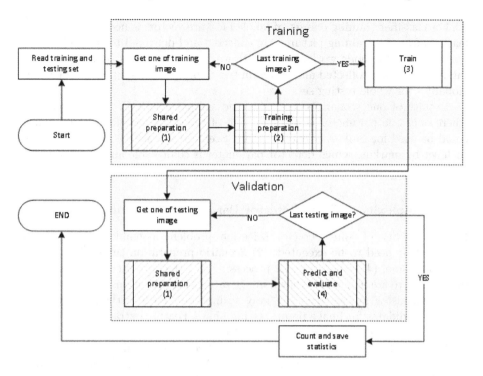

Fig. 2. Scenario execution schema

Application creates number of scenario computing i-th value v using formula:
$v = min + i * step$ until $v <= max$. Then each scenario is executed; detailed
results for each tested image are stored for each scenario in separate file and
global results such as true positive, false positive, true negative and false negative count are stored in one result file for template. Basing on this template
result file ROC is created.

3.1 Scenario Description

Scenario consists of two general modes: training and validation. In each mode so
called chains are executed in a way shown on Fig 2. Chain is a series of operation
executed one after another. Shared preparation chain mark by (1) on Fig. 2 is
common for both modes, while training preparation (2) and test (3) chains are
specific for training mode, but predict and evaluation chain (4) is specific for
validation mode. Shared preparation chain should contain operation of image
preparation and feature extraction. This is crucial for proper work of scenario
execution. Training preparation chain is destined for labeling pixels in a way
specific for classification algorithm. Different algorithms could demand different
way of labeling, e.g. SVM needs two or more classes, while Nearest Neighbour
just one class.

For classifier training one set of labeled feature vectors is needed. Thats why features from all training pictures are collected and delivered to train phase. In testing phase methods works in opposite way: each picture is considered separately; results are collected and stored in two tribes: locally - for each image and globally - for whole testing set.

Results of one scenario constitutes one sample for ROC. Set of scenarios where only one parameter is differ in each of them creates whole ROC, which could be used for rank different detectors. Scenario template is a description of such set by marking which detector parameter is changeable and rank in which changes are made.

3.2 Complexity of Joint Detector Parameter Selection Problem

For complexity of joint detector selection problem assembles: (1) Scenario amount that need to be executed, (2) Scenario preparation time, (3) Scenario execution time, (4) scenario result processing, (5) the size of training and testing set. There are many well-known image processing operations like smoothing, denoising, histogram equalization, gray scaling, etc. Each of them was considered as candidates for image preprocessing phase. Feature extracting using local neighborhood descriptors could be a mixture of several methods as: SURF [1], ORB [7], BRISK [8], FREAK [9]. If only 2 component mixture will be considered number of mixtures will reach 15. SVM [2], Nearest Neighbor [4] and CART [3] was tested as classifiers. Multiplying this quantities gives 180 different detectors. For usefulness analysis about 20 variant for each detector should be tested, which gives 3600 scenarios to execute.

Execution time of one scenario depend of its component complexity and fluctuates in range from 15min. to 30min.

Output of joint detector is a set of point predicted as joint centers. There are several operation that need to be done: (1) labeling results as positive or negative (2) count positive and negative results and (3) processing it to form proper for AUC analysis. Only 100 images in evaluation set will produce much over 360 000 results for AUC analysis and will take 1200h (assuming only 20min for avg. scenario execution time). On table 1 there are complexity analisys for results presented in this paper. It is good example showing the problem with big amount of variants, long execution time and massive result set, which are generated in joint detector selection problem.

Table 1. Detector and scenario count for results presented in this paper

Classifier	Detector	Scenario / detector	Scenarios count	Avg. exec time	Whole exec. Time	Result count
SVM	6	26	156	15min	39h	900
NN	4	15	60	25min	25h	1000
CART	4	10	40	30min	20h	1200
		Sum:	412	-	137h	3100

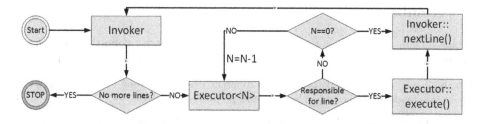

Fig. 3. Schema of implementation generic chain of responsibility pattern

3.3 Implementation

Implementation of MEDUSA Script was based on two programming techniques: Chain of Responsibility pattern [10] and Meta-function idiom [11] used as a novel tool for generic chain implementation. Each specialization of generic meta-function class is responsible for another MEDUSA Script function implementation. Details of the way how it works is shown on Fig. 3.

Invoker reads script file and parses each line, which is recognised as one command. Then invoker calls meta-function [11] named Executor with the gratest number in chain. Executor decieces if it is responsible for current command and: (1) Execute this command or (2) calls the next meta-function element in chain. After current line execution steering returns to invoker. Thats why adding new functions is not complicated and the working amount could be decreased.

4   Descriptors Comparison Method

In joint detector problem pixel descriptors or theirs mixtures forms feature vectors fitting classifiers. As was mentioned in this paper results are based on SVM classifier. The parameters selection methods, its values and prove that it is better solution for classification was presented in [12] and it is based on AUC analysis. The preprocessing phase was limited to gray scaling only. Used parameters of SVM are: $C = 25$, RBF kernel with $\gamma = 0.85$. 4-class classification was used with one positive class and 3 negative, with penalty weights: [10,0.1,0.1,0.1].

Classification results could be divided in four groups: true positive, true negative, false positive and false negative. Because the general measure of classifiers based on all this quantities was described in [12] for this phase of research authors limits their attention to false positive value named as false detections.

The measure that was used for make a descriptor rank base on false detection in evaluation image set. False detection count could be computed in two places of detector: just after classification and after clusterization. The necessary condition claims that in every image from evaluation set there has to be minimum one positive detection of joint. For each variant false detection count (FDC) is computed. The best variant is one with smallest FDC percentage.

5   Results

Below (Table 2) there is a comparison of fulfilling necessity condition and false detection percentage for detectors with different mixture of SURF descriptor:

Table 2. Comparison of detectors with different descriptors used as feature vector. Tests was made with the same classifier and preprocessing phase.

	Descriptor or mixture	Window sizes	mixture weights	Necessity fullfilling	FDC percentage
1	SURF	80	-	YES	100,0
2	SURF	200	-	YES	60,4
3	SURF	350	-	NO	-
4	2 x SURF	[200, 80]	1:1	YES	31,8
5	2 x SURF	[350, 80]	1:1	YES	44,0
6	3 x SURF	[200, 80, 40]	1:1:1	YES	34,0
7	3 x SURF	[260, 200, 80]	1:1:1	YES	23,1
8	3 x SURF	[350,200,80] v1	1:1:1	YES	23,0
9	3 x SURF	[350,200,80] v2	1:1:1	YES	23,0

All images have 960x720 resolution, but were cropped to 740 x 490 10 px size for cutting off metadata (like USG device settings, Hospital name, etc.) inprinted in image by USG device. SURF was used with orientation computation turned on, window size was a parameter that influence for results was investigated; remaining parameters : number of pyramid octaves and number of octave layers within each octaves have nearly zero influence for classification, and hessian threshold is used in SURF detector not by descriptor computing.

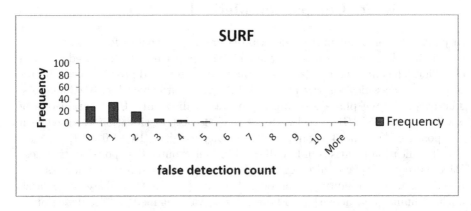

Fig. 4. Histogram of false detection count in case of one SURF descriptor feature vector

Implementing one SURF descriptor as feature vector allows neccessity ciondition fullfilling but with high (60-100) percentage of false detection count. It means that in 60-100% of images there are at least one false detection. Applying 2 x SURF mixture decreses FDC to 30-50% and 3 x SURF to level 20-30%. The

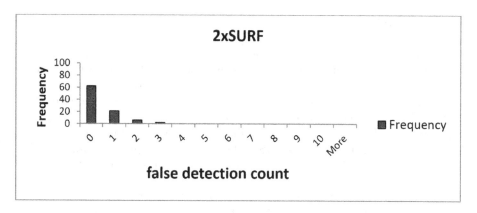

Fig. 5. Histogram of false detection count in case of two SURF descriptor feature vector with windows sizes: 200 and 80 pixels

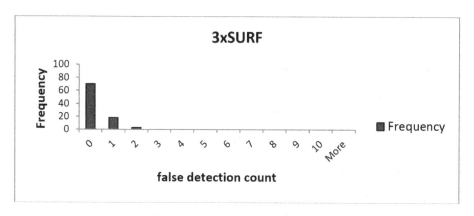

Fig. 6. Histogram of false detection count in case of three SURF descriptors feature vector with win-dows sizes: 350, 200 and 80 pixels

best and very similar features vectors are 3 x SURF with window size parameters got from set 350,200,80 and 260,200,80. Other parameters of SURF doesn't have big influence for FDC- two descritpors differencing with SURF paranmeters shown in line 8 and 9 in Table 2 has exactly the same FDC values.

On Fig. 4, 5 and 6 there are histograms presenting false detection count for one of each class of mixture.

On histograms above there could be seen another regularity: not only FDC decreases with descriptor count in mixture growth but also maximum count of false detection on images decreases: in case of 1 SURF feature vector there are images with 5 false detections while in case of 3 SURF mixture in only one image there was 2 false detection in the rest one or none.

6   Conclusions

A new approach to automated selection of parameters, components and modes of a learning joint detector, and its implementation as MEDUSA Script has been described. New, improved results point to a triple SURF descriptor as the best choice of components of the feature vector. Using a mixture of descriptor, the triple SURF improved final results in two ways: (1) Decreased the number of image samples that contain false detections, and (2) in the samples that contain false detections, their count is reduced. The presented approach eliminates the need for time consuming experimentation in configuring the joint detector.

Acknowledgments. The research leading to these results has received funding from the Polish-Norwegian Research Programme operated by the National Centre for Research and Development under the Norwegian Financial Mechanism 2009-2014 in the frame of Project Contract No. Pol-Nor/204256/16/2013.

References

1. Bay, H., Tuytelaars, T., Van Gool, L.: SURF: speeded up robust features. In: 9th European Conference on Computer Vision, Gratz, Austria (2006)
2. Chang, C.-C., Lin, C.-J.: LIBSVM: A Library for Support Vector Machines. National Taiwan University, Tai-pei, 4 March 2013
3. Breiman, L., Friedman J.H., Olshen R.A., Stone C.J.: Classification and regression trees. Chapman, Wadsworth
4. Muja, M., Lowe D.G.: Fast approximate nearest neighbors with automatic algorithm configuration. In: International Conference on Computer Vision Theory and Applications (VISAPP), Lisbon, Portugal (2009)
5. McNally, E.G.: Ultrasound of the small joints of the hands and feet: current status. Skeletal Radiology **37**(2), 99–113 (2008)
6. Till, D.J., Hand, R.J.: A Simple Generalisation of the Area Under the ROC Curve for Multiple Class Classification Problems. Machine Learning **45**, 171–186 (2012)
7. Rublee, E., Rabaud, V., Kurt Konolige, K., Bradski., G.R.: ORB: an efficient alternative to SIFT or SURF. In: ICCV, pp. 2564–2571, Barcelona, Spain (2011)
8. Leutenegger, S., Chli, M., Siegwart, R.: BRISK: binary robust invariant scalable keypoints. In: ICCV, pp. 2548–2555 (2011)
9. Alahi, A., Ortiz, R., Vandergheyns, P.: FREAK: Fast Retina Keypoint. EPFL Lausanne, Switzerland (2012)
10. Gamma, E., Helm, R., Johnson, R., Vissides, J.: Design patterns: abstraction and reuse of object-oriented design. In: ECOP 1993, Kaiserslautern, Germany, July 1993
11. Abrahams, D., Gurtovoy, A.: Chapter 2.2: Metafunction. C++ Template Metaprogramming, pp 15–17. Addison-Wesley (2005)
12. Wereszczyński, K., Segen, J., Kulbacki, M., Mielnik, P., Fojcik, M., Wojciechowski, K.: Identifying a joint in medical ultrasound images using trained classifiers. In: Chmielewski, L.J., Kozera, R., Shin, B.-S., Wojciechowski, K. (eds.) ICCVG 2014. LNCS, vol. 8671, pp. 626–635. Springer, Heidelberg (2014)
13. Project MEDUSA. http://eeagrants.org/project-portal/project/PL12-0015

Automatic Markers' Influence Calculation for Facial Animation Based on Performance Capture

Damian Pęszor[1,2]([✉]), Konrad Wojciechowski[2], and Marzena Wojciechowska[2]

[1] Institute of Informatics, The Silesian University of Technology,
Akademicka 16, 44-100 Gliwice, Poland
[2] Polish-Japanese Academy of Information Technology,
Koszykowa 86, 02-008 Warsaw, Poland

Abstract. Marker-based performance capture is a technique that enables acquisition of expression and mimicry of human face. This data can be used to propel facial animation system, be it bone driven or similarly dependant on position of points in space. Every model that is to be animated has to be analyzed in order to select level of influence each marker has over each vertex of said model. This process can be quite tedious if done manually. In this paper we present an approach for automatic calculation of markers' influence based on position of vertex on human face's surface obtained by acquisition using structured light-based scanner or similar approach.

Keywords: Marker influence · Vertex weight · Facial animation · Performance capture · Face scan · Bone driven animation · Fiducial points

1 Introduction

Performance capture is a technique of acquiring data about deformation of surface of actor's face in order to capture his mimicry for later use in facial animation. While there are many specific methods to implement performance capture, most of them are based on tracking reflective markers that are either painted on face's surface or attached to it. Once obtained, performance capture data is used to create realistic facial animation. There are few techniques to create facial animation, each having different flaws and benefits. Texture based animation, while quite common in two dimensional applications, looks poorly in three dimensional environment. Physiological models based on tissue, muscle and skeleton of human face give very good results, but level of complexity and thus computational cost are big enough to make this approach unsuitable for most real time animation. Blending of different poses has low computational cost and is applicable to 3D environment, but is model-specific and therefore requires huge amount of work for every animated mesh. Bone driven animation is most commonly used because of low computational and memory cost. This technique is able to imitate subtle transformations of facial skin and is usable for not only single mesh but also those with similar morphology. The biggest

© Springer International Publishing Switzerland 2015
N.T. Nguyen et al. (Eds.): ACIIDS 2015, Part II, LNAI 9012, pp. 287–296, 2015.
DOI: 10.1007/978-3-319-15705-4_28

problem with bone driven animation is the need to assign a weight for each vertex-bone pair that will represent influence this bone has over given vertex. Since every marker can be considered the end of the bone, this issue can also be seen as defining specific marker's influence over vertices. If markers are positioned on facial surface with anthropometric features in mind rather than in simple grid arrangement, one can find fiducial points on animated model. Those points can then be used to obtain vertices' weights that are specific for animated model, thus eliminating both major disadvantages of bone driven animation.

2   Facial Mesh

The method presented in this paper is designed to calculate influence of markers placed on human face. It can be, however, used on different surfaces that have underlying structure, given that points on this surface can be detected similarly to fiducial points on facial mesh or can be manually selected. In case of artificial models of human face, the quality of the model determines if it can be analyzed the same way as realistic model used in this research. In case of low quality models (ones that do not preserve realistic curvatures of facial surface) or models that are representation of damaged, deformed, artistic or non-human faces, manual detection of fiducial points might be needed.

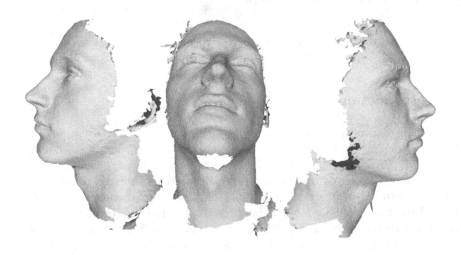

Fig. 1. Mesh as acquired by 3dMDface System

2.1   Mesh Acquisition

While artistic representations of human face are often simplified and therefore easy to animate, those that are realistic can prove to be a difficult subject to animation. The more realistic the face, the easier it is to spot parts of model that move differently than natural mimicry. Therefore, for the purpose of this paper, meshes of real faces have been acquired and analyzed. Depending on technique used (stereo vision, structured light, time of flight, etc.), various type of errors can be introduced in acquired mesh. Lens distortion, lighting differences, interference from another camera can all lead to errors. Faces used in this research were obtained using 3dMDface system, a structured light based scanner, at The Institute of Theoretical and Applied Informatics of Polish Academy of Sciences in Gliwice. Being structured light based, this equipment produces errors on surfaces that are highly reflective, such as hair, glossy skin or eyes, due to incorrect determination of depth. Most of expressive facial area is however well preserved and therefore can be used as a basis for animation. Eight different actors were scanned for a total of forty models representing neutral facial expression, sample of which are presented in Fig. 2.

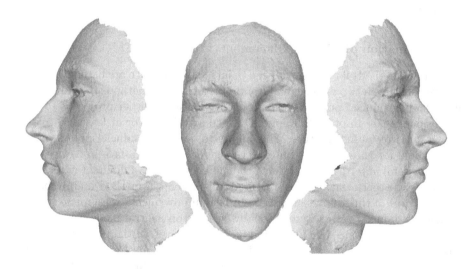

Fig. 2. Example of preprocessed mesh

2.2   Mesh Preprocessing

Meshes obtained using structured light or other methods are prone to various types of errors. First of all, those meshes often contain more than single continuous surface, therefore any surfaces out of face's expressive area should be removed. It is typical for the surface that has biggest number of vertices to be the one that contains facial expressive area, so only this surface has to be preserved. Oftentimes scanned models also contain topological errors like non-manifold edges and vertices which have to be removed in order to correctly estimate fiducial points. Furthermore, due to the noise present, edges of the analyzed surface might be composed of vertices that do not represent skin affected by mimicry. Those can be safely removed in order to make sure that they are not affected by facial expressions' animation. Depending on how face is oriented to the scanner, facial hair and other factors, it is probable, that resulting mesh will contain gaps or holes. Those need to be filled with respect to preserving curvatures of human head, for which [1] was used.

3   Model Division

In order to calculate influence of each marker over each vertex, one has to be able to distinguish markers that affect given vertex from those which do not. All vertices of expressive facial area lie between fiducial points that can be found on face structure, therefore finding those fiducial points that correlate to marker placement is essential to calculate markers' influence on every vertex. In some cases, such as testing marker arrangement prior to performance capture session, manual selection of fiducial points can be useful. In most cases, however, this process can be automated.

3.1   Estimation of Marker Placement

Most of the methods for finding fiducial points on human face are based on colour features present in two dimensional frontal pictures. Few methods that are capable of finding fiducial points in three dimensional mesh either project it to two dimensional space in order to apply 2D based algorithms, or at least use colour data instead of mesh structure. This allows to find some fiducial points but is not suitable for those points that describe structure of the face in a way that could be useful in terms of retargeting the mimicry to face of different actor than the one whose performance was captured. [2] presents a method mainly based on curvatures of three dimensional mesh. Few improvements were made to this method; all fiducial points were manually selected on neutral model to compensate for points that are difficult to find using curvatures. Initial two dimensional window of 96x96mm used for estimation of nose tip of mesh composed of vertices $V = (v_1, v_2, ..., v_n)$ has been replaced with sphere. This sphere is located in neutral face's nose tip p_{nt} of radius r such that:

$$r = \frac{\left\| p_{nt} - \frac{\left(\sum_{i=1}^{n} v_i \right)}{n} \right\|}{2} l \tag{1}$$

Next, nose tip was differentiated from surroundings using gradient based method by traversing from vertices closest to centroid of the model to farthest. This allowed to place the sphere in farthest point from centroid while reducing it's radius 5 times. At this point curvature estimation based on paraboidal fitting [3] is much more likely to correctly point to nose tip if it's calculated using more than one ring neighbourhood. Similar approach was used to find other fiducial points. Points that do not have specific curvature can be found using their correspondent position on neutral model with relation to those fiducial points which were already found. Position of marker m_j with normal N on analyzed mesh can then be found as a vertex v_i such that:

$$\underset{i}{\arg\min} \frac{\|v_i - m_j\|((v_i - m_j) \cdot N)}{\|v_i - m_j\|\|N\|} \tag{2}$$

3.2 Triangular Division of the Model

Having found fiducial points one can divide the model into triangles based on those found points. All fiducial points should be connected into a topologically correct mesh with few rules in mind. First, every triangle should represent possibly flat area - this way all edges can be extended into planes that will bisect facial model easily. Second, each triangle should be possibly equilateral in order to avoid uneven distribution of influence. Third, triangles should be possibly small, so distance from marker to vertex that is affected by said marker will be small as well. Since surface of the mesh is not coplanar with triangles connecting markers, there is no clear association between vertex and triangles near it. Obtaining such association, however, is needed in order to limit the number of markers that have influence over specific vertex to those that are near it. Those markers represent surfaces that are affected by same transformation as the one sampled in analyzed vertex. Two methods were designed to obtain this association.

Geometric method is based on spatial coordinates of analyzed vertex with relation to marker's position. Each markers' based triangle represents a plane in three dimensional space. Edge of triangle therefore represents intersections of two planes. Using spherical linear interpolation [4] of normals, one can obtain a third plane intersecting through the edge with same angle to both of triangles' planes. Each vertex lies on either side of this plane and eventually - inside of pyramid based on edges of one of triangles. In some cases, e.g. in vertices lying on cheeks, vertex can be inside two or more pyramids - one correct, and others on far side of the face. Distance between the triangle's plane (alternatively, it's centroid) and vertex can then be used to verify which triangle should be associated with specific vertex. This reveals the main problem of the method - the pyramids might extend beyond expressive area thus resulting in non-expressive surface being dependant on markers' movement. The remedy for that is to cut out of the model any non-expressive areas or use graph-based method instead.

Graph-based method uses mesh structure to obtain vertex-triangle association. Each marker is associated with vertex that is topologically connected to every other marker through some other vertices. One can therefore assume that with marker arrangement based on anthropometric features, shortest path between markers corresponds to smallest distance. Using A* [5] algorithm to obtain paths between neighboring markers results in finding edge of markers' based triangle on actual surface model similarly to extending planes in geometric method. Instead of testing for side of plane, breadth first search is used to decide on which side of the path vertex is located. Due to the nature of triangular mesh, one has to select one of middle vertices of the edge's path and start the search from this vertex to the direction of opposite marker. All vertices found between three edges' path can be associated to triangle composed of those edges. This method ensures that no vertices from outside of expressive area are associated with markers. The main flaw of this method is that it is dependant on topology of the mesh rather than surface itself, which can lead to problems in case of structured light induced errors or markers placed with low density.

Using graph method to cut outer edge of expressive area before using geometric method proves to give best results. Graph method can also be used to improve estimation for those vertices that are contained by two or more pyramids in geometric method.

4   Influence Calculation

Each vertex is positioned on surface in three-dimensional space. Each part of surface is based on triangle and therefore each vertex can be projected onto this triangle. Similarly to the method used to divide the model, one can define a plane as extension of triangle's edge by using normal to the triangle's surface. In all cases except the edge of the expressive area, there will be a neighboring triangle on the other side of analyzed edge. Since both triangles can have different orientation, there is a portion of space that is placed between triangles' normals' based planes. To cover for that, neighboring triangles' normals are interpolated using spherical linear interpolation, so that plane based on interpolated normal would split space equally for both triangles.

Having three planes containing the space, each described by a point (marker) $\{A, B, C\}$ lying on it and it's unit normal n_i, a point of intersection p_{int} of all three can be found.

$$p_{int} = \begin{vmatrix} n_{00}n_{10}n_{20} \\ n_{01}n_{11}n_{21} \\ n_{02}n_{12}n_{22} \end{vmatrix}^{-1} [(A \cdot n_0)(n_1 \times n_2) + (B \cdot n_1)(n_2 \times n_0) + (C \cdot n_2)(n_0 \times n_1)] \quad (3)$$

A line between point p_{int} on top of pyramid and vertex v_i inside of it can be extended to intersect with the plane of triangle that is a basis of said pyramid and has normal n and a marker-related point A lying on it. The intersection found is projection v_i' of v_i onto the base triangle.

$$v_i' = (v_i - p_{int}) \frac{(A - p_{int}) \cdot n}{(v_i - p_{int}) \cdot n} + p_{int} \quad (4)$$

Having found this point on triangle's surface, barycentric coordinates are calculated with relation to triangle's vertices.

$$a = \frac{(y_B - y_C)(x_{v'} - x_C) + (x_C - x_B)(y_{v'} - y_C)}{(y_B - y_C)(x_A - X_C) + (x_C - x_B)(y_A - y_C)}$$
$$b = \frac{(y_C - y_A)(x_{v'} - x_C) + (x_A - x_C)(y_{v'} - y_C)}{(y_B - y_C)(x_A - X_C) + (x_C - x_B)(y_A - y_C)} \quad (5)$$
$$c = 1 - a - b$$

Barycentric coordinates could be identified with influence each of surrounding three markers has over the vertex. This, however, would result in differences between vertices that are near of each other, but are assigned to different triangles due to the edge separating them. Those differences, visible as sharp, unwanted features of the model, need to be smoothed. It is also worth to note, that in most cases, especially when the marker arrangement is not dense, markers on neighboring triangles do affect how skin is deformated inside of the analyzed triangle, so those also need to be taken into account.

Surface of triangle is therefore based on it's three vertices and three another points. This is exactly how cubic Bézier triangle is constructed. [6] Surface of cubic Bézier triangle composed of n points $p_i | i \in \mathbb{N} \wedge i \in [0, n-1]$ each having barycentric coordinates of a, b, and c related to triangle's vertices α^2, β^2 and γ^2 respectively and influenced by control points $\alpha\beta$, $\alpha\gamma$ and $\beta\gamma$ is described with following equation:

$$p_i(a,b,c) = (\alpha a + \beta b + \gamma c)^2 = \alpha^2 a^2 + \beta^2 b^2 + \gamma^2 c^2 + 2\alpha\beta ab + 2\alpha\gamma ac + 2\beta\gamma bc \quad (6)$$

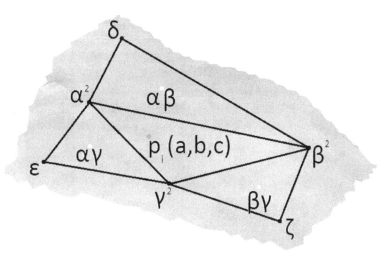

Fig. 3. Example of surface. Black dots represent markers, white dots represent additional control points of Bézier triangle and grey dot represents example point

Since projected barycentric coordinates, vertices of triangle and control points are all known, one could assume that equation can be used directly. This is however not the case, because appropriate control points (vertices of neighboring triangles) will not result in obtaining surface expressed by points v_i due to the difference between points' projection and points themselves. Instead of directly calculating change in barycentric coordinates that would cover for that, another method is used. First, barycentric coordinates are assumed to be correct and control points are estimated using nonlinear least squares fitting [7] in order to find a surface approximating actual points' positions. Once obtained, each control point weighted by barycentric coordinates is replaced with position of third vertex of neighboring triangle $(\delta, \epsilon, \zeta)$ with new weight. Similarly, weight of each triangle's vertex is calculated:

$$
\begin{aligned}
A &= a^2 \\
B &= b^2 \\
C &= c^2 \\
D &= \frac{2\alpha\beta ab}{\delta} \\
E &= \frac{2\alpha\gamma ac}{\epsilon} \\
F &= \frac{2\beta\gamma bc}{\zeta}
\end{aligned}
\tag{7}
$$

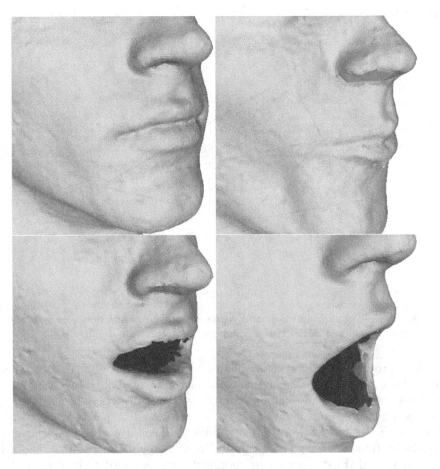

Fig. 4. Surface with calculated markers' influence, original and distorted by movement of markers

This can be used as weighting term for corresponding markers, so the difference between marker's position in neutral frame and in current frame f can be expressed as:

$$p_{i,f}(A, B, C, D, E, F) = p_{i,0}(A, B, C, D, E, F) + A(\alpha_i - \alpha_0)$$
$$+B(\beta_i - \beta_0) + C(\gamma_i - \gamma_0) + D(\delta_i - \delta_0) + E(\epsilon_i - \epsilon_0) + F(\zeta_i - \zeta_0) \tag{8}$$

Various ways of applying offset using e.g. normal to surface can be used instead to further enhance wrinkles and similar details of facial surface.

5   Exemplary Results

The results of using weights obtained by automatic calculation of influence presented in this paper can be seen in Fig. 4. One can notice that using weights obtained in this matter produces unrealistically smooth surfaces (as seen on cheek in bottom-right picture) or a valley-like surface in the position of marker (as seen in top-right picture), which do not accurately correspond to how real skin reacts. This can be however compensated by reducing animation impact through using appropriate initial vertices' positions. This is also a good way to reproduce wrinkles and similar features.

6   Conclusion and Further Works

The method presented in this paper allows to automatically estimate markers' influence over vertices of human facial model. While this seems to result in acceptable weights estimates, it is still clear that using simple offset, as typical in bone-driven animation, will not produce satisfactory results in presenting skin deformations. A different approach, based on distortions dependant on interpolated normals could possibly be applied in order to improve the way distortions are animated, which will be part of future research. Still though, this work provides easy and fast way to estimate how surface is affected by changes in markers' positions and therefore is useful in both further research in animation automatization and practical applications. Apart of producing animation itself, this can be also useful to test markers' positions on facial surface and it's effects on possible expressions on mesh used before comitting to expensive performance capture procedure without need to manually select weights for each possible marker arrangement.

Acknowledgments. This project has been supported by the National Centre for Research and Development, Poland.(project INNOTECH In-Tech ID 182645 "Nowe technologie wysokorozdzielczej akwizycji i animacji mimiki twarzy." - "New technologies of high resolution acquisition and animation of facial mimicry.")

References

1. Zhao, W., Gao, S., Lin, H.: A Robust Hole-Filling Algorithm for Triangular Mesh. The Visual Computer: International Journal of Computer Graphics **23**(12), 987–997 (2007)
2. Gupta, S., Markey, M.K., Bovik, A.C.: Anthropometric 3D Face Recognition. International Journal of Computer Vision **90**(3), 331–349 (2010)
3. Surazhsky, T., Magid, E., Soldea, O., Elber, G., Rivlin, E.: A comparison of gaussian and mean curvatures estimation methods on triangular meshes. In: Proceedings of IEEE International Conference on Robotics and Automation, vol. **1**, pp. 1021–1026 (2003)
4. Shoemake K.: Animating rotation with quaternion curves. In: SIGGRAPH 1985 Proceedings of the 12th Annual Conference on Computer Graphics and Interactive Techniques, pp. 245–254 (July 1985)
5. Hart, P.E., Nilsson, N.J., Raphael, B.: A Formal Basis for the Heuristic Determination of Minimum Cost Paths. IEEE Transactions on Systems Science and Cybernetics **4**(2), 100–107 (1968)
6. Farin, G.: Curves and Surfaces for CAGD, 5th edn. Academic Press ISBN: 1-55860-737-4
7. Bates, D.M., Watts, D.G.: "Nonlinear Regression Analysis and Its Applications". Wiley, New York (1988). ISBN 978-0-471-81643-0

Automated Analysis of Images from Confocal Laser Scanning Microscopy Applied to Observation of Calcium Channel Subunits in Nerve Cell Model Line Subjected Electroporation and Calcium

Julita Kulbacka[1](✉), Marek Kulbacki[2], Jakub Segen[2], Anna Choromańska[1],
Jolanta Saczko[1], Magda Dubińska-Magiera[3], and Małgorzata Kotulska[4]

[1] Department of Medical Biochemistry, Medical University,
Chałubińskiego 10, 50-367 Wroclaw, Poland
Julita.Kulbacka@umed.wroc.pl
[2] Polish-Japanese Academy of Information Technology,
Koszykowa 86, 02-008 Warsaw, Poland
[3] Department of Animal Developmental Biology,
Institute of Experimental Biology, University of Wroclaw, Wroclaw, Poland
[4] Department of Biomedical Engineering, Wroclaw
University of Technology, Wybrzeze Wyspianskiego 27, 50-370 Wroclaw, Poland

Abstract. We assess a possibility of applying automated image analysis to immunofluorescence microphotographs from confocal laser scanning microscopy (CLSM). Several modes of automated analysis were tested to inspect differentiation of voltage dependent calcium channel subunits from a model of nerve cells PC12 (rat pheochromocytoma) subjected to electroporation (EP), with regard to extracellular calcium level. The objective of the experiments was evaluating sensitivity of the channel expression to the presence of calcium and electroporation voltage. For this purpose non-selective nanopores of a controlled conductivity were generated in the cell membrane using electroporation, at physiological or increased extracellular calcium concentrations. Introduction of Ca^{2+} into the cells was possible through electropores and physiological voltage-dependent calcium channels. Two subunits of calcium channel (α1H and α1G) were immunofluorescentically stained and the automated analysis of changes in cellular morphology was performed, based on comparative assessment of the fluorescence signal. The automated analysis allowed apparently higher observation capabilities. The results showed morphological changes in the channel subunits and higher expression of the channel, following its exposition to electric field or calcium.

Keywords: Electroporation · Calcium channel subunits · Fluorescence image analysis

1 Introduction

Automatic analysis of images obtained from immunodetection studies has long been a major problem for researchers [1]. The development in the cell biology techniques

© Springer International Publishing Switzerland 2015
N.T. Nguyen et al. (Eds.): ACIIDS 2015, Part II, LNAI 9012, pp. 297–306, 2015.
DOI: 10.1007/978-3-319-15705-4_29

discloses that fluorescence images are critical data for cell analysis, and therefore increasingly used in quantitative tests. The area of our present interests is cell image analysis algorithms, including: fluorescent microscopy images stained with selected antibodies, nuclei disorders detection and data association to the corresponding controls.

There are several studies that point to automated or semi-automated methods for immuno-stained cells analysis. Wahlby et al. [2] measured the concentration of at least six different antigens in each cell semi-quantitatively, using sequential immunofluorescence staining and image analysis techniques. Zhaozheng et al. [3] evaluated cell segmentation on the restored irradiance signals by simple thresholding. The experimental results validated that high quality cell segmentation can be achieved by this approach. However, to the best of our knowledge, no automatic analysis methods are available for a comparative analysis of immunostained images.

In our study, we applied an automated image analysis methods to the model of neural cells pheochromocytoma cell line (PC12 cells). PC12 is a standard model system to study neuronal differentiation. The main aim of current research was quantification of overloading intracellular calcium effects. The electroporation (EP) method was applied for generation of non-selective nanopores in cellular membrane under conditions of physiological or no extracellular calcium concentrations. EP enabled Ca^{2+} introduction into the cell through temporary electropores and physiological voltage-dependent calcium channels. Voltage-dependent Ca channels play an important role in regulating cellular Ca^{2+} concentration in excitable cells [4]. These Ca^{2+} channels not only are central in controlling neuro-transmitter release, excitation–contraction coupling, and excitation–secretion coupling, they are also involved in gene expression and neuronal migration. The α1 subunit forms the Ca^{2+} channel pore and can bind Ca^{2+} channel blockers [5,6,7]. The influence of electroporation on calcium channels expression has not been studied before.

After in vitro experiments, viability of cells was estimated and α1H and α1G subunits of the calcium channel were immunofluorescentically stained and visualized in confocal microscopy. The immunofluorescent images were subjected to automatic image analysis by selected algorithms.

2   Materials and Methods

2.1   Cell culture

The studies were performed on PC-12 (ATCC®CRL-1721™) cell line derived from a pheochromocytoma of the rat adrenal medulla that have an embryonic origin from the neural crest that has a mixture of neuroblastic cells and eosinophilic cells. The cells were grown in RPMI 1640 medium (Biochrom, Polgen, Poland), supplemented by 10% fetal bovine serum (FBS, Lonza, Poland) and supplemented by antibiotics (antibiotic-antimycotic solution, Lonza). For the experiments, the cells were removed by trypsinization (Trypsin 0.025 % and EDTA 0.02 %; Sigma) and washed with PBS (Sigma). The cells were maintained in a humidified atmosphere at 37°C and 5 % CO_2.

2.2   Electroporation Protocol

For electroporation experiments cells were prepared by prior trypsinization and gentle centrifugation 500×g. Then cells were suspended in cuvettes in low conductivity (0.12 S/m) electroporation buffer (10 mM KH_2PO_4/K_2HPO_4, 1 mM $MgCl_2$, 250 mM sucrose, pH 7.4) with or without 2.5 mM calcium solution. As electrodes, two aluminum parallel plates were used, 4 mm apart (BTX Harvard Apparatus). Microsecond electroporation (EP) was performed with ECM 830 Square Wave Electroporation System (manufactured by BTX Harvard Apparatus). The electric field was applied in the series of amplitudes: 500 and 1000 V/cm, 8 impulses in a sequence, 100 µs each, at the frequency of 1 Hz. Suspension of cells was electroporated in volume of 800 µl. Additionally experiments were performed with addition of 2mM calcium chloride (CaCl, Sigma Aldrich). After exposition to electrical pulses cells were maintained for 10 min in 37°C, then cells were designed for viability or CLSM study.

2.3   Viability Assay

Cellular viability was determined by the MTT assay. MTT (3-(4,5-dimethylthiazol-2-yl)-2,5-diphenyltetrazolium bromide, Sigma, Poland) assay was performed 24 or 48 hours after the end of experiments to evaluate cells mitochondrial dehydrogenase activity as a viability marker. The cells viability assay was performed according to the manufacturer's protocol (In Vitro Toxicology Assay, Sigma). The absorbance was measured at 570 nm using multiwell plate reader (EnSpire Multimode Reader, Perkin Elmer). Each experiment was performed in 3 independent repetitions. Mean values and standard deviations of all results were calculated. The final results were expressed as the percentage of mitochondrial function relative to untreated control cells.

2.4   Immunofluorescence – CLSM Study

The PC12 cells after EP and EP with Ca^{2+} were seeded on microscopic cover glasses in Petri dishes (35 mm) and after 24 hours were fixed in 4% paraformaldehyde. T-type voltage- gated Ca^{2+}- channel (subunit alpha-1G and alpha-1H) was immunofluorescently stained. The primary calcium channels antibodies were used: T-type Ca^{2+} alpha-1H (Santa Cruz Biotechnology, USA) and alpha 1-G (Sigma Aldrich, Poland). The primary antibodies were visualized with rabbit polyclonal antibodies in concentration 1:100. The secondary antibody goat anti-mouse IgG TRITC (for 60 min. at room temperature; 1:50; Sigma-Aldrich) were applied. DNA was stained with DAPI contained in fluorescent mounting medium (Roth). For imaging, Olympus FluoView FV1000 confocal laser scanning microscope (Olympus) was used. The images were recorded by employing Plan-Apochromat 60x oil-immersion objective.

2.5   Image Processing and Analysis

Based on visual analysis it is possible to distinguish visual differences between 12 fields from one case (Fig.3). In each field we have different parameters of voltage activating the subunits of calcium in the presence or absence of calcium ions. Fluorescent intensity within the cytoplasm is a factor that indicates similarities and differences between different examples (experiment repetitions) of the same field and similarities and differences between different fields which repeat between selected examples. The intensity of cytoplasm and nuclei is a signifier of the similarities between different cases of the same field (Fig.3), and differences between electrical field strengths ($\alpha 1$H and $\alpha 1$G expression under conditions of EP) just based on distribution of fluorescence intensity within the cytoplasm. We would like to extract the region of cytoplasm automatically, such that neither the nucleus nor the background areas are included, so that isolated cytoplasm regions could be analyzed and compared. We used RGB TIFF images of fluorescently stained cells. We constructed the following chain of image processing operations that begins with a composite input image and ends with a histogram of the cytoplasm region using Fiji [8] and R [9] for each image in the following steps:

(I)   Separation of a composite input image by splitting channels with stained calcium channel subunits and nuclei. It gives two pictures - red channel (calcium channel subunits – mainly cytoplasm) and blue channel (nuclei);

(II)   Conversion all separated images to 8 bit gray scale;

(III)   Constructing mask images for removing the background noise and the   nucleus regions from a cytoplasm image. This operation comprises smoothing by Gaussian filtering, formation of a binary image by segmentation of a gray scale image using Fiji operations *Smoothing, Mean white threshold* and cleaning the binary image using rank filters, including the *Minimum* filter and *Fill Holes* operations from Fiji. The last operation has an effect similar to morphological operations, closing small holes and removing spots;

(IV)   Extraction of connected regions (cells) using a connected components algorithm on the result of Step (III). It gives a numerical information such as the area, mean and standard deviation;

(V)   Removing the background and the nucleus regions from a gray scale cytoplasm image, using the binary images extracted in Step (III) as masks. This is done by AND-ing the gray scale image with the binary image from cytoplasm, then AND-ing the result with the inverted binary image of nuclei.

(VI)   Generation of a histogram from the image result of Step (V).

We separated the objects from the background using pixel intensity thresholding. We tested Fiji implementations of Huang [10] Intermodes [11], Li [12], MaxEntropy [13], MinError(I) [14], Moments [15], Otsu [16], Percentile [17], Shanbhang [18], Triangle [19] and Yen [20] methods on the set of 142 immunofluorescence microphotographs. Ultimately we chosen iterative procedure based on isodata algorithm [21] as most suitable for our data. Fiji implementation of the algorithm divides the image into object and background by taking the initial threshold, computes the average of averages of pixels at or below the threshold and increments the process until the threshold is larger than the composite average.

A histogram generated for each target image informs only about intensity of calcium channel subunits area without the noise from background and nuclei. For cell segmentation we used Particle Analysis (Fig. 1.(6)) plugin that enables PALM/STORM 2D/3D/4D

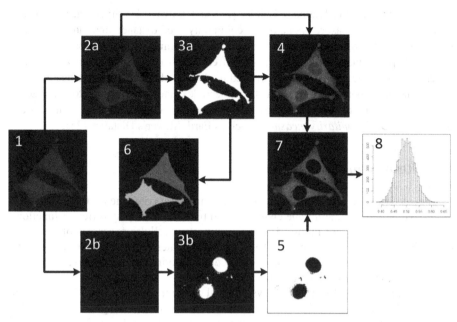

Fig. 1. Chain of selected operations for ROI preparation process and histogram generation: input image (1); extracted channels with cytoplasm (2a) and nuclei (2b), applied threshold and filters (3a,3b, 5); separated background (4); segmented cells' areas (6); cytoplasm without nuclei (7); histogram from the ROI (8)

particle detection and image reconstruction [22] and for each segmented cell calculates an area of each cytoplasm (A_C) and corresponding nuclei (A_N). We created factor Area Ratio $AR = \frac{A_N}{A_C}$ that informs about proportions between nuclei and cytoplasm size between 0 and 1. The influence of electroporation causes decreasing cytoplasm in relation to nuclei. For example, we obtained the following ratio values for alpha-1G subunit in case of cells incubated with calcium solution:

Table 1. The Area Ratio between cytoplasm and nuclei for alpha-1G staining

Electrical field intensity	Area ratio
0 V/cm	0,256
500 V/cm	0,292
1000 V/cm	0,370

In order to describe the overall brightness content of gray-scale immunofluores-cence microphotographs, one dimensional distribution of image intensities represented by histogram was applied. To find dissimilarity between two histograms $A=\{a_i\}$ and $B=\{b_j\}$, $j=i$ we used R package *bin-by-bin dissimilarity measures* from Utility Functions for R Histograms based on binary *ground distance* d_{ij} for bin i and j with threshold depending on bin size. The dissimilarity between pairs of images selected from 12 fields was tested using various methods: Minkowski-form distance,

Histogram intersection, Kullback-Leibler [23] and Jeffrey [24] divergence. Kullback-Leiber divergence measure is one of the commonly used distance measure that is used for computing the dissimilarity score between histograms. The comparison enabled us to give the preliminary information about differences between histograms representing different fields. The main drawback of used methods is that they take into account only correspondences between bins with the same index and are sensitive to bin size. In order to obtain more reliable results from histograms comparison we plan to study other *cross-bin dissimilarity measures,* such as Earth Mover's Distance [25].

3   Results and Discussion

Voltage-activated T-type Ca^{2+} channels are an important component of the large array of voltage-dependent membrane channels used by neurons to play different functions. T-type Ca^{2+} currents are involved in many important cellular processes of the cardiovascular and nervous systems [26]. Our study indicates that electroporation process and calcium ions concentrations in external cellular environment may have a regulatory effect on the calcium channel subunits.

The influence of EP and EP in physiological calcium concentration (2mM) on murine pheochromocytoma cells was performed (Fig.2, viability assay after 24 and 48 h of incubation). Microsecond EP can induce "nanopores" or releases in the cell outer membrane. Thus all ions and active particles can migrate or be transported inside the cells. We used this effect to load nervous cells with calcium ions. EP method has been previously applied by other researchers [27, 28] and in our previous study to enhanced drug delivery [29, 30]. Here we observed that EP alone was NOT cytotoxic neither, in zero, nor in physiological calcium concentrations. We can observe the significant decrease of cellular viability (ca. 24% of control cells) after the highest electric field intensity (1000 V/cm) in the presence of calcium ions after 24 hour incubation. The longer time of incubation (48h) resulted in increase of cellular viability up to 48%, which can be a sign of cellular recovery.

Additionally, some data indicate that permeabilization of the cell plasma membrane by intense pulsation of electrical field may be accompanied by prolonged inhibition of voltage-gated (VG) currents through Na^+, Ca^{2+}, and K^+ channels [31, 32].

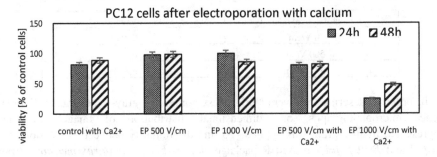

Fig. 2. The influence of microsecond electroporation on PC12 pheochromocytoma cells after 24 and 48 hour incubation

The most interesting point was evaluation of calcium channels expression after electroporation with or without calcium. The results are presented in Fig. 3. Both determined Ca^{2+} channels subunits: α1H and α1G are voltage-dependent. The results indicate a slight increase of fluorescence signal in both calcium channel subunits after electroporation without calcium after 500 V/cm. After 1000 V/cm the increased fluorescent signal was detected for α1H subunit in the absence of calcium ions. For the cells treated with electroporation in the presence of calcium ions we could observe the fluorescence signal increasing proportionally to increasing electrical field intensity in case of both subunits: α1H and α1G.

The visible morphological changes were detected after electroporation without and with Ca^{2+}. Cells volume reduced and cytoplasm shrunk, which was an effect of electropermeabilization.

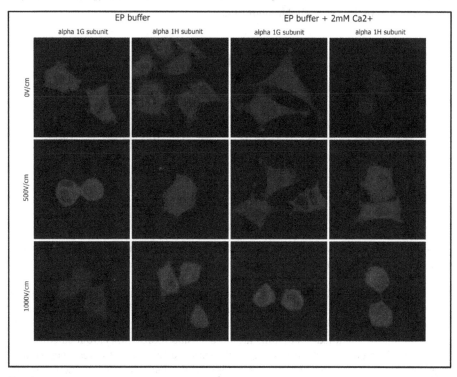

Fig. 3. The immunofluorescent detection of calcium channels subunits (α1H and α1G) in PC12 cells after electroporation in 0 mM of Ca^{2+} concentration and in physiological calcium suspension

The more detailed analysis of the fluorescence images is presented in Fig.4 A and B. The results were generated individually for each experiment. In Fig. 4 only two sets of histograms are presented. The analysis allowed us to observe similarity of histograms between each other, for the same therapeutic cases. Histograms were generated as the result of average fluorescence intensity. As we can see the image analysis indicated similar dependence between control and EP treated or EP+Ca^{2+}, cells as it is presented in Fig.3. However these results are more viable because the analysis is not descriptive as in case of standard "manual" immunofluorescence examination. Here we have the results calculated from minimum three experiments for a single sample. This evaluation provides the opportunity to verify the repeatability of experiments and the observed differences. The automatic image analysis presents that electroporation alone induced an increase of calcium channels expression, in case of α1H and α1G after 500 V/cm. In case of stronger electric field intensity the effect was more intense for α1H subunit and inversed in case of α1G. The increased expression of both subunits can be a cellular response to electropermeabilization. Both analyzed subunits are a part of voltage-dependent calcium channel, which indicates that EP may have a simulative effect on activity and expression of this type of channel. However, when electric field intensity increase up to 1000 V/cm the expression of α1G subunits was lower. This may be related with the absence of calcium ions in EP buffer which is in the extracellular environment.

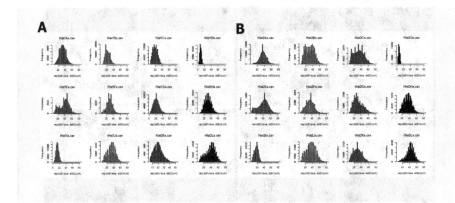

Fig. 4. The fluorescence intensity presented as histograms obtained from immunofluorescence images, A) results for example No. 1; B) results for example No. 2

Cells incubated in the presence of calcium solution (physiological concentration) responded with a slight decrease of the florescent signal (Fig. 3) and this effect is also visible on histograms (Fig. 4). When cells are electroporated in the presence of calcium solution the obtained histograms are broader and more points are located on the side of the light colors, which is compatible with the increased fluorescence signal. Thus, we can indicate that when cells recognize the lack of calcium then calcium channel subunits α1H and α1G tend to be overexpressed. Electroporation alone also induces the increase of calcium subunits expression, in particular the α1H subunit. At the physiological Ca^{2+} concentration the expression of both subunits increases with the rising electrical field intensity. These observations were not so evident when fluorescent microphotographs were analyzed in a traditional manner. More detailed image analysis processing was required, which could be provided by automated methods. Using this approach we could conclude that electroporation in physiological conditions stimulated expression of voltage-dependent calcium channels subunits and affect cellular morphology with significant decrease of cell viability.

Acknowledgements. This work was supported by National Science Center; grant No. 2011/01/D/NZ4/01255 (J. Kulbacka), Statutory Funds of Wroclaw Medical University, The Foundation the Medical University and partially by COST Action TD1104.

References

1. Kanade, T., et al.: Cell image analysis: Algorithms, system and applications. In: 2011 IEEE Workshop on Applications of Computer Vision (WACV), IEEE (2011)
2. Wählby, C., Erlandsson, F., Bengtsson, E., Zetterberg, A.: Sequential immunofluorescence staining and image analysis for detection of large numbers of antigens in individual cell nuclei. Cytometry **47**(1), 32–41 (2002)
3. Yin, Z., Su, H., Ker, E., Li, M., Li, H.: Cell-Sensitive Microscopy Imaging for Cell Image Segmentation. In: Golland, P., Hata, N., Barillot, C., Hornegger, J., Howe, R. (eds.) MICCAI 2014, Part I. LNCS, vol. 8673, pp. 41–48. Springer, Heidelberg (2014)
4. Liu, H., Felix, R., Gurnett, C., De Waard, M., Witcher, D.R., Campbell, K.P.: Expression and Subunit Interaction of Voltage-Dependent Ca2+ Channels in PC12 Cells. The Journal of Neuroscience **16**(23), 7557–7565 (1996)
5. Beam, K.G., Adams, B.A., Niidome, T., Numa, S.: Tanabe T Function of a truncated dihydropyridine receptor as both voltage sensor and calcium channel. Nature **360**, 169–171 (1992)
6. Ghosh, A., Ginty, D.D., Bading, H., Greenberg, M.E.: Calcium regulation of gene expression in neuronal cells. J. Neurobiol. **25**, 294–303 (1994)
7. Dunlap, K.: Luebke JI, Turner TJ Exocytotic Ca2+ channels in mammalian central neurons. Trends. Neurosci. **18**, 89–96 (1995)
8. Schindelin, J., Arganda-Carreras, I., Frise, E., Kaynig, V., Longair, M., Pietzsch, T., Preibisch, S., Rueden, C., Saalfeld, S., Schmid, B., Tinevez, J.Y., White, D.J., Hartenstein, V., Eliceiri, K., Tomancak, P., Cardona, A.: Fiji: an open-source platform for biological-image analysis. Nature Methods **9**(7), 676–682 (2012). doi:10.1038/nmeth.2019
9. R Development Core Team. R: A language and environment for statistical computing. R Foundation for Statistical Computing, Vienna, Austria. ISBN 3-900051-07-0, URL (2008). http://www.R-project.org
10. Huang, L.-K., Wang, M.-J.J.: Image thresholding by minimizing the measure of fuzziness. Pattern Recognition **28**(1), 41–51 (1995)
11. Prewitt, J.M.S., Mendelsohn, M.L.: The analysis of cell images. Annals of the New York Academy of Sciences **128**, 1035–1053 (1966)
12. Li, C.H., Lee, C.K.: Minimum Cross Entropy Thresholding. Pattern Recognition **26**(4), 617–625 (1993)
13. Kapur, J.N., Sahoo, P.K., Wong, A.C.K.: A New Method for Gray-Level Picture Thresholding Using the Entropy of the Histogram. Graphical Models and Image Processing **29**(3), 273–285 (1985)
14. Kittler, J., Illingworth, J.: Minimum error thresholding. Pattern Recognition **19** (1986)
15. Tsai, W.: Moment-preserving thresholding: a new approach. Computer Vision, Graphics, and Image Processing **29**, 377–393 (1985)
16. Otsu, N.: A threshold selection method from gray-level histograms. IEEE Trans. Sys., Man Cyber. **9**, 62–66 (1979). doi:10.1109/TSMC.1979.4310076
17. Doyle, W.: Operation useful for similarity-invariant pattern recognition. Journal of the Association for Computing Machinery **9**, 259–267 (1962). doi:10.1145/321119.321123

18. Shanbhag, A.G.: Utilization of information measure as a means of image thresholding. Graph. Models Image Process. (Academic Press, Inc.) 56(5), 414–419, ISSN 1049-9652 (1994)
19. Zack, G.W., Rogers, W.E., Latt, S.A.: Automatic measurement of sister chromatid exchange frequency. J. Histochem. Cytochem 25(7), 741–53. PMID 70454 (1977)
20. Yen, J.C., Chang, F.J., Chang, S.: A New Criterion for Automatic Multilevel Thresholding. IEEE Trans. on Image Processing 4(3), 370–378, ISSN 1057-7149, (1995). doi:10.1109/83.366472
21. Ridler, T.W., Calvard, S.: Picture thresholding using an iterative selection method. IEEE Transactions on Systems, Man and Cybernetics 8, 630–632 (1978)
22. Henriques, R., Lelek, M., Fornasiero, E.F., Valtorta, F.: Christophe Zimmer & Musa M Mhlanga. QuickPALM: 3D real-time photoactivation nanoscopy image processing in ImageJ. Nature Methods 7, 339–340. (2010). doi:10.1038/nmeth0510-339
23. Kullback, S.: Information Theory and Statistics. Dover, New York (1968)
24. Puzicha, J., Hofmann, T., Buhmann, J.M.: Non-parametric similarity measures for unsupervised texture segmentation and image retrieval. In: Proceedings of CVPR (June 1997)
25. Rubner, Y., Tomasi, C., Guibas, L.J.: A metric for distributions with applications to image databases. In: IEEE International Conference on Computer Vision, pp. 59–66 (January 1998)
26. Klockner, U., Lee, J.H., Cribbs, L.L., Daud, A., Hescheler, J., Pereverzev, A., Perez-Reyes, E., Schneider, T.: Comparison of the Ca2+ currents induced by expression of three cloned α1 subunits, α1G, α1H and α1I, of low-voltage-activated T-type Ca2+ channels. Eur J Neurosci 11, 4171–4178 (1999)
27. Tang, L.: Electroporation for drug and gene delivery in cancer therapy. Curr. Drug. Metab. 14(3), 271 (2013)
28. Gibot, L., Wasungu, L., Teissié, J., Rols, M.P.: Antitumor drug delivery in multicellular spheroids by electropermeabilization. J. Control Release 28, 167(2), 138–47 (2013)
29. Kulbacka, J., Daczewska, M., Dubińska-Magiera, M., Choromańska, A., Rembiałkowska, N., Surowiak, P., Kulbacki, M., Kotulska, M., Saczko, J.: Doxorubicin delivery enhanced by electroporation to gastrointestinal adenocarcinoma cells with P-gp overexpression. Bioelectrochemistry. pii: S1567–5394(14) 00065-6 (April 4, 2014)
30. Kulbacka, J., Kotulska, M., Rembiałkowska, N., Choromańska, A., Kamińska, I., Garbiec, A., Rossowska, J., Daczewska, M., Jachimska, B., Saczko, J.: Cellular stress induced by photodynamic reaction with CoTPPS and MnTMPyPCl5 in combination with electroporation in human colon adenocarcinoma cell lines (LoVo and LoVoDX). Cell Stress Chaperones 18(6), 719–731 (2013)
31. Nesin, V., Bowman, A.M., Xiao, S., Pakhomov, A.G.: Cell permeabilization and inhibition of voltage-gated Ca(2+) and Na(+) channel currents by nanosecond pulsed electric field. Bioelectromagnetics 33(5), 394–404 (2012)
32. Pakhomov, A.G., Kolb, J., White, J., Shevin, R., Pakhomova, O.N., Schoenbach, K.S.: Membrane effects of ultrashort (nanosecond) electric stimuli. Society for Neuroscience 37th Annual Meeting; Nov. 2–7, 2007; San Diego, CA. 2007a. 2007 Neuroscience Meeting Planner CD-ROM, Presentation Number: 317.14

Registration of Ultrasound Images for Automated Assessment of Synovitis Activity

Jakub Segen[1], Marek Kulbacki[1(✉)], and Kamil Wereszczyński[1,2]

[1] Polish-Japanese Academy of Information Technology,
Koszykowa 86, 02-008 Warszawa, Poland
mk@pja.edu.pl
[2] Institute of Informatics, Silesian University of Technology,
Akademicka 16, 44-100 Gliwice, Poland

Abstract. Ultrasound images of joints are used by doctors to assess a degree of synovitis activity, in diagnosis and treatment of rheumatoid arthritis. Research on automation of synovitis assessment from ultrasound images is being conducted, with objectives of lowering medical costs and improving patients care. Analysis of synovitis area in an image should be done relative to the joint and bones, therefore the joint and bones must be located in the initial step. An approach is proposed for locating joint and bones, by registering structural descriptions of the joint region. A preliminary result is presented that includes a description of a registration method that iteratively improves the registration quality, and its application example based on synthetic data.

Keywords: Medical ultrasound images · Image registration · Structural image description · Synovitis activity · Automated diagnosis

1 Introduction

Medical ultrasound imaging is a useful tool in detection and monitoring of synovitis, which is the inflammation of the synovial membrane that covers a joint. Synovitis is most often caused by rheumatoid arthritis. Ultrasound examination of joints provides good synovitis detection modality, comparable with MRI, less costly and more available. Recent articles by Zufferey et al. [1], Kunkel et al. [2] and van der Stadt, et al. [3] describe the use of ultrasound in and scoring of synovitis. The utrasound images are examined by medical experts to assess a presence and degree of synovitis. Automating this process is desirable; the exam would become less expensive, more available and free of subjective discrepancies in scoring. A project MEDUSA [4] is being conducted in Poland and Norway towards this goal.

The appearance of synovitis in an ultrasound image is described [6], and [5], as an image region that represents an hypoechoic area inside the joint, which means an area that produces less ultrasound signal reflections than the surrounding region, and therefore it appears darker. Since synovitis is located within the joint area, distances from an examined region to the joint and the bones are important as features for evaluating the

© Springer International Publishing Switzerland 2015
N.T. Nguyen et al. (Eds.): ACIIDS 2015, Part II, LNAI 9012, pp. 307–316, 2015.
DOI: 10.1007/978-3-319-15705-4_30

likelihood that this region represents synovitis. In addition, the measurement of the width of the hypoechoic region near the joint, in a direction normal to the bone, is used during the expert assessment of synovitis [6]. Therefore, such measurements also can be expected to be helpful in the automated analysis. To use such geometric measurements relative to the joint and bones, the regions that represent the joint and the bones in the ultrasound image must be located. Reliably locating the joint and bone areas in an ultrasound image is the first step of our proposed approach to the automated evaluation of synovitis. The remaining steps of the proposed approach will include finding the hypoechoic areas in the image, making a range of measurements on these areas that include the measurements relative to the joint and bones, constructing a decision function to determine which areas represent synovitis activity, and a function that computes the numerical assessment of the synovitis activity using a scale 0-3. The following part of this paper describes the current status of the development of a method for automatically locating the joint and bone regions in an ultrasound image of a joint.

2   Locating Joint and Bone Regions in Ultrasound Joint Images

A human can quickly learn to find joints and bones in ultrasound images of finger joints, but finding these image regions automatically is not trivial. Visual examination of images of finger joints, as well as the results of our ongoing work on developing detectors of joint and bones, based on local image neighborhood, has led us to conclude that such detectors, considered in isolation, will not be sufficient to reliably locate the joint and bone regions. Therefore, we are taking the approach of identifying a number of image features as a group, rather than individually, using not only the properties resulting from the analysis of a local neighborhood, but also the geometric relations within this group. It is expected, that this approach will result in a significantly more reliable detection of individual features than individual and isolated neighborhood based detectors. This approach uses the individual detectors, but does not rely solely on their output, and instead considers the outputs of all detectors in different parts of the image, which are mapped to a structural model, that has been priorly constructed by a learning method. The mapping part of this approach is conceptually related to recognition-by-components   theory proposed by Biederman [7].
We call the process of mapping detector outputs to the parts of the model image-to-model registration, since it can be considered   an extension of image-to-image registration [8] methods.

3   Registration of Ultrasound Images of Finger Joints

The role of image registration in a sinovitis detector is to identify the parts in ultrasound images such as bone and joint that will be used to guide the search for a possible inflammation region. The registration task is formulated as a problem of finding a correspondence between a set of image features and a structural model, that includes a set of parts, description of the part properties, and descriptions of geometric relations within the set the parts. The set of image features consists of the regions in an image that are found as a result of an application of a group of feature detectors such as joint, bone, skin and possibly tendon detectors. While, in isolation, such individual detectors are imperfect, the registration method, that combines their results with learned relational constraints, has a high likelihood of correctly identifying the parts. The registration method is based on an objective, or cost, function, that includes global and local rigid transformations. It proceeds by searching for a correspondence between the model and the image feature set that minimizes the cost. The models are inferred from a set of training images with annotations, by a combined supervised and unsupervised learning. The following subsections describe the feature detectors, the model structure, the process of mapping features to a model, and model learning.

3.1   Joint Detector

A learning approach to detecting a joint location in ultrasound images is being developed, using point feature descriptors. The training and test sets consist of images with the joint regions identified. An image point feature descriptor such as SURF or SIFT is used as the feature vector for a classifier. A pixel classifier is constructed, by training multiple simple classifiers, including k-nearest neighbor, nearest descriptor cluster, and SVM, from which a composite classifier is constructed using an ensemble learning method such as boosting. To increase the computational efficiency, the pixels where a descriptor is computed are initially screened by applying an image point feature detector. The final joint detector is the result of clustering the pixels classified as "joint region". The joint detector is described in more detail in a separate paper by Wereszczynski et al. [9].

3.2   Bone and Skin Detectors

To identify bones, skin and possibly tendons in an ultrasound image, a general detector for linear forms is being developed. The detector is a trainable classifier, which is applied to a stack or vector of images (VOI). The component images of the VOI are the results of passing the input image through a bank of filters, which are preselected to enhance the linear characteristics of an image, such as edges and ridges. These filters include a Gaussian smoothing filter, the first derivative filter (gradient operator), the second derivative filter, the Laplacian, and threshold operators.

To apply, for example, a bone detector to a pixel to test if this pixel is a part of a bone, each component of the VOI is sampled along the sampling line, a line which crosses the tested pixel, and is approximately perpendicular to the estimated orientation of the bone. An equal number of samples are taken on each side of the pixel. The resulting values are assembled in a sample array, whose rows correspond to the components of VOI and columns to the sample positions along the sampling line. The sample array, treated as a vector, is used as the input to classifiers and learning estimators. The rows and columns of the sample array may be removed or added as a part of the classifier's feature selection process being developed.

One type of a linear, hyperplane classifier, is shown in Fig. 1. The classifier is represented by a vector of convolution kernels that correspond to the component images of VOI, and associated weight factors. The classification function is a weighted sum of normalized convolutions of the component images and their corresponding component kernels. The kernels and weights of the classifier are computed from a training set of ultrasound images using supervised learning methods. This form of a classifier is especially easy to implement within an image processing toolkit, such as the ITK toolkit [12] which is used in the project MEDUSA.

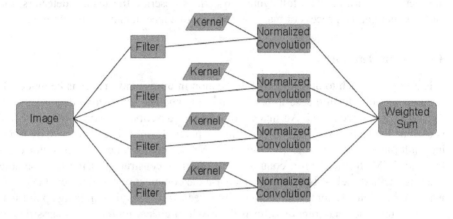

Fig. 1. A linear classifier

Fig. 2a shows an example of ultrasound image data to which classifiers are applied in order to detect linear forms such as bone or skin.   Fig. 2b shows the same ultrasound image, with added annotations. The regions of skin and bone are marked with lines, the area of the joint and the region of synovitis activity are enclosed inside closed curve and polygon.

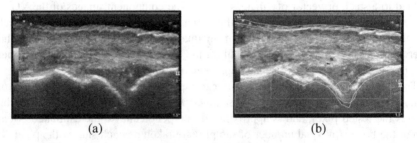

(a)   (b)

Fig. 2. Ultrasound Image of a finger joint without (a) and with (b) annotations. Brown line (top) marks the skin, green lines mark the bones, blue polygon surrounds the joint, pink closed curve outlines the synovitis region

Figure 3 is an illustration of an image processing module implemented in the ITK toolkit. The Input image is smoothed in X and Y directions and passed to Gradient, 2nd Derivative filters. Independently, a 2d Gaussian smoothing filter is applied to the

Input image. The 2nd Derivative X and Y images are the input to a Laplacian filter. In addition the gradient, 2nd Derivative and Laplacian images are input to threshold filters, and a ZeroCrossing filter is applied to the gradient X and Y images. The Cascade submodule implements the linear classifier shown in Fig. 1. The filter marked Primitive Bone Detector is an Interest Operator for linear forms.

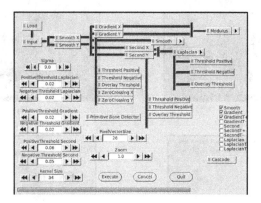

Fig. 3. Layout of an image processing module in ITK toolkit

The results of the applying the Smoothing filter, Gradient-Y filter and 2nd Derivative-Y filter to an ultrasound image in Fig. 4a, are shown in Fig. 4b, Fig. 4c and Fig. 4d.

Fig. 4e shows the results of a Lower Threshold filter applied to 2nd Derivative image. Fig. 4f shows the results of a Zero Crossing filter applied to the Gradient-Y image. An Interest Operator for linear forms can be built by applying logical AND to the results of gradient zero-crossing and negative threshold of the second-derivative. Applying the logical AND operation to the images of Fig. 4e and Fig. 4f results in the Interest Operator image shown in Fig. 4g. The best values for filter window sizes, thresholds and weights will be determined using cross-validation. Development of the skin and bone detectors is in progress - the results will be reported when it is completed.

3.3   Structural Description of the Joint Region

The features extracted from an image are used to form the primitive parts for a structural description of a joint region. There are three types of a primitive part: a "joint", "skin" and "bone". The joint has a geometric form of a point, both skin and bone have a form of an oriented line segment. The structural description is composed from such primitive parts placed at specific locations, and orientations for the skin and bone. The primitive parts are extracted either from the annotations in training and test images or obtained through the image analysis. Using the annotations, the joint is at the image location marked as the joint, and skin and bone lines are obtained by segmenting into straight line segments the annotation curves used to annotate these features. The image analysis extraction is applied to the test images and to images that are not annotated. The joint is constructed directly from the output of the joint detector.

The skin and bone line segments are built from the responses of the skin and bone detectors, by assembling the detected pixels into pixel chains to which line segments are fitted using the piecewise linear approximation, such as [10]. We will refer to the primitive parts in a structural description as nodes.

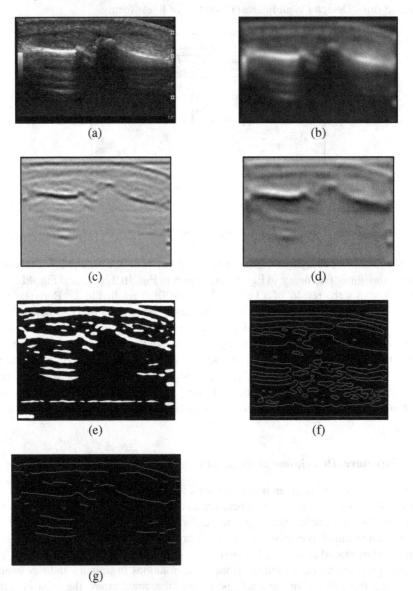

Fig. 4. The results of the applying filters: (a) ultrasound input image; (b) Result of the smoothing filter; (c) Result of 2nd Derivative-Y filter; (d) Result of Gradient-Y filter; (e) Result of Lower Threshold filter applied to 2nd Derivative-Y image; (f) Result of Zero Crossing Filter applied to the Gradient-Y image; (g) Results of the Interest Operator for linear forms

3.4   Ultrasound Image Registration by Matching Structural Descriptions

Image registration, within the scope of this paper, means bringing to correspondence the features of two images, using a transformation that is a superposition of a rigid planar transformation and a set of local deformations. The registration can be achieved by matching structural descriptions obtained either from image annotations or through image analysis.

Given two structural descriptions, a reference R and a target X, a matching problem is defined as finding a function Map from the nodes of X to the nodes of R, and a planar rigid transformation T, which minimize the objective function Q, defined below. Multiple nodes of X can be mapped to the same node of R, and there may be nodes of R to which no node of X is mapped, in other words the function Map is generally neither injective nor surjective. Also, some of the nodes of X may remain unmapped, in which case they will be considered to be mapped to a *null* node. A null node is added to R as the node R_0.

$$Q(X, R, T, Map) = \sum_{i=1}^{n} d^2(X_i, T(R_{Map(i)})) + mC_R \tag{1}$$

n and m are the numbers of nodes in the target and the reference structures, respectively, C_R is a regularization coefficient and $d^2(x, r)$ is the squared distance between the nodes x and r, which is defined in the following subsection.

The goal of the matching process is finding Map and T that give the minimum of Q, that is

$$Q^{Opt}(X, R) = min_{\{Map, T\}} Q(X, R, T, Map) \tag{2}$$

The local deformations mentioned earlier do not need to be applied to evaluate Q, but are used implicitly, to compute distances d^2 in equation 1. The value Q^{Opt} can be treated as a squared distance between the target and the reference structures, and used in supervised and unsupervised learning. The best values of the constants - the coefficient C_R, and the distance from a target node to the reference null node $d(x, R_0)$, can be obtained through a process of supervised learning, using cross-validation.

Node Translation Vector and Distance between Nodes

To complete the specification of Q, the node distance function d^2 is described. A distance between two joint type nodes is their Euclidean distance. A distance between a joint and non-joint node is infinite. A distance between a non-null node and a null node is a constant D_n. A distance between two line-segment type nodes, a reference segment R and a target X, is illustrated in Figure 5, and defined as follows. The midpoint of R is projected onto a line extension of X, resulting in a vector of projection V_p. The segment R is translated in parallel to X, by a vector V_s, such that V_s is the minimal length vector that moves a projection of R onto the line extending X, denoted R_p, such that R_p is entirely contained in X, or X is contained in the R_p. The vector sum of V_p and V_s is the node translation vector V_t:

$$Vt = Vp + Vs \qquad (3)$$

The squared distance between line segments R and X is

$$d^2(X,R) = Vt^t \cdot Vt = \|Vt\|^2 \qquad (4)$$

The node translation vector V_t is used by the minimization search method described below.

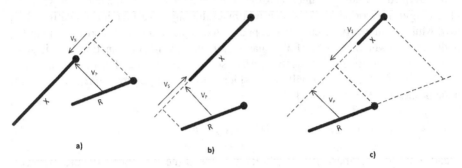

Fig. 5. Node translation vector components VP and VS for three cases a), b) and c)

The Minimization Search Method

The minimization search uses a version of the RANSAC method [11]. To match a reference structural description R with a target description X, the method examines the pairs of pairs of line segment nodes (r, x), where r is a pair of nodes from R and x is a pair of nodes from X. For each pair (r, x):

1. an initial rigid transformation (rotation and translation on a plane) T_i is calculated, such that $T_i(r)$ has an orientation angle and center position averaged from the nodes of x.
2. the reference description R is transformed by T_i to R', $R' = T_i(R)$, by transforming each node of R.
3. the mapping *Map* between X and R is computed by finding for each node x in X a corresponding node r' in R', which is nearest to x in distance d (Equation 4).
4. a rotation angle from r' to x is calculated for each node x and its corresponding node r', given by *Map*; an average rotation angle A is computed from these node rotation angles, and R' is rotated by the angle A, into R'', where $R'' = Rotation(R',A)$. The current rigid transformation T_c, such that $R'' = T_c(R)$ is computed and the current best score Q_b is set to infinity.
5. the translation for R'' which finds a local minimum of Q is computed by the following iteration loop:
    (a) For each node x in X and its corresponding (as given by *Map*) node r'' in R'', compute the translation vector V_t and the squared distance $d^2(x,r'')$;
    (b) Compute the vector sum V_g of all V_t node translation vectors, and the value of $Q_c = Q(X, R, Tc, Map)$, using the squared distances from the step (a);

(c) If $Q_b - Q_c <$ *epsilon*, then terminate the loop, otherwise set $Q_b = Q_c$;
(d) Update the current rigid transformation T_c by adding the vector V_g to T_c translation component, update R'' by translating it by the vector V_g and go to Step (a).
6. Select the pair of node pairs for which the above iteration results in the smallest value of Q_b, and return the corresponding *Map* and T_c as the results of the minimization search.

An example of structure matching using the minimization search method applied to a synthetic data is shown in Fig. 6. Figure 6a shows the target structure X and Figure 6b shows a reference structure R. Figure 6c shows target X overlaid with the transformed reference in red color,   for the node pairs $X:[0, 2] - X$, and $R:[0, 1]$, with local minimum $Q = 63.3$. Figure 6d shows the improvement for the pairs $X:[4, 5]$, and $R:[0, 1]$ with $Q = 51.8$. Figure 6e shows the final result for the pairs $X:[4, 7]$, and $R:[0, 1]$ with $Q = 34.9$.

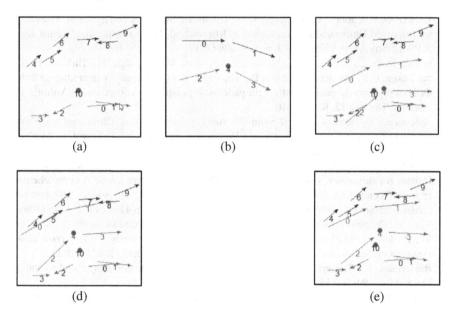

(a)   (b)   (c)

(d)   (e)

Fig. 6. a) Target, b) Reference, c) initial match, d) improved match, e) final match

4   Conclusions

A proposed approach to automated assessment of synovitis activity requires a solution to registering representations of ultrasound images of joints. A method for ultrasound image registration by matching structural descriptions, composed of detected skin, bone, and joint features, has been described and illustrated with a synthetic example. The described method uses node locations and orientations, but does not use skin and bone labels returned by the detector. A planned improvement will include the use of such labels. Other improvements or extensions, such as a scale transformation, or inclusion in Equation 1 the differences of node relations as a measure of geometric incoherence, may be added as a result of upcoming tests on labeled ultrasound images.

Acknowledgements. The research leading to these results has received funding from the Polish-Norwegian Research Programme operated by the National Centre for Research and Development under the Norwegian Financial Mechanism 2009-2014 in the frame of Project Contract No. Pol-Nor/204256/16/2013.

References

1. Zufferey, P., Tamborrini, G., Gabay, C., Krebs, A., Kyburz, D., Michel, B., Moser, U., Villiger, P.M., So, A., Ziswiler, H.R.: Recommendations for the use of ultrasound in rheumatoid arthritis: literature review and SONAR score experience. Swiss Med Wkly. **143**, w13861 (2013)
2. Kunkel, G.A., Cannon, G.W., Clegg, D.O.: Combined Structural and Synovial Assessment for Improved Ultrasound Discrimination of Rheumatoid, Osteoarthritic, and Normal Joints A Pilot Study. Open Rheumatol J. **6**, 199–206 (2012)
3. van de Stadt, L.A., Bos, W.H., Meursinge Reynders, M., Wieringa, H., Turkstra, F., van der Laken, C.J., van Schaardenburg, D.: The value of ultrasonography in predicting arthritis in auto-antibody positive arthralgia patients: a prospective cohort study. Arthritis Research & Therapy **12**, R98 (2010)
4. Automated Assessment of Joint Synovitis Activity from Medical Ultrasound and Power Doppler Examinations using Image Processing and Machine Learning Methods. http://eeagrants.org/project-portal/project/PL12-0015
5. Backhaus, M., Kamradt, T., Sandrock, D., Loreck, D., Fritz, J., Wolf, K.J., Raber, H., Hamm, B., Burmester, G.R., Bollow, M.: Arthritis of the finger joints: A comprehensive approach comparing conventional radiography, scintigraphy, ultrasound, and contrast-enhanced magnetic resonance imaging. Arthritis & Rheumatism **42**(6), 1232–1245 (1999)
6. Vlad, V., Berghea, F., Libianu, S., Balanescu, A., Bojinca, V., Constantinescu, C., Abobului, M., Predeteanu, D., Ionescu, R.: Ultrasound in rheumatoid arthritis - volar versus dorsal synovitis evaluation and scoring. BMC Musculoskeletal Disorders **12**, 124 (2011)
7. Biederman, I.: Recognition-by-components: a theory of human image understanding. Psychol Rev. **94**(2), 115–147 (1987)
8. Sotiras, A., Davatzikos, C., Paragios, N.: Deformable Medical Image Registration: A Survey. IEEE Trans Med Imaging. **32**(7), 1153–1190 (2013)
9. Wereszczyński, K., Segen, J., Kulbacki, M., Mielnik, P., Fojcik, M., Wojciechowski, K.: Identifying a joint in medical ultrasound images using trained classifiers. In: Chmielewski, L.J., Kozera, R., Shin, B.-S., Wojciechowski, K. (eds.) ICCVG 2014. LNCS, vol. 8671, pp. 626–635. Springer, Heidelberg (2014)
10. Dobkin, D.P., Levy, S.V.F., Thurston, W.P., Wilks, A.R.: Contour Tracing by Piecewise Linear Approximations. ACM Transactions on Graphics **9**(4), 389–423 (1990)
11. Fischler, M.A., Bolles, R.C.: Random Sample Consensus: A Paradigm for Model Fitting with Applications to Image Analysis and Automated Cartography. Comm. of the ACM **24**(6), 381–395 (1981)
12. The Insight Segmentation and Registration Toolkit: www.itk.org

Quantifying Chaotic Behavior in Treadmill Walking

Henryk Josiński[2]([✉]), Agnieszka Michalczuk[1], Adam Świtoński[1,2],
Romualda Mucha[3], and Konrad Wojciechowski[1]

[1] Polish-Japanese Academy of Information Technology,
Aleja Legionów 2, 41-902 Bytom, Poland
{amichalczuk,aswitonski,kwojciechowski}@pjwstk.edu.pl
[2] Institute of Informatics, Silesian University of Technology,
Akademicka 16, 44-100 Gliwice, Poland
{Henryk.Josinski,Adam.Switonski}@polsl.pl
[3] Medical University of Silesia, Batorego 15, 41-902 Bytom, Poland
romam28@wp.pl

Abstract. The authors describe an example of application of nonlinear time series analysis directed at identifying the presence of deterministic chaos in human motion data by means of the largest Lyapunov exponent (LLE). The research aimed at determination of the influence of gait speed on the LLE value with a view to verification of the belief that slower walking leads to increased stability characterized by smaller LLE value. Analyses were focused on the time series representing hip flexion/extension angle, knee flexion/extension angle and dorsiflexion/plantarflexion dimension of the ankle. Gait sequences were recorded in the Human Motion Laboratory (HML) of the Polish-Japanese Academy of Information Technology in Bytom by means of the Vicon system. Application of the AC5000M treadmill allowed recordings in three variants: at the preferred walking speed (PWS) of each subject, at 80% of the PWS and at 120% of the PWS. According to the recommendations from the literature the LLE value was estimated twice for every time series: as the short-term LLE_1 for the first stride and as the long-term LLE_{4-10} over a fixed interval between the fourth and the tenth stride. In the latter case it was confirmed that the LLE value increases with walking speed for both limbs.

Keywords: Nonlinear time series analysis · Phase space reconstruction · Deterministic chaos · Human motion analysis

1 Introduction

Dynamics properties of a system can be determined on the basis of its model (provided that it is known) consisting of differential or difference equations or through analysis of experimental data collected as result of system observation.

© Springer International Publishing Switzerland 2015
N.T. Nguyen et al. (Eds.): ACIIDS 2015, Part II, LNAI 9012, pp. 317–326, 2015.
DOI: 10.1007/978-3-319-15705-4_31

The state of a dynamical system at a given instant of time can be represented by a point in the phase space spanned by the state variables of the system. Many nonlinear or infinite-dimensional dynamical systems exhibit chaotic behavior. The presence of deterministic chaos is characterized by extreme sensitivity to initial conditions. This hallmark means that initially nearby points can evolve quickly into very different states. In case of analysis of experimental data, fundamental components of the process of determining existence of chaos in a signal represented by a time series are phase space reconstruction and subsequent estimation of the Lyapunov exponents which quantify the average exponential rate of divergence of initially nearby phase space trajectories [1]. Thus, a positive value of the largest Lyapunov exponent (LLE) implies chaotic behavior.

Chaoticity was observed in a variety of systems from several areas including, among others, meteorology, physics, engineering, economics and biology. From among biomedical signals EEG, ECG and gait kinematic data are worthy to note.

The chaotic characteristics of the ECG signals (Lyapunov exponents spectrum and correlation dimension) were incorporated to the set of features for the purpose of biometric individual identification [2] but, first of all, chaos theory has been applied to the analysis of electrocardiogram for examination of cardiac disorders [3].

Chaos is also present in epileptic EEG signals. Brain activity during seizure differs greatly from that of normal state which can be observed as a decrease in chaoticity in the minutes before the seizure. Thus, analysis of the changes of the LLE allows the detection and prediction of the incoming epileptic seizure [4], [5].

Methods for estimating LLE from experimental data provide a promising means of directly quantifying local dynamic stability (LDS) during locomotion [6], that is to say, the degree of resilience of gait control to infinitesimally small perturbations that occur naturally during walking and are manifested as natural kinematic variability [7], [8]. These disturbances, resulting in stride-to-stride[1] differences in kinematic measurements, are attenuated in time – at least within the current stride and possibly across subsequent strides – by the neuro-controller and musculoskeletal system in order to maintain a stable walking pattern and the LLE can be used to quantify the exponential attenuation of variability between neighboring kinematic trajectories [9]. As far as running is concerned, the influence of both speed and use of a leg prosthesis on the dynamic stability expressed by means of the LLE for subjects with and without unilateral transtibial amputations was studied in [10].

One of the other approaches quantifying stability from experimental data was based on the Floquet multipliers and was used for post-polio patients [11]. However, this method, in contrast to the LLE, requires the assumption of strict periodicity of human walking, whereas humans do not walk in an exactly periodic manner [12]. The decrease of walking stability of elderly subjects in lateral plane was reported in [13] on the basis of analysis of variability of the centre of gravity using the approximate entropy (ApEn) technique. Both approaches – LLE and

---

[1] A stride is defined as a full cycle of limb movement – from a heel-strike to heel-strike again.

ApEn – were used to quantify local stability and measure variability, respectively, in the anterior cruciate ligament deficient knee during walking [14], [15].

The research described in the present paper aimed at determination of the influence of gait speed on the LLE value with a view to verification of the belief that slower walking leads to increased stability which is characterized by smaller LLE value.

The organization of this paper is as follows – section 2 contains a brief description of the method used for the purpose of human motion time series analysis. Section 3 deals with procedure of the experimental research along with its results. The conclusions are formulated in section 4.

2   Method of the Time Series Analysis

Nonlinear time series analysis methods enable the determination of characteristic invariants such as the LLE of a particular system solely by analyzing the time course of one of its variables [16]. Nevertheless, identification of chaotic behavior based on experimental data is a multistage process. The first step constitutes a phase space reconstruction. On the basis of Takens' embedding theorem [17] the phase space can be reconstructed using time-delayed measurements of a single observed signal in form of a time series. Reconstruction consists in viewing a time series $x_k = x(k\tau_s), k = 1, \ldots, N$ in a Euclidean space $\mathbb{R}^m$, where m is the embedding dimension and τ_s is the sampling time [18]. Each m-dimensional embedding vector is formed as $\mathbf{x}_k = \left[x_k, x_{t+\tau}, x_{t+2\tau}, \ldots, x_{t+(m-1)\tau} \right]^T$, where τ is the delay time. The selection of τ and m is important for the sake of reconstruction quality. It is worthwhile to mention that the properties associated with the system's dynamics (*inter alia* Lyapunov exponents) are preserved in the new phase space.

Time delay τ was calculated from the first local minimum of the mutual information function (MI). Mutual information between x_t and $x_{t+\tau}$ is a measure of how much information can be predicted about one time series point given full information about the other [1]. Hence, given that the time delay τ was determined from the first MI minimum, $x_{t+\tau}$ adds the largest amount of information. Assuming that the range of values in a time series was partitioned into j intervals of equal length, the mutual information function $I(\tau)$ can be computed according to the following formula:

$$I(\tau) = \sum_{h=1}^{j} \sum_{k=1}^{j} P_{h,k}(\tau) \log_2 \left(\frac{P_{h,k}(\tau)}{P_h P_k} \right) \tag{1}$$

where h and k are indices of intervals, P_h, P_k denote the probabilities that x_t assumes a value within the h-th, k-th interval, respectively, and $P_{h,k}(\tau)$ is the joint probability that x_t belongs to the h-th interval and $x_{t+\tau}$ is taken from the k-th interval.

The minimal embedding dimension m that is required to fully resolve the structure of the system in the reconstructed phase space was found by the method of "False Nearest Neighbors" (FNN) [19], which is based on the following assumption constituting the condition of no self-intersections – if the attractor (i.e., a set of states towards which neighboring states asymptotically approach in the course of dynamic evolution [20]) is to be reconstructed successfully in $\mathbb{R}^m$, then all points that are close in $\mathbb{R}^m$ should also be sufficiently close in $\mathbb{R}^{m+1}$ [18]. A point which does not satisfy this condition is a "false" neighbor. Its neighborhood did not result from the dynamics of the system but from the projection issues. The number of such points is computed for increasing embedding dimension until the percentage of "false" neighbors is below a given threshold.

The mean divergence between neighboring trajectories in the phase space at time t is described by the following formula:

$$d(t) = De^{\lambda_1 t} \tag{2}$$

where D is the initial separation between neighboring points and λ_1 is the LLE [6]. The Rosenstein algorithm [21] estimates the LLE on the basis of the appropriately reconstructed attractor locating for each point on the attractor its nearest neighbor on adjacent orbit and computing the divergence between successive pairs of points along the trajectories. On the basis of the formula (2) the Euclidean distance $d_j(i)$ between the j-th pair of nearest neighbors after i time steps of the length equal to Δt and the LLE are linked in the following way:

$$\ln[d_j(i)] \approx \lambda_1 \cdot (i \cdot \Delta t) + \ln[D_j] \tag{3}$$

Considering all pairs of nearest neighbors a set of parallel lines with slope equal to the LLE can be defined on the basis of formula (3). Hence, after averaging ($\langle \cdot \rangle$) the logarithmic divergence of the neighboring trajectories after i time steps over all values of j:

$$\lambda_1 \cdot (i \cdot \Delta t) \approx \left\langle \ln\left[\frac{d_j(i)}{D_j}\right] \right\rangle \tag{4}$$

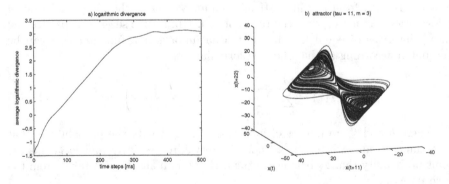

Fig. 1. a) Average logarithmic divergence for the Lorenz system as a function of time, b) reconstructed 3D attractor of the Lorenz system

the LLE is estimated as the slope of the linear best-fit line to the average logarithmic divergence of the neighboring trajectories across the given time span. The average divergence as a function of time computed for the Lorenz system [22] and the reconstructed 3D attractor are presented in Fig. 1. The curve cannot continuously grow with time, because the attractor is bounded in the phase space. Therefore, the linear best-fit to the divergence must be performed on the precisely determined linear region [8].

3   Experimental Research

The research consisted in analysis of gait sequences which were recorded in the Human Motion Laboratory (HML) of the Polish-Japanese Academy of Information Technology [23] by means of the Vicon Motion Kinematics Acquisition and Analysis System. The Vicon system is equipped with 10 NIR (Near InfraRed) cameras recording the movement of an actor wearing a special suit with attached markers (the *motion capture* process). Positions of the markers in consecutive time instants constitute basis for reconstruction of their 3D coordinates. Application of the AC5000M treadmill allowed recordings in three variants: at the preferred walking speed (PWS) of each subject (denoted as "Normal"), at 80% of the PWS ("Slower") and at 120% of the PWS ("Faster"). Three sequences of continuous walking of length of several dozen seconds were recorded with a frequency of 100 Hz at given walking speed for every person. Four healthy subjects (denoted as B0156, B0238, B0244 and B0245) participated in the experiments. Their walking speeds are presented in Table 1.

Table 1. Three variants of walking speed [m/s]

	B0156	B0238	B0244	B0245
Slower	0.85	0.80	1.43	0.71
Normal	1.07	1.03	1.79	0.89
Faster	1.29	1.25	2.15	1.07

A stride interval (i.e., the time elapsed between subsequent ipsilateral heel strikes) varies across subjects and speeds. Mean values of a single stride interval for every subject are included in Table 2.

Table 2. Mean values of a single stride interval [s]

	B0156	B0238	B0244	B0245
Slower	1.52	1.30	1.07	1.39
Normal	1.20	1.15	0.96	1.22
Faster	1.13	1.07	0.88	1.16

A single recorded time series represented movements at the given joint in one of the following planes: *sagittal* (*lateral*), *frontal* (*coronal*) and *transverse* (*horizontal*), which divide body into left/right, anterior/posterior (front/back) and superior/inferior (upper/lower) parts, respectively. However, for the comparative purposes the analyses were focused on the time series related to the movements in the sagittal plane. The considered data cover dorsiflexion/plantarflexion angle of the ankle, knee flexion/extension angle and hip flexion/extension angle (Fig. 2a).

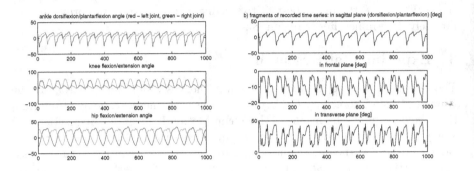

Fig. 2. Initial fragments of recorded time series: a) movements of all 3 joints in sagittal plane, b) movements of left ankle joint in all 3 planes

Local dynamic stability is assumed to describe how a subject responds to small initial differences in kinematics over the course of 10 strides [24]. Hence, according to the recommendations from the literature [6] the LLE value was estimated twice for every time series: as the short-term LLE_1 for the first stride and as the long-term LLE_{4-10} over time interval between the fourth and the tenth stride. The short-term divergence estimates the local stability immediately after a perturbation, whereas the long-term divergence has been adopted empirically. Limits of each stride were exactly demarcated due to precisely marked occurrences of the "heel-strike" event. The parameters of the phase space reconstruction (τ and m) were determined by the MI and FNN methods for each time series separately. The threshold of the "false" neighbors acceptability was equal to 1%. The values of τ varied from 3 to 48, whereas 13, 17 and 20 were used most frequently. As far as the embedding dimension m is concerned, selected values belonged to the set of $\{4, 5, 6, 7\}$ and 5 was the predominant value (used for circa 57% of time series), which was confirmed by other researchers [6], [7], [9]. Fig. 3 illustrates results of successive stages of the LLE computation for the time series from Fig. 2b representing the dorsiflexion/plantarflexion angle of the B0244 subject's left ankle at the preferred walking speed: a) mutual information, b) percentage of false nearest neighbors, c) 3D projection of the reconstructed attractor, d) average logarithmic divergence (vertical lines delineate the region across which estimates of LLE_1 and LLE_{4-10} were calculated). All computations were performed using MATLAB.

Values of both short-term and long-term largest Lyapunov exponents (LLE_1 and LLE_{4-10}, respectively) for dorsiflexion/plantarflexion angle of the ankle, hip flexion/extension angle and knee flexion/extension angle, averaged over all subjects, are presented in upper, middle and lower rows of Fig. 4, respectively,

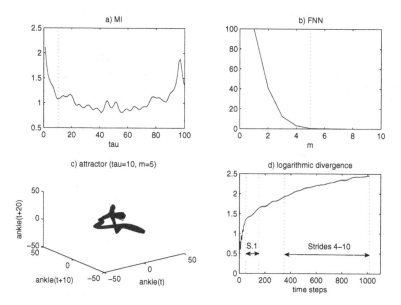

Fig. 3. Stages of the LLE computation for the dorsiflexion/plantarflexion angle: a) mutual information, b) percentage of false nearest neighbors, c) 3D projection of the reconstructed 5D attractor, d) average logarithmic divergence

taking into account three variants of walking speed and both (left/right) sides of the human body.

All types of ankle, hip and knee movements in the sagittal plane are characterized by the positive LLE values which detect and quantify the presence of deterministic chaos. It is worthwhile to mention that larger short-term values indicate greater sensitivity to local perturbations during the time needed for 1 step. This remark allows to justify the fact that the smallest mean LLE_1 value occurs at the preferred walking speed which is natural, comfortable and requires the least muscular effort. This observation concerns all considered right side joints as well as left knee joint and for both exceptions (left ankle joint and left hip joint) difference between mean LLE_1 values at slower and normal velocities is rather marginal. Mean long-term value which increases with walking speed for both limbs without exceptions confirms the assumption that slower walking leads to increased stability characterized by smaller LLE_{4-10} value. Besides, both ankle joints are distinguished by the smallest LLE values in all variants of walking speed. However, the results may be influenced by the fact that subjects were recorded during walking on treadmill.

In summary, local dynamic stability is influenced by walking speed with different contributions from considered joints. The above mentioned observations are mostly consistent with results presented in [9], [10].

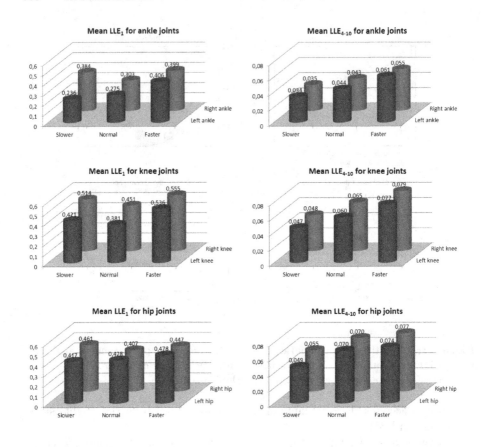

Fig. 4. Mean values of both short-term and long-term largest Lyapunov exponents: for the dorsiflexion/plantarflexion angle of the ankle (upper row), for the knee flexion/extension angle (middle row) and for the hip flexion/extension angle (lower row)

4   Conclusion

The authors described an example of identification of the presence of chaotic behavior in human motion data based on phase space reconstruction and estimation of the LLE. The presented approach was used to characterize local dynamic stability of walking. The results are consistent with outcomes reported in the literature. Quantifying dynamic stability during walking is important for assessing people who have a greater risk of falling. For this reason, the applied procedure of a time series analysis will be extended by incorporation of other measures, such as correlation dimension and approximate entropy, in the hope that it will constitute support for assessment of gait disorders (among others resulting from Parkinson's disease, stroke, osteoarthritis of the hip or osteoarthritis of the spine).

Acknowledgments. This work was supported by the project DEC-2011/01/B/ST6/06988 from the Polish National Science Centre.

References

1. Henry, B., Lovell, N., Camacho, F.: Nonlinear dynamics time series analysis. In: Akay, M.(ed.) Nonlinear Biomedical Signal Processing: Dynamic Analysis and Modeling, vol. 2, pp. 1–39. Wiley Online Library (2012) (published online)
2. Chen, C.-K., Lin, C.-L., Chiu, Y.-M.: Individual identification based on chaotic electrocardiogram signals. In: 6th IEEE Conference on Industrial Electronics and Applications, pp. 1765–1770 (2011)
3. Cohen, M.E.: Chaos. Wiley Encyclopedia of Biomedical Engineering (2006)
4. Osowski, S., Świderski, B., Cichocki, A., Rysz, A.: Epileptic seizure characterization by Lyapunov exponent of EEG signal. COMPEL: The International Journal for Computation and Mathematics in Electrical and Electronic Engineering **26**(5), 1276–1287 (2007)
5. Mormann, F., Andrzejak, R.G., Elger, C.E., Lehnertz, K.: Seizure prediction: the long and winding road. Brain **130**, 314–333 (2007)
6. Dingwell, J.B., Cusumano, J.P.: Nonlinear time series analysis of normal and pathological human walking. Chaos **10**(4), 848–863 (2000)
7. Dingwell, J.B., Marin, L.C.: Kinematic variability and local dynamic stability of upper body motions when walking at different speeds. Journal of Biomechanics **39**, 444–452 (2006)
8. Terrier, P., Deriaz, O.: Non-linear dynamics of human locomotion: effects of rhythmic auditory cueing on local dynamic stability. Frontiers in Physiology **4**, 1–13 (2013)
9. England, S.A., Granata, K.P.: The influence of gait speed on local dynamic stability of walking. Gait & Posture **25**, 172–178 (2007)
10. Look, N., Arellano, C.J., Grabowski, A.M., McDermott, W.J., Kram, R., Bradley, E.: Dynamic stability of running: The effects of speed and leg amputations on the maximal lyapunov exponent. Chaos **23**, 043131 (2013)
11. Hurmuzlu, Y., Basdogan, C., Stoianovici, D.: Kinematics and Dynamic Stability of the Locomotion of Post-Polio Patients. Journal of Biomechanical Engineering **118**(3), 405–411 (1996)
12. Dingwell, J.B., Kang, H.G.: Differences Between Local and Orbital Dynamic Stability During Human Walking. Journal of Biomechanical Engineering **129**(4), 586–593 (2007)
13. Arif, M., Ohtaki, Y., Nagatomi, R., Inooka, H.: Estimation of the Effect of Cadence on Gait Stability in Young and Elderly People using Approximate Entropy Technique. Measurement Science Review **4**(2), 29–40 (2004)
14. Stergiou, N., Moraiti, C., Giakas, G., Ristanis, S., Georgoulis, A.D.: The effect of the walking speed on the stability of the anterior cruciate ligament deficient knee. Clinical Biomechanics **19**, 957–963 (2004)
15. Georgoulis, A.D., Moraiti, C., Ristanis, S., Stergiou, N.: A novel approach to measure variability in the anterior cruciate ligament deficient knee during walking: the use of the approximate entropy in orthopaedics. Journal of Clinical Monitoring and Computing **20**, 11–18 (2006)
16. Perc, M.: The dynamics of human gait. European Journal of Physics **26**, 525–534 (2005)

17. Takens, F.: Detecting Strange Attractor in Turbulence. Lecture Nodes in Mathematics, vol. **898**, pp. 366–381 (1981)
18. Kugiumtzis, D.: State Space Reconstruction Parameters in the Analysis of Chaotic Time Series - the Role of the Time Window Length. Physica D: Nonlinear Phenomena **95**(1), 13–28 (1996)
19. Kennel, M.B., Brown, R., Abarbanel, H.D.I.: Determining embedding dimension for phase-space reconstruction using a geometrical construction. Physical Review A **45**(6), 3403–3411 (1992)
20. Weisstein, E.W.: Attractor From MathWorld - A Wolfram Web Resource. http://mathworld.wolfram.com/Attractor.html
21. Rosenstein, M.T., Collins, J.J., De Luca, C.J.: A practical method for calculating largest Lyapunov exponents from small data sets. Physica D **65**, 117–134 (1993)
22. Awrejcewicz, J., Mosdorf, R.: Numerical Analysis of Some Problems of Chaotic Dynamics. WNT (2003)
23. Webpage of the Human Motion Laboratory of the Polish-Japanese Academy of Information Technology. http://hm.pjwstk.edu.pl
24. Toebes, M.J.P., Hoozemans, M.J.M., Furrer, R., Dekker, J., van Dieën, J.H.: Local dynamic stability and variability of gait are associated with fall history in elderly subjects. Gait & Posture **36**, 527–531 (2012)

Augmented Reality and 3D Media

Human Detection from Omnidirectional Camera Using Feature Tracking and Motion Segmentation

Joko Hariyono, Van-Dung Hoang, and Kang-Hyun Jo[✉]

Graduate School of Electrical Engineering, University of Ulsan,
Ulsan 680–749, Korea
{joko,hvzung}@islab.ulsan.ac.kr, acejo@ulsan.ac.kr

Abstract. This paper proposes a motion segmentation method on images which are captured by an omnidirectional camera. A simple unwrapping method is performed to convert an omnidirectional image into a panoramic image. Two consecutive panoramic images are used for motion analysis. Corner features are extracted from the image, and their locations are defined in local patches by dividing an image into grid cells. Then, each feature in previous frame is tracked to find its corresponding in the current frame. The affine transformation is performed using three corresponding features. The regions of moving object are detected as transformed objects which are different from the previously registered background. Morphological processing is applied for smoothing the motion region. Histogram vertical projection and boundary saliency are applied to segmenting the motion. Finally, the proposed motion segmentation method is used for human detection in omnidirectional images. The performance result shown the best detection rate is 97.25% at 0.3 false positive rate.

Keywords: Human detection · Omnidirectional camera · Feature tracking · Motion segmentation

1 Introduction

Human detection is one of the essential tasks for understanding environment in robotic and autonomous navigation. It is more challenging to detect human or pedestrian, in order to avoid an accident and control locomotion of the vehicle. Thus it can be implementing for autonomous driving and vision based driving assistance system.

Over last decade, the question of how to detect human in the image has been thoroughly investigated [1, 6, 8, 9]. Due to the random influence, such as scene structure, variation of clothes, occlusion, the problem remains challenging and continues to attract research. Simple and applicable methods also attract research in real-time mobile robot applications. Human detection methods for mobile robot have been actively developed. Gavrila et al. [2] employed hierarchical shape matching to find pedestrian candidates from moving vehicle. Local descriptor is proposed for object recognition and image retrieval. Some authors presented methods for human detection using HOG

© Springer International Publishing Switzerland 2015
N.T. Nguyen et al. (Eds.): ACIIDS 2015, Part II, LNAI 9012, pp. 329–338, 2015.
DOI: 10.1007/978-3-319-15705-4_32

and SVM, such as [2, 6, 13, 14]. The HOG features are calculated by taking orientation histograms of edge intensity in a local region. The HOG features are extracted from all locations of a dense grid on an image region. The HOG features are fed to SVM to classify objects. [11, 12, 15] tried to use the omnidirectional sensor for detected human from mobile robot application.

Accurate estimation of the vehicle ego-motion relative to the road and environment is a key component for autonomous driving [3]. Liu et al. [4] tried to measure the camera ego-motion using Kanade-Lucas-Tomasi (KLT) optical flow tracker. Corresponding features are obtained in consecutive two omnidirectional images. The motion analysis of feature points is used. The camera ego-motion was calculated based on an affine transformation of two consecutive images, where corner features were tracked by KLT optical flow tracker [5]. Using corner feature for tracking with only one affine transformation model, the detecting moving objects resulted in a problem. It could not represent the whole background changes. In our previous work [7], the method using each affine transformation of local pixel groups was proposed to detect moving objects. Grid windows-based is used for KLT tracker by tracking each local sector of the input image. In the other work [8] motion analysis of moving objects using optical flow are employed. Using two consecutive images, the motion distances of pixel groups are calculated. Based on those calculations, the independence motions of objects are segmented by subtracts the pixel with its corresponding. Then moving objects are segmented from the motion caused by the camera ego-motion.

The paper presents a method to detect human in omnidirectional images based on motion analysis and segmentation. This work proposed a motion segmentation method in images which are obtained from the omnidirectional camera A simple unwrapping method is proposed to converted omnidirectional images into panoramic image. Two consecutive panoramic images are used for motion analysis. Corner features are extracted using [5] method. Those features are defined in local patch by divided images into grid cells. Then optical flow tracker is applied by tracking each feature in the previous frame to find corresponding feature in the current frame. Three corresponding features are used to compute affine transformation. It is performed according to each location of corresponding cells in the consecutive images. Motion regions are detected as object movements which are different from the previously registered background. Morphological process is applied for smoothing detected motion regions. In order to localize regions, the histogram vertical projection and boundary saliency are applied. The HOG features are extracted on the candidate region and classified using linear Support Vector Machine (SVM). The HOG feature vectors are used as input of linear SVM to classify the given input into pedestrian/non-pedestrian. Fig.1 shows the overview of the pedestrian detection algorithm for driver assistance system.

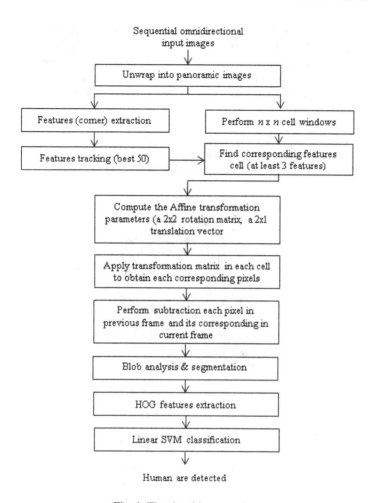

Fig. 1. The algorithm overview

2   The Omnidirectional Camera System

This section presents the omnidirectional camera system which is used in this work. The omnidirectional camera mounted on the mobile robot [16] as shown in Fig. 2. The camera consists of the perspective camera and the hyperboloid mirror. It captures an image reflecting from the mirror so that the image obtains reflective scene, as shown in Fig. 3. Actually those images are very useful, because it shows in 360 degree field of view. However some pre-processing is needed. For this task, unwrapping into panoramic image is performed. It will take a little bit computational cost, however it is needed for easier to analyze pattern of motion.

A simple transformation is used for unwrapping from omnidirectional into panoramic images. Calibration parameters such as focal length, mirror equation and projection plane are not needed. The necessary parameters for this unwrapping are center and radius of the projected circles from both mirror borders, inner and outer. Actually the calculated pixels in the omnidirectional camera image will not corresponding exactly, one to one, to the pixels of projected image in panoramic view. So, sub pixel anti-aliasing methods should be used. The bilinear interpolation method will be used that may lead to aliasing in case the omnidirectional image is under-sampled.

Fig. 2. An omnidirectional camera mounted on the mobile robot

3   Feature Extraction and Tracking

To analyze the motion on the image, features are important things which are needed to extract from the image. Our motion feature is motivated by the fact that strong cues exist in the movements of different body parts when the human is walking. They include the motion of two legs, those between two parts of an arm, or those between a leg and an arm. They provide useful cues to identify the walking motion.

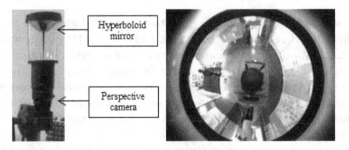

Fig. 3. The structure of omnidirectional vision and its image

Corner features are used in this work as the interest point to perform the task for feature tracking. Using Shi and Tomassi [5] method, fifty best corner features are extracted in each image. Then, all the features are defined within patches (cells) which generate by divided the image using n x n cell windows.

Two consecutive images are compared and tracked corresponding features from previous frame to the current frame. The feature which located in a group cell is used by method from [9] to find the motion distance of each pixel in a group of cells. The motion distance d in x and $y-$ axis by of feature cell in the previous frame $g_{t-1}(i,j)$ is obtained by finding most similar cell $g_t(i,j)$ in the current frame. The affine parameter is calculated by the least square method using three corresponding features in those two consecutive frames. The camera ego-motion compensation is obtained by subtracting frame difference on the tracked corresponding pixel cells. Since the camera ego-motion is applied, regions of moving objects are segmented. Fig.4 shows moving object segmentation is obtained.

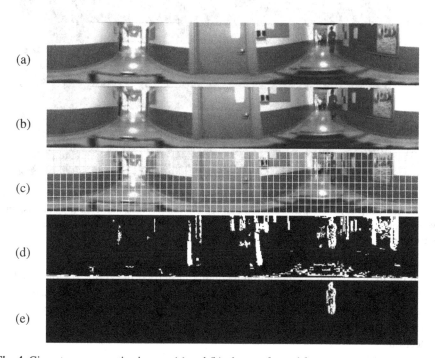

Fig. 4. Given two consecutive images (a) and (b), then performed features extraction and grid cell windows on first image (c) Frame difference result (d) Ego-motion compensation result (e)

4   Motion Segmentation

The motion segmentation algorithm is devised for accurately locates bounding boxes of the motion in the different image. Each pixel output from frame difference with the ego-motion compensation cannot show clearly as silhouette. It just gives information of motion area from moving object. Those areas are applied morphological process to obtain region of moving objects and noise removal.

The fact that humans usually appear in upright positions, and conclude that segmenting the scene into vertical strips is sufficient most of the time [17]. Then, detected moving objects are represented by the position in width in x axis. Using projection histogram h_x by pixel voting vertically project image intensities into x − coordinate.

We detect moving object based on the constraint of moving object existence that the bins of histogram in moving object area must be higher than a threshold and the width of these bins should be higher than a threshold. Fig. 5 shows the motion segmentation process.

(a)

(b)

(c)

Fig. 5. The region, result from ego-motion compensation for moving object detection (a). Then, morphological process are performed on that region, the result is shown in (b) The histogram vertical projection is performed with specific threshold, so that the region of moving object is localized (c).

Adopting the region segmentation technique proposed in [9], the region is defined using boundary saliency. It measures the horizontal difference of data density in the local neighborhood. The local maxima correspond to where maximal change in data density occurs. They are candidates for region boundaries of human in moving object detection.

5   Experimental Results

Our robot system moved in the corridor. The omnidirectional camera captured more than 3,000 sequent images with more than 6,800 people. The System is evaluated using those images. Proposed algorithm was programmed in MATLAB and executed on a Pentium 3.40 GHz, 64-bit operating system with 8 GB RAM.

The reliability of our system is evaluated using the proposed cell window based flow estimation whether it is still visible at several levels. Several sizes of cell window are tested for optical flow tracking. It determines the relative distance movement of a feature in the consecutive images. So that we consider using the flow field window tracking which is still accurate for varied sizes of landmark and object on the image. As a counterweight parameter, computational cost was considered for performance balancing. Table 1 shows the miss detection rate and computational cost of several different window sizes. The result shows that the cell window size 6x6 is lowest on the miss detection rate, but slowest in computational cost. The cell window size 8x8 is selected based on lower in the miss rate and faster computational speed.

Table 1. The miss rates of various cell window sizes

Cell window size	False positive rate	Computational Cost (fps)
6 x 6	0.031	9.67
8 x 8	0.032	9.82
10 x 10	0.034	9.81
12 x 12	0.038	9.87

For the human detection stage, the original HOG proposed by Dalal and Triggs is implemented. The HOG features are extracted from 16×16 pixels of the candidate image region. Candidate images are obtained from motion segmentation results. The first, Sobel filter is performed to obtain the gradient orientations from each pixel in this local region. The local region is divided into small cells with cell size is 4×4 pixels. Histograms of gradients orientation with eight orientations are calculated from each of the local cells. Then, the total number of HOG features becomes $128 = 8 \times (4 \times 4)$ and they constitute a HOG feature vector.

For the training process, the person INRIA datasets [2] are used. These images were used for positive samples. The negative samples were collected from images of mountain, airplane, building, sky etc. The number of sample images is 3,000. From these images, 1,000 person images and 2,000 negative samples were used as training samples to determine the parameters of the linear SVM.

When the original HOG is implemented using sequential images dataset. Due to the quality of panoramic images, the recognition rate for test dataset is 95.33% at 0.4 false positive rates. Then, the proposed method is applied using combination of the camera ego-motion compensation (EMC) and the HOG feature. The HOG feature vectors were extracted from locations of the candidate image. Then, the feature vectors were used as input of the linear SVM. The selected subsets were evaluated by cross validation. Also the recognition rates of the constructed classifier using test samples are evaluated.

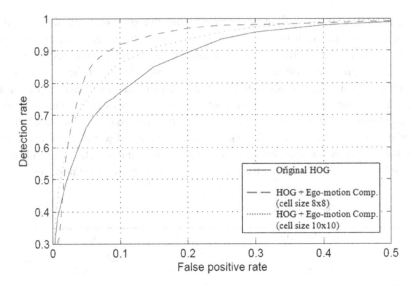

Fig. 6. Comparison result when the proposed method tested and the original HOG

The relation between the detection rates and the number of false positive rate are shown in Fig. 6. The best recognition rate 97.25 % was obtained at 0.3 false positive rates, while original HOG obtain lower. It means that higher detection rate with smaller false positives rate is obtained. Table 2 shows the computational cost also reduces five and seven times better for HOG and EMC which the size of cell 8 x 8 and 10 x 10 respectively. Several results of detected humans are shown in the Fig.7.

Fig. 7. Successful detection results

Table 2. Computational cost comparison of human detection system when the omnidirectional image is performed only using original HOG and combination HOG + proposed motion segmentation method using cell windows size 8x8 and 10x10 (in fps)

Original HOG	HOG + EMC (8 x 8)	HOG + EMC (10 x 10)
0.87	4.39	6.25

6   Conclusion

The paper presents a method to detect human in omnidirectional images based on motion analysis and segmentation. This work proposed a motion segmentation method in images which are obtained from the omnidirectional camera. An unwrapping method is performed to converted omnidirectional images into panoramic image. Two consecutive panoramic images are used for motion analysis. Corner features are extracted from the image. Those features are defined in local patch by divided images into grid cells. Then optical flow tracker is applied by tracking each feature in the previous frame to find corresponding feature in the current frame. Three corresponding features are used to compute affine transformation. It is performed according to each location of corresponding cells in the consecutive images. Motion regions are detected as object movements which are different from the previously registered background. Morphological process is applied for smoothing detected motion regions. In order to localize regions, the histogram vertical projection and boundary saliency are applied. The HOG features are extracted on the candidate region and classified using linear SVM. The HOG feature vectors are used as input of linear SVM to classify the given input into pedestrian/non-pedestrian. The best recognition rate 97.25 % was obtained at 0.3 false positive rates. The computational cost also reduces seven times better when the proposed motion segmentation is used for detecting moving object region as a human candidate.

In the future work, real-time omnidirectional camera application for pedestrian detection is an interesting topic for our improvement.

References

1. Gavrila, D.M., Munder, S.: Multi-cue Pedestrian Detection and Tracking from a Moving Vehicle. International Journal of Computer Vision 73(1), 41–59 (2007)
2. Dalal, N., Triggs, B.: Histograms of oriented gradients for human detection. In: IEEE Conference on Computer Vision and Pattern Recognition (CVPR) (2005)
3. Vassallo, R.F., Santos-Victor, J., Schneebeli, H.: A general approach for egomotion estimation with omnidirectional images. In: Proceedings of the Third Workshop on Omnidirectional Vision (2002)
4. Liu, H., Dong, N., Zha, H.: Omni-directional Vision based Human Motion Detection for Autonomous Mobile Robots. Systems Man and Cybernetics 3, 2236–2241 (2005)

5. Tomasi, C., Kanade, T.: Detection and tracking of point features. In: Proceedings of Fourteenth International Conference on Pattern Recognition, vol. 2, p. 1433 (1998)
6. Hoang, V.D., Vavilin, A., Jo, K.H.: Fast human detection based on parallelogram haar-like feature. In: The 38th Annual Conference of The IEEE Industrial Electronics Society, Montreal, pp. 4220–4225 (2012)
7. Hariyono, J., Wahyono, D.C., Jo, K.H.: Accuracy enhancement of omnidirectional camera calibration for structure from motion. In: International Conference on Control, Automation and Systems, Korea (2013)
8. Hariyono, J., Hoang, V.D., Jo, K.H.: Moving Object Localization using Optical Flow for Pedestrian Detection from a Moving Vehicle. The Scientific World Journal **2014** (2014). http://dx.doi.org/10.1155/2014/196415
9. Hariyono, J., Hoang, V.D., Jo, K.H.: Motion segmentation using optical flow for pedestrian detection from moving vehicle. In: 6th International Conference on Computational Collective Intelligent Technologies and Application, Seoul (2014)
10. Dollar, P., Wojek, C., Schiele, B., Perona, P.: Pedestrian Detection: An Evaluation of the State of the Art. IEEE Transactions on Pattern Analysis and Machine Intelligence **34**(4), 743–761 (2012)
11. Wang, M.L., Lin, H.Y.: Object recognition from omnidirectional visual sensing for mobile robot applications. In: IEEE International Conference on Systems, Man and Cybernetics (2009)
12. Arican, Z., Frossard, P.: OMNISIFT: scale invariant features in omnidirectional images. In: IEEE Int. Conf. on Image Processing (2010)
13. Hoang, V.D., Le, M.H., Jo, K.H.: Hybrid cascade boosting machine using variant scale blocks based HOG features for pedestrian detection. Neurocomputing **135**, 357–366 (2014)
14. Hoang, V.-D., Hernandez, D.C., Jo, K.-H.: Partially obscured human detection based on component detectors using multiple feature descriptors. In: Huang, D.-S., Bevilacqua, V., Premaratne, P. (eds.) ICIC 2014. LNCS, vol. 8588, pp. 338–344. Springer, Heidelberg (2014)
15. Kang, S., Roh, A., Nam, B., Hong, H.: People detection method using GPUs for a mobile robot with an omnidirectional camera. Optical Engineering **50**(12), 127204 (2011)
16. Hariyono, J., Hoang, V.-D., Jo, K.-H.: Human detection from mobile omnidirectional camera using ego-motion compensated. In: Nguyen, N.T., Attachoo, B., Trawiński, B., Somboonviwat, K. (eds.) ACIIDS 2014, Part I. LNCS, vol. 8397, pp. 553–560. Springer, Heidelberg (2014)
17. Roh, C.H., Lee, W.B.: Development of a 3d tangible-serious game for attention improvement. International Journal of Intelligent Information and Database Systems **8**(2), 85–96 (2014). doi:10.1504/IJIIDS.2014.063253

Automatic Fast Detection of Anchorperson Shots in Temporally Aggregated TV News Videos

Kazimierz Choroś[(✉)]

Department of Information Systems, Wrocław University of Technology,
Wybrzeże Wyspiańskiego 27, 50-370 Wrocław, Poland
`kazimierz.choros@pwr.edu.pl`

Abstract. The temporal aggregation method applied for sports news videos detects and aggregates two kinds of shots: sequences of long shots, mainly studio shots unsuitable for content-based indexing, and sports player shots adequate for sports categorization. Hereby, it significantly reduces the number of frames analyzed in content-based indexing of TV sports news. The tests have shown that applying the temporal aggregation method it was possible to reject about half of video frames and despite this almost all sports scenes reported in TV sports news have been indexed. The paper examines the influence of the temporal aggregation on the detection of anchorperson shots in news videos. The TV news video editing is similar to that of TV sports news although news shots are longer in average than sports player shots. The interviews, statements, and commentaries are more significant in news than in sports news for content-based analyses because these statements are not necessarily spoken by an anchorman, so they are usually important informative parts of TV news. The experiments carried out on TV news and described in the paper have shown that anchorperson shots as well as interview shots may be more easily and faster selected when TV news videos are temporally aggregated. These experiments were performed in the Automatic Video Indexer AVI.

Keywords: Content-based video indexing · Temporal aggregation · Video indexing strategies · Video categorization · Video structure · TV news analyses · Broadcast news · Anchorperson shots · Interview shots · AVI indexer

1   Introduction

The methods of content-based video indexing are applied to an automatic processing of television broadcast, mainly to TV news and TV sports news. For effective retrieval of video data in very huge video data bases more and more sophisticated indexing and retrieval methods are being developed. The main goal of content-based video indexing of broadcast videos can be to ensure an effective retrieval of special events or special people, of official statements or political polemics and commentaries, etc. In the case of TV sports news the goal is to detect players and games, or to detect reports of special sports categories. Due to the automatic categorization of sports events, i.e. the automatic detection of the sports disciplines of reported events videos can be automatically indexed and then retrieved. The retrieval of individual

© Springer International Publishing Switzerland 2015
N.T. Nguyen et al. (Eds.): ACIIDS 2015, Part II, LNAI 9012, pp. 339–348, 2015.
DOI: 10.1007/978-3-319-15705-4_33

sports news and sports highlights such as the best or actual games, tournaments, matches, contests, races, cups, etc., special player behaviours or actions like penalties, jumps, or race finishes, etc. in a desirable sports discipline becomes more effective. The other goal could be also the detection of advertisement billboards and banners, authentic emotion detection of audience, and so on.

If the content-based indexing method needs to analyze all frames of a video such a procedure becomes extremely time consuming. So, it is preferred that the indexing is limited to the key-frames, to only one or to a few for every shot or even for every video scene. To apply such a strategy we need to recognize the structure of a video. The shots are detected due to very effective temporal segmentation methods. Unfortunately, the detection of scenes is not so easy to perform.

A scene is defined as a group of consecutive shots sharing similar visual properties and having a semantic correlation – following the classical rule of unity of time, place, and action. One of the method of scene detection is the temporal aggregation method [1]. This method groups shots taking into account only shot lengths. Such an approach is very fast. The temporal aggregation method has been already tested for TV sports news leading to the detection of player scenes. A player scene is a scene presenting the sports game, i.e. a given scene was recorded on the sports fields such as playgrounds, tennis courts, sports hall, swimming polls, ski jumps, etc. All other non-player shots and scenes usually recorded in a TV studio such as commentaries, interviews charts, tables, announcements of future games, discussions of decisions of sports associations, etc. are called studio shots or studio scenes. Studio shots are slightly useful for video categorization (detection of sports discipline) and therefore can be rejected. It was observed that the studio scenes may be even two thirds of TV sports news. This rejection of non-player scenes before starting content analyses creates an opportunity to reduce significantly computing time and conduct these analyses more efficiently.

Generally different video genres have different editing style. The specific nature of videos has an important influence on the efficiency of temporal segmentation methods. It has been tested in the experiments performed in [2]. The efficiency of segmentation methods was analyzed for five different categories of movies: TV talk-show, documentary movie, animal video, action & adventure, and pop music video. The segmentation parameters should be suitable to the specificity of the videos.

In the case of TV news video editing is similar to that of TV sports news but shots are longer in average. Then the statements and commentaries can be more significant in news than in sports news for content-based analyses because these statements are not spoken by anchorman but also by politicians. The detection of politicians is important and may be realized using for example face detection methods.

In this paper the usefulness of temporal aggregation method in detection of the main structural units of TV news is verified.

The paper is organized as follows. The next section describes some related work in the area of an automatic anchorperson shots detection. The main idea of the temporal aggregation method is presented in the third section. The detection of pseudo-scenes using temporal aggregation is outlined in the forth section. The fifth section presents the experimental results of the detection of the main structural units of TV news obtained in the AVI Indexer. The final conclusions and the future research work areas are discussed in the last sixth section.

2   Related Work

Much research has been carried out in the area of automatic recognition of video content and of visual information indexing and retrieval [3–6]. Traditional textual techniques frequently applied for videos are not sufficient for nowadays video archive browsers. The effective methods of the automatic categorization of a huge amount of broadcast news videos would be highly desirable. Most of proposed methods require the detection of the structure of videos being indexed [7].

Anchor/non-anchor shots are frequently used as a starting point for the automatic recognition of a video structure. Anchorperson shot detection is still a challenging and important stage of news video analysis and indexing. Recent years, many algorithms have been proposed to detect anchorperson shots. Because we observe the very high similarity between anchor shots (very static sequences of frames, small changes, the same repeated background) one of the approaches of an anchor shot detection is based on template matching. Whereas the other methods are based on different specific properties of anchor shots. In the first group of methods a set of predefined models of an anchor should be defined and then, they are matched against all frames in a news video, in order to detect potential anchor shots. The second group of an anchor shot detection methods is mainly based on clustering. Unfortunately, the proposed methods are very time-consuming because they require complex analyses of a great number of video frames.

The high values of recall and precision for anchorperson detection have been obtained in the experiments on 10 news videos [8]. The news videos were firstly as usually segmented into shots by a four-threshold method. Then the key frames were extracted from each shot. The anchorperson detection was conducted from these key frames by using a clustering-based method based on a statistical distance of Pearson's correlation coefficient.

The new method presented in [9] can be also used for dynamic studio background and multiple anchorpersons. It is based on spatio-temporal slice analysis. This method proposes to extract two different diagonal spatio-temporal slices and divide them into three portions. Then all slices from two sliding windows obtained from each shot are classified to get the candidate anchor shots. And finally, the real anchor shots are detected using structure tensor. The experiments carried out on news programs of seven different styles confirmed the effectiveness of this method.

The algorithm described in [10] analyzes audio, frame and face information to identify the content. These three elements are independently processed during the cluster analysis and then jointly in a compositional mining phase. The temporal features of the anchorpersons for finding the speaking person that appears most often in the same scene are used to differentiate the role played by the detected people in the video. Significant values of precision and recall have been obtained in the experiments carried out for broadcast news coming from eight different TV channels.

A novel anchor shot detection method proposed in [11] detects an anchorperson cost-effectively by reducing the search space. It is achieved by using skin colour and face detectors, as well as support vector data descriptions with non-negative matrix factorization.

It is observed that the most frequent speaker is the anchorman [12]. An anchor speaks many times during the programme, so the anchorperson shots are distributed all along the programme timeline. This observation leads to the selection of the speaker who most likely is the anchorman. It is assumed that a speaker clustering process labels all the speakers present in the video and associates them to temporal segments of the content. However, there are some obvious drawbacks, because a shot with a reporter (interview shots) or with a politician (statement shots) frequently found in news can be erroneously recognized as an anchor shot.

Another observation in a large database [13] draws much attention to interview scenes. In many interview scenes an interviewer and an interviewee recursively appear. A technique called interview clustering method based on face similarity can be applied to merge these interview units.

May be the strategy should be that all statement shots, i.e. anchor, reporter, interview shots should be detected by the same procedure and then using face detection method we should try to identify a person speaking.

Video analyses discussed in the related papers as well as in this research are the methods using visual features only. There are also audio-visual approaches analyzing not only visual information but also audio (see for example [14, 15]).

3 Temporal Segmentation and Aggregation in the AVI Indexer

The Automatic Video Indexer AVI [16] is a research system designed to develop new tools and techniques of automatic video content-based indexing for retrieval systems, mainly based on the video structure analyses [17] and using the temporal aggregation method [1]. The standard process of automatic content-based analysis and video indexing is composed of several stages. Usually it starts with a temporal segmentation resulting in the segmentation of a movie into small units called video shots. Shots can be grouped to make scenes, and then key-frame or key-frames for every scene can be selected for further analyses. In the case of TV sports news every scene is categorized using such strategies as: detection of playing fields, of superimposed text like player or team names, identification of player faces, detection of lines typical for a given playing field and for a given sports discipline, recognition of player and audience emotions, and also detection of sports objects specific for a given sports category. Whereas in the case of TV news scenes can be categorized basing on the people or place detection using face detection or object detection.

The detection of video scenes facilities the optimization of indexing process. The automatic categorization of news videos will be less time consuming if the analyzed video material is limited only to scenes the most adequate for content-based analyses like player scenes in TV sports news or official statements in TV news. The temporal aggregation method implemented in the AVI Indexer is applied for a video structure detection. The method detects and aggregates long anchorman shots. The shots are grouped into scenes basing on the length of the shots as a sufficient sole criterion.

The temporal aggregation method has two main advantages. First of all it detects player scenes, therefore the most informative parts of sports news videos. Then, it significantly reduces video material analyzed in content-based indexing of TV sports news because it permits to limit indexing process only to player scenes. Globally, the length of all player scenes is significantly lower than the length of all studio shots.

The temporal aggregation is specified by three values: minimum shot length as well as lower and upper limits representing the length range for the most informative shots. The values of these parameters should be determined taking into account specific editing style of a video and its high-level structure.

Formally, the temporal aggregation process is defined as follows [18]:

- single frame detected as a shot is aggregated to the next shot,

  if $(L(shot_i) == 1$ [frame]) then $L(shot_{i+1}) = L(shot_{i+1}) + 1$;

  $LS = LS - 1$;

  where $L(shot_i)$ is the length [measured in frames] of the detected shot i and $shot_{i+1}$ is a next shot on a timeline and LS is a number of shots detected;

- very short shots should be aggregated till their aggregated length attains a certain value Min_Shot_Length,

  while $((L(shot_i) < MIN\_Shot\_Length)$ and $(L(shot_{i+1}) < MIN\_Shot\_Length))$ do

  $$\{ L(shot_i) = L(shot_i) + L(shot_{i+1});$$
  $$LS = LS - 1; \}$$

- all long consecutive shots should be aggregated because these shots seem to be useless in further content analyses and categorization of sports events,

  while $((L(shot_i) > MAX\_Shot\_Length)$ and $(L(shot_{i+1}) > MAX\_Shot\_Length))$ do

  $\{ L(shot_i) = L(shot_i) + L(shot_{i+1});$

  $LS = LS - 1; \}$

- after aggregation all shots of the length between two a priori defined maximum and minimum values should remain unchanged – these shots are very probably the most informative shots for further content-based analyses.

4   Scene Detection by Temporal Aggregation

The temporal aggregation method enables us to detect and to aggregate two kinds of shots: sequences of long studio shots unsuitable in the case of sports news for content-based indexing and player scene shots adequate for sports categorization. It has two main advantages. First of all it detects player scenes, therefore the most informative sports news units of videos. Then, it significantly reduces video material analyzed in content-based indexing of TV sports news because it permits to limit indexing process only to player scenes.

The temporal aggregation is specified by three values: minimum shot length as well as lower and upper limits representing the length range for the most informative shots. The values of these parameters was determined basing on the previous analyses of TV sports news and the analyses of their high-level structure.

Very short shots including single frames are relatively very frequent. Generally very short shots of one or several frames are detected in case of dissolve effects or they are simply wrong detections. The causes of false detections may be different [19]. Most frequently it is due to very dynamic movements of players during the game, very dynamic movements of objects just in front of a camera, changes (lights, content) in advertising banners near the player fields, very dynamic movements of a camera during the game, light flashes during games or interviews. These extremely short shots resulting from temporal segmentation are joined with the next shot in a video. So, the first two steps of the temporal aggregation of shots also leads to the significant reduction of false cuts incorrectly detected during temporal segmentation.

The result of the temporal aggregation are the sequences of shots of the lengths between two a priori defined values. All these sequences are separated by only one very long studio shot. Such a sequence of shots is treated as a pseudo-scene (Fig. 1).

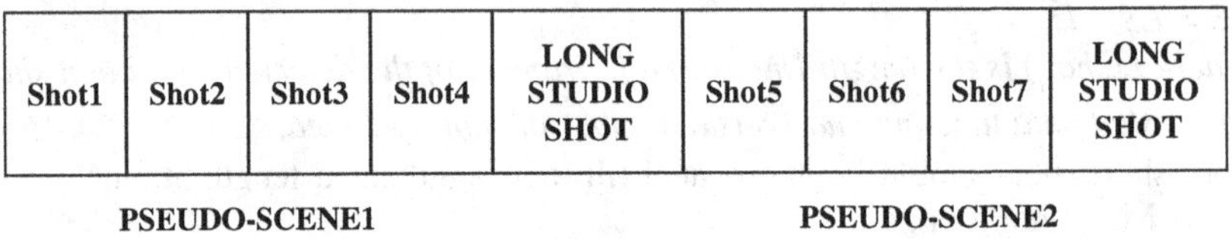

Shot1	Shot2	Shot3	Shot4	LONG STUDIO SHOT	Shot5	Shot6	Shot7	LONG STUDIO SHOT

            PSEUDO-SCENE1   PSEUDO-SCENE2

Fig. 1. Pseudo-scene set as the result of the temporal aggregation of shots

5   Temporal Aggregation of TV News in the AVI Indexer

The method of temporal aggregation of news videos has been applied in the experiments performed in the AVI Indexer. Six editions of the TV News „Teleexpress" used in the experiments have been broadcasted in the first national Polish TV channel (TVP1). Their characteristics before and after temporal aggregation are presented in Table 1. The „Teleexpress" is broadcasted every day and is of 15 minutes. This TV program is mainly dedicated to young people. It is dynamically edited, it is very fast paced with very quickly uttered anchor comments. So, the dynamics of the „Teleexpress" News can be comparable to the dynamics of players scenes in TV sports news. However the number of topics and events reported in the news is usually much greater than in the sports news. The question is whether the temporal aggregation method is as effective as in the case of sports videos.

Table 1. Characteristics before and after temporal aggregation of the six „Teleexpress" News videos broadcasted in March 2014: 2014-03-03, 2014-03-05, 2014-03-06, 2014-03-08, 2014-03-09, and 2014-03-11

	Video 1	Video 2	Video 3	Video 4	Video 5	Video 6	Average
Length [sec.]	907	900	899	894	893	905	900
After temporal segmentation							
Total number of shots before aggregation	520	461	492	465	523	453	486
Number of shots of less than 15 frames	310	260	277	274	304	260	281
Number of shots of a single frame	261	201	222	205	223	181	216
Real number of anchor shots	13	12	13	14	13	13	13

Table 1. (*Continued*)

	After temporal aggregation						
Total number of shots after aggregation	176	178	173	159	197	169	175
Number of shots tipped to be reports (45<=length<=305)	161	164	157	144	182	153	160
Percentage of frames in the report shots in the video [%]	73	69	72	66	73	70	71
Number of long aggregated shots (>305 frames)	14	14	16	15	14	15	15

The results of the shot aggregation of the „Teleexpress" News videos are presented in the Table 2. The temporal aggregation has been applied with such parameters that only shots of the duration not lower than 45 frames (MIN_Shot_Length) and not greater than 305 frames (MAX_Shot_Length) have been not aggregated. These are shots of the length from 2 to 12 seconds ± 5 frames of tolerance. For each analyzed news video 50 longest shots are included and classified.

Table 2. The longest shots in the "Teleexpress" News videos after the temporal aggregation

L – Length of a shot   A – Anchor
T – Type (category) of a shot   C – Chart
                                     F – Final Animation
                                     I – Intro
                                     R – Report
                                     S – Statement, Reporter, or Interview

	L	T	L	T	L	T	L	T	L	T	L	T
1.	918	A+R	1122	A+R+S	768	A	913	A+R	828	A+S	1217	A+S
2.	627	R+A	886	A	499	A	843	R+A	450	A	538	A
3.	501	A	637	A	467	A	630	A	442	A	480	A
4.	436	A	577	R+A	465	A	569	A	433	A	474	A
5.	430	A	538	R+A	432	R	553	A	424	A	464	A
6.	417	A	535	A	427	A	536	A	422	C	431	A
7.	406	A	409	A	410	A	522	A	407	A	427	A
8.	404	A	409	S	383	A	465	R+A	404	A	410	A
9.	398	A	389	A	377	A	442	A	396	A	405	A
10.	382	A	361	S+R	334	R	406	A	390	A	395	A
11.	377	A	346	A	333	A	406	A	372	A	379	R
12.	347	R	315	A	327	S	354	R	371	R+A	368	R+S
13.	342	S	309	A	313	R	346	A	371	A	350	A
14.	327	A	306	F	311	A	336	A	326	R	342	S
15.	300	F	293	A	311	A	309	R	281	R	330	F

Table 2. (*Continued*)

16.	290	R	290	A	310	S	299	R	273	S	284	
17.	271	R	281	I	297	A	282	F	229	R	260	S
18.	259	S	248	R	294	A	278	R	205	S	259	S
19.	242	R	232	R	283	S	261	S	205	R	255	S
20.	240	R	224	S	277	S	259	R	196	R	255	S
21.	238	R	220	R	268	S	259	R	189	I+A	248	A
22.	221	S	209	R	246	F	255	S	188	R	236	S
23.	219	R	190	R	240	S	238	R	177	S	225	C
24.	195	S	184	R	235	R	224	S	166	S	224	S
25.	194	S	178	R	234	S	212	R	165	R	223	R
26.	185	R	177	S	214	R	185	I+A	161	R	212	R
27.	183	S	174	S	210	R	182	R	155	R	206	S
28.	182	R	165	S	208	S	171	R	149	R	206	R
29.	178	R	160	S	204	R	167	R	145	R	204	S
30.	176	R	153	R	194	R	167	A	135	S	191	R
31.	176	R	153	R	193	R	166	R	133	R	190	R
32.	166	R	148	R	191	R	164	R	129	R	186	R
33.	165	R	147	R	189	R	161	S	128	R	181	R
34.	163	R	146	R	184	S	159	R	127	R	176	R
35.	162	R	146	R	184	R	157	R	126	R	162	R
36.	162	R	143	R	177	R	154	R	124	R	162	R
37.	160	R	143	R	175	R	151	R	124	R	155	R
38.	157	R	136	R	165	R	146	R	123	R	152	R
39.	157	R	132	R	161	S	143	R	123	R	151	R
40.	157	A	132	R	152	R	142	R	122	R	151	S
41.	155	R	131	R	151	R	139	R	121	R	150	R
42.	153	R	130	S+R	149	R	136	R	121	R	150	R
43.	153	R	129	R	147	R	135	R	121	R	147	R
44.	152	R	123	R	143	R	134	R	121	R	139	R
45.	150	R	123	R	142	I	133	R	121	C	139	R
46.	148	R	122	R	136	R	133	R	120	R	138	S
47.	146	R	120	R	134	R	131	R	117	R	133	R
48.	144	S	120	R	126	R	127	R	115	R	132	R
49.	143	R	119	R	126	R	126	R	113	R	132	R
50.	140	R	117	R	123	R	125	R	113	R	131	R

After the application of the temporal aggregation the anchor shots are still the longest shots in news. Although, it should be noticed that the statement, reporter, or interview shots are also at the beginning of the ranking of longest shots in news videos. They are almost as frequent (7 shots) in long shots as report shots (8 shots). So, the shot aggregation facilitates the detection of speaking person shots. The analysis of long aggregated shots in the tested news videos is presented in Table 3.

Table 3. Analysis of long aggregated shots

	Video 1	Video 2	Video 3	Video 4	Video 5	Video 6	Average
Number of aggregated long shots	14	14	16	15	14	15	14.67
Real number of all anchor shots	13	12	13	14	13	13	13.00
Number of anchor shots in long shots	12	11	11	13	12	11	11.67
Percentage of detected anchor shots [%]	92.31	91.67	84.62	92.86	92.31	84.62	89.77
Number of report shots	1	0	3	2	1	1	1.33
Percentage of report shots in long shots [%]	7.14	0	18.75	13.33	7.14	6.67	8.84

Between all 88 long aggregated shots there are 70 anchor shots and 7 other speaking people shots, eight report shots, one chart shot, and one final animation. The report shots represent only about 9 % of all aggregated long shots. To detect faster anchor shots it is desirable to reduce the video space by using temporal aggregation.

6   Final Conclusions and Remarks

There are many methods of content-based video indexing. Many of them adapted to the content analysis of TV news videos are based on video structure. The detection of anchorperson shots in TV news videos is very important because these shots are usually treated as shots setting in video timelines the limits of events or group of events reported in broadcasted news.

The temporal aggregation can be successfully applied to reduce the video space analysed in content-base indexing without disturbing an anchor detection process. Furthermore, the temporal aggregation can be also used to select not only anchor shots but also candidate shots for statement, reporter, or interview shots. The results of tests performed in the AVI Indexer have confirmed that the temporal aggregation facilitates the automatic parsing of video structure of news videos.

All candidate shots for anchor shots as well as for statement, reporter, or interview shots can be then analysed using usually proposed approaches based mainly on face detection and recognition. In further research the tests on more reach and diverse news video will be performed.

References

1. Choroś, K.: Temporal aggregation of video shots in TV sports news for detection and categorization of player scenes. In: Bădică, C., Nguyen, N.T., Brezovan, M. (eds.) ICCCI 2013. LNCS, vol. 8083, pp. 487–497. Springer, Heidelberg (2013)

2. Choroś, K., Gonet, M.: Effectiveness of video segmentation techniques for different categories of videos. In: New Trends in Multimedia and Network Information Systems, pp. 34–45. IOS Press, Amsterdam (2008)

3. Money, A.G., Agius, H.: Video summarisation: a conceptual framework and survey of the state of the art. J. of Visual Communication and Image Representation **19**, 121–143 (2008)

4. Hu, W., Xie, N., Li, L., Zeng, X., Maybank, S.: A survey on visual content-based video indexing and retrieval. IEEE Transactions on Systems, Man, and Cybernetics, Part C: Applications and Reviews **41**(6), 797–819 (2011)

5. Del Fabro, M., Böszörmenyi, L.: State-of-the-art and future challenges in video scene detection: a survey. Multimedia Syst. **19**(5), 427–454 (2013)

6. Asghar, M.N., Hussain, F., Manton, R.: Video indexing: a survey. Int. J. of Computer and Information Technology **3**(1), 148–169 (2014)

7. Kompatsiaris, Y., Mérialdo, B., Lian, S. (eds.): TV Content Analysis: Techniques and Applications. CRC Press, Boca Raton (2012)

8. Ji, P., Cao, L., Zhang, X., Zhang, L., Wu, W.: News videos anchor person detection by shot clustering. Neurocomputing **123**, 86–99 (2014)

9. Zheng, F., Li, S., Wu, H., Feng, J.: Anchor shot detection with diverse style backgrounds based on spatial-temporal slice analysis. In: Boll, S., Tian, Q., Zhang, L., Zhang, Z., Chen, Y.-P.P. (eds.) MMM 2010. LNCS, vol. 5916, pp. 676–682. Springer, Heidelberg (2010)

10. Broilo, M., Basso, A., De Natale, F.G.: Unsupervised anchorpersons differentiation in news video. In: Proc. of the 9th International Workshop on Content-Based Multimedia Indexing (CBMI), pp. 115–120. IEEE (2011)

11. Lee, H., Yu, J., Im, Y., Gil, J.M., Park, D.: A unified scheme of shot boundary detection and anchor shot detection in news video story parsing. Multimedia Tools and Applications **51**(3), 1127–1145 (2011)

12. Montagnuolo, M., Messina, A., Borgotallo, R.: Automatic segmentation, aggregation and indexing of multimodal news information from television and the Internet. Int. J. of Information Studies **1**(3), 200–211 (2010)

13. Dong, Y., Qin, G., Xiao, G., Lian, S., Chang, X.: Advanced news video parsing via visual characteristics of anchorperson scenes. Telecommunication Systems **54**(3), 247–263 (2013)

14. El Khoury, E., Sénac, C., Joly, P.: Audiovisual diarization of people in video content. Multimedia Tools and Applications **68**(3), 747–775 (2014)

15. Qu, B., Vallet, F., Carrive, J., Gravier, G.: Content-based inference of hierarchical structural grammar for recurrent TV programs using multiple sequence alignment. In: Proc. of the IEEE International Conference on Multimedia and Expo, ICME, pp. 1–6 (2014)

16. Choroś, K.: Video structure analysis and content-based indexing in the automatic video indexer AVI. In: Nguyen, N.T., Zgrzywa, A., Czyżewski, A. (eds.) Advances in Multimedia and Network Information System Technologies. AISC, vol. 80, pp. 79–90. Springer, Heidelberg (2010)

17. Choroś, K.: Video structure analysis for content-based indexing and categorisation of TV sports news. Int. J. of Intelligent Information and Database Systems **6**(5), 451–465 (2012)

18. Choroś, K.: Automatic detection of headlines in temporally aggregated TV sports news videos. In: Proc. of the 8th International Symposium on Image and Signal Processing and Analysis (ISPA), pp. 147–152. IEEE (2013)

19. Choroś, K.: False and miss detections in temporal segmentation of TV sports news videos – causes and remedies. In: Zgrzywa, A., Choroś, K., Siemiński, A. (eds.) New Research in Multimedia and Internet Systems. AISC, vol. 314, pp. 35–46. Springer, Heidelberg (2014)

Combined Motion Estimation and Tracking Control for Autonomous Navigation

Van-Dung Hoang and Kang-Hyun Jo[(⊠)]

School of Electrical Engineering, University of Ulsan, Ulsan 680-749, Korea
hvzung@islab.ulsan.ac.kr, acejo@ulsan.ac.kr

Abstract. This paper presents an assisting system for autonomous vehicle in outdoor environment. The system consists of several modules, which includes localization estimation, tracking control for navigation. In order to provide the position of a vehicle, which supports for future navigation, a motion estimation method based on nonholonomic constraint is presented. The method simplifies solution to 3D motion estimation using minimum set of parameters of geometric constraint. To reduce the number of parameters for accelerating computational speed, it is supposed that the vehicle moves following the model of Ackermann steering constraint, which requires constraints between rotation and translation components. An edge matching method based on omnidirectional vision is used to estimate the oriented heading of the vehicle motion. The advantage of the omnidirectional camera is that allows tracking landmarks in large rotation angle, which is utilized for high accuracy estimating. Finally, a stable and robust control method is used for motion tracking and control to navigate vehicle. The simulation and experimental results demonstrate the effectiveness in accuracy of the proposed method under variety of terrains in outdoor environments.

Keywords: Autonomous navigation · Intelligent transportation · Path planning · Geometric constraint · Motion estimation · Tracking control

1 Introduction

In recent years, automatic navigation systems have been developed and applied into many researches on robotics, autonomous navigation, and other industry applications. Many methods have been proposed for localization, navigation, visual odometry, which have been applied in modern intelligent systems, especially intelligent transportation, surveillance systems. Unmanned ground vehicle navigation becomes an important research area in various applications. In autonomous navigation, first, it requires to provide a full path of road network for vehicle motion. That information will be fed to automatic navigation part for driving vehicle. To navigate correctly, the current position of vehicle moving is required to supports for vehicle navigation. Therefore, a method for estimating the localization of vehicle becomes an important part of online autonomous navigation. Finally, to give a decision for navigation, the system should estimate the error between the current positions of a vehicle with regard to the planned path. In the motion estimation field, combining of vision system

© Springer International Publishing Switzerland 2015
N.T. Nguyen et al. (Eds.): ACIIDS 2015, Part II, LNAI 9012, pp. 349–358, 2015.
DOI: 10.1007/978-3-319-15705-4_34

and other electro-magnetic devices has been considered as a solution for the accumulated error problem. The early research on vision-based odometry used a single perspective camera [1, 2]. Because of field of view limitations, some author groups proposed methods using an omnidirectional camera. The basic principles of these approaches are corresponding features and epipolar geometry constraints. Some other groups have integrated multiple magnetic sensors [3, 4]. The authors proposed the method using rotation multiple 2D laser rangefinders (LRF) for constructing maps with almost planar motion assumption.

The tracking control of an autonomous vehicle following the predefined trajectory is a challenging task because of the nonholonomic constraint, nonlinear characteristics of autonomous device and affecting of kinematic noises and conditional environment. Up today, there have been many contributions focusing to deal with the problem of trajectory tracking control [5, 6]. That method has been applied successful on reality autonomous robot. Yutaka *et al.* in [7] proposed a stable tracking control law for nonholonomic motion constraint of vehicles based on the Lyapunov function. That proposed method was successfully implemented on reality mobile robot.

The objective of this paper is to develop an efficient application for online localization estimation and autonomous tracking control. They provide a trajectory of autonomous vehicle motion for tracking control navigator. The method using omnidirectional camera is proposed for 3D motion estimation. This method is expanded based on the ideal of the car-like structured motion model (CSMM) in planar motion assumption. Finally, the tracking control method is applied and estimated appropriate parameters for coverage and stable global trajectory tracking on vehicle motion in outdoor scenes.

2   System Overview

This section presents a configuration of experimental system for motion estimation and tracking control, which assists for navigation of unmanned ground vehicle in outdoor environment. The general flowchart of the proposed system is shown in Fig.1. Due to limited space of paper, the task path planning, we are presented in another literature, or the reader can refer to [8, 9] for more detail. In this paper we only focus on motion estimation and trajectory tracking control.

The intelligent electronic vehicle equipped with the omnidirectional camera, stereo perspective cameras, GPS receiver, LRF and industrial computer with Wi-Fi connection. The direction of omnidirectional image was defined at the first frame, collinear with the directional head of the vehicle. The parameters of omnidirectional image are 1280×960pixesl resolution, center point (646, 460), inner circular radius 150 pixels, and outer circular radius 475 pixels. The parameters omnidirectional camera system is calibration using toolbox [10]. The SICK LMS-291 laser worked under the conditional parameters of angular range 180 degrees, angular resolution 0.5 degrees, and maximum distance measurement 80m. The extrinsic relative parameters between LRF device and camera system is calibrated using method in [11].

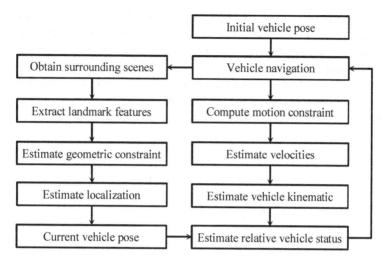

Fig. 1. The flowchart of autonomous vehicle navigation

3   Nonholonomic Based Geometric Constraint

The visual odometry system is composed of consecutive image pair constraints. Those constraints are analyzed based on the epipolar constraint using the essential matrix. Fig. 2 shows a 3D point P with respect to two correspondence reprojection rays of r and r' from the focal point of the hyperboloid mirror to P. The rays of r and r' are observed from two camera poses, the constraint can be described as follows:

$$r'^T Er = 0 \tag{1}$$

where the essential matrix $E=[T]_\times R$, is computed based on the translation vector T and rotation matrix R (*yaw, pitch, roll*).

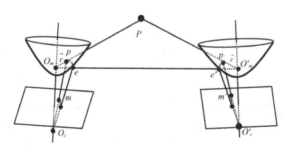

Fig. 2. Geometry constraint based on sequent omnidirectional images

There are several methods, which have been successfully applied to solve problem of epipolar geometric, such as the eight-point [2]. The computational time of RANSAC outlier filter is dependent on the number of corresponding points, which are required for representing geometrical constraint. Therefore, it is important to minimize required corresponding point for reducing computational time. The CSMM constraint is following the Ackermann steering constraint, as depicted in Fig.3 (a). This paper presents a method based on expanding the Ackermann steering constraint for 3D motion as depicted in Fig. 3(b).

As presented in [12], motion estimation task is equivalent to recover five components of transformation. The expression of scale translation on T_X, T_Y, T_Z and rotation components are shown in (2) and (3).

$$T = \begin{bmatrix} \cos(\alpha/2)\cos\beta \\ \sin(\alpha/2) \\ \cos(\alpha/2)\sin\beta \end{bmatrix} \qquad (2)$$

$$R = \begin{bmatrix} \cos\alpha\cos\beta & -\sin\alpha\cos\beta & \sin\beta \\ \sin\alpha & \cos\alpha & 0 \\ -\cos\alpha\sin\beta & \sin\alpha\sin\beta & \cos\beta \end{bmatrix} \qquad (3)$$

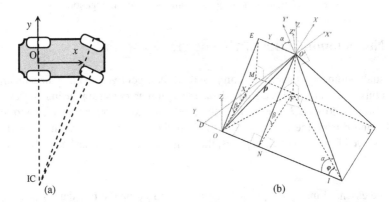

(a)   (b)

Fig. 3. Relation between vehicle poses under nonholonomic: (a) Ackermann steering constraint, (b) 3D motion model

Therefore, to extract the translation and rotation components, it is required to discover two variables β, α. In this paper, the vision-like compass is used to estimate oriented rotation of motion. It provides the capture of transformation on the horizontal scene in front of vehicle. In small period of sequent motion, the yaw angle α is approximated as oriented rotation. Final solution requires one corresponding image point for estimating β.

In this scene, we take advantage of omnidirectional image to estimate orientation rotation. The special phenomenon of omnidirectional camera is that when camera moves on a straight, the scenes on mirror at two areas of front and rear of vehicle are slowly and separately change onto two sides of mirror. However, the scenes on the image are uniformly changed when the camera rotates. Omnidirectional image is used as a vision-like compass for estimating rotation of vehicle. The corresponding features in the front and rear of vehicle are matched by the Chamfer method [13]. The Chamfer edge matching for estimation the rotation angle was implemented [14].

4   One Point Solution

According to result of previous section, the α angle is computed. This subsection presents the method for solving the geometric constraint problem for motion estimation. The result of α is substituted to (2) and (3) for essential matrix as follows:

$$
E = \begin{bmatrix}
-c_1 s_2 sin(\beta) - c_2 s_1 sin(\beta) & -c_1^2 sin(\beta) + s_1 s_2 sin(\beta) & s_1 cos(\beta) \\
2c_1 c_2 cos(\beta) sin(\beta) & -2c_1 s_2 cos(\beta) sin(\beta) & c_1 sin(\beta)^2 - c_1 cos(\beta)^2 \\
c_1 s_2 cos(\beta) - c_2 s_1 cos(\beta) & c_1^2 cos(\beta) + s_1 s_2 cos(\beta) & -s_1 sin(\beta)
\end{bmatrix} \quad (4)
$$

where parameters c_1, c_2, s_1, s_2 are $cos(\alpha/2)$, $cos(\alpha)$, $sin(\alpha/2)$, $sin(\alpha)$, respectively.

The essential matrix with the corresponding rays of $r = [x,y,z]^T$ and $r' = [x',y',z']^T$ from the two sequent images are substituted to (1), presented as follows:

$$
\begin{aligned}
f(\beta) = &x'(-x(c_1 s_2 sin(\beta) + c_2 s_1 sin(\beta)) + 2yc_1 c_2 cos(\beta) sin(\beta) + z(c_1 s_2 cos(\beta) \\
&- c_2 s_1 cos(\beta))) + z'(xs_1 cos(\beta) - y(c_1 cos(\beta)^2 - c_1 sin(\beta)^2) - zs_1 sin(\beta)) \\
&- y'(xc_1^2 (sin(\beta) - s_1 s_2 sin(\beta)) + 2yc_1 s_2 cos(\beta) sin(\beta) - z(c_1^2 cos(\beta) + s_1 s_2 cos(\beta)))
\end{aligned} \quad (5)
$$

Solution of β is recovered by solving (7) with parameters of (c_1, c_2, s_1, s_2) from vision-like compass, $[x,y,z]^T$ and $[x',y',z']^T$ from corresponding rays. The solution is depend only one variable, that mean it requires at least one corresponding point for solving equation. The Symbolic Math Toolbox of Matlab is used to solve this equation.

The SIFT feature method [15] is used to detect the keypoint and matching corresponding points between pair consecutive images. The method in [10] is used to calibrate the camera system and construct the reflection rays from the focal point of mirror to world points following spherical model, as depicted in Fig. 5.

$$
g(m) = \begin{bmatrix}
u(\xi + \sqrt{1 + (1 - \xi^2)(u^2 + v^2)}) / (u^2 + v^2 + \gamma^2) \\
v(\xi + \sqrt{1 + (1 - \xi^2)(u^2 + v^2)}) / (u^2 + v^2 + \gamma^2) \\
\xi / \gamma - \gamma(\xi + \sqrt{1 + (1 - \xi^2)(u^2 + v^2)}) / (u^2 + v^2 + \gamma^2)
\end{bmatrix} \quad (6)
$$

where γ is considered as the focal length of camera system, ξ is the deviation from origin of omnidirectional mirror to origin of the spherical model. The point (u, v) is transformed of image point to the center point (u_0, v_0) of omnidirectional image.

(a)   (b)

Fig. 4. Geometric constraint: (a) omnidirectional image, (c) spherical image model

5   Tracking Control for Auto Navigation

The motion of a vehicle is operated based on the linear velocity and the angular velocity. Let the linear and angular velocities of the vehicle are v and w, respectively. The kinematics of vehicle is formulated by:

$$
\dot{P} = \begin{bmatrix} \dot{x} \\ \dot{y} \\ \dot{\alpha} \end{bmatrix} = \begin{bmatrix} \cos\alpha & 0 \\ \sin\alpha & 0 \\ 0 & 1 \end{bmatrix} \begin{bmatrix} v \\ w \end{bmatrix} \tag{7}
$$

Let $P_c = [x_c, y_c, \alpha_c]^T$ and $P_n = [x_n, y_n, \alpha_n]^T$ be the current pose of vehicle and the expected pose of a vehicle in next step (known as reference position) that we want to tracking, respectively. The relative pose tracking between the current position and the expected position of the vehicle is represented as follows:

$$
P_r = \begin{bmatrix} x_r \\ y_r \\ \alpha_r \end{bmatrix} = \begin{bmatrix} \cos\alpha_c & \sin\alpha_c & 0 \\ -\sin\alpha_c & \cos\alpha_c & 0 \\ 0 & 0 & 1 \end{bmatrix} \begin{bmatrix} x_n - x_c \\ y_n - y_c \\ \alpha_n - \alpha_c \end{bmatrix} \tag{8}
$$

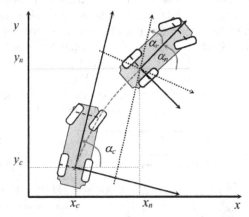

Fig. 5. The motion tracking model of autonomous vehicle

The specific control low using the Lyapunov optimization theory, which was proposed by [7], is used estimation the target linear and angular velocity as follows:

$$
q_r = \begin{bmatrix} v \\ w \end{bmatrix} = \begin{bmatrix} v_n \cos\alpha_r + K_x x_r \\ w_n + v_n (K_y y_r + K_\alpha \sin\alpha_r) \end{bmatrix} \tag{9}
$$

where K_x, K_y and K_α are the positive coefficients, which are identified by experiment such that the system is convergence and stability.

6   Evaluation Results

The proposed method is evaluated using synthetic data to demonstrate the performance of our approach. This method is compared with other methods, such as the eight-point. A fully geometrical constraint based eight-point method is proposed to estimate the 6DOF motion. Therefore, it will result in highly accurate, but requires more computational time than others require.

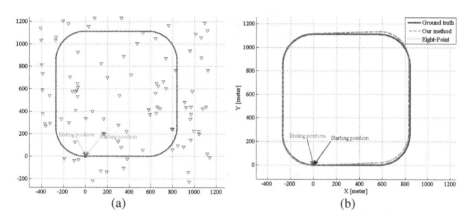

(a)   (b)

Fig. 6. Simulation result: (a) hypothesis trajectory and landmarks, (b) top-view results

The synthetic data were generated so that vehicle 3D motion follows the nonholonomic CSMM. The trajectory consists of 400 frames and the set of world landmarks were generated by randomization, as depicted in Fig. 7(a). The positions of observed landmarks at each frame were added noise. The landmark in front and rear of motion direction is tracked to estimate the oriented rotation of vehicle motion. Fig. 7(b) shows some results. Fig. 8 represents the localization errors. Although, eight-point method is the fully constraint method, but it is also encountered the accumulative error. Our method gives the best results due to the motion following CSMM constraint. The means and standard deviations of error are shown in table 1.

Table 1. Estimated localization errors comparision

Method	Max	Mean	Standard deviation
Eight-Point	27.09	15.75	6.56
Our method	20.76	11.34	6.23

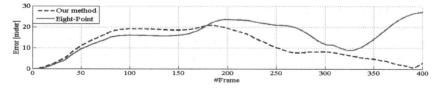

Fig. 7. Comparison results

In experiment, the system was carried the electric vehicle with the omnidirectional camera mounted on the roof, the GPS receiver, IMU and the laser device mounted on the bumper. The vehicle moved on the terrain with distance about 587 meters, which consists of 285 images and laser scans. To evaluate the methods, the values of ground truth were measured by an average of GPS information at markers on the road. The results were plotted on a Google image for comparison, as depicted in Fig. 9.

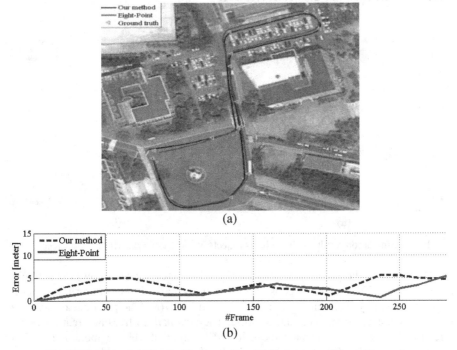

(a)

(b)

Fig. 8. Motion estimation result of real data: (a) Trajectories, (b) position errors

Fig. 10(a) shows some example results of path planning from the source (S) to the destination (D). This result is also the minimum distance-cost, which is computed using [16]. The trajectory for motion is denoted by red color from the source S to the destination D with the minimal cost to travel is around 1,226m. In the task of trajectory tracking control for vehicle navigation, we simulate and evaluate using road data at our campus. Fig. 11(b) shows the result of the global trajectory tracking control of vehicle travel. The control parameters for simulation are defined by tried and tested $K_x=10/s$, $K_y=0.365$ rad/m^2, $K_\alpha=0.35 rad/m$, the sampling time is 0.05s. The vehicle velocity is limited at 25m/s. The bigger K_x can reach convergence trajectory faster, but it will not appropriate with the sampling time of vehicle control. The shorter distance for tracking control is also making convergence to reference trajectory faster, however it causes of unsuitable velocities and unstable trajectory tracking control vehicle.

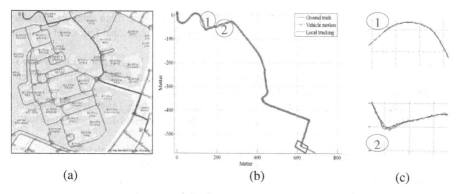

(a)   (b)   (c)

Fig. 9. Simulation results of path planning and motion tracking control: (a) Path planning result, (b) trajectory tracking control, (c) some special terrain positions

7   Conclusions

This paper presented the assistant system for autonomous vehicle, which consist of two contributions: motion estimation, and convergent trajectory tracking control. First stage presented the method based on simplified solution of 3D motion estimation using nonholonomic CSMM supposition. The method was expanding on the planar CSMM to minimal set of geometric constraint parameters, which requires constraints between rotation and translation. The proposed method uses oriented motion with only one corresponding point of image for estimating vehicle motion. The advantage of the omnidirectional camera is that allows tracking landmarks in long travel supporting for correct the motion estimation, especially in large rotation. Second, the tracking control method based on the Lyapunov function is proposed applying to evaluate the trajectory tracking control of vehicle motion in outdoor scene by appropriately parameters setting. The simulation and experimental results demonstrate the effectiveness in accuracy of the proposed method under variety of terrains in outdoor environments. The proposed model can be applied in particle road terrain of outdoor environment.

Acknowledgment. This work was supported by the National Research Foundation of Korea (NRF) Grant funded by the Korean Government (MOE) (NRF2013R1A1A2009984).

References

1. Royer, E., Lhuillier, M., Dhome, M., Lavest, J.-M.: Monocular Vision for Mobile Robot Localization and Autonomous Navigation. Int. J. Comput. Vis. **74**, 237–260 (2007)
2. Nistér, D., Naroditsky, O., Bergen, J.: Visual odometry for ground vehicle applications. Journal of Field Robotics **23**, 3–20 (2006)

3. Kim, S., Yoon, K., Lee, D., Lee, M.: The localization of a mobile robot using a pseudolite ultrasonic system and a dead reckoning integrated system. International Journal of Control, Automation and Systems **9**, 339–347 (2011)
4. Suzuki, T., Kitamura, M., Amano, Y., Hashizume, T.: 6-DOF localization for a mobile robot using outdoor 3D voxel maps. In: IEEE/RSJ International Conference on Intelligent Robots and Systems (IROS), pp. 5737–5743 (2010)
5. Broderick, J.A., Tilbury, D.M., Atkins, E.M.: Characterizing Energy Usage of a Commercially Available Ground Robot: Method and Results. Journal of Field Robotics **31**, 441–454 (2014)
6. Do, K.D.: Bounded controllers for global path tracking control of unicycle-type mobile robots. Robotics and Autonomous Systems **61**, 775–784 (2013)
7. Kanayama, Y., Kimura, Y., Miyazaki, F., Noguchi, T.: A stable tracking control method for an autonomous mobile robot. In: IEEE International Conference on Robotics and Automation pp. 384–389. IEEE (1990)
8. Mnih, V., Hinton, G.E.: Learning to detect roads in high-resolution aerial images. In: Daniilidis, K., Maragos, P., Paragios, N. (eds.) ECCV 2010, Part VI. LNCS, vol. 6316, pp. 210–223. Springer, Heidelberg (2010)
9. Chai, D., Forstner, W., Lafarge, F.: Recovering line-networks in images by junction-point processes. In: Computer Vision and Pattern Recognition (CVPR), pp. 1894–1901 (2013)
10. Mei, C., Rives, P.: Single view point omnidirectional camera calibration from planar grids. In: IEEE International Conference on Robotics and Automation (ICRA), pp. 3945–3950 (2007)
11. Hoang, V.-D., Cáceres Hernández, D., Jo, K.-H.: Simple and efficient method for calibration of a camera and 2D laser rangefinder. In: Nguyen, N.T., Attachoo, B., Trawiński, B., Somboonviwat, K. (eds.) ACIIDS 2014, Part I. LNCS, vol. 8397, pp. 561–570. Springer, Heidelberg (2014)
12. Hoang, V.-D., Hernández, D.C., Le, M.-H., Jo, K.-H.: 3D motion estimation based on pitch and azimuth from respective camera and laser rangefinder sensing. In: IEEE/RSJ International Conference on Intelligent Robots and Systems (IROS), pp. 735–740, 2013
13. Barrow, H.G., Tenenbaum, J.M., Bolles, R.C., Wolf, H.C.: Parametric correspondence and chamfer matching: two new techniques for image matching. In: 5th International Joint Conference on Artificial Intelligence, vol. 2, pp. 659–663. Morgan Kaufmann Publishers Inc., Cambridge, USA (1977)
14. Le, M.-H., Hoang, V.-D., Vavilin, A., Jo, K.-H.: One-point-plus for 5-DOF localization of vehicle-mounted omnidirectional camera in long-range motion. International Journal of Control, Automation and Systems **11**, 1018–1027 (2013)
15. Lowe, D.: Distinctive Image Features from Scale-Invariant Keypoints. Int. J. Comput. Vis. **60**, 91–110 (2004)
16. Hoang, V.-D., Jo, K.-H.: Path planning for autonomous vehicle based on heuristic searching using online images, Vietnam Journal of Computer Science, 1–12. doi:10.1007/s40595-014-0035-4

Eye Tracking in Gesture
Based User Interfaces Usability Testing

Jerzy M. Szymański[1], Janusz Sobecki[1(✉)], Piotr Chynał[1], and Jędrzej Anisiewicz[2]

[1] Wrocław University of Technology, Wybrzeże Wyspiańskiego 27,
50-370 Wrocław, Poland
{jerzy.szymanski,janusz.sobecki,piotr.chynal}@pwr.edu.pl
[2] Aduma S.A., Klecińska 125, 54-413 Wrocław, Poland
j.anisiewicz@aduma.pl

Abstract. In the paper we present a method for usability evaluation of kinetic gesture based user interfaces with application of eye tracking. First we present the problems of the kinetic interaction and the application of eye tracking in usability studies. Then we present our method of the usability verification with applied eye tracking, by performing a sample experiment, then its results and discussion.

Keywords: Kinetic interfaces · Usability · Kinetic application evaluation · Eye tracking

1 Introduction

In recent years various interaction methods and user interfaces have been introduced. We have mobile devices with touch screens, we have gesture based interaction kinetic user interfaces, and new technologies are being introduced at a dynamic rate. However one thing still does not change, we need to evaluate usability of such interfaces to provide the best experience for the end user.

The ISO9241-11 norm defines usability as "extent to which a product can be used by specified users to achieve specified goals with effectiveness, efficiency and satisfaction in a specified context of use" [6]. There are many well-known techniques for the usability verification (for example focus groups, interviews, observations, surveys, etc.), and new methods are being introduced to match the evolution of user interfaces.

One of the most interesting usability testing techniques is eye tracking [4], [7]. This method enables to track the movement of user gaze on the screen, using a special device called eye tracker. In the result of such test we receive graphical reports of where users were looking during performing tasks in the application. This provides data for effectiveness and efficiency analysis. It may have however some disadvantages, such as head immobility during eye tracking, using a variety of invasive devices, a relatively high price of commercially available eye trackers and a difficult calibration [2], [7]. However, it provides very valuable information for usability studies and enables a thorough evaluation of a particular application with participation of it users.

N.T. Nguyen et al. (Eds.): ACIIDS 2015, Part II, LNAI 9012, pp. 359–366, 2015.
DOI: 10.1007/978-3-319-15705-4_35

In our previous research we have performed eye tracking usability research of mobile applications [3], so the main purpose of this study was to verify the known eye tracking method in the considerably new, gesture based kinetic interaction environment, by creating a specific apparatus setup. We decided to use eye-tracker in a new application to check the validity of such studies. For this purpose we carried out a short experiment to verify what results we could get.

We decided to carry out research on the simple application – interactive shopping gallery plan, which shows the map of each floor and the list of shops, restaurants etc. located on it. The selection of elements in this application is done by moving the pointer, using hand gestures, over the element and waiting for at least 3 seconds. We have invited users that were familiar with gesture interaction and also those who were completely new to such method of interaction. Gesture based interaction is quite often used in many types of Augmented Reality (AR) applications [5], so building the methodology for verification of usability of such systems is of increasing importance nowadays.

The construction of this paper is following - firstly we present the tools we have applied in the usability evaluation and eye tracking, next we describe the experiment. In the next section we show the results of the eye tracking experiments and in the last section we present the summary and future works.

2   Eye Tracking in Kinetic User Interfaces

Since Microsoft introduced its Kinect system in 2011, it was only a matter of time that system developers would introduce gesture based user interfaces that used this technology. Various studies have been conducted on how to design such interaction [1], [8]; however none of them has focused on usability of such interfaces.

Eye tracking evaluation of large screen kinetic user interfaces provides many challenges. One of the most significant is the user necessity to move the body quite a lot, which may very likely cause the loss of the tracking of user's eye by the eye tracking equipment. Another challenge in such research is how to set up the eye tracking apparatus because we work with a large display and not with a computer screen.

The best solution to overcome these challenges would be to use head mounted eye tracker or special glasses that enable eye tracking. However methods with head-mounted eye trackers also have drawbacks. This type of devices is still very costly and not so popular. Data obtained from them are more difficult to analyze and often there is a need to manually mark areas in frame-by-frame mode. Also it is important that standard, stationary eye trackers could be used in real environment, without asking users for participation in tests and setting up devices on their heads. Therefore, we needed to create our own method for the purpose of this experiment.

In our research we used Tobii X2-60 eyetracker. It is a very small, mobile, 60 Hz binocular eye tracking research system. It is connected to a computer using USB cable. Its main purpose is to perform eye tracking studies on laptops and mobile devices. Its precision and eye tracking accuracy is highest at a distance of 65 cm (26"), but it can work even when user is at a distance of 90 cm from the device. Knowing the above factors and the fact that thanks to its small size, X2-60 eye tracker can be mounted anywhere, we decided to try to use it in kinetic interface evaluation.

3   The Experiment

We selected pretty simple application "Sky Tower Shopping Centre Information" (Figure 1) with kinetic user interface. The main goal of the application was to deliver the users the list of shops, restaurants and other places that are situated in the Sky Tower Shopping Centre, as well as their position on the plan of each of the three floors.

Fig. 1. Tested application: Sky Tower Shopping Centre Information

Five users took part in the experiment – three male and two female users. They were all employees of Aduma company, however none of them were familiar with the application. They were aged from 22 to 25, because kinetic interfaces are generally used by younger people, so this way we could emulate the target group for this application. Because the application was pretty easy to use only three different tasks were prepared. The experiment was carried out using Tobii eyetracker and recorded using Tobii Studio application. Below we will describe the experiment and its results in more details.

4   Plan of the Experiment

The experiment devices and the position of the users were configured as shown in the Figure 2 and Figure 3. The application was displayed on the 48" screen - A, the application input device was the MS Kinect device – B placed on the top of the screen A, the Tobii X2-60 eyetracker was mounted on the top of the standard photo tripod – C, and finally the user – D, was staying in one position using his or her left hand to control the application (this is because most of the interactive elements of the application were situated on the left side of the screen and usage of the right hand would interfere with the eyetracker – it would hindered the direct view of the user eyes).

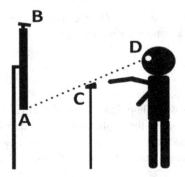

Fig. 2. Schema of the experiment equipment configuration. A: LCD display, B: depth sensor, C: eyetracker, D: participant's gaze line.

Fig. 3. Photo from the usability test

The tasks that the users should complete using the tested application:

a) Assuming that you are on the level "0" in front of the "McDonald's" restaurant, find the nearest fashion shop.

b) Assuming that you are on the level 2 find the nearest grocery.

c) Give the number of tracks in the bowling center

Users were asked to stand at the fixed position in front of the screen and the Tobii eye tracker (see position D at the Figure 2). Then we started the experiment procedure with Tobii Studio, which begins with eye tracker calibration, and after the successful calibration we started the actual tasks. To do this, the initial screen of the application was displayed, where each user was asked to raise his or her hand to start the kinetic interaction. Then we verified if user was able to control the application with left hand. In the following step the moderator read the tasks and afterwards, the user proceed to their realization. After finishing the task, which was declared by the user, or after specified amount of time elapsed (3 minutes), the moderator read the next task or ended the session.

After the session was finished the user was asked to fill out the post-test questionnaire that contained the following questions: were the tasks formed clearly, did you feel lost during the test, what was the cause of the possible troubled during the test, was the application simple and intuitive, did you enjoy working with the application, do you have any other comments.

Figure 2 presents the photo from the test. We can see the LCD screen, MS Kinect on the top of the screen, the Tobii X2-60 eyetracker on the tripod and one of the laboratory staff members testing the equipment configuration.

5   Results of the Experiment

Only one participant had problem with Task 3. All others tasks were accomplished successfully by all participants. The time and gazepoints completeness results of test are presented in Table 1.

Table 1. Tasks time and gazepoints completeness data

	Task 1 [s]	Task 2 [s]	Task 3 [s]	Recorded gazepoints
Person 1	112	160,3	-	25%
Person 2	88,6	53,4	123,9	93%
Person 3	87,4	35,4	40,8	81%
Person 4	35,4	123,8	34,3	87%
Person 5	175,9	85,2	29,9	23%
Average	**99,9**	**85,2**	**57,2**	**61,8%**

There were some difficulties observed during experiment such as: problems with movement sensor accuracy, loosing hand tracking or physical difficulties arising from the need to maintain a relatively constant position. Our observations are presented in Table 2.

Table 2. Kinetic interface usability tests issues

	Observations and remarks
User comfort	Rather high, decrease at loosing hand tracking and during cursor movement problems
Eye tracking data precision	Eye tracking data from 61,8% of time, reason in participants movements and eyetracker hiding by arm
Calibration	The calibration process was rather trouble-free and fast
Interaction	Some problems with interaction. Long-time of cursor select confirmation. Right or left handed factor.
Other remarks	The study was real life situation simulation of shopping centre kinetic application usage. Participants take part in usual usability test enhanced with eye tracking.

Cursor is animated, so user can see progress of choosing. Participants reported, that this time was too long for them, and they had problem with keeping their arm in stable position for that time.

Detailed analysis of standard usability test is out of the scope of this paper, thus we wanted to focus on combining gaze tracking and kinetic interfaces.

Basis analysis of gaze tracking test contains gaze plot and heat maps. Our gaze tracking installation with kinetic user interface allowed us to conduct that kind of analysis. In Figure 4 sample gaze plot and heat map from Task 1 is presented. We can see that mostly used parts of the application were left and center. The obtained data does not differ from data obtained using standard gaze tracking methods. It is possible to perform standard gaze tracking analysis.

Based on our application structure we have defined few "Areas of Interest". We measured participants gaze activity in following areas: Left Menu, Interactive Map, Main Logo, Floor Changing Menu, Shop Logo and Shop Description.

Fig. 4. Sample gaze plot and heat map

Figure 5 shows time to first gaze fixation mean (with fixations minimum time 15ms) for all participants. We can see, that firstly noticed elements are Left Menu, Interactive Map and Floor Changing Menu. Other areas are noticed tens of seconds later. Without gaze tracking combined with kinetic installation we would not know, that Logos and Descriptions are invisible for users for about a minute. It could be very important in real-life situations, where that kind of applications are placed in big shopping centers and user involvement for such long time may be more difficult.

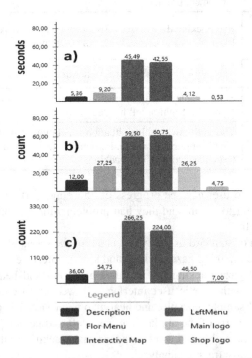

Fig. 5. Factors indicating interest about areas: a) total visit duration mean, b) visit count mean, c) fixation count mean

Usage of gaze tracker gave us opportunity to analyze data about number of visits in defined areas, time spend in that areas and number of fixation occurred in each of them. In Figure 6 we can see that the most popular (in terms of visits, time spend and fixations) are areas with Interactive Map and Left Menu. Least popular are specific information about shop – Description and Shop Logo.

The obtained results allowed us to precisely determine the usability of the tested application. We were able to view which elements participants noticed, which were not visible to them, and basically we got the same amount of information regarding usability of this application as we would get during eye tracking evaluation of a website or other application controlled with a cursor and keyboard.

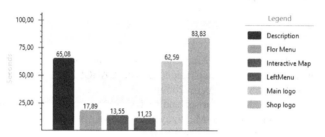

Fig. 6. Time to first fixation mean by defined areas

6   Summary and Future Works

In this paper we have presented a work-in-progress method of eye tracking usability studies of gesture based kinetic user interfaces applied in large screen displays. We have conducted a preliminary verification in form of a test with users. We managed to configure the application as well as eye tracker to handle successfully the experiment and to perform further analysis. We obtained reliable eyetracking data from all users taking part in the experiment and we were able to perform the application usability analysis using the Tobii Studio software. Still our research environment with the Tobii X2-60 eye tracker proved to be sufficient for this experiment. However in future works we will try to perform similar research with mobile eye tracking glasses. This will eliminate the problems that we had with participants' hands interrupting the tracking of their gaze.

Eye tracking evaluation of this application gave us a better insight into the process of gesture based kinetic interaction. During the experiment, we have observed a rather obvious cause and effect relation between the eye fixation and the consequent selection of the option (active point). This discovery may be used for example in building the application controlled by eye fixation. We have already tested such applications [2] and they prove to be quite effective, however, their quality depends on the quality of the eye tracking device, which may be pretty expensive (even 30 times more expensive than MS Kinect).

In conclusion our experiment showed that it is possible to use eye tracking for gesture based interface usability evaluation and that it is an effective method for such research. Also, we have discovered that eye tracking for gesture based application was pretty much similar to the eye tracking of applications with "standard" types of interaction and only problems were caused by users' movement. This can be over-come with a mobile – glasses eye tracker.

As for the future works we will try to analyze the kinetic interaction more tho-roughly. We will try to implement a mechanism that could recognize and record each gesture for further analysis. In addition it is possible to perform kinetic interaction with more than one pointer on the screen, so it is also worth to examine.

Acknowledgements. The research was partially supported by the European Commission under the 7th Framework Programme, Coordination and Support Action, Grant Agreement Number 316097, ENGINE - European research centre of Network intelliGence for INnovation En-hancement (http://engine.pwr.edu.pl/).

References

1. Bruegger, P., Hirsbrunner, B.: Kinetic user interface: interaction through motion for perva-sive computing systems. In: Stephanidis, C. (ed.) UAHCI 2009, Part II. LNCS, vol. 5615, pp. 297–306. Springer, Heidelberg (2009)
2. Chynał, P., Sobecki, J.: Comparison and analysis of the eye pointing methods and applica-tions. In: Pan, J.-S., Chen, S.-M., Nguyen, N.T. (eds.) ICCCI 2010, Part I. LNCS, vol. 6421, pp. 30–38. Springer, Heidelberg (2010)
3. Chynał, P., Szymański, J.M., Sobecki, J.: Using eyetracking in a mobile applications usa-bility testing. In: Pan, J.-S., Chen, S.-M., Nguyen, N.T. (eds.) ACIIDS 2012, Part III. LNCS, vol. 7198, pp. 178–186. Springer, Heidelberg (2012)
4. Duchowski, A.T.: Eye tracking methodology: Theory and practice, pp. 205–300. Springer-Verlag Ltd., London (2003)
5. Hayes, G.: 16 Top Augmenter Reality Business Models. A weblog by Gary Hayes posted at August 14 2009. Downloaded in October 2014 from http://www.personalizemedia.com/ 16-top-augmented-reality-business-models/
6. International Standard ISO 9241-11. Ergonomic requirements for office work with visual display terminals (VDTs) – Part 11: Guidance on Usability. ISO (1997)
7. Mohamed, A.O., Perreira Da Silva, M., Courbolay, V.: A history of eye gaze tracking (2007). http://hal.archivesouvertes.fr/docs/00/21/59/67/PDF/Rapport_interne_1.pdf (March 10 2014)
8. Pallotta, V., Bruegger, P., Hirsbrunner, B.: Kinetic User Interfaces: Physical Embodied In-teraction with Mobile Pervasive Computing Systems. IDEA Group Publishing (2008)

Maximization of AR Effectiveness as a Didactic Tool with Affective States Recognition

Piotr Hrebieniuk[1](✉) and Zbigniew Wantuła[2]

[1] Cohesiva, Wroclaw, Poland
piotr.hrebieniuk@gmail.com
[2] ADUMA S.A., Wroclaw, Poland
zbysto@gmail.com

Abstract. Some research shows that affective state of learners is not less important for didactic effectiveness than their cognitive state. This paper describes an attempt to stimulate some affective states in order to boost a didactic process quality by using augmented reality systems as a stimuli. Results of some research about AR effectiveness in didactic process were analyzed, and they show that augmented reality as it is isn't sufficient to entail and preserve adequate affective states in pupils. Subsequently, there were shown examples AR applications that stimulate specific affective states crucial for optimal didactic process. Some Augmented Reality application mechanisms were suggested, that corresponds with those affective states beneficial for didactics optimization. Finally, users of those applications were examined, and the results were analyzed in order to suggest a guidelines for designing AR applications, which by using affective stimulation are best suited for didactic process.

Keywords: Augmented reality · Affective states · AR effectiveness

1   The Role of Affective States in Didactic Process

In the age of accelerating technological changes, and naturally following social and civilization changes, education is becoming a key subject in public discussion and is often perceived as a key to our future in the reality of accelerating changes [1]. We can find lots of initiatives that focus on redefining modern education objectives, mainly by shifting accents from knowledge acquisition to new ways if its assimilation, creativity and flexibility for change [2].

Irrespective of new ways of education, the technology evolution enables us to create completely new tools, and to gather information and conclude about methodology and mechanics of education: didactics. Research made by Yerkes and Dodson at the beginning of previous century [3] became a fundament of broader considerations about didactics optimization. The law of Yerkes&Dodson waited until 50s' of XX century to be transposed on the field of psychology, when it has been interpreted as a relation between effectiveness of performing tasks and a level of arousal.

© Springer International Publishing Switzerland 2015
N.T. Nguyen et al. (Eds.): ACIIDS 2015, Part II, LNAI 9012, pp. 367–376, 2015.
DOI: 10.1007/978-3-319-15705-4_36

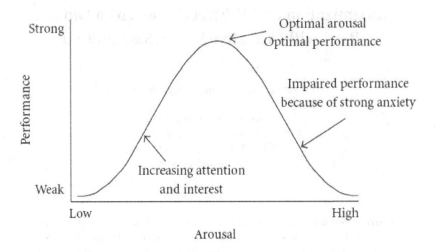

Fig. 1. Hebbian version of Yerkes-Dodson curve for single task, with hyperarousal impact on performance

Further research [4] has shown, that the optimal level of arousal for performing tasks differs depending on complexity of the task. Tasks more challenging intellectually, that require more focus, involve less arousal to maintain concentration - too high arousal causes concentration disorders, and finally anxiety. On the other hand, tasks that require more stamina and consistence may be performed better with higher level of arousal, which increases motivation. Very important thesis presented in the same collective work, states that too much arousal has negative impact on task completion success, regardless of its nature and complexity (as depicted on picture 2).

In this work, when we say "task" we mean completion of a didactic exercise. As we said, modern considerations about learning process show, that creativity is a key to success in didactics - it's generally desired and most useful in grown up life. So far education was focusing on teaching to adapt to our habitat, mainly by mastering skills in following well defined and described directions, in fixed conditions. In the era in which changes go faster than our adaptability allows us to adapt, it's creativity that holds the key to success. In his work, Fredrickson [5] refines the impact of arousal on task completion. She suggests five affects of high arousal - joy, interest, contentment pride and love - as a set of emotions maximizing creativity during task execution. He postulates, that those affects let people boost their focus on performed tasks and diversifies their thoughts. Those deliveries allow us to believe, that placing pupils in an affective state with high arousal, and maintaining level of arousal appropriate to performed task can be crucial to increase the quality of effective didactic process.

2   Application of Augmented Reality

In their work about AR usage in education [6] Panteva and Ivanova touch the topic of stimulating affective states in didactic process by displaying object in Augmented Reality. During their research, the authors were showing consecutive 3D objects in augmented reality to the pupils, and then observed their reactions. Each of those models was quite similar, as all of them depicted wild animals, that children could not have met in nature. Surveying teachers and making observations of kids reactions, they presented a list of emotions (affects) with percentage occurents among the kids while interacting with AR objects. Those affects where selected arbitrary, and didn't reflect any known classification or affective model.

The results are interesting, especially in the context of desired high arousal affects, that in this work are joy, enthusiasm, excitement. Throughout the experiment recognition rate of those affects differed for particular 3D models. For 5 consequently shown AR models percentage value of subjectively felt affects rated, respectively for excitement: 11%, 26%, 21%, 26%, 21%, for enthusiasm: 16%, 26%, 16%, 21%, 37%, and for joy: 0%, 5%, 5%, 11%, 21%. Authors clearly stressed, that all of AR models were similar, that is didn't varied in impressiveness, ways of interaction, their behaviour, complexity or educational factor. For this conditions, for a series of AR models, we'd intuitively expect a downward trend, or optionally - when time interval would be very short - constant level of pupils arousal. According to Panateva and Ivanova, AR has a positive effect on education, and can greatly diverse school classes while at the same time boosting attractiveness of didactic process. On the other hand, varied and unpredictable level of arousal for similar AR models tells us, that typical usage of AR applications in education isn't sufficient, as keeping arousal of students on appropriate level is a key to fully controlled and effective teaching. Moreover, the variability of measured arousal is unintuitive, which we interpret as a sign of complexity of interaction between human and AR applications in context of affective processing.

Augmented Reality systems are gaining popularity in the space of solutions for Retail market, interactive multimedia installations at the museums, and all interactive surroundings for the youngest technology users. Although for a couple of years people began to get used to this kind of attractions, thus AR lost its biggest success factor - the "WOW effect". We've observed that it's caused mainly by simplified and primitive UX of AR applications. Display of bare and merely interactive 3D models isn't enough to raise awareness and excitement, as users are accustomed to the idea of mashing real-life and virtual objects. Nevertheless, we think that appropriately designed AR applications can become a huge value, especially in modern didactic processes, in which boosting students' creativity plays a key role.

The purpose of this paper is to examine Aduma AR applications on the field of didactics and to confirm, that we can manipulate affects characterized by high arousal. Therefore, we think we can use those applications to directly increase efficiency of didactic processes. As we said before, stimulating a learner isn't enough to achieve this goal, so we suggest some AR patterns, that enables us to graduate the level of arousal. This way we show, that specific AR applications can directly influence users' arousal, hence they're a great tool to various types of didactic tasks. We should mention, that showing a mechanism for gradation of induced arousal in AR applications is not the solution for choosing optimal levels of arousal, neither for valuating those levels and assigning them to optimal state to fulfill didactic tasks. We also don't grade complexity of those tasks - we will simply pick two tasks which nature obviously varies from each other.

3   Affective Models

To conduct our experiment it is crucial to choose affective model, which we will use to classify affective states and to measure level of arousal induced by interactive AR applications.

One of the most popular of affective models, especially in marketing research, is P.A.D., emotions description model described by A. Mehrabian [7]. This model consists of the assumption, that every emotional state can be described by a vector in 3-dimension space. Those dimensions are three emotional components: pleasure, arousal and dominance (hence PAD). This model is often applied for consumer marketing studies and construction of artificial emotions [8]. Although, as Mehrabian points out, this model is extremely useful to measure emotional traits, i.e. temperament, author shows that unitary components of emotional trait are temporary and dynamic, thus it's hard to average and compare their values. Trait, on the other hand, is a kind of emotional tendency, that is already averaged.

The other popular affective model proposed by Russel [9], is a circumplex model of affect. In this model affects are interpreted as a combination of arousal and pleasure in 2-dimensional space. On figure 2 is depicted a graphic representation of this model, which also shows individual affective states, according to our common understanding. The X axis stands for valence, respectively unpleasantness (left side of axis) and pleasure (right side). The dimension of arousal is scaled on Y axis, with highest arousal in the upper part of the plot. For instance, calm affect is induced by moderate pleasure and extremely low arousal. The excitement on the other hand means extreme pleasure and high arousal. So according to Russel's model we can grade affects in respect of arousal and valence. Russel's model captures 28 different affective states.

The model was confirmed in many experimental researches. Analysis of this model has shown, that assumed affective states categorization has intercultural application - experiments was conducted in Poland, England, Estonia, Greece and others [10]. Moreover, many research of human brain nature, for instance neuroimaging fMRI had shown, that Russel's model is biologically meaningful, because affective information computing in space of valence and arousal can be easily referred to brain's reward system, while arousal's role in affective computation referred to activity of brain's excitement system [11].

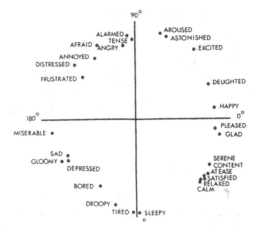

Fig. 2. Russel's circumplex model

4   Assumptions and Experiment

For the purposes of this research it's important to select one affective state, which we want to modulate. This affect has to be induced by very high arousal. For the didactic process to succeed, it's desirable for the affect to be subjectively pleasure. Moreover, because of our limited capabilities in measuring affective states of AR applications users, chosen emotion (with high arousal and pleasure) has to be intuitively undestandable and easily noticeable and distinguishable.

Those outlines indicates that the best solution is to use Russel's circumplex model of affects as a referential model, especially because:

– Excitement is characterized by high arousal, and it's arousal level we want to measure. Moreover, excitement is also a pleasant affect.
– In 28-valued space of Russel's model excitement is distinctive enough, so we wouldn't have much difficulties in identification and classification of affects. The "arousal", which is close to excitement can be treated as convergent with excitement.

Summarizing, based on Russel's circumplex model of affects we've chosen excitement to be the measured affect, and it's excitement that should be modulated, and variation that should be observed during conducted experiment.

Basing on our experience and observations on Augmented Reality systems for kids, we assumed, that the greatest factor of excitement is the level of interaction between AR application and its users. We've observed, that kids tend to stay longer with interactive AR, and most likely they tend to play better, when object can react to their actions. We didn't have any particular observations when it comes to excitement grading, or adjusting specific AR apps to tasks difficulty level.

To conduct our research we needed two AR installations, each one used to enhance different didactic process. We wanted to inspect easier task, that requires more consistency and stamina, and the other one challenging intellectually. This way, we could relate to diversified Yerkes&Dodson law of arousal impact on performance. Each of the two experiments described below were conducted in two versions: basic, which less interaction, and extended, where users had to interact more. Experiment 1 addresses lower arousal need for optimal completion, while Experiment 2 needed higher level of arousal.

4.1   Experiment 1: Catch the Butterflies!

The first installation we used was Augmented Reality based mobile application, called "Catch the butterflies!". The task here was to get more knowledge about different kinds of butterflies. Each species differentiated by colour, shape, size and original habitat. We divided the application into two independent versions.

Basic version of application - the one with less interaction - was very simple. Kids had to assimilate some knowledge about particular butterfly species. Then, if they could name the animal and answer the teacher's question about some basic facts about them, they could finally see the butterflies in AR, flying around them in a augmented space. Movement of the smartphone around vertical axis allowed them to see movement of butterflies all around the room.

In extended version of this application, the educational aspect remained unchanged. However, interaction with 3D models changed vastly. From now, kids could not only watch butterflies, but also catch them in the net, by finger tapping the screen. With each tap the net swooshed: user saw movement of the net, and heard characteristic sound. The experience stopped when kid caught all of butterflies in augmented space.

Fig. 3. Examining "Catch the butterflies!" mobile AR app

4.2   Experiment 2: Family Square with AR

The second installation we used was Family square - a comprehensive product that combines a fun zone for kids with a place of rest for parents. We used a couple of technology solutions, i.e. interactive floor and multitouch desk with a set of didactic and funny applications. Additionally, we capture kids movement on top of interactive floor, and augment it on large LCD screen hanging next to the square.

Basic version of our experiment assumed very simple usage of AR technology. The kids were playing a game of guessing the animals on interactive floor. Each time a player guesses correctly what animal makes played sound (squeak, roar, etc), he could see the animal on a screen in Augmented Reality.

We extended the application, by adding interaction to AR objects. Now kids could influence the object by making body gestures. For example, raising their hands made 3D animal shown on the screen make a sound, moving around the floor caused it to walk, and when sited, the animal went back to its stable position.

Fig. 4. Aduma's Family Square

The first experiment was definitely more of an intellectual challenge - kids had to focus to gain systematic knowledge about butterflies. Therefore, we classify this task as requiring lower arousal. Second task was different - it lasted longer, and required kids to engage more. According to Yerkes&Dodson law, we need more arousal to boost efficiency when fulfilling this kind of task.

5   Methodology

For both cases of different level of expected optimal arousal (Experiment 1 and Experiment 2) we asked parents/custodians about their impressions of kids affects. We picked three extreme affects from Russel's model, that differs significantly: excitement (high pleasure, extreme arousal), boredom (very low arousal, moderate pleasure), and frustration (low pleasure, high arousal). Our choice was made solely on intuitive emotions recognition, as kids' emotional state was to be judged arbitrarily by adults. All of those affects are easy to identify, so that their distinction would be easy irrespective of the way we choose to measure them. Moreover, this set is suitable for arousal tracking - the two most expected affects with high arousal (frustration and excitement) differ extremely by the pleasure factor, while boredom's pleasure factor is almost neutral. Parents were asked to tell if their kid seemed bored, frustrated or excited each time he/she was interacting with the 3D model. Parents were observing their kids during interaction with the system and judging their state every time a new AR object occurred. In each variant (basic version of application, extended interactive version) kids were playing with 4 different AR objects - different kinds of butterflies, different animals.

6   Results and Interpretation

We examined a group of 20 kids for Butterfly catching game, and a group of 42 kids for family square case. For both cases of different level of expected optimal arousal (Experiment 1 and Experiment 2) we asked parents/custodians about their impressions of kids affects. We picked three extreme affects from Russel's model, that differens significantly: excitement (high pleasure, extreme arousal), boredom (very low arousal, moderate pleasure), and frustration (low pleasure, high arousal). Parents were asked to grade if their kid seem bored, frustrated or excited in three-valued scale, each time the kid was interacting with new 3D model. In each variant (basic version of application, extended, more interactive version) kids were playing with 4 consequent AR objects - different kinds of butterflies, different animals.

Table 1.   Results of survey for Experiment 1

Experiment 1 - Catch the butterflies					
App variant	Basic			Extended	
Arousal	low	medium	high	low	Medium	high
Object1	0%	20%	80%	0%	10%	90%
Object2	0%	30%	70%	0%	30%	70%
Object3	0%	50%	50%	0%	30%	70%
Object4	10%	40%	50%	10%	20%	70%

In the basic scenario kids were gradually getting bored. Even though they were astonished by first object, arousal began to drop, what could have been seen as boredom rise. New butterflies were similar, so application wasn't giving them any new experience.

When we implemented our interactive extension, the kids could influence those models. The arousal level stayed stable - parents generally rated it as high, while kids were playing quietly and seemed interested. We noticed, that interacting with AR objects not only extends fun, but redefines the experience. Possibility of influencing AR objects is probably thoroughly changing primal association of performed action. As this interpretation goes beyond the scope of this work, we settle on noticing, that the arousal level stayed stable during user's interaction with the app.

In the basic scenario kids quickly started to feel frustrated, as they couldn't do anything with AR objects. They were amazed by AR at the first time, but this high level of arousal was hard to maintain with high pleasure factor at the same time, hence parents rated their mood as aroused and unpleasant, that is frustrated. As mentioned in previous research, too high arousal is hard to reduce, therefore sometimes the reaction of examined is pleasure decrease, as they can't maintain both high pleasure and arousal too long. When we implemented extended version of the app, kids have had some problems with catching up with gestures at the first time. The task was just too difficult. We could see this confusion as a relatively low excitement level, compared to basic scenario. With second and further models on the other hand, kids already knew how to interact with the model. They seemed excited all the time, as they used their bodies very intensively - they high arousal and pleasure were maintained. The key difference compared to Experiment 1, that let us avoid the pleasure/arousal value decrease, was the intensity of interaction, that is the extent of body movement.

Table 2. Results of survey for Experiment 2

Experiment 2 - Family square						
App variant	Basic			Extended		
Arousal	low	medium	high	low	medium	high
Object1	0.0%	4.8%	95.2%	0.0%	11.9%	88.1%
Object2	0.0%	9.5%	90.5%	0.0%	7.1%	92.9%

7   Conclusions

Our experiment showed, that the key to maintaining adequate level of arousal in didactic process supported by Augmented Reality applications, is to engage learners in interaction with AR models. Precisely speaking, people get excited when they can influence what is being augmented. Augmented objects reacting to our behavior significantly increase intensity of UX, without disrupting the feeling of pleasure. Moreover, picking two completely different ways of interaction with AR, gave us two different levels of average excitement. We can conclude, that people seem to excite more, if the interaction is more engaging. Tasks requiring lower level of arousal tend to fit well with non-intrusive, well known and broadly used interfaces, like smartphone in Experiment 1. Those that goes better with intensive stimulation are good for more extravagance interfaces, which requires us to move our body more.

References

1. Kurzweil, R.: The Age of Spiritual Machines. Viking Press (1999)
2. Craft, A.: Creativity and Education Futures: Learning in a Digital Age. Trentham Books Ltd. (2010)
3. Yerkes, R.M., Dodson, J.D.: The relation of strength of stimulus to rapidity of habit-formation (1908)
4. Diamond, D.M., Campbell, A.M., Park, C.R., Halonen, J., Zoladz, P.R.: The Temporal Dynamics Model of Emotional Memory Processing: A Synthesis on the Neurobiological Basis of Stress-Induced Amnesia, Flashbulb and Traumatic Memories, and the Yerkes-Dodson Law, 28 March 2007
5. Fredrickson, B.: What good are positive emotions? Journal of General Psychology 2(3) (1998)
6. Panteva, P., Ivanowa, M.: Exploration on the Affective States and Learning During an Augmented Reality Session. JADLET Journal of Advanced Distributed Learning Technology (2013)
7. Mehrabian, A.: Pleasure-Arousal-Dominance: a general framework for describing and measuring individual differences in temperament. Current Psychology 14 (1996)
8. Ratneshwar, S., Mick, D.G.: The why of consumption: contemporary perspectives on consumer motives (2003)
9. Russel, J.A.: A circumplex model of affect. Journal of Personality and Social Psychology (1980)
10. Russell, J.A., Lewicka M., Niit, T.: A cross-cultural study of a circumplex model of affect. Journal of Personality and Social Psychology (1989)
11. Posner, J., Russell, J.A., Peterson, B.S.: The circumplex model of affect: An integrative approach to affective neuroscience, cognitive development, and psychopathology. Development and Psychopathology 17(3) (2005)

A Method of the Dynamic Generation
of an Infinite Terrain in a Virtual 3D Space

Kazimierz Choroś[(✉)] and Jacek Topolski

Department of Information Systems, Wrocław University of Technology,
Wybrzeże Wyspiańskiego 27, 50-370 Wrocław, Poland
{kazimierz.choros,jacek.topolski}@pwr.edu.pl

Abstract. The paper presents a method of generating infinite environment including various ecosystems specified by the user. The ecosystems in virtual
worlds might be generated with different set of textures and by using different
formulas to generate shape of the landscape which after all will blend smoothly
between each other. This includes intelligent spreading of flora and fauna along
the areas, and other area-specific stuff. Rendering a large terrain in real-time
strictly imposes a lot of algorithms for optimizations to obtain the highest frame
per second rate possible. Therefore, it is obligatory to simplify hardly visible
elements to reduce complexity of the scene. This results in creating a patched
terrain where each patch can be parameterizable to lower its quality. In the
method presented in the paper patches will be generated on the CPU side in a
separate thread to eliminate stuttering during calculations and then final data
will be sent to GPU. Calculating the patches on CPU creates an opportunity to
edit them to provide more details if needed.

Keywords: Virtual 3D spaces · Infinite terrains · Real-time rendering · Scene
complexity · Terrain patch generation

1 Introduction

Rendering of the 3D graphics involves enormous calculations that need to be done on
a computer. It is now possible to render realistic terrain visualizations hardly distinguished from a photograph in just a few hours on an average user PC. However, there
is still some limitation on real-time rendering, especially in computer games or interactive visualizations where current devices impose on saving resources for also other
things like animations, artificial intelligence, or game logic. Furthermore, as the time
elapses the bigger and more complex areas for sightseeing are in demand.

In the context of computer games or terrain visualisations every object is composed
of many points called vertices forming their shapes, which are later overlaid with
miscellaneous colours and texture coordinates. Unlike 2D spaces, which are able to
represent only planar structures, 3D space can represent real world very intuitively by
simply adding depth as a third component. Therefore, this model is generally chosen
to visualise photorealistic and interactive spaces. The model space is the space in
which points' coordinates are in relative positions for the object itself, so when we

© Springer International Publishing Switzerland 2015
N.T. Nguyen et al. (Eds.): ACIIDS 2015, Part II, LNAI 9012, pp. 377–387, 2015.
DOI: 10.1007/978-3-319-15705-4_37

create an object by specifying its vertices it is done in the model space. After placing the object into the virtual world these coordinates become world space coordinates, although, without any transformations they are the same. However, we can use world space in which we can simply move the vertices to different coordinates. If we want to move two objects to different positions (without defining two models) we have to transform each object with different world matrices. A world matrix is a homogeneous 4x4 matrix (in 3D space) which describes the transformation of coordinates relative to the world origin. The most significant advantage of such matrices is that we can store different transformations in a simple 4x4 matrix.

View space is an auxiliary space which converts all of the object coordinates to the position relative to the view of camera. To calculate the matrix which converts world space into view space we take the world transformation matrix of camera and invert it. After the inversion we simply multiply all of the vertices by this matrix.

Before flattening the 3D space onto the screen we need to transform all of the vertices to the projection space which stores the whole space in a cube. To display on the screen the defined vertices for a model we have to go through the world, view and projection space which basically a consequence of multiplication of the transformation matrices. So, we need a world-view-projection matrix which transforms the model vertices from model space to the coordinates on the 2D screen.

In this paper a method of the dynamic generation of an infinite terrain in a virtual 3D space will be presented. The method uses a modification of the Voronoi diagram.

The paper is organized as follows. The next section describes related work in the area of the generation of infinite terrains in 3D virtual worlds. The main idea of the Voronoi diagrams is outlined in the third section. The method of the dynamic generation of an infinite terrain based on the modified Voronoi diagram is presented in the forth section. The fifth section presents the implementation and experimental test results. The final conclusions are discussed in the last sixth section.

2   Related Work

The infinite terrain means we don't delimit the explorable virtual world by impassable borders (a valley bounded with mountains or an island bounded with water), but we want to allow visitors to go whatever distance they wish without blocking their path. An infinity definition is something that can not be currently achieved, mostly because of the technical reasons as the numbers, memory, or CPU have their limits. Hence, the infinity must be somehow simulated. Some approaches loop the virtual world by using an abstraction of torus model called wraparound [1], because it's impossible to reach the end of the torus. The advantage of torus is that it gives a feeling as if it were some kind of planet that may be encircled, but we have to solve the issues with welding the begin-end edge of the generated world. However, even than the landscape is actually finite.

There are not many papers related to the infinite terrain generation, because of its inconvenience and uncommon applications. Usually, a finite landscape is created with the fixed size, although very large. They are easier to maintain and process, because we get the complete information of the base elevation that may be used to logically distribute terrain features and special objects, whereas an infinite terrain implies generating its content partially, thus it limits our knowledge about the landscape.

The possibility of generating an infinite terrain appeared once the procedural algorithms, like noise generation, have been invented. A very detailed paper referring to the infinity in the context of 3D terrain generation is the Dollins dissertation [2]. The author describes an approach using a hash-based quadtree algorithm to manage terrain parts. All of the parts are procedurally generated on-the-fly without saving the content to a file, so it doesn't allow for runtime terrain deformations, because whenever the visitor comes back to the last visited place it is generated once again.

Unfortunately, the presented approach does not have any mechanism which would allow us to generate every part by using another procedural algorithm, so it results in lack of diversity for the landscape shape, making the visitors bored pretty fast. Furthermore, the whole terrain is in fact overlaid with only one texture.

Another approach has been used, among others, in the paper [3]. The authors have used a technique called a projected grid, which is frequently used for sea or ocean rendering, because it allows to create surface which is expanded up to the horizon using a grid with fixed number of vertices. A projected grid is in fact computed on the GPU, so it is quite hard to incorporate a logic there which would allow us to provide mechanism for creating various areas that are textured and shaped differently. Current hardware, usually together with the hardware tessellation, allows producing very complex planetary scenes by dealing with the wraparound in a different way. This is called a planet rendering. The common problem with planet rendering is keeping up with the real size of a planet. It requires working with large distances and details, but by involving fractals it is possible to create very realistic landscape [4].

Another way of simulating the infinity is to simply generate the subsequent areas before approaching the end of the current landscape. This approach has been used in the Minecraft game [5] and fully relays on the procedural generation as the terrain is completely infinite without any intended repetitions.

A procedural generation [6] means that the content is created algorithmically rather than manually. Procedural approaches have applications in numerous situations like generating landscapes, vegetation, buildings, fire or water and give the opportunities to make large differential worlds in any scale, whereas manual approach requires time proportionally to the size of the landscape. When generating terrain manually artists have great control over the look, but even then they make extensive use of procedural tools to overlay areas with fine natural looking details. In our research we are going to choose the approach based on a full landscape infinity (without wraparound or planet rendering), therefore it is not possible to create a terrain manually or by using some existing maps, so, we have to rely on the procedural generation techniques. There are a lot of procedural generation techniques [7] which differ in the visual aspect, computational complexity and what they are used for.

Systems which introduce a landscape generation have grown into quite impressive number. The current trend is that most of the implementations like [8] or [9] use GPU for generating a terrain geometry, because it doesn't require transmitting a lot of data from CPU to GPU. The geomorphing technique is much more efficient and easier when done on GPU, then we don't have to use vertex buffers which are usually created as dynamic buffers degrading its performance and using a lot of memory.

3   Voronoi Diagram

The Voronoi diagram is a way of dividing space into regions (called simply Voronoi cells). Each cell contains a point called seed, site, or generator. Cell is created by calculating distance using a distance function, that is each point in the cell has the lowest distance to the seed creating the cell than to any other.

Let S be a non-empty set from Euclidean space with a distance function d. Let P be a set with all of the seeds and k, j are indices of the seeds. The equation (1) produces set of points creating the Voronoi cell for the given k-seed

$$R_k = \{x \in X \,|\, d(x, P_k) \leq d(x, P_j), \forall j \neq k\} \tag{1}$$

When the distance function is Euclidean distance it basically means that two neighbour points have a straight line perpendicular to the segment connecting these points. The straight line is then drawn on the halfway of the segment until crossing another straight line created from another two neighbouring points. When the distance function is e.g. Manhattan distance the straight lines become lines with some kind of step. There are naturally other types of distance functions, nonetheless, the Euclidean distance is the most frequently used.

The Voronoi diagram has a great number of applications. Among others, there are applications in mining for estimating the reserves of minerals, in climatology for calculating the rainfall for an area as well as in ecology, architecture, or even machine learning. In [10] the authors have widely described the practical applications of the diagram. In our method it will used to simulate spreading area types over the landscape. There are few algorithms for producing the Voronoi diagram. We will use some kind of brute force approach as we need to find out also the points values in the cells, not only the edges. Furthermore, we will need to modify the algorithm to apply some parameters that allow us to control the diagram in a more flexible way.

The approach we will use works in the way that for the given position (in our case we use 2D space) we search for the cell for which the seed point is the closest to the given position. Therefore, we have to iterate over the neighbouring seeds to calculate the distances. The seeds are in random positions, although they are determined, which basically means that for the same given position the algorithm uses the same seeds positions. Using the Euclidean distance it is not necessary to use the original equation as we only need to compare distances, thus we can optimize it using squared Euclidean distance which doesn't apply the final square root to the result.

4   Proposed Solution for an Infinite Terrain

Despite the advantages of using GPU for generating terrain geometry our solution will use CPU as current CPUs are fast enough to generate the data for landscape in runtime and by using few manners it is possible to reduce the data so much that it is small enough to send them to the GPU when crossing terrain areas. Moreover, there is still a lack of mechanisms that handle different ecosystems in a flexible way, for which it is much harder to create logic on the GPU side.

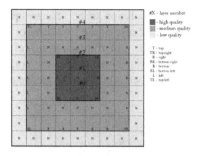

Fig. 1. Terrain quads organization L={2,2,1}

It is impossible to maintain whole mesh within the same detail, therefore entire terrain is organized into quads of the same size. Every quad has its own quality parameter which is used in the LOD effect. It also contains an additional value for being able to remove the cracks (T-vertices) on the terrain – with this value we can determine the position of the quad (whether it is on the left, right, some of the corners, etc.), and thus we generate the grid with modified vertices on edges to fit them to another quad with different quality level. The quads are organized in a simple matrix preserving real positions (Figure 1).

Obviously, every quad consists of the data with all of the vertices necessary to form a shape. The structure of a vertex is based on three elements – position, normal vector, and weights of areas. The latter is an array containing values specifying how much of the given area is on the vertex. We can think of it as we would have a transition between e.g. a desert, a grassy area, and some rocky area, so at some point we can notice that the area is a bit sandy, a bit grassy, and a bit rocky, thus we can define weights for them like: 33.3% of sand texture, 33.3% of grass texture, and 33.3% of rock texture. Those values are then put into the weight array.

All of the mentioned data is also stored in a file containing whole terrain. Having such file creates an opportunity to load the generated quads from a file, thus we are able to deform the terrain in runtime, because they will not be generated every time.

To enable a flexibility and possibility to investigate the best configuration for quality and level number we will use as input information an array of the layers L={A, B, C, ...} where A, B, C, ... ∈ Z and the number correlates to the number of layers for the given quality. Therefore, the A is number of layers for the highest quality level, B is the number of layers for medium quality level, etc. Moreover, we will use another parameter for the quad size, but it won't interfere with our algorithm to update the quad array.

At an initial position we are in the centre quad of the terrain having some quads around us. Obviously, we have to update the visible quads and its quality when the position of camera changes. To avoid frequent updates (forcing sending data to GPU) we will update the quads when moving to another quad, that is crossing a quad at the top edge results in loading new quads in the first row of the matrix and freeing memory from the quads at the last row. We need to update the parameters of remained quads like quality and position parameter in the matrix. We can notice (Figure 2) that it is possible to obtain a generic solution to be able to update the quads with such variability using the L array. It's because of the similarity of the squares.

The only changes are the numbers of right, bottom, left, and top quads (that are equal to each other). As there is always one quad at the corner we can provide a simple equation to determine that number, which is a sequence of odd numbers.

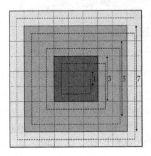

Fig. 2. Dependencies between different layers and its positions. It also presents a spiral approach for assigning the values from output array from the algorithm for determining position parameters for quads for a given layer index.

We also need to keep in mind that for the quads that are not on the edges with another quality level, we want to put there a value indicating it should not be modified. The algorithm of determining position parameters for quads is handling this situation. The algorithm produces an array with quad positions for every layer sequentially. In the case of Figure 2 we will have an array:

R, BR, B, BL, L, TL, T, TP,
N, N, N, N, N, N, N, N, N, N, N, N, N, N, N, N, N, (16 times N)
R, R, R, R, R, BR, B, B, B, B, B, BL, L, L, L, L, L, TL, T, T, T, T, T, T, TR,
N, N, N, N, N, N, N, N, N, N, N, N, N, N, N, N, N, N, N, N, N, N, N, N, N, N, N, N,
   N, N, N, N, N (32 times N)

Having such an array we use the outward spiral method [1] for iterating over the quad matrix to assign the parameters to its corresponding quads. The complexity of the both algorithms are O(n) and O(n • k) respectively, where n is the number of repetitions for repeatable parts of the layer and k is layer number, we also need to incorporate the complexity of spiral iterating which is $O(n^2)$. Nonetheless, we should not worry about it, because the quad matrix size will be small enough to overlook the performance impact. The outward spiral method has already been used for assigning quad parameters. It will also be used to render them.

What we want to achieve is a font-to-back rendering [11], which is in fact achieved directly using the spiral method. The front-to-back rendering means that we are rendering the objects closer to the camera first. This may cause overlaying other objects which will eliminate call to the pixel shader saving rendering time. Therefore, in our method we render the closest quads earlier going to the further quads later.

To support a variety of areas we will use the modified Voronoi diagram. The Voronoi cell will be identified as an area. Such algorithm should handle flexible parameters to be able to affect areas in some ways producing results for our needs. These parameters will be area size, speed of latitude change (which will affect type of areas), and dissolution. The first parameter is the easiest one and affects only the cell size in the diagram. Our algorithm will be able to produce all of the cells in the same size without any random variations, however after introducing the other parameters we will get areas that will indirectly differ in size. The second parameter will affect the type of an area. On the high latitude there are more arctics and tundras whereas on the lower latitude we have more tropicals. This is what we are going to achieve. The last parameter is some kind of dispersion of the cells. By using higher values for the dissolution we can eliminate the straight connectivity line between groups of cells with two different types which produces more diverse and less predictable world.

To produce borders of areas we use the algorithm of Voronoi as it produces uniform colours for the cells as well as the cells' shapes are different which makes the landscape look more attractive. Hence, we can discretize the colours to such number that it's equal to the number of area types. In fact, at this point we have enough information to render the terrain with different ecosystems, but basing only on Gouraud interpolation between only two vertices creates very sharp edges, so we have to smooth them somehow. To make smooth transition between two areas Gaussian blur is used. However, the algorithm needs to produce results in weight arrays, thus we can not threat the output as if it were a pure 2D image but a 3D with N depth where N is a number of area types. Moreover, we have to be able to generate the input Voronoi further than rendered terrain to make the proper blur on the edges of visible terrain, e.g. if we want to have transition within 10 vertices, we need to extend our Voronoi map by 20 (10 vertices from two sides) in height and width.

The easiest parameter to implement is cell size, which requires just diving the input data (x and y) by some size factor. The bigger the factor is the bigger the cell will be.

The speed of latitude change is a bit more complex. Firstly, we must think about a function to specify the factor used for changing area type according to the given position. A good choice would be some basic linear function to make the zones with similar sizes. We also need to take into consideration that our terrain is infinite, thus we have to define what happens when we reach the last possible zone. There are two possible ways – after reaching the last zone we can start from the first zone, so having a tropics-tundra-arctic zones, after the arctic there will be tropics. However, it's pretty unrealistic, so we will choose another approach which in fact exists in real life, so it will be tropics-tundra-arctic-arctic-tundra-tropics. We are using arctics (and tropics) twice – this lets us provide very basic function (2) for obtaining the mentioned factor for area type, where x is the position value from the axis along the latitude and S is the speed of latitude change. In the case we don't want twice as big arctic and tropics areas we can specify smaller range of latitude where it may exists.

$$f(x) = \begin{cases} \frac{x}{S} & \text{if } x < S \\ -\frac{x}{S} + 2 & \text{if } x \geq S \end{cases} \tag{2}$$

Before processing the position value by the function (2), we need to prepare the input position in such a way that it makes the function periodical. The easiest way is to use modulo $x = |x| \bmod (S \cdot 2)$. Having such function we can apply it to the Voronoi algorithm as a step to perform latitude change, where the x is the seed's horizontal/vertical position. Therefore, the output of the f(x) function will be the output of the method generating the Voronoi diagram.

Last parameter, a very important factor to improve the look of a landscape, is a dissolution which makes the areas spread irregularly. This is very cheap feature, because it only requires clamping the f(x) values to the range of [S – d; S + d], where d is the factor of dissolution.

Every vertex consists of array with area weights, so, we are in a very comfortable situation for applying terrain shape. We can basically multiply the generated height value by the appropriate weight (for the same area type). Unfortunately, in current design it must be done for all areas, so if an area is very far away from the given vertex and its weight value is equal to 0, we still have to multiply it.

However, this concept is very flexible as we can choose whatever 2D generator we wish like Perlin noise, Simplex noise, or whatever other noise-based stuff (or even not noise-based ones). What is more, we can choose different generators for different area types, even if they don't fit on the edges, because the connection on the edges is dealt by the use of the Gaussian blur at earlier stage. Furthermore, if a game has a diverse terrain geometry with different ecosystems it probably precalculates the mesh before rendering, but we can generate such terrain geometry on-the-fly without precalculating anything before rendering.

The strength of the Gaussian blur has pretty big affect on the areas edges, because with small kernel size the border will be much sharper which can sometimes produce unexpected results when there is a big difference between heights from two different generators. Obviously, it can be eliminated by increasing the kernel size, although it increases calculation time. One can say that after generating the weights it won't have any impact on the rendering process, but unfortunately it impacts the performance quite a lot indirectly. When the kernel size is bigger it produces more small weights which then must be processed in the pixel shader.

We used the Gaussian blur because it is very convenient way of dealing with the connectivity problem. For instance, box blur does not work well, because it produces sharp edges in some cases on its own, but we rather want to eliminate them. In fact, every blur which produces smooth results might be used here.

Texturing is mostly based on the vertex weights. This way we deal with texturing areas in different way, although we may use also other types of texturing like height-based or slope-based. By using the weights every vertex can process different textures independently by having an array of textures, array texture or some atlas, but it has disadvantages as well, mainly in performance. Moreover, every texture might have its own chain of mipmaps. By using texture array we can easily determine what texture should be sampled on a given vertex.

5   Tests and Results

Using this approach we may easily prepare landscape with various biomes that can blend smoothly between each other. To add another area type we need to implement one interface containing methods for returning area placement (latitude) and height for given position of a vertex. After that we have to modify our pixel shader to include the new area – this step is one of the main drawbacks of our system, but it may be solved by using Hammes approach [12] for texturing. Then we just add that implemented object to the engine via one method and the engine is capable of doing all of the work to spread the areas with given parameters.

The approach is effective because GPU does not need to recompute the geometry every frame, however FPS may drop pretty fast when many area types are visible at once because we have to sample a lot of textures. This especially happens when we use big kernel size for blurring because it produces a lot of small weights and causes sampling few textures for most of pixels. In the Table 1 we present performance statistics for generating one quad with size 65x65.

Table 1. Steps needed to generate and send a quad to GPU

Step	Action	Milliseconds
1.	Creating Voronoi map with area types	4.456
2.	Performing Gaussian blur for smoothing areas borders (kernel size = 20)	9.065
3.	Preparing structure with data for a quad (assumed height to be 0 to make it independent from noise generator)	2.608
4.	Saving quad to a file	3.044
5.	Sending quad data to GPU	0.354
	Total	19.527

As we can see generating a complete quad takes 19.527 milliseconds, that means having a terrain with 17 layers and 20 as Gauss kernel size it will generate all quads (1089) in around 21 seconds, but we need to keep in mind that when moving over terrain we only need to generate quads at one edge (or two – if moving diagonally), that means we need to generate 33 (or 65) quads which results in only 0.644 sec (or 1.269 sec). This gives us an opportunity to move over terrain quite quickly, so it can support various vehicles if used as game engine. Most of the time is taken by performing Gaussian blur, so we must choose the most efficient implementation, which is performing horizontal and vertical blur independently. When a quad is generated we save it to a file, so when we come back to already visited area the quad is loaded in around 1.156 ms. Unfortunately, this method requires a lot of memory (Figure 3) and can be a bottleneck for some situations. Hopefully, nowadays we may stop generating new layers at around 500MB memory usage and the generated terrain will be large enough for most cases.

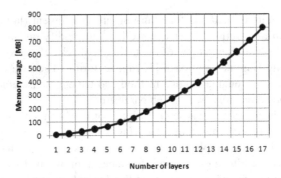

Fig. 3. Memory usage depending on the number of layers

6   Conclusions

This method provides comfortable way of dealing with various ecosystems on an infinite terrain. We showed a way for determining an ecosystem landscape shape and texturing as well as exposed information necessary for spreading flora and fauna very easily. Our engine also supports parameterization for placing the ecosystems in a more plausible ways, that is including latitude, area size, or dispersion of the areas. To achieve that we used modified Voronoi diagram along with Gaussian blur. Moreover, we support different quad sizes and variable number of terrain layers, for which, we can easily specify geometry quality. We also presented an uncommon way for rendering order by using spiral order rendering instead of very popular quad trees.

Despite of many functionalities and advantages our engine has also some inconveniences. Main drawback is that we have to modify pixel shader in order to include newly added area type, although including new area type is very easy. Another disadvantage is that we can not apply adaptive mesh for terrain geometry, it's because our engine is texturing terrain according to vertices information and adaptive algorithm may produce on flat areas only few vertices. However, both of the problems can be solved by using Hammes solution [12] for texturing terrain (merging the textures on CPU and sending them to GPU in order to cover the terrain mesh). Another undesirable situation refers to memory usage, we support terrain LOD, but even with low quad quality we are still holding all of the vertices in memory in order to not update too many quads on the GPU when moving over terrain, we only switch between predefined index buffers to provide correct LOD. In order to reduce the use of memory we need to go for a GPU-based implementation or sacrifice some speed of generating quads, that is generating that many vertices as we currently need to draw a quad but it would probably remove the possibility to move over terrain quickly.

References

1. Wolf, M.J.P. (ed.): The Medium of the Video Game. University of Texas Press (2001)
2. Dollins, S.C.: Modelling for the Plausible Emulation of Large Worlds. Ph.D. thesis, Brown University (2002). http://cs.brown.edu/~scd/world/dollins-thesis.pdf
3. Schneider, J., Boldte, T., Westermann, R.: Real-time editing, synthesis, and rendering of infinite landscapes on GPUs. In: Proceedings of Vision, Modeling, and Visualization 2006, pp. 145–152. IOS Press (2006)
4. Cozzi, P., Ring, K.: 3D Engine Design for Virtual Globes. CRC Press (2011)
5. Minecraft Blueprints: Step By Step Guide For Building Houses & Other Structures. Minecraft Books (2014)
6. Raffe, W.L., Zambetta, F., Li, X.: A survey of procedural terrain generation techniques using evolutionary algorithms. In: Proceedings of the IEEE Congress on Evolutionary Computation (CEC), pp. 1–8 (2012)
7. Hendrikx, M., Meijer, S., Van Der Velden, J., Iosup, A.: Procedural content generation for games: A survey. ACM Transactions on Multimedia Computing, Communications, and Applications (TOMCCAP) 9(1), 1–22 (2013). Article 1
8. Livny, Y., Kogan, Z., El-Sana, J.: Seamless patches for GPU-based terrain rendering. The Visual Computer 25(3), 197–208 (2009)
9. Losasso, F., Hoppe, H.: Geometry clipmaps: terrain rendering using nested regular grids. ACM Transactions on Graphics 23(3), 769–776 (2004)
10. Okabe, A., Boots, B., Sugihara, K., Chiu, S.N.: Spatial Tessellations: Concepts and Applications of Voronoi Diagrams, vol. 501. John Wiley and Sons, New York (2009)
11. Fernando, R., Haines, E., Sweeney, T.: GPU Gems: Programming Techniques, Tips and Tricks for Real-Time Graphics. Addison-Wesley Professional (2004)
12. Hammes, J.: Modeling of ecosystems as a data source for real-time terrain rendering. In: Westort, C.Y. (ed.) DEM 2001. LNCS, vol. 2181, pp. 98–111. Springer, Heidelberg (2001)

Accuracy Evaluation of a Linear Positioning System for Light Field Capture

Suren Vagharshakyan[✉], Ahmed Durmush, Olli Suominen,
Robert Bregovic, and Atanas Gotchev

Tampere University of Technology, Tampere, Finland
{suren.vagharshakyan,ahmed.durmush,olli.j.suominen,
robert.bregovic,atanas.gotchev}@tut.fi

Abstract. In this paper a method has been proposed for estimating the positions of a moving camera attached to a linear positioning system (LPS). By comparing the estimated camera positions with the expected positions, which were calculated based on the LPS specifications, the manufacturer specified accuracy of the system, can be verified. Having this data, one can more accurately model the light field sampling process. The overall approach is illustrated on an in-house assembled LPS.

1 Introduction

In order to properly capture the light field, significant number of densely positioned cameras are needed, e.g. a camera array. For static scenes, such camera array can be replaced by a system in which a single camera is precisely positioned on the camera array plane. A motorized linear positioning system (LPS) which can position a camera on a plane with high precision (sub-pixel resolution) has been built so that it is possible to capture images from different viewpoints. Thus, the spatial resolution of the light field captured by moving a single camera is higher than a light field, which would be captured by an array of similar cameras. In the following section we will describe in detail the built LPS and present manufacturer provided specifications for it. In Section 1.2 we describe the well-known pinhole camera model together with modeling of the lens distortions. In Section 1.3 we explain the mathematical model of the light field capturing process using an LPS. In Section 2 we describe the methodology of estimating model parameters for the LPS, which will allow us to evaluate the accuracy specifications provided by the manufacturer. Experimental results presented in Section 3 show the importance of evaluating the working accuracy of the LPS based on proposed methods in comparison to standard methods.

1.1 LPS Specifications

The in-house assembled LPS is composed of a hard alloy aluminum base, precision lead screw, anti-backlash nut and a stepper motor. The hard alloy aluminum base provides the structural stability to the system. The precision lead screw with an anti-backlash nut

© Springer International Publishing Switzerland 2015
N.T. Nguyen et al. (Eds.): ACIIDS 2015, Part II, LNAI 9012, pp. 388–397, 2015.
DOI: 10.1007/978-3-319-15705-4_38

converts rotary motion of the stepper motor to precise linear movements. Backlash describes the loss of motion due to gaps between the mechanical parts and it directly influences the positioning accuracy. Therefore, anti-backlash mechanisms are used in precision required applications. Furthermore, stepper motors used in the LPS have built-in encoders that provide closed loop servo like operation and increase the precision of the system.

The specifications of the LPS based on the characteristics of the mechanical and electrical parts are summarized in Table 1. The accuracy is measured by the deviation of the actual position from the desired position along the whole travel distance whereas repeatability is the measure of a system's consistency to achieve identical results. Straight-line accuracy is the measure of deviation from straight line along the motion axis.

Table 1. Specifications provided by the manufacturer

Accuracy	±20μm
Precision (Repeatability)	4μm
Straight Line Accuracy	38μm
Maximum Linear Speed	20 mm/s
Maximum Payload	20 kg
X Axis Travel Distance	1524mm
Y Axis Travel Distance	1016mm

1.2 Pinhole Camera with Lens Distortions Image Formation Model

A view captured by a camera is formed by projecting 3D points of the scene on the image plane of the camera according to the transform

$$sm = A[R|t]M$$

$$m = \begin{bmatrix} u \\ v \\ 1 \end{bmatrix}, A = \begin{bmatrix} f_x & 0 & c_x \\ 0 & f_y & c_y \\ 0 & 0 & 1 \end{bmatrix}, R = \begin{bmatrix} r_{11} & r_{12} & r_{13} \\ r_{21} & r_{22} & r_{23} \\ r_{31} & r_{32} & r_{33} \end{bmatrix}, t = \begin{bmatrix} t_1 \\ t_2 \\ t_3 \end{bmatrix}, M = \begin{bmatrix} X \\ Y \\ Z \\ 1 \end{bmatrix}$$

where (X,Y,Z) is the 3D position of a scene point expressed through its homogeneous coordinates M, $f = (f_x, f_y)$ and $c = (c_x, c_y)$ are the focal lengths and principal point coordinates measured in pixels, respectively, r_{ij} represents the camera rotation matrix and (t_1, t_2, t_3) are coordinates of the camera's optical center. Moreover, (f, c) and (R, t) are respectively the intrinsic and extrinsic camera parameters. For camera rotation there exists a more compact representation based only on three free parameters, however the matrix form will be kept for convenience. The pair (u, v) is the 2D coordinate of the projected point as a pixel position on the image plane. The result of the perspective projection is written in homogeneous coordinates m with a scaling factor s.

The pinhole camera model describes the perspective projections for a given camera. For real world cameras, another important problem is to model the distortions produced by camera lenses. Lens radial and tangential distortions at the image plane can be formalized using Brown's distortion model [1]

$$x^* = x(1 + k_1 r^2 + k_2 r^4 + k_3 r^6) + 2p_1 xy + p_2(r^2 + 2x^2)$$

$$y^* = y(1 + k_1 r^2 + k_2 r^4 + k_3 r^6) + p_1(r^2 + 2y^2) + 2p_2 xy$$

where $(k_1, k_2, k_3, p_1, p_2)$ are radial and tangential distortion coefficients, (x, y) are undistorted coordinates at image plane $\left(r = \sqrt{x^2 + y^2}\right)$ and (x^*, y^*) are final coordinates taking into account lens distortions. Hereafter, the distortion transform is denoted as D. The model can be extended by adding higher order polynomial terms of r and assuming also fractional terms of r.

The image formation procedure is formalized by Eq. 1. The matrix F combines the extrinsic matrix (R, t), the distortion-modelling transform and the intrinsic matrix (see Fig. 1). For brevity, the set of intrinsic parameters with lens distortion parameters are denoted as $I = (f, c, k_1, k_2, k_3, p_1, p_2)$.

$$\begin{bmatrix} u \\ v \end{bmatrix} = F_{R,t,I}\left(\begin{bmatrix} x \\ y \\ z \end{bmatrix}\right) \tag{1}$$

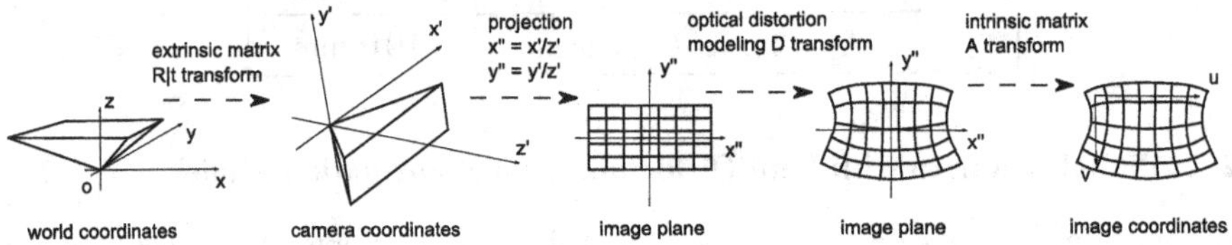

Fig. 1. Mathematical model of image formation

1.3   Model of the Motorized LPS

In our model of the motorized LPS, it is assumed that a single specific camera moves in space along a line with fixed step size d and takes photos of a stationary 3D scene. Using the mathematical model of the capturing function presented before, at each movement step, the camera projection can be described as function $F_{R,t_i,I}$, where parameters R, I are fixed during the movement, while the positions t_i are linearly changing, $t_i = l_0 + (id)l_n$, $i = 1, \dots, K$. Here, l_0 is vector to the start of the line over which the camera is moving and l_n is a normalized direction vector of the line as shown in Fig 2.

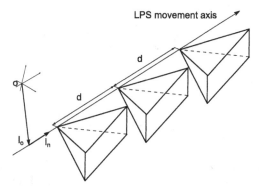

Fig. 2. Parameterization of the motorized LPS

2   Estimation

2.1   Estimation of Parameters for a Single Camera

Extraction of camera parameters can be broken down to estimating the 3D position, rotation and distortion coefficients of the camera. For that purpose, for a given camera setup, projections of 3D points are measured and then the inverse problem is solved to estimate R, t, I parameters. In other words, for a given set of points in space $M_k = \begin{bmatrix} x_k \\ y_k \\ z_k \end{bmatrix}, k = 1, ..., N$ and their corresponding projections $m_k = \begin{bmatrix} u_k \\ v_k \end{bmatrix}, k = 1, ..., N,$

$$\underset{R,t,I}{\text{argmin}} \sum_{k=1}^{N} \left\| F_{R,t,I}(M_k) - m_k \right\|^2$$

is found. When all parameters are unknown and the function F is nonlinear, the solution should be obtained through nonlinear optimization algorithms. For that purpose, we use camera estimation algorithms implemented in the OpenCV library [2] based on [3, 4].

2.2   Motorized LPS Parameter Estimation

Estimation of the motorized LPS parameters refers here to estimating the movement precision over the LPS movement axis for specific fixed camera with unknown intrinsic parameters. Methods for unconstrained estimation of the camera locations at each capture step do not provide valuable results. Particularly, results obtained through estimation of the unconstrained locations with common intrinsic parameters and lens distortion coefficients [2, 3], i.e.

$$\underset{\substack{I,R_i,t_i,\\i=1,\dots,K}}{\operatorname{argmin}} \sum_{i=1}^{K} \sum_{k=1}^{N} \left\| F_{R_i,t_i,I}(M_k) - m_{k,i} \right\|^2 \tag{2}$$

or independent estimation of the locations

$$\underset{R_i,t_i}{\operatorname{argmin}} \sum_{k=1}^{N} \left\| F_{R_i,t_i,I^*}(M_k) - m_{k,i} \right\|^2, i = 1,\dots,K \tag{3}$$

using beforehand independently estimated lens distortions and intrinsic ters $I^* = (f^*, c^*, k_1^*, k_2^*, k_3^*, p_1^*, p_2^*)$, are not precise enough. They give only a rough estimation of the camera movement through space.

In our proposed approach we suggest considering a linear dependence between the camera positions. In this case the problem can be formulated as

$$\underset{R,l_0,l_n}{\operatorname{argmin}} \sum_{i=1}^{K} \sum_{k=1}^{N} \left\| F_{R,l_0+d_i l_n, I^*}(M_k) - m_{k,i} \right\|^2 \tag{4}$$

where the rotation denoted by R is common for all positions, l_0, l_n define a line in the space, $\{d_i\}_{i=1,\dots,K}$ is the distribution of positions over that line, and I^* denotes the intrinsic parameters and lens distortions coefficients which are estimated beforehand using (2). The minimization problem is solved by using the Levenberg–Marquardt algorithm [5]. It is a non-linear minimization algorithm and therefore providing good initial estimates for R, l_0, l_n improves the estimation performance. In particular, initial estimates for the minimization algorithm (4) can be obtained by using independently estimated 3D position $t_i, i = 1,\dots,K$ and rotation $R_i, i = 1,\dots,K$ of each camera position by minimizing (3). The initial common rotation estimation R is the mean of all rotations $R_i, i = 1,\dots,K$ and l_0, l_n are the least square solution of

$$\begin{bmatrix} t_1 \\ \vdots \\ t_K \end{bmatrix} = [l_0 \ l_n] \begin{bmatrix} 1 & \cdots & 1 \\ d & \cdots & Kd \end{bmatrix} \tag{5}$$

where d is a predefined uniform step size. Alternatively, l_0, l_n can be found based on principle component analysis for fitting a line to points $t_k, k = 1,\dots,N$ [6].

After finding solution for (4), $d_i, i = 1,\dots,K$ are estimated such that

$$\underset{d_i}{\operatorname{argmin}} \sum_{k=1}^{N} \left\| F_{R,l_0+d_i l_n, I^*}(M_k) - m_{k,i} \right\|^2, i = 1,\dots,K. \tag{6}$$

In fact, the interesting values are $d_i, i = 1,\dots,K$, which allow estimating or verifying the precision of the movement over the LPS compared to the intended uniform step size d. The proposed hybrid algorithm can be summarized as follows:

Input: K number of views, N number of observed fixed points of the scene and $M_k, k = 1,\dots,N$ their 3D coordinates in the space, $m_{k,i}, k = 1,\dots,N, i = 1,\dots,N$

corresponding projection image coordinates in pixels of i-th point in k-th view, d motorized LPS movement step size, and $I^* = f^*, c^*, k_1^*, k_2^*, k_3^*, p_1^*, p_2^*$ previously estimated intrinsic parameters and lens distortion coefficients.

1. Find R_i rotation matrix and t_i position $(i = 1, ..., K)$ of the camera for each view by solving (3).
2. Calculate mean rotation matrix R based on rotation matrices R_i. Calculate least square solution of (5) for l_0, l_n (normalize l_n, if it is necessary).
3. Using already found R, l_0, l_n as initial values, update them by solving (4)
4. Solve (6) to find d_i, using grid search method for each $i = 1, ..., N$ independently.
5. Repeat step 3, 4 until $err = \sum_{i=1}^{K} \sum_{k=1}^{N} \left\| F_{P_i}(M_k) - m_{k,i} \right\|^2 < T$, where T is the predefined convergence tolerance.

3   Experimental Results

3.1   Experiment with Synthetic Data

To evaluate the proposed algorithm, we generate a synthetic dataset based on the mathematical model presented in sections 1.2 and 1.3. The dataset is generated by projecting synthetic chessboard corners (Fig. 4(a)) to virtual cameras located along the line and adding noise to projected coordinates to model inaccuracy of the feature detection algorithms. The inaccuracy is assumed to be less than 1px in absolute value along each axis. A realistic set of LPS parameters is used in projection calculation, but lens distortions are not considered. For camera motion we consider a scenario where consecutive step size varies as a sine wave Fig. 3 (ground truth). As seen in the figure, estimated consecutive distances are close to the ground truth even in the presence of noise, which shows that the proposed algorithm is robust to inaccuracy in feature point detection. This makes our approach suitable for LPS accuracy evaluation in a real world setting.

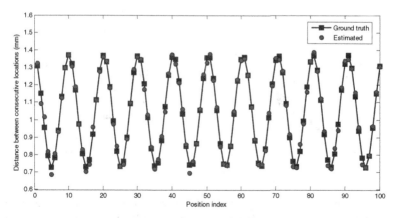

Fig. 3. Comparison of the estimated locations against the ground truth for synthetic dataset

3.2   Experiment with Real Data

To form a set of 3D points, we use a chessboard texture on a plane such that each inner corner of the chessboard defines a point in 3D space as shown in Fig 4(a). Corner detection algorithm presented in OpenCV allows detecting corners with sub pixel precision in image coordinates, see Fig 4(b). For the given LPS, a nominal step size of 1 mm is used for camera motion and a chessboard is captured by the camera at each step for later processing and movement precision estimation, see Fig. 5.

(a)   (b)

Fig. 4. (a) Origin of the 3D space positioned at the edge corner of a chessboard and all inner points shown as red dots related to selected axis. (b) Detected image coordinates of inner points of a chessboard (18x30 inner points).

Fig. 5. Example of views taken with 11mm distance using LPS. Original data contain 400 views of size 1980x1080px taken with 1mm distance.

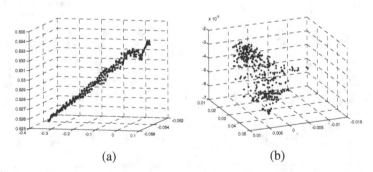

(a)   (b)

Fig. 6. 3D coordinates of estimated positions (a) in meters and rotations (b) in radians obtained by solving Eq. 2 of the camera at 400 locations over the LPS movement axis with 1mm step size

Solving Eq. 2 is not trivial because of the large number of unknowns. For example, in the case of $N = 32$ (chessboard with 4x8 inner corners) and $K = 400$ (K is the number of images), it is necessary to estimate 9 intrinsic parameters and $9K$ extrinsic parameters based on NK measurements, in other words, to solve a nonlinear optimization problem with 3609 unknowns based on 12800 measurements. Due to the large number of unknowns, the method is too slow in practice – for 30 iterations of minimization it takes about three hours of computation. Estimated location results are presented in Fig. 6, and the estimated intrinsic parameters are:

$$f = 9m, c_x = 962.68px, c_y = 539.13px, k_1 = -0.233, k_2 = 0.247, k_3 = -0.173, p_1 = 0.0028, p_2 = -0.0035.$$

Beforehand estimation of the lens distortion coefficients independent from LPS can be done by using only e.g. 6 arbitrary positioned images of the chessboard, similar to the one shown in Fig. 4(b). For a better estimation, it is necessary to have images where a large part of the field of view is occupied by a chessboard with a large number of inner corners. In that case, in Eq. 2 one will have a large number of measurements (3240) with relatively small number of unknowns (69). Then one can effectively solve Eq. 2 using the Levenberg–Marquardt algorithm. The obtained results are:

$$f = 8.18m, c_x = 951.59px, c_y = 551.08px, k_1 = -0.169, k_2 = 0.111, k_3 = -0.02, p_1 = -0.0007, p_2 = -0.00002.$$

After determining the lens distortion coefficients and camera intrinsic parameters, the camera positions and rotations are calculated by solving Eq. 3 and the results are shown in Fig. 7. They are quite similar to the results shown in Fig. 6, however, their calculation takes only a few seconds. Unfortunately, the estimation quality is not high enough to allow making conclusions about the LPS working accuracy.

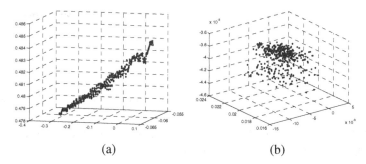

(a)   (b)

Fig. 7. 3D coordinates of estimated positions (a) and rotations (b) of the camera obtained by solving Eq. 3, at 400 locations over the LPS movement axis with 1mm step size

The proposed constraint in Eq. 4 together with Eq. 6 facilitates getting meaningful results about the accuracy of the LPS movement as shown in Fig. 8. For the given LPS, maximum estimated positioning misalignment of the LPS has an absolute value less than 0.04mm. Together with joint camera rotation estimation for all positions, results provide LPS model parameters for further processing and properly interpreting sampled light field data. We also noticed that the precision of the lens distortion coefficient estimation highly affects the camera location estimation. However, the results clearly show that the proposed method is able to detect incorrect and unexpected camera movement by the LPS. This is illustrated in Fig. 9 by adding outliers in the original dataset.

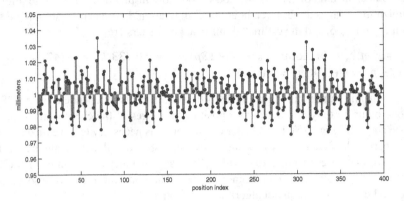

Fig. 8. Distances between estimated consecutive locations $(d_i - d_{i+1})$ of the camera over the line in space. They are calculated based on the proposed method from a data set containing images with 1 mm step size along the LPS movement axis.

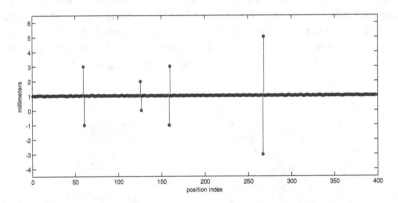

Fig. 9. Similar results as in Fig. 7 for same dataset, but with several outliers added artificially by replacing images in dataset, which are clearly visible from the results

4   Conclusion

A motorized LPS allowing very fine light field sampling has been constructed and a method for its precision verification has been developed. Proposed calibration algorithm also provides estimates of the camera lens distortions and camera rotation during capture. Having this data, one can model the light field sampling process with a higher accuracy. As a future work we plan to extend capturing process from linear capturing system to a planar capturing system and attempt to provide a more advanced estimation procedure for evaluating the accuracy of a 2D positioning system.

References

1. Brown, D.C.: Decentering distortion of lenses. Photogrammetric Engineering **32**(3), 444–462 (1966)
2. OpenCV (Open Source Computer Vision Library). www.opencv.org
3. Bouguet, J.-Y.: Camera Calibration Toolbox for Matlab. www.vision.caltech.edu/bouguetj /calib_doc/
4. Zhang, Z.: A Flexible New Technique for Camera Calibration. IEEE Transactions on Pattern Analysis and Machine Intelligence **22**(11), 1330–1334 (2000)
5. Marquardt, D.: An Algorithm for Least-Squares Estimation of Nonlinear Parameters. SIAM Journal on Applied Mathematics **11**(2), 431–441 (1963)
6. Hawkins, D.M.: On the Investigation of Alternative Regressions by Principal Component Analysis. Journal of the Royal Statistical Society, Series C **22**(3), 275–286 (1973)

A System for Real-Time Passenger Monitoring System for Bus Rapid Transit System

Jonathan Samuel Lumentut[1], Fergyanto E. Gunawan[2]($^{(\boxtimes)}$),
Wiedjaja Atmadja[3], and Bahtiar S. Abbas[3]

[1] School of Computer Science, Bina Nusantara University, Jakarta 11480, Indonesia
[2] Binus Graduate Programs, Bina Nusantara University, Jakarta 11480, Indonesia
f.e.gunawan@gmail.com
[3] Faculty of Engineering, Bina Nusantara University, Jakarta 11480, Indonesia

Abstract. TransJakarta is a BRT-based mass transportation system operating in Jakarta, the capital of Indonesia. Currently, the system delivers rather low level of service; as a result, the number of passengers is rather low in comparison to that of the other BRT-based systems. In this work, we propose a design of the passenger counting system for BRT-based system, which is very important for the BRT fleet management to increase the system level of service. The counting system is established by deploying computer vision techniques. A few algorithms are evaluated and the Adaptive Median Filtering with the sampling rate of 13 produces the highest level of precision and recall.

Keywords: Computer vision · Passenger counting system · Background substraction · Bus rapid system

1 Introduction

Traffic congestion is a major issue faced by many large cities across the globe. Jakarta, the capital of Republic of Indonesia, also deals with the problem in the daily basis. The congestion has caused the city setback in various sectors. In the economic sector, the predicted loss due the congestion is about Rp 65 trillion per year or about 5 billion USD. Within this number, about Rp 35 trillion per year or about 2.8 billion USD are due to the lost in the vehicle operation [1]. The congestion also has negative impacts in health, environment, and social sectors.

Metropolitan cities require structured mass transportation systems to lessen the level of congestion [2]. The system can be based on rail system or on bus system. The bus system is often called as the Bus Rapid Transit (BRT) system. The rail-based system has larger capacity, but requires longer development time and higher cost. In the other side, the BRT system has smaller capacity, but requires shorter development time and lower cost.

The adoption rate of the BRT system is higher than that of the rail-based system since the last decade. The adoption rates of the both systems are shown in Fig. 1.

© Springer International Publishing Switzerland 2015
N.T. Nguyen et al. (Eds.): ACIIDS 2015, Part II, LNAI 9012, pp. 398–407, 2015.
DOI: 10.1007/978-3-319-15705-4_39

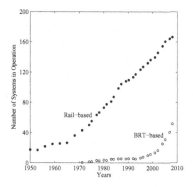

Fig. 1. The number of development of the train-based and BRT-based public transportation systems across the globe [3]

Jakarta also adopts the BRT system and was firstly constructed in 2004. The system is called TransJakarta BRT. By 2014, TransJakarta already has 12 corridors with 180 km busway length in total. Table 1 lists the monthly number of passengers of TransJakarta in 2013.

Table 1. The Number of Monthly Passengers of TransJakarta on a number of corridors in 2013

Corridor	Jan	Feb	Mar	Apr	May	Jun
1	1722650	1753109	1753109	2144327	2208102	2240811
2	608449	607270	701911	656851	678055	726560
3	651398	695467	821056	793286	799330	844527
4	519565	519701	576392	580012	575861	573580
6	661385	652530	726278	726278	724563	703151
9	984900	1012136	1145375	1127795	1122989	1143005

Corridor	Jul	Aug	Sep	Oct	Nov	Dec
1	226959	1856363	2217651	2185972	2199600	2367065
2	712579	674294	730465	712963	708227	769027
3	797685	751904	862652	846839	841850	873630
4	583415	519188	646062	646242	650763	668191
6	691517	598530	708459	708459	731415	757512
9	1164793	1020026	1210573	1235085	1218115	1251486

Source: http://transjakarta.co.id/

The transportation system can be divided into two sub-systems: the supply-side system and the demand-side system. The transportation system should be designed to balance the capacity of the both sub-systems. If the level of demand is higher, then the level of service of the system will drop. If the level of the supply is higher, the system may be not operating efficiently. For this reason, monitoring the two sides of the transportation system is important for designing a cost-effective system.

To operate the transportation system efficiently and effectively, we need to monitor the both sides of the system. In the case of the BRT system, the fleet of buses is the main constituent of the supply side; meanwhile, the BRT passengers are the main constituent of the demand side. The bus fleet can now be monitored using various technologies such the Global Positioning System (GPS) [4–7]. However, the passenger monitoring system is extremely limited. This work intends to propose a monitoring system for the passenger using computer vision techniques.

Many previous studies have been performed to develop methodology to track moving objects using computer vision. In this article, those methods will be further developed for monitoring BRT passengers.

Reference [8] develops method to track the movement of many human objects in real-time utilizing the Conditional Density Propagation algorithm. Three improvement were made in his work: the use of an effective template for the human form using self-organizing map; the use of the hidden Markov model for modeling the dynamic of the human shape; and the use of a competition rule to separate a person from others. In addition, Ref. [9] study passenger detection using the characteristics of the head area. They intend to track the pedestrian movement. The tracking is achieved in two steps: applying the background difference algorithm, dynamic threshold algorithm, and the method of morphological processing to filter the image noise; and applying head matching algorithm using a mask template.

In addition, Ref. [10] develops an algorithm for estimating the volume of a passenger flow. Reference [11] discuss on tracking human motion in outdoor environment. Reference [12]develop real-time video processing on Android platform.

In this work, we develop a method for counting the number of passengers in a BRT station. The method is developed using computer vision techniques where the passengers are detected and tracked by background subtraction algorithms. We evaluate the following algorithms: Pixel-based Adaptive Segmenter algorithm [13], Godbehere-Mitsukawa-Goldberg algorithm [14], and approximated median filtering [15].

2   Methods

2.1   Design Consideration of Bus Rapid Transit System

The Bus Rapid Transit (BRT) system is the bus-in-steroid system. The system eliminates aspects that slowing down the traditional bus system in serving large number of passengers. The system delivers high performance by deploying a number of design considerations [16].

Those design considerations mainly are: the BRT bus lanes are designed to be located on the high-demand segments, off-vehicle fare collection, the busway is separated and physically protected from the mixed traffic.

The use of the dedicated lane is a very important design consideration. And, the lane is placed in the road center to minimize conflict with the mixed traffic, pedestrians spilling into the roadway, man-powered vehicles, and other conflict-ing factors. The off-vehicle fare collection, instead of the on-vehicle fare collection of the traditional bus system, is important to increase the passenger transfer rate.

In addition to those main design aspects, the BRT stations are designed on the median of the road and are used to share by both directions of service; see Fig. 2. Thus, the required number of stations is less and it can save cost with fewer stations. However, with this design, the BRT requires special buses which can serve passengers from the two sides of the bus.

Fig. 2. A typical design and location of the TransJakarta BRT station

2.2   Proposed Monitoring Strategy

Figure 2 shows a typical TransJakarta BRT station. The station is clearly con-structed by considering the design factors discussed in Sec. 2.1. Passengers arrive on the station by using the elevated stair. The TransJakarta BRT stations tend to be narrow but long. Thus, the station can serve more than one bus simulta-neously.

The proposed real-time monitoring system for such station design is shown in Fig. 3. The system consists of a central processing unit, a number of image acquisition units (CCTVs or IP cameras), wireless data communication unit, and server. The acquisition unit records the moving images of the movement of passengers across the access point. The recorded images will be sent to a central processing unit where the passenger counting will be performed. The algorithms in the CPU should be able to count the remaining passengers in the station. The data including the associated time-stamp data, will then be submitted to a server via a wireless network.

2.3   Passenger Counting Using Background Subtraction Algorithms

In the following, we will discuss three computer vision algorithms for background subtraction, which are evaluated for the purpose of passenger counting. For the detail exposition, the reader is advised to consult Ref. [15].

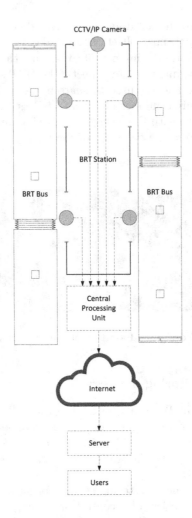

Fig. 3. The proposed passenger monitoring system for TransJakarta BRT station

In the following, we use notations: $I_t^c(x, y)$ to denote the value of channel c of the pixel at location (x, y) at time t and $B_t^c(x, y)$ is the background model.

Godbehere-Mitsukawa-Goldberg. The Godbehere-Mitsukawa-Goldberg (GMG) algorithm combines statistical background image estimation, per-pixel Bayesian segmentation, and an approximate solution to the multi target tracking problem using a bank of Kalman filters and Gale-Shapley matching. A heuristic confidence model enables selective filtering of tracks based on dynamic data. Figure 4 describes the GMG algorithm.

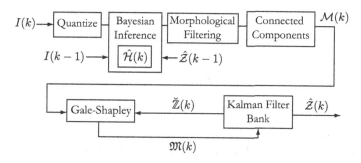

Fig. 4. GMG algorithm block diagram. An image $I(k)$ is quantized in color-space, and compared against the statistical background image model, $\hat{\mathcal{H}}(k)$, to generate a posterior probability image. This image is filtered with morphological operations and then segmented into a set of bounding boxes, $\mathcal{M}(k)$, by the connected components algorithm. The Kalman filter bank maintains a set of tracked visitors $\hat{Z}(k)$, and has predicted bounding boxes for time k, $\check{Z}(k)$. The Gale-Shapley matching algorithm pairs elements of $\mathcal{M}(k)$ with $\check{Z}(k)$; these pairs are then used to update the Kalman Filter bank. The result is $\hat{Z}(k)$, the collection of pixels identified as foreground. This, along with image $I(k)$, is used to update the background image model of $\hat{\mathcal{H}}(k)$. This step selectively updates only the pixels identified as background [14].

Adaptive Median Filtering. The median filtering algorithm is actually derived from the non-recursive techniques median filtering [15]. Reference [17] propose a recursive technique that is easy to estimate the median. This technique is also used for background modeling at traffic monitoring. In the algorithm, the estimated median is added by 1 if the input pixel is larger than the previos estimated median. Inversely, the median is subtracted. Mathematically, it is written:

$$B_{t+1}^{C} = \begin{cases} B_t^C + 1 & \text{if} \quad I_t^C > B_t^C \\ B_t^C - 1 & \text{if} \quad I_t^C < B_t^C \\ B_t^C & \text{if} \quad I_t^C = B_t^C \end{cases} \tag{1}$$

Pixed-based Adaptive Segmenter. Pixel-Based Adaptive Segmenter or called PBAS is a non-parametric background modeling technique [13]. The background is modeled on the basis of the observed pixel value history. The foreground decision depends on a decision threshold. The background update is based on a learning parameter. This algorithm uses dynamic per-pixels state variables and introduce dynamic controllers for each of them. The PBAS algorithm is summarized in the flowchart in Fig. 5. For the detail, we advice the reader to consult Ref. [13].

2.4   Experimental Setup

Figure 6 shows a typical access point of a typical BRT station. We draw the red virtual line and use the line to separate the region into two regions: Region A and Region B. Region A is located to the left of the line. We attach a camera to the

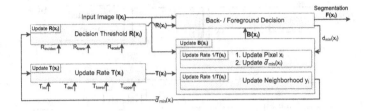

Fig. 5. PBAS algorithm block diagram. Variable $I(x_i)$ denotes the current pixel value, $B(x_i)$ is the background model, $R(x_i)$ is the per-pixel threshold, $T(x_i)$ is the learning parameter, $F(x_i)$ is the foreground segmentation mask [13].

top of that access point and record any passenger movement around the access point. When a BRT bus arrives at the station, the bus door will be align to the access point. The bus door will be in Region A. To board the bus, the passenger should walk across the red line and enter the bus. In the experiment, we record the activity in this access point for a duration of 30 minutes. We should note that in practice, passengers often wait for the bus along the red line, resulted in the counting error.

Fig. 6. An access point to a typical BRT station. The image recording system is located on the top. The red is a virtual line which is constructed to separate Region A, to the left of the line, and Region B, to the right of the line.

3   Implementation and Results

This section reports the empirical data collected during verification of the proposed method. The verification is performed as the following. In the experiment, the three algorithms are utilized and the results of each algorithms are reported.

The experimental results are tabulated in Table 2 for the PBAS, GMG, and AMF algorithms. The other results are tabulated in Table 3 for the AMF algorithm with varying the sampling rate. We should note that in this experiment the recorded image is rather low in resolution and contains high level of noise. As the results, the algorithms PBAS and GMG are not able to recall any object crossing the line and only detect the noise. In addition, the both algorithms require rather long processing time. Only the AMF algorithm can identify the object with reasonable level of accuracy and precision.

On the basis of the previous results, we further investigate the perform of the AMF algorithm by varying the sampling rate. We vary the sampling rate as 1, 3, 5, 9, and 13. The results are presented in Table III, and it indicates that the precision of the AMF algorithm tends to be strongly affected by the sampling rate. The precision level increases as the sampling rate increasing.

Table 2. The Experimental Results for PBAS, GMG, and AMF algorithms. The AMF uses the sampling rate of 1.

Parameters	PBAS Algorithm		GMG Algorithm		AMF Algorithm	
	$A \to B$	$B \to A$	$A \to B$	$B \to A$	$A \to B$	$B \to A$
The number of object captured by the algorithm	19	16	66	80	9	6
The number of correct object captured by the algorithm	0	0	0	1	2	3
The number of the actual object	3	4	3	4	3	4
Recall	0	0	0	0.25	0.67	0.75
Precision	0	0	0	0.25	0.22	0.50

Table 3. The Effect of the Sampling Rate to the Counting using AMF Algorithm

Parameters	Sampling Rate = 5		Sampling Rate = 9		Sampling Rate = 13	
	$A \to B$	$B \to A$	$A \to B$	$B \to A$	$A \to B$	$B \to A$
The number of object captured by the algorithm	11	10	12	10	12	9
The number of correct object captured by the algorithm	2	4	2	4	2	4
The number of the actual object	3	4	3	4	3	4
Recall	0.67	1.00	0.67	1.00	0.67	0.44
Precision	0.18	0.40	0.17	0.40	0.67	1.00

4   Conclusion

This research proposes a system for counting the number of passengers waiting in a BRT station. The system is designed to provide the real-time data of the number of passengers. This data is very important for the BRT fleet management. The passenger counting system utilizes computer vision technique to track the movement of passengers across the station access point. In the image of the access point, a virtual line is drew, and the passenger is counted upon crossing the line. The image can be recorded using any recording devices such as IP TV, mobile TV, or CCTV. In the designed system, the recorded image will be sent to a central processing unit for processing the passenger counting on each access point, to determine the number of the total passengers in the station, and to submit the data to a dedicated server. In the current development, only one access point is considered. The developed system is used to count the passenger crossing the access point. Then, the result is compared to the manual counting. The empirical findings suggest that the AMF algorithm provides the best results in term of the accuracy in comparison to the PBAS and GMG algorithms. The result of the AMF algorithm can also be improved by increasing the sampling rate. The required time for analysis is also lower for the AMF algorithm.

Acknowledgement. We offer our highest appreciation to Directorate General Higher Education of Republic of Indonesia for their support via the National Competitive Research Grant No.: 017.A/DRIC/V/2013.

References

1. Hartawan, T.: Macet terus, jakarta rugi rp 65 triliun per tahun (2013). http://www.tempo.co/read/news/2013/03/24/214468984/Macet-Terus-Jakarta-Rugi-Rp-65-Triliun-per-Tahun (retrieved on December 2013)
2. Morichi, S.: Long-term strategy for transport system in asian megacities. Journal of the Eastern Asia Society for Transportation Studies **6**, 1–22 (2005)
3. Campo, C.: Bus rapid transit: Theory and practice in the united states and abroad. Master's thesis, School of civil and environmental engineering, Georgia Institute of Technology (2010)
4. Chandra, F.Y., Gunawan, F.E., Glann, G., Gunawan, A.A.: Improving the accuracy of real-time traffic data gathered by the floating car data method. In: The 2nd International Conference on Information and Communication Technology (IColCT), Bandung, Indonesia (2014)
5. Gunawan, F.E.: Real-time traffic monitoring system. In: International Conference and Advanced Informatics: Concepts, Theory and Applications (ICAICTA), Bandung, Indonesia (2014)
6. Gunawan, F.E., Chandra, F.Y.: Optimal averaging time for predicting traffic velocity using floating car data technique for advanced traveler information system. Procedia - Social and Behavioral Sciences **138**, 566–575 (2014)
7. Gunawan, F.E., Chandra, F.Y.: Real-Time Traffic Monitoring System using Floating Car method. Lambert Academic Publishing (2014b)

8. Kang, H.-G., Kim, D.: Real-time multiple tracking using competitive condensation. Pattern Recognition **38**, 1045–1058 (2005)
9. Wang, B., Chen, Z., Wang, J., Zhang, L.: A fast passenger detection algorithm based on head area characteristics. In: International Workshop on Information and Electronic Engineering (IWIEE), Harbin, China, pp. 184–188. Elseiver (2012)
10. Zhou, D.: Bus passenger recognition and track of video sequence. Journal of Multimedia **8**, 262–269 (2013)
11. Haritaoglu, I., Davis, L.: W4: Real-time surveillance of people and their activities. IEEE Transactions on Pattern Analysis and Machine Intelligence **22**, 809–830 (2000)
12. Saipullah, K., Anuar, A., Ismail, N., Soo, Y.: Real-time video processing using native programming on android platform. In: IEEE 8th International Colloquium on Signal Processing and Its Applications (2012)
13. Hofmann, M., Tiefenbacher, P., Rigoll, G.: Background segmentation with feedback: The pixel-based adaptive segmenter. In: 2012 IEEE Computer Society Conference on Computer Vision and Pattern Recognition Workshops (CVPRW), pp. 38–43 (2012)
14. Godbehere, A., Matsukawa, A., Goldberg, K.: Visual tracking of human visitors under variable-lighting conditions for a responsive audio art installation. In: American Control Conference (ACC), 4305–4312 (2012)
15. Parks, D., Fels, S.: Evaluation of background subtraction algorithms with postprocessing. In: IEEE Fifth International Conference on Advanced Video and Signal Based Surveillance 2008, AVSS 2008, pp. 192–199 (2008)
16. Weinstock, A., Hook, W., Replogle, M., Cruz, R.: Recapturing global leadership in bus rapid transit: A survey of select u.s. cities. Technical report, Institute for Transportation and Development Policy (2011)
17. McFarlane, N., Schofield, C.: Segmentation and tracking of piglets in images. Machine Vision and Applications **8**(3), 187–193 (1995)

Cloud Based Solutions

Cloud and m-Learning:
Longitudinal Case Study of Faculty of Informatics and Management, University of Hradec Kralove

Ivana Simonova and Petra Poulova[✉]

Faculty of Informatics and Management, University of Hradec Králové,
Hradec Králové, Czech Republic
{ivana.simonova,petra.poulova}@uhk.cz

Abstract. Shift has been detected in the ICT-enhanced teaching/learning, particularly from e-learning to cloud and m-learning. For education, it offers new possibilities to structure and perform learning processes. That was the reason why a case study was prepared summarizing how this process was carried out at the Faculty of Informatics and Management (FIM), monitoring (1) learners´ knowledge reached with/without the ICT support, (2) the impact of ICT in the instruction tailored to learners´ preferences, (3) evaluating the work with virtual desktops and (4) learners´ readiness for m-learning, mainly what mobile devices they own, which of them they use for personal purposes, whether they have adequate skills developed, and whether mobile devices could be implemented in the process of instruction. Despite no statistically significant differences were detected in researches 1 and 2, learners expressed high satisfaction with using virtual desktops, wide possession of mobile devices and good skills in using them for personal activities.

Keywords: Case study · Cloud computing · Database · Education · e-learning · m-learning

1 Introduction

The net-generation, e-/i-generation, digital natives and other attributes describe features which define characteristics of current young people. Lately, they have been called the C-generation, or generation C, where C stands for cloud.

Different approaches to setting the Generation C definition are available, e.g. Pickett places Generation C between 1982 and 1996 [1], while Strauss and Howe set them in the Millennial period (i.e. born the interval between 1982 and 2001). Pickett adds Generation C is a group of individuals who share a similar state of mind, certain personality traits, values, attitudes, interests, lifestyles, they turn to the Internet naturally and extensively and are very Web 2.0-savvy [1], i.e. all of them being digital natives defined by Prensky [2].

Cloud computing is a 'model for enabling convenient, on-demand network access to a shared pool of configurable computing resources (e.g., networks, servers, storage,

© Springer International Publishing Switzerland 2015
N.T. Nguyen et al. (Eds.): ACIIDS 2015, Part II, LNAI 9012, pp. 411–420, 2015.
DOI: 10.1007/978-3-319-15705-4_40

applications, and services) that can be rapidly provisioned and released with minimal management effort or service provider interaction' [3]. Cloud computing can be used in many fields of human activities including educators, institutions, individual students, to support particular teaching and learning experience, to organize software availability, it enables access to software applications, hardware, data, and computer processing power on the Web, which is preferred to installing software onto one computer or server. For education, it offers new possibilities to structure and perform learning processes, thus providing potential impact on education. In teaching, the interoperability, transferability, security and privacy, backup and perpetuity, denial of service and content issues are the most appreciated features [4]. As Awodele et al. state cloud computing, combined with web-based applications and services enables to achieve higher levels of productivity, innovation, and effectiveness as well as the access to software applications, hardware, data, and computer processing power on the Web, rather than installing software onto one's computer or server. The educators can exploit new web software applications for learning purposes thus supporting innovation in the use of new technologies for learning with minimal investment. For students, cloud applications can add richness and variety to their learning experience, collaboration with other students after regular school schedule. [5] And, educators are now seeking efficient solutions for online collaboration and online teaching for overcoming the current limitations. Cloud computing could be one approach supporting educators and students on their journey towards an efficient, network based, online learning environment. [6]

The contemporary development of information and knowledge society depends on technology which has long been based on the increasing performance of devices and their miniaturization at the same time, thus providing a number of benefits to its users, reduce equipment, costs and maintenance [7]. And, cloud solutions allow the teaching, research and development to be more efficient, particularly in:

- the economy of financial resources (to pay for what is used);
- elasticity of use, given the possibility of using initial the small services;
- increased availability, e.g. Google offers approximately 100% for educational applications;
- end-user satisfaction because the applications accessed in the cloud include the latest tools to be used without being purchased, installed and maintained;
- collaboration possibilities;
- data storage   for free use, accessible from any time and place and through any type of mobile device; and,
- learning management systems (LMS) such as Blackboard or Moodle can be hosted there [8].

Offering on-demand infrastructure, applications, and support services in a cloud could drive down the total cost and capital invested for IT in higher education, facilitate a transparent matching of IT cost, demand, and funding, scale IT, foster further IT standardization, accelerate the time to market by reducing IT supply bottlenecks, and increasing the interoperability between disjoined technologies in or between organizations [9].

Reflecting the technical and technological development, the process of transformation in delivery of education changed providing new transformation tools, i.e. from using non-portable devices (e-instruction/e-learning), followed by portable, handheld ones (m-instruction/m-learning) and mobile services, and thus using latest mobile devices and requiring new skills from the users (both teachers and learners) [10].

Although mobile devices and technologies have been around for a long time period, their impact on university education is recent, and the estimated impact is even more clouded, difficult to be imagined [11]. Reflecting the fact that mobile phones (smartphones) have become of mass use and tablets are on the rise (as verified e.g. in our survey and many others), these probably will be the leading tools for the future couple of years, both for personal use and education purposes.

Below an example of Faculty of Informatics and Management (FIM), University of Hradec Kralove (UHK), Czech Republic is presented. The study comprises data which have been collected since 2009/10 academic year. It is structured into several phases where following research questions were answered:

1. Do students reach better knowledge when taught in the ICT-enhanced way compared to traditional methods without ICT support? (Research 1)
2. Do students reach better knowledge if their learning preferences are reflected in the process of instruction? (Research 2)
3. Do students appreciate the possibility to use virtual desktops for remote access to virtual classrooms equipped with software not available to anybody anywhere? (Research 3)
4. Are students ready for cloud m-learning? (Research 4)

Reflecting the technology development, both e-learning (research questions 1 and 2) and m-learning (research questions 3 and 4) are in the focus of attention.

2   Research Design

As the Faculty of Informatics and Management comprises the research activities in the field of Informatics, as well as it functions as the educational institution preparing professionals in the area, two main fields can be detected in the FIM research: (1) professional, focusing on Informatics, Information- and Knowledge Management-related problems, and (2) education-focused, dealing with branch didactics, i.e. methodology how to teach the learning content efficiently. Currently, the focus on latest technologies has been emphasized.

2.1   Course of Research

Being the leader in the Czech Republic, the process of ICT implementation into education started in 1997 at FIM and spread widely after 2000, when the LMS WebCT/Blackboard was used. Currently, approximately 250 online courses supporting the complete scope of subjects are available to students (e-subjects), either to enhance the teaching/learning process, or to be used in the distance form. This phase of FIM development is reflected in research question 1 (see above) where the ICT-enhanced process of teaching/learning was researched; and in research question 2 trying to verify whether the individualization if the ICT-enhanced process of instruction results in learners´ better knowledge.

Since 2012/13 academic year the virtual desktops have been available to students, mainly for work with software not providing free/open access (e.g. MS SQL Server, Enterprise Architect). Virtual classrooms were formed enabling students to work from any place any time. On the other side, despite been future IT specialists and the work with virtual desktop was intuitive to a large extent, new skills were required from learners. Thus possible problems, both technical and learning content-related were analyzed within research question 3.

Since 2013/14 the Blackboard Mobile Learn™ version 4.0 for Apple and Android devices has been piloted. Research question 4 investigates answer to what types of sources of information students use for their higher education (university study), focusing in detail on the availability of mobile devices for students.

2.2 Research Sample

Four research samples were included in the case study, i.e. totally 1,408 participants, as displayed in table 1.

Table 1. Amounts of respondents

Research	Started research	Finished research
1	772	678
2	400	324
3		203
4		203
Total		1,408

2.3 Research Methods

Data were collected from pedagogical experiments (research questions 1 and 2) and by questionnaires (research questions 3 and 4) and processed by NCSS2007 statistic software applying t-test, Mann-Whitney and Kolgomorov-Smirnov tests and frequency analyses for data collected by questionnaires.

In research 1 the pedagogical experiment was applied with didactic pre-test detecting the entrance knowledge, post-test of learners´ final knowledge after the instruction enhanced/non-enhanced by ICT and retention test after three months (post-test2) in three profile subjects (Database Systems, Management, IT English), interview and observation [12].

In research 2 the pedagogical experiment with didactic pre-test and post-test was applied in three groups as follows:

- Students in the experimental group 1 (group LCI) were offered such study materials, exercises, assignments, ways of communication and other activities which suit their individual learning styles. The selection was made electronically by an application which automatically generates the "offer", i.e. it provides each student with types of materials appropriate to his/her learning style.
- Students in experimental group 2 have access to all types of materials (CG – content general) and the process of selection is the matter of individual decision (group CG).
- Students in the control group 3 (group K) study under traditional conditions, when their course is designed according to the teacher's style of instruction which they are made to accept.

The process of instruction in all groups was tracked by the LMS [13].

In researches 3 and 4 the questionnaire consisting of 12 statements was applied. Respondents expressed their experience, opinions and evaluation on the six-level scale (1 – full agreement, 6 – total disagreement), and proposals in the form of open answers were welcomed on how to improve the current state. The monitored fields of interest were (1) to collect learners´ feedback on the possibility to use virtual desktops and (2) what mobile devices they use so that we could consider whether they are ready for cloud m-learning.

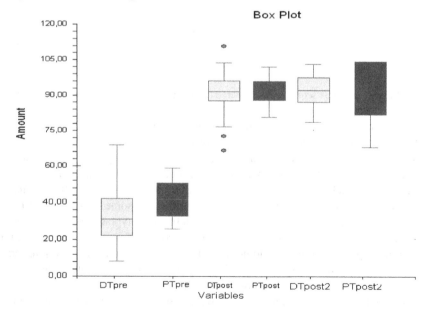

Fig. 1. Test scores in pre-test, post-test and post-test2

3   Results

In research 1, the hypothesis (*There exist statistically significant differences in learners performance in the ITC-enhanced process of instruction compared to the traditional one*) was falsified, as test scores detected in the experimental and control groups were very close (figure 1, DT: ICT-enhanced instruction, PT: non-ICT-enhanced instruction; pre: pretest; post: posttest).

In research 2, both hypotheses (***H1:*** *Students reach higher increase in knowledge if the process of instruction is adjusted to their learning style (LCI group 1) in comparison to the process reflecting teacher´s style of instruction (group K) and* ***H2:*** *Students reach higher increase in knowledge if they can study independently using all types of provided study materials (CG group) in comparison to the process reflecting teacher´s style of instruction (group K)*) were falsified as the differences in test results were not statistically significant (figure 2, dif K versus LCI, dif K versus CG).

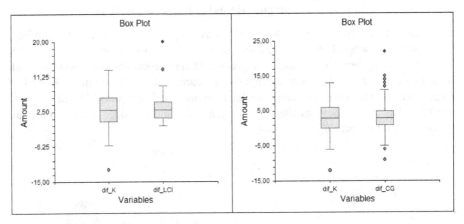

Fig. 2. Differences in test scores in K compared to LCI groups and K compared to CG groups

Results in both pedagogical experiments in researches 1 and 2 led us to the conclusion the statistically significant contribution of ICT applied in the process of instruction at the Faculty of informatics and Management, University of Hradec Kralove, was not proved. The reason might be that research participants are students of IT study programs (bachelor and master Applied Informatics and Information Management study programs). Being of 20 – 25 years old (with some exceptions of older students in part-time programs), they can be defined as typical digital natives or e-/i-generation who consider the use of ICT in the process of instruction standard, not any type of support. They mostly do not need to be trained in the use of ICT for education purposes, having a high level of intuition developed and professional knowledge and skills under the development. Above all, research 2 proved that despite they had individual learning preferences, they were flexible enough to get knowledge, either the process of instruction was tailored to their preferences, or not.

In research 3, learners´ feedback was collected after two-year long (i.e. four-semester) use of virtual desktops. Users expressed their high satisfaction with this approach calculated as 1.8 value from six-level scale (1 – full agreement, 6 – total disagreement, as described above). Despite some technical problems appeared at the beginning evaluated as 2.3 in the first semester, final value reached 1.4 value.

These results, been supported by high quality equipment and conditions at FIM (hardware, software, support etc.) and continuous training of teachers lecturing other than IT-relating subjects (e.g. English, German, Math, Accounting, Management, Psychology, Ethics etc.) entitled us to approach to wider use of ICT in education. Following latest trends, mobile devices were implemented in the concept. The pilot project using LMS Blackboard Mobile Learn™ version 4.0 started in 2013/14. This version supports iOS6+, i.e. iPhone 3GS, iPad 2+, IPad mini, iPod Touch 4+ and Android OS 2.3+) was carried in February – May 2014. Mobile devices enhanced independent foreign language learning (English for IT students), as it was considered natural from the learners´ point of view. Parts of learning content in IT English courses were available for students who mainly used smartphones and tablets for this purpose. Before this pilot project started, research 4 on availability of mobile devices and their use for personal and educational purposes was carried out. The collected data showed:

- Sources of information respondents use within university study: online courses (e-subjects) in LMS are the most frequently used source. (92 % of respondents), personal attendance of lectures (85 %), materials from the Internet for free (77 %), materials from the FIM web page (there is a file with study texts where teachers uploaded the materials before online courses were widely used, and numerous teachers still use both this way and the online courses) and discussion groups (both 72 %), Wikipedia (42 %), half of students borrow textbooks (53 %), but only one third buys them (31 %), Facebook is used by more than half of respondents (58 %), Google+ and LinkedIn are not frequently used (11 %, 1 %).
- Mobile devices respondents own: notebooks (88 %), TV (67 %), smartphones (61 %), mobile phones (52 %), PC (52 %), mp3 player (49 %), DVD player (39 %), radio (30 %), Hi-fi (27 %), tablets (24 %), game console (13 %), netbooks (10 %).
- Mobile devices respondents use for entertainment: notebooks (81 %), TV (54 %), smartphones (49 %), mp2 player (29 %), mobile phone (28 %), radio (18 %), DVD player and tablet (16 %), HI-FI (12 %), game console (10 %), netbook (7%).
- Mobile devices respondents use for private communication: notebook (79 %), smartphone (59 %), mobile phone (55 %), PC (45 %), tablet (13 %), netbook (8 %); and personal contact (96 %).
- Mobile devices respondents use for university study (in all subjects, for all activities except searching for information): notebook (87 %), smartphone (43 %), PC (42 %), mobile phone (18 %), tablet (18 %), netbook (7 %).

Data are summarized in figure 3.

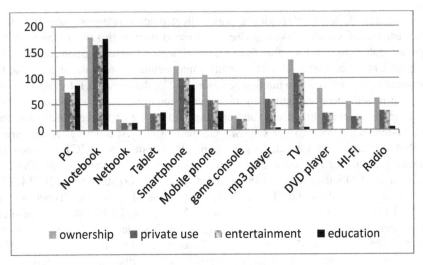

Fig. 3. Mobile devices used for various purposes

4   Cloud M-learning Concept

Two features are clearly seen from these data: (1) mobile devices are widely used by students and (2) as they are competent in private use, it will be easier (if possible at all) to learn how to use them for education. Together with results in researches 1 and 2, the following concept of cloud m-learning was designed:

- Lectures will be read in the classroom. Students can attend them either face-to-face, watch them online or later on.
- For seminars students will be divided into several groups of 20. They can work either form laboratory in the FIM building, or to use the virtual desktop services and participate from home or any other place and time.
- This choice can be approached from non-portable computer, notebook, netbook, tablet and smartphone. Consultations with teacher can be carried both personally and in online/offline communication.

This concept will be used in the subject of Database Systems II, were approximately 150 students are enrolled.

5   Conclusions

Information and communication technologies help substantially in education (despite not proved in knowledge increase as in our researches, but they definitely support motivation and make approach to learning easier and open [14].

M-learning, based on learners´/teachers´ ownership of mobile devices and mastering skills on how to use them for education purposes, including their willingness to cover relating financial requirements for services used, is a natural solution for the net generation of digital natives [2] and generation C [1] as it puts together favorite learning aims (i.e. mobile devices) and methods of constructivism, connectivism, collaborative active learning and others.

Numerous surveys have been published but didactic-related ones are still missing. So, it is the challenge for future research activities. As Benson and Morgan state, mobile devices have become a default way to accessing the Internet [15]. The proliferation of smart mobile devices and easily accessible Internet have had a significant impact on higher education (HE). Virtual Learning Environments (VLEs) have been deployed by HE institutions. Advanced capabilities of mobile devices and better connectivity have led to a growing number of the student population accessing VLEs through their smart phones, downloading and working with lecture slides on iPads instead of paper notes, participating in VLE discussion forums through their phones, etc. With the growing number of mobile devices in the hands of the younger population, it is only a matter of time before HE students will be expecting wireless access to learning materials to complement and/or replace current Internet-based VLEs.

Above all, ethical problems relating to the use of mobile devices and cloud computing – these are the hot topics as well. The ethical feeling of non-/contacting the teacher anytime anywhere is connected to general behavior of the digital natives and their good manners. And, they, being allowed to feel free from childish age, disrupt this gentle border line easily. Despite this and as most results show, vast majority of both students and teachers are ready for efficient use of mobile devices within cloud computing.

Acknowledgment. The paper is supported by the SPEV project.

References

1. Pickett, P.: Who Is Generation C? Characteristics of Generation C. How Can You Categorize These Digital Natives? About.com (2012). http://jobsearchtech.about.com/od/techindustrybasics/a/Generation_C.htm
2. Prensky, M.: Digital natives, digital immigrants. On the Horizon. MCB University Press, vol. 9, No. 5, (2001)
3. National Institute of Standards and Technology (NIST), NIST Cloud Computing Program (2010). http://www.nist.gov/itl/cloud/
4. Oberer, B., Erkollar, A.: Cloud eLearning: Transforming Education Through Cloud Technology: Preliminaries for Generation C. In: Lam, P. (ed.) Proceedings of the 7th International Conference on eLearning. Chinese University of Hong Kong, Hong Kong (2012)
5. Awodele, O., Kuyoro, S.O., Adejumobi, A.K., Awe, O., Makanju, O.: Citadel ELearning: A New Dimension to Learning System. World of Computer Science and Information Technology Journal (WCSIT) 1(3), 71–78 (2011)
6. Bondarev, V., Ossyka, A., Mghawish, A.: Integrated Environment for Developing Learning Lectures. International Journal of Computer Science & Information Technology (IJCSIT) 3(6), 119–128 (2011)

7. Cloud computing in education, IITE Policy Brief, UNESCO Institute for Information Technologies in Education (2010)
8. Isăilă, N.: Cloud computing in Education, Knowledge horizons – economics, vol. 6, no. 2, pp. 100–103 (2014)
9. Katz, R., Goldstein, P., Yanosky, R. Demystifying Cloud Computing For Higher Education. Educause Center for Applied Research. Research Bulletin **2009**(19), Boulder: ECAR, (2009)
10. Pieri, M., Diamantini, D.: From e-learning to mobile learning: New opportunities. In: Mohamed, A. (ed.) Mobile Learning. Transforming the delivery of education and training, pp. 183–194. AU Press, Athabasca (2009)
11. Kukulska-Hulme, A., Jones, C. The next generation: design and the infrastructure for learning in mobile and network world. In: Olofsson, A.D., Lindberg, J.O. (eds.). Informed design of educational technologies in higher education: enhanced learning and teaching, pp. 57–78 (2011)
12. Poulova, P., Simonova, I., Janecka, P. Didactic Approaches to ICT-Enhanced Teaching and Learning. In International conference on e-learning and e-technologies in education (ICEEE), p. 37–42. IEEE Press, Lodz, (2012)
13. Poulova, P., Simonova, I. On-line Process of Instruction Reflecting Learning Styles Respondents' feedback. In: IEEE Global engineering education conference (EDUCON), pp. 573–580. IEEE Press, Berlin, (2013)
14. Poulova, P., Simonova, I. Didactic Reflection of Learning Preferences in IT and Managerial Fields of Study. In: 12th International conference on information technology based higher education and training (ITHET 2013). IEEE Press, Antalya, (2013)
15. Benson, V., Morgan, S.: Student experience and ubiquitous learning in higher education: Impact of wireless and cloud applications. Creative Education **4**(8), 1–5 (2013)

Granular-Rule Extraction to Simplify Data

Reza Mashinchi[1], Ali Selamat[1,3(✉)], Suhaimi Ibrahim[2], and Ondrej Krejcar[3]

[1] Faculty of Computing, Universiti Teknologi Malaysia, 81310 Johor Baharu,
Johor, Malaysia
r_mashinchi@yahoo.com, aselamat@utm.my
[2] Advanced Informatics School, Universiti Teknologi Malaysia,
54100 Kuala Lumpur, Malaysia
suhaimiibrahim@utm.my
[3] Faculty of Informatics and Management, Center for Basic and Applied
Research, University of Hradec Kralove, Rokitanskeho 62,
500 03 Hradec Kralove, Czech Republic
ondrej@krejcar.org

Abstract. Granulation simplifies the data to better understand its complexity. It comforts this understanding by extracting the structure of data, essentially in big data or cloud computing scales. It can extract a simple granular-rules set from a complex data set. Granulation is associated with theory of fuzzy information granulation, which can be supported by fuzzy C-mean clustering. However, intersections of fuzzy clusters create redundant granular-rules. This paper proposes a granular-rules extraction method to simplify a data set into a granular-rule set with unique granular-rules. It performs based on two stages to construct and prune the granular-rules. We use four data sets to reveal the results, i.e., wine, servo, iris, and concrete compressive strength. The results reveal the ability of proposed method to simplify data sets by 58% to 91%.

Keywords: Fuzzy information granulation · Fuzzy C-mean clustering · Membership function · Simplification · Granular-rule

1 Introduction

Granulation simplifies information as human mind does. Mind maps the real world complexity into simpler computable parts. Granulation approaches the same way as an art of problem solving [1]. The world is surrounded with flood of information, and concretely, we need intelligent systems to encounter it. L.A. Zadeh [2] addresses them as information revolution and intelligent system revolution. To reach the mind model, information needs granular representation, and conversely, granular information requires intelligent systems to compute granularity. In other words, granular computing survives both information and intelligent systems.

Theory of information granulation (TIG) plays a key role in granulation. Since theory of fuzzy information granulation (TFIG) established the TIG [2-4], various approaches have been developed under the model of granular computing [5, 6]. Besides TFIG, abstraction and theory of hierarchal planning study the granulation.

© Springer International Publishing Switzerland 2015
N.T. Nguyen et al. (Eds.): ACIIDS 2015, Part II, LNAI 9012, pp. 421–429, 2015.
DOI: 10.1007/978-3-319-15705-4_41

Theory of abstraction [7] processes information to show the relevant and irrelevant details. It formulates similarly to the rough sets theory [8]; however, it represents the necessary terms of granulation. It reflects the understanding and representing the world in humans based on grain seizes, and it abstracts information as their current interests. Theory of hierarchal planning associates with studies on granulating the plan [9] and theory of quotient space. The latter is based on hierarchal description and problem representation [10], which visualizes the granulation as a structural problem-solving approach.

The core idea of granulation is simplification; however, it creates ambiguity. As show in Figure 1, the intersected areas create ambiguity as every data point may belong to any fuzzy granules. Consequently, the granular-rule extraction comes with redundant granular-rules. This paper proposes a method to prune the redundancies.

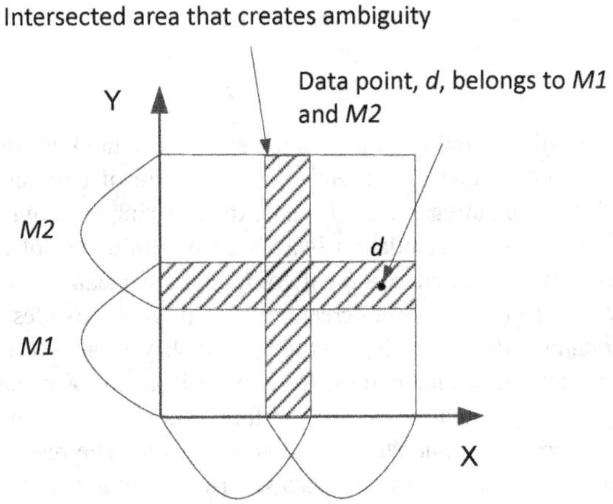

Fig. 1. Ambiguity areas in fuzzy granulation

The organization of this paper is as follows. Section 2 explains the granulation, and subsection 2.1 distinguishes the fuzzy information granulation. Section 3 introduces the proposed fuzzy granulation method that extracts the granular-rules. Section 4 gives the results, and Section 5 reaches the conclusion.

2   Granulation

Granulation is a decomposition of the whole to parts based on indistinguishability, similarity, proximity, or functionality. An aim of granulation is simplification, where in turn, it ignores the extras. It is similar in human, when one encounters flood of information [11-13].

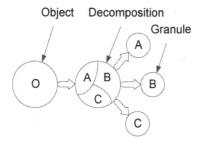

Fig. 2. A general model for information granulation

Fig. 2 shows the general concept of granulation that can be supported by clustering mechanisms. Equation 1 defines the granulation process given by [14].

$$X \rightarrow (\varphi_1, \varphi_2, \varphi_i) \tag{1}$$

Where X is data set and φ_i is information granule. Each granule is a tangible entity that has well-defined semantics. This definition gives the general scheme; though, granulation involves several models, e.g., interval analysis, random sets, probabilistic sets, and rough sets. However, fuzzy set, rough set, and quotient space widely appeals in the information granulation. The model of fuzzy information granulation is based on fuzzy set theory of information granulation. This model is used for two reasons: its trend in the literature, and, the ability of fuzzy modeling to hybrid with other computational models.

2.1 Fuzzy Information Granulation

The fuzziness is the specification of human to granulate the information. Fuzzy information granulation decomposes an object to parts based on fuzzy measurements. Fundamental concepts of linguistic variables and if-then rules established it to spark the granular computing. The main studies associate with classification and clustering fields. Studies related to classification describe the classifiers as rules to express the corresponding fuzzy granules. To name a few, Zhou et al. [15] gives a general framework for classification rule mining; Yao et al. [16] introduce the classification rules by granular computing; and, Yang and Fusheng [17] apply fuzzy information granulation to machine learning for the classification problem. Studies related to clustering applied fuzzy clustering. To name a few, Pedrycz et al. [18] encode and decode the numeric data using fuzzy clustering to obtain granules; Panoutsos et al. [19] study the granular feature to linguistic interpretability of fuzz logic rule base; and, Kwak et al. [20] apply fuzzy granulation to design linguistic model realized by context-based fuzzy C-mean clustering. In addition, studies related to evolutionary computation applies evolutionary optimization to define the granules. Reformat and Pedrycz [14] aims at covering the most of data point, while defining the size of granules as small as possible. In this paper, we propose granulation method based on fuzzy information granulation by applying the fuzzy C-mean clustering; while the granulation extracts the granular-rules equivalent with original data set.

3   Proposed Fuzzy Granulation Method

This section proposes the granulation method. It transforms the crisp data set into fuzzy granular-rule set – henceforth, we denote it as G-rule. The algorithm performs granular rule extraction based on fuzzy clusters obtained by fuzzy C-mean clustering algorithm. It includes three main phases as shown in Fig. 3.

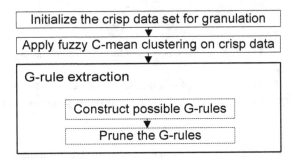

Fig. 3. Proposed granulation method

First, it receives the fuzzy clusters as granules; and second, it constructs all possible G-rules from the crisp data set. The latter stage finds the possible rules based on intersected parts of the granules. Third, it eliminates the repeated and the multi-output G-rules. The G-rule extraction stage includes two steps as shown in Fig. 4.

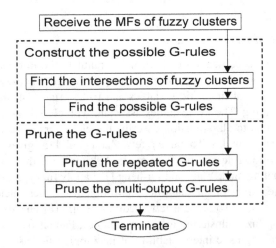

Fig. 4. Steps of extracting the G-rules

Fig. 4 shows the steps of constructing and, then, pruning the G-rules. The first step derives possible G-rules from crisp data by finding every crisp data within the intersecting parts of fuzzy clusters. Then, the second step prunes possible G-rule by eliminating the extra G-rules. It eliminates two types of G-rules as cross-output and repeated. Fig. 5 and Fig. 6 give the algorithms of first and second steps, respectively.

Function: Finds the intersections between fuzzy clusters
Input: Fuzzy clusters
Output: G-rules in intersected area between fuzzy clusters
Begin
$Inter_{f.clust}$ = FIND_intersetecd.fuzzy.clusters ( );
While ($Inter_{f.clust}$)
    $Inter_{area}$ = FIND_intersected.area ( );
End
//Find the possible G-rules//
$Data_{shared}$ = FIND_data_in_$Inter_{area}$ ( );
While ($Data_{shared}$)
    Switch ($Data_{shared}^i$)
        Case 1: $Data_{shared}^i \in Cluster_c$
            Possible_G-rules = CONSTRACT_G-rule ($Cluster_c$);
        Case n: belongs to cluster_n
            Possible_G-rules = CONSTRAC_G-rule ($Cluster_n$);
        Case (n+1): belongs to
            Possible_G-rules = CONSTRAC_G-rule ($Cluster_c$ to $Cluster_n$);
    End
    i = i ++;
End

Fig. 5. Algorithm of constructing possible G-rules

Function: Prune the G-rules
Input: Constructed G-rules of intersected areas
Output: Unique G-rules of intersected areas
Begin
While (Possible_G-rule)
    //Prune the repeated G-rules//
    $Grule_{repeated}$ = FIND_repeated.G-rule ( );
    For (i = 1 to (n − 1))
        KILL ($G.rule_{repeated}^i$);
    End
    //Prune the multi-output G-rules//
    $Grule_{multi}$ = FIND_multi_output_G-rules ( );
    $Freq_{most}$ = FIND_highest_frequency_of_each_multi-output_G-rule ( );
    For (j = 1 to n)
        If ($freq_{multi}^j < freq_{most}^j$) then KILL ($G.rule_{multi}^j$);
    End
End

Fig. 6. Algorithm of pruning the G-rules

The first step (Fig. 5) finds the intersected parts of the clusters; and then, it finds the crisp data that belong to the intersecting parts. First, it loads the fuzzy clusters from fuzzy C-mean clustering using the alpha cuts, and then it evaluates the status of each crisp data to its corresponding cluster. A data intersects if it belongs to more than one cluster. In other words, two clusters contain shared data when they overlap. Once the intersected data are found, then it constructs the possible G-rules by associating every shared data to belonging clusters.

The second step (Fig. 6) prunes the redundant G-rules. First, it eliminates the repeated G-rules; and second, it eliminated the multi-output G-rules by surviving the most frequent G-rules. The frequently of every multi-output G-rule is the repetition of its input part. Thus, the most repeated G-rules are survived by highest frequency. Consequently, it eliminated the remained G-gules from the rule base. Eventually, the extracted G-rule base has been originated from the crisp data set.

4   Results

This section gives the result of proposed granulation method. We performed fuzzy C-mean clustering over the following four data sets: wine, servo, iris, and concrete compressive strength (C.C.S) – from University California Irvine (UCI) repository (http://archive.ics.uci.edu/ml). For wine, servo, and iris data sets, we limited the number of variable to five of them to conform the representation of results as given in Table 4 in appendix.

We applied fuzzy C-mean clustering based on two clusters to initiate the proposed granulation method (see Fig. 3). Table 1 gives the MFs of obtained clusters represented in triangular fuzzy numbers $\tilde{A}$ as $(\alpha_1, \alpha_2, \alpha_3)$. Each cluster is, then, used to perform the G-rule extraction.

Table 1. Fuzzy clusters to build the G-rules

Data set	Variable	Cluster1	Cluster2
Wine	1	(11.41,12.42,16.63)	(12.42,16.63,14.38)
	2	(0.74,1.9,3.9)	(1.9,3.9,5.8)
	3	(1.71,2.15,2.54)	(2.15,2.54,3.22)
	4	(70,94,113)	(94,113,164.11)
	5	(0.89,2,2.8,5)	(2,2.8,3.88)
Servo	1	(1,3,4)	(3,4,5)
	2	(1,3,4)	(3,4,5)
	3	(3,5,6)	(5,6,6)
	4	(1,2,4)	(2,4,5)
	5	(0.1312,1.3,3.7)	(1.3,3.7,7.1001)
Iris	1	(4.5,5.2,6.6)	(5.2,6.6,7.9)
	2	(2.8,3.5,4.4)	(3.5,4.4,4.4)
	3	(1.6,5,6.9)	(5,6.9,6.9)
	4	(0.2,1.8,2.5)	(1.8,2.5,2.5)
	5	$\tilde{1}$	$\tilde{2}$
C. C. S.	1	(102,214.9,387)	(214.9,387,540)
	2	(0,24,172.6)	(24,172.6,359.4)
	3	(0,24.5,142.3)	(24.5,142.3,200.1)
	4	(121.8,161.9,188)	(161.9,188,247)
	5	(0,1.9,10.8)	(1.9,10.8,32.2)
	6	(801,916,1046.9)	(916,1046.9,1145)
	7	(594,690,824)	(690,624,992.6)
	8	(1,56,270)	(56,270,365)
	9	(2.33,23.35,51.86)	(23.35,51.86,82.6)

We performed the proposed granulation in two parts. First, we constructed possible G-rules by finding the intersected part between two MFs for each variable. Second, we eliminated the repeated G-rules, and the G-rules with multiple outputs. Table 2 gives the results; where, 1646 G-rules were eliminated in wine, servo, and iris data sets, and, 44263 G-rules were eliminated in C.C.S data set. We computed the elimination rate, and we revealed that the pruning procedure decreases the number of G-rules for at least 98%. Each rule is, then, represented in linguistic terms as given in Table 3. Respectively, H and L indicate the high and low representations.

Table 2. Number of G-rules before and after purining

G-rules	Data set			
	Wine	Servo	Iris	C. C. S.
Possible	1662	1662	1030	44487
Pruned	16	16	16	225

Table 3. Linguistic G-rules after prunning, with L for low and H for high

G-rule	Data set			
	Wine	Servo	Iris	C. C. S.
1	HLLHL	HLLHL	LLLLM	225 G-rules
2	HLHHL	HLHHL	LLLHM	
3	LLLLL	LLLLL	LHLLL	
4	LLLHL	LLLHL	LHHLL	
5	LHLLL	LHLLL	LHLHM	
6	LHLHL	LHLHL	LHHHM	
7	HLLLL	HLLLL	HHLLM	
8	HHLLL	HHLLL	HHLHM	
9	HHLHL	HHLHL	HHHLM	
10	LLHLL	LLHLL	HHHHH	
11	LLHLH	LLHLH	HLLLM	
12	LHLHL	LHLHL	HLLHM	
13	LHHHL	LHHHL	LLHLM	
14	HLHLL	HLHLL	LLHHM	
15	HHHLL	HHHLL	HLHLM	
16	HHHHL	HHHHL	HLHHM	

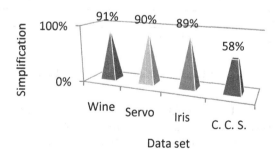

Fig. 7. Impact of granulation to simplify the crisp data

R. Mashinchi et al.

We computed the ability of simplification in Fig. 7 to summarize the results of the proposed method. We compared the number of obtained G-rules against the number of data in each data set. It shows that wine, servo, iris, and C.C.S. respectively consist of 178, 167, 150 and 1030 data, and concretely 16, 16, 16, and 255 G-rules. Therefore, we observe that the proposed granulation method significantly simplifies the crisp data sets.

We can assert that the proposed graduation method potentially abate the complexity of big data, as it significantly simplifies the data set. Fuzzy C-mean clustering in the proposed method initiates the simplification by finding the interconnections between data; and concretely, we can assert that the proposed granulation extracts the structure of big data. This is an essential to encounter big data, or could computing scale, to understand the data prior to hardware extensions; which in result, it reduces the processing costs.

5   Conclusion

This paper proposed a granulation method to extract a set of granular rules from a crisp data set. Simplification of data was the core idea to the proposed method along with the theory of fuzzy information graduation (TFIG). The proposed method consists of two stages, with two steps in each stage. The first stage constructs all granular rules by: finding the intersection of fuzzy clusters, and finding possible granular rules. The second stage prunes the extras from the constructed granular rules by: pruning the repeated granular rules, and pruning multi-output granular rules. We applied the proposed method on following data sets: wine, servo, iris, and concrete compressive strength. We performed fuzzy C-mean clustering (FCM) with two clusters to initiate the granular-rule extraction. The results were represented in linguistic form, and, it showed significantly decreases the number of extracted granular rules in compared with number of data in original data set. This shows the impact of proposed method to simplify data; where, it could simplify data from 91% to 58% for wine, servo, iris, and concrete compressive strength. One can increase the rate of simplification in future works by investigating the optimal size of granules.

A   Appendix

Table 4. Attributes of each data that are used in this paper

Data set	Variables				
Iris	Sepal length	Sepal width	Sepal petal length	Petal width	Classes
Wine	Alcohol	Malic acid	Ash	Magnesium	Flavanoids
Servo	Amplifier	Motor	lead screw/nut	Sliding carriage	Classes

Acknowledgements. The Universiti Teknologi Malaysia (UTM) and Ministry of Education Malaysia under research university grants 00M19, 01G72 and 4F550 are hereby acknowledged for some of the facilities that were utilized during the course of this research work. This work and the contribution were also supported by project "Smart Solutions for Ubiquitous Computing Environments" FIM, University of Hradec Kralove, Czech Republic.

References

1. Yao, Y.: The art of granular computing. In: Kryszkiewicz, M., Peters, J.F., Rybiński, H., Skowron, A. (eds.) RSEISP 2007. LNCS (LNAI), vol. 4585, pp. 101–112. Springer, Heidelberg (2007)
2. Zadeh, L.A.: Toward a theory of fuzzy information granulation and its centrality in human reasoning and fuzzy logic. Fuzzy sets and systems **90**, 111–127 (1997)
3. Zadeh, L.A.: Outline of a new approach to the analysis of complex systems and decision processes. IEEE Transactions on Systems, Man and Cybernetics, 28–44 (1973)
4. Zadeh, L.A.: Toward a theory of fuzzy systems (1969)
5. Reza Mashinchi, M., Selamat, A.: An improvement on genetic-based learning method for fuzzy artificial neural networks. Applied Soft Computing **9**, 1208–1216 (2009)
6. Sanchez, M.A., Castillo, O., Castro, J.R.: Information granule formation via the concept of uncertainty-based information with Interval Type-2 Fuzzy Sets representation and Takagi–Sugeno–Kang consequents optimized with Cuckoo search. Applied Soft Computing
7. Giunchiglia, F., Walsh, T.: A theory of abstraction. Artificial Intelligence **57**, 323–389 (1992)
8. Pawlak, Z.: Rough sets. International Journal of Computer & Information Sciences **11**, 341–356 (1982)
9. Knoblock, C.: Generating abstraction hierarchies: An automated approach to reducing search in planning, vol. 214. Springer (1993)
10. Zhang, B., Zhang, L.: Theory and applications of problem solving. Elsevier Science Inc. (1992)
11. Mashinchi, M.H., Mashinchi, M.R., Shamsuddin, S.M.H.: A Genetic Algorithm Approach for Solving Fuzzy Linear and Quadratic Equations. World Academy of Science, Engineering and Technology **28** (2007)
12. Mashinchi, M.R., Mashinchi, M.H., Selamat, A.: New Approach for Language Identification Based on DNA Computing. BIOCOMP, pp. 748–752 (2007)
13. Mashinchi, M.H., Orgun, M.A., Mashinchi, M.R.: A least square approach for the detection and removal of outliers for fuzzy linear regressions. In: 2010 Second World Congress on Nature and Biologically Inspired Computing (NaBIC), pp. 134–139. IEEE (2010)
14. Reformat, M., Pedrycz, W.: Evolutionary optimization of information granules. In: IFSA World Congress and 20th NAFIPS International Conference, 2001. Joint 9th, vol. 4, pp. 2035–2040. IEEE (2001)
15. Shi, Z.-H.Z.Y.J., Chen, F.: A general neural framework for classification rule mining. International Journal of Computers, Systems, and Signals **1**, 154–168 (2000)
16. Yao, J.T., Yao, Y.Y.: Induction of classification rules by granular computing. In: Alpigini, J.J., Peters, J.F., Skowron, A., Zhong, N. (eds.) RSCTC 2002. LNCS (LNAI), vol. 2475, pp. 331–338. Springer, Heidelberg (2002)
17. Li, Y., Yu, F.: Optimized fuzzy information granulation based machine learning classification. In: 2010 Seventh International Conference on Fuzzy Systems and Knowledge Discovery (FSKD), vol. 1, pp. 259–263. IEEE (2010)
18. Pedrycz, W., de Oliveira, J.V.: A development of fuzzy encoding and decoding through fuzzy clustering. IEEE Transactions on Instrumentation and Measurement **57**, 829–837 (2008)
19. Panoutsos, G., Mahfouf, M., Mills, G.H., Brown, B.H.: A generic framework for enhancing the interpretability Of granular computing-based information. Intelligent Systems (IS), 2010 5th IEEE International Conference, pp. 19–24. IEEE (2010)
20. Kwak, K.-C., Pedrycz, W.: A design of genetically oriented linguistic model with the aid of fuzzy granulation. 2010 IEEE International Conference on Fuzzy Systems (FUZZ), pp. 1–6. IEEE (2010)

Wireless Positioning as a Cloud Based Service

Juraj Machaj[(⊠)] and Peter Brida

Faculty of Electrical Engineering, Department of Telecommunications and Multimedia,
University of Zilina, Univerzitna 1, 010 26 Zilina, Slovakia
{juraj.machaj,peter.brida}@fel.uniza.sk

Abstract. In the recent time seamless positioning in hybrid environment is becoming hot topic for large number of researchers. In this paper we will present approach for seamless positioning using modular positioning system implemented as a cloud service. The idea of the proposed system is to provide position estimates in both indoor and outdoor environments by using different positioning modules. Currently we have implemented three positioning modules – GPS module, Wi-Fi module and GSM module. Each of the modules have different role in the system. GPS should provide positioning information mainly in outdoor environment, Wi-Fi module should be used mainly in indoor environment and GSM module should provide position estimates in case that GPS and Wi-Fi positioning modules are not able to estimate position of the device. For example due to low number of received signals or poor quality of signals. In the proposed system positioning GSM and Wi-Fi modules utilize fingerprinting approach for the position estimation.

Keywords: Seamless positioning · Localization · Fingerprinting · Wi-Fi · GSM

1   Introduction

In the past years many different positioning systems emerged. This is mainly caused by the rising popularity of smart devices and increased development of Location Based Services [1]. Lately situation in providing position information becomes more complex and traditional GNSS based positioning is not sufficient any more, since users are looking to use LBS in various environments e.g. dense urban environment and indoor environment. Problem with GNSS positioning systems arise especially in indoor environment, where radio signals from satellites propagates thru the walls and thus their power is extremely low. Another phenomenon negatively affecting GNSS positioning systems in complex environments is multipath propagation, which can cause significant errors in position estimates.

Due to the fact that GNSS systems cannot be used in the indoor environment, new positioning systems based on radio networks emerged in the last few years. In these positioning systems signals from radio networks deployed in the buildings are used to estimate the position of a mobile device in the indoor environment. These systems can utilize different wireless networks technologies like ZigBee [2], Bluetooth [3,4], Wi-Fi [4-7] and GSM [8-11]. However each of these technologies has its pros and cons.

© Springer International Publishing Switzerland 2015
N.T. Nguyen et al. (Eds.): ACIIDS 2015, Part II, LNAI 9012, pp. 430–439, 2015.
DOI: 10.1007/978-3-319-15705-4_42

In our work we decided to implement to our system positioning based on Wi-Fi and on GSM. Main reasons to implement positioning based on these two technologies are their ubiquitous deployment and the fact that receivers for these networks are implemented almost in all new smart devices.

In this paper we will propose the modular localization system which will be implemented as a cloud service from the user point of view. In the proposed system we utilize fingerprinting approach where calibration measurements are stored in the database and used to estimate the position of mobile device based on comparison of signals in the radio map database and signals measured by the device. Users only need internet connection and to install application on their devices. No other modifications to the device are needed to use the positioning service. This means that our implementation can be described as Software as a Service (SaaS) cloud implementation.

The rest of a paper will be divided as follows; in the Section 2 the related work in the area of positioning and basics of fingerprinting positioning approach will be presented. In Section 3 the implementation of proposed positioning system will be described. Experimental scenarios used for testing of the proposed positioning system will be described and achieved results will be discussed in Section 4. Section 5 will conclude the paper and provide some information about future work.

2   Related Work and Fingerprinting Positioning

2.1   Positioning in GSM Networks

Currently the GSM network is ubiquitously deployed over the world and thus can be assumed as a basic communication infrastructure for wireless communication. Based on this fact it seems to be ideal candidate for use in wireless positioning systems. The simplest algorithm used position estimation in the GSM network is CoO (Cell of Origin). In this algorithm position of mobile device is estimated as the position of the BTS (Base Transceiver Station) with the highest received signal power. Advanced positioning algorithms used in the GSM network can be divided in three groups:

- distance based positioning,
- angle based positioning,
- GSM fingerprinting.

Algorithms from the group of distance based positioning are based on the assumption that positioning service provider knows both position of the BTS and its transmit power. In such case it is possible to estimate the distance between BTS and mobile device from the measured RSS (Received Signal Strength) using most appropriate radio signal propagation model [8].

In contrast when angle based positioning algorithms are used only information about the position of the BTS is needed. Antenna configuration has a high impact on the resolution and accuracy of measured AoA (Angle of Arrival) [9]. Further negative effects on the accuracy of the AoA measurement can be caused by NLoS (Non-Line-of-Sight) conditions and multipath propagation. Angle based positioning in combination with measurements of RTT (Round Trip Time) is widely used in the ECoO (Enhanced Cell of Origin) [10] positioning algorithm. This algorithm estimates position of the mobile device within given sector of an area covered by the BTS.

Currently the most appropriate way to provide positioning service in GSM networks seems to be use of algorithms from the fingerprinting localization framework. In the GSM fingerprinting [11] position of mobile device can be estimated by comparison of RSS values measured by the mobile device from BTSs in the area with RSS values stored at the localization server in the database called radio map. Fingerprinting framework will be described in detail in the subsection 2.3.

2.2   Positioning in Wi-Fi Networks

In the recent time deployment of Wi-Fi networks become extremely popular in both urban areas and in indoor environments. Currently almost all new smart devices are equipped with built-in Wi-Fi transmitters. Similarly to GSM network the process of positioning in Wi-Fi networks can be executed in different ways [11]. The most common frameworks for positioning based on Wi-Fi are following:

- distance or angle based positioning (multilateration),
- Wi-Fi fingerprinting.

Although more simple positioning algorithms can also be used in Wi-Fi networks, e.g. CoO (Cell of Origin) and propagation modeling; these are not popular due to low accuracy of estimated position [12]. Even thou propagation modeling is very similar to fingerprinting approach. The main difference is that radio propagation models are used to create radio map database instead of real measurements. Advantage is that complex calibration process is removed; however such system achieves significantly lower accuracy of position estimates.

Distance or angle between transmitter and receiver is used in multilateration positioning. To perform AoA measurements in Wi-Fi network it is necessary to use MIMO (Multiple-Input and Multiple-Output) technology and LoS conditions are required [13].

On the other hand it is possible to use RSS, ToA (Time of Arrival) or RTT measurements to estimate distance between transmitter and receiver [14]. When it comes to the accuracy of distance estimate, all aforementioned methods have their own drawbacks. Latencies, which falsify time stamps, can highly affect the accuracy of distance estimated using RTT measurements [15]. Typical variation of the time measurement can be 5μs which means 1500m error when converted to distance estimate. Another error source is represented by clock drift of RTT measurements [16]. When distance estimate is computed from RSS the accuracy is given mainly by the precision of propagation model used for the distance estimation [17].

In the recent time empirical fingerprinting localization framework become extremely popular in the Wi-Fi networks. The process of position estimation is the same as in GSM network. Mobile device measure RSS from APs (Access Points) in the communication range and compare the measures RSS values with data stored in radio map database in order to estimate its position. The main advantage of the fingerprinting positioning seems be to immunity to multipath phenomenon. The fact that there is no need to know exact positions of AP can be also considered as advantage when compared to lateration based positioning. Measurements of RSS in the Wi-Fi networks are performed on beacon signals; therefore measurements are not affected by the implemented adaptive power regulation.

2.3   Fingerprinting Localization Framework

In comparison to other positioning frameworks the fingerprinting seems to achieve the best accuracy. This fact is even more obvious in the areas with a strong multipath propagation and NLoS conditions. Therefore we decided to use fingerprinting positioning in our system. Operation of the fingerprinting positioning system can be divided into two independent phases – calibration phase and positioning phase [5].

2.3.1   Calibration Phase

Calibration phase, sometimes also called offline phase, represent essential step in the fingerprinting localization. It has to be performed before the deployment of positioning service in order to create radio map database. This database is commonly stored in the localization server, which is used to estimate position of mobile devices.

Localization area is divided into small cells during the calibration phase and each cell is characterized by the reference point [5] as can be seen in Figure 1. RSS values from all APs within range are measured at each reference point. Measured RSS data (fingerprints) are sent to the localization server and stored in the radio map database. Data in the database are stored as vectors that can be described as:

$$S_j = (\alpha_1,...,\alpha_{N_j} ,c_j,\theta_j) \qquad j = 1,2,...,M , \qquad (1)$$

where N_j is the number of APs heard at the j-th reference point, M is the number of reference points, α_i are RSS values, c_j represent coordinates of j-th reference point and parameter vector θ_j can contain any additional information that may be used in the localization phase.

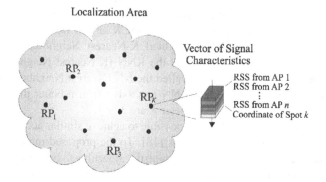

Fig. 1. Radio map principle

2.3.2   Localization Phase

Positions of mobile devices are estimated during the localization phase, in some lite-rature also called online phase. In this phase mobile device measures RSS values from all APs within the communication range and send them to the localization server. The server uses algorithm to compare measured RSS data to the fingerprints stored in the database in order to estimate position of mobile device. Algorithms that can be used for position estimation are commonly divided into two groups – deterministic and probabilistic.

In the probabilistic framework the basic assumption is that position of mobile de-vice can be modeled as a random vector. The location candidate γ is chosen if its posterior probability is the highest [5]. The decision rule uses Bayes' theorem:

$$P(c_i|S) = \frac{P(S|c_i)P(c_i)}{P(S)} , \qquad (2)$$

where posterior probability $P(c_i|S)$ is a function of likelihood $P(S|c_i)$, prior probability $P(c_i)$ and observed evidence $P(S) = \sum_i P(S|c_i)P(c_i)$, vector S represents the ob-served RSS values during online phase and c_i stands for i-th location candidate, i.e. reference point (RP).

On the other hand the basic assumptions for deterministic algorithms are that RSS values at the receiver are not random and are dependant on the position of a mobile device [9]. Thus data in the radio map that are most similar to measured data should be measured at position of a mobile device. The position estimate cam be estimated computed using:

$$\hat{x} = \sum_{i=1}^{M} \omega_i \cdot c_i \left/ \sum_{i=1}^{M} \omega_i \right. , \qquad (3)$$

where ω_i is a non-negative weighting factor [5]. Weights are commonly calculated as the inverted value of the distance between RSS vectors from the positioning phase and radio map database. The most common way to calculate the distance is use of the Euclidian distance, but other distance metrics can be used with similar results [19].

The estimator of the formula (3) that keeps the K largest weights and sets the oth-ers to zero is called the WKNN (Weighted K-Nearest Neighbours) method [10]. WKNN with all weights $\omega_i = 1$ is called the KNN (K-Nearest Neighbours) method. The simplest method, where $K = 1$, is called the NN (Nearest Neighbour) method [7]. In [5] it was found that WKNN and KNN methods perform better than the NN me-thod, particularly when values of parameter K are 3 or 4.

In the literature it is shown that it is possible to achieve similar accuracy using both probabilistic and deterministic algorithms [18]. To the proposed modular positioning system we decided to implement deterministic algorithms, because in contrast to probabilistic algorithms there is no need of accurate statistical model.

3   Proposed Positioning Service

We proposed the modular positioning system as an integrated set of components that provide localization service for its users. The proposed system is able to simultaneously provide services to multiple users. The proposed positioning system can be described as centralized, network based system with device assistance. Localization server implemented in a cloud is responsible for position estimation process. The idea is to perform all computations at the cloud server operating in the network side of the system. In order to get necessary information for position estimation process the cloud server sends request for measurements to the mobile device.

Since the proposed system communicates with the existing infrastructure without need of any modification it can be considered as fully autonomous. System is currently able to automatically switch between indoor and outdoor mode based on data measured by mobile device. Algorithm responsible for the process of switching is called Modular Localization Algorithm (MLA) [20].

The basic assumption considered in the proposed system is, that combination of GPS (Global Positioning System), GSM and Wi-Fi localization system allows ubiquitous positioning in hybrid (indoor and outdoor) environment (see Figure 2). Since standard smart devices are equipped all necessary hardware needed by these technologies we consider these as a basic modules of the system. The proposed system is easily extendible by implementation of new modules based on other technologies e.g. Bluetooth, Zig-Bee etc.

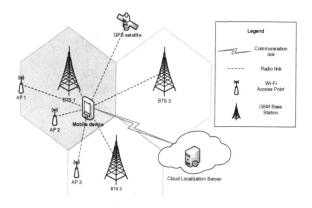

Fig. 2. Example of environment

The proposed system consists of several components with mutual communication and their own responsibilities. The basic components of the system are mobile devices, existing network infrastructure and the cloud localization server.

Localization requests and signal information from GSP, GSM and Wi-Fi are processed by the proposed MLA algorithm. The MLA algorithm has to select the most appropriate platform and return estimated position. The flowchart in Figure 3 depicts the operation of MLA algorithm in detail. To estimate the position MLA util-ize deterministic fingerprinting localization framework. Currently we have implemented NN, KNN and WKNN algorithms, which were described in the previous section.

The operation of the MLA is as follows; firstly the algorithm checks GPS availa-bility. If position can be determined using GPS, algorithms returns the position to the mobile device. However, in some environments GPS may not be available, mobile device may not have GPS hardware or there will be low number of visible satellites. In such case the algorithm utilize Wi-Fi or GSM measurements. Algorithm calculates number of transmitters with RSS 10dBm above the receiver sensitivity for both Wi-Fi (NAP) and GSM (NBTS) networks. If NAP is higher than or equal to 3, Wi-Fi based positioning is performed. We decide to set the threshold to 3 transmitters (APs or BTSs); because minimum number of signals needed to unique definition of position in 2D space is 3. If Wi-Fi based localization fails e.g. there are no similar radio map vectors due to non-mapped area or if NAP < 3, GSM based localization is performed.

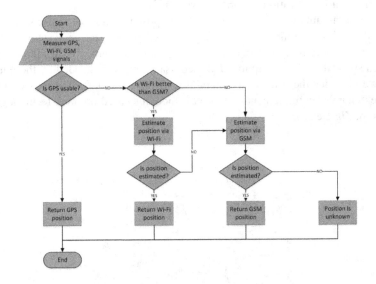

Fig. 3. Flowchart of modular localization algorithm

In the proposed algorithm the RSS thresholds were set to -90dBm for Wi-Fi signals and -103dBm for GSM signals. As stated before these thresholds are 10dB above the minimum sensitivity of devices, most devices provides minimum sensitivity of -100dBm for Wi-Fi and -113dBm for GSM signals. We decided to implement the thre-shold based on previous results published in [21]. These results show that low RSS values may have negative impact on the accuracy of position estimates. It is caused by the fact that these values are more significantly affected by RSS fluctuations.

4   Testing Scenario and Achieved Results

To evaluate performance of the proposed cloud based positioning service we performed measurements in the real world conditions. The measurements were performed in the campus of the University of Zilina in both indoor and outdoor environments. Radio map in the area was created in the area using both GSM and Wi-Fi measurements. It is important to note that the area of the campus of University of Zilina cannot be characterized as dense urban environments, thus lower number of wireless transmitters was detected.

Calibration phase was performed in the evening hours, when lower number of moving obstacles (moving people and vehicles) was presented in the area. In the indoor environment measurements were performed on 937 RPs for both GSM and Wi-Fi, however due to low number of detected signals only 547 RPs was used for Wi-Fi positioning. In the indoor environment measurements were performed at 50 RPs for GSM and 43 for Wi-Fi network. In average 5 BTS per RP and 15 APs per RP were detected in the radio map.

During the positioning phase position of mobile device was estimated in total 260 times, 177 times in the outdoor environment and 83 times in the indoor environment. During the positioning phase the GPS receiver at the mobile device was turned off to simulate worst case scenario, when no GPS signals can be received and position must be estimated by the radio network. Measurements were performed with Smartphone running on Android OS and equipped with both GSM and Wi-Fi transmitters. Achieved positioning error is shown in the Figure 4.

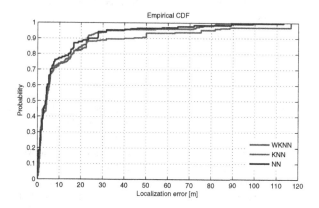

Fig. 4. CDF of positioning error in hybrid environment

From the figure it can be seen that positioning error was lower than 5m in 50% and less than 10m in more than 70% of positioning estimates for all three implemented algorithms. It can also be seen that WKNN and NN algorithms outperformed KNN algorithm, this can be seen especially for the higher probabilities. It is important to note that Wi-Fi positioning was preferred by the MLA algorithm to estimate position of mobile device. Wi-Fi signals were used for 78% position estimates in outdoor and for 51% of position estimates in indoor environment.

It is also important to note that localization errors achieved in the outdoor environment were significantly higher compared to indoor environment. In the indoor environment the highest localization error was less than 9 m. This is partially caused by the fact that campus of the University of Zilina is further from the dense urban environment and thus relatively low number of transmitters was detected.

5   Conclusion and Future Work

In the paper the modular positioning system implemented as a cloud service was proposed and described. Functionality of the proposed solution was evaluated using experiments in the real world conditions. From the achieved results it seems that proposed solution is able to provide ubiquitous positioning in hybrid environment. The accuracy of the system can be increased by using GPS positioning, especially in outdoor environment.

The main scope of the future work will be to improve the performance of proposed solution from the accuracy point of view and development of the system which will not be affected by the use of different devices. Another hot topic should be improvement of decision algorithm for modular positioning system, which will provide more sophisticated measurements of signal properties.

Acknowledgment. This work was partially supported by the Slovak VEGA grant agency, Project No. 1/0394/1 and by project: University Science Park of the University of Zilina (ITMS: 26220220184) supported by the Research & Development Operational Program funded by the European Regional Development Fund.

References

1. Cerny, M., Penhaker, M.: Wireless body sensor network in health maintenance systems. Electronics and Electrical Engineering 115(9), 113–116 (2011)
2. Xing, B., Geng, Z., Han, L., Du, S.: Intelligent alarm positioning system based on zigbee wireless networks. In: 6th International Conference on Intelligent Networks and Intelligent Systems, pp. 278–281 (2013) ISBN: 978-1-4799-2808-8
3. Chawathe, S.: Low-latency indoor localization using bluetooth beacons. In: 12th International IEEE Conference on Intelligent Transportation Systems, ITSC 2009, pp. 1–7 (2009)
4. Behan, M., Krejcar, O.: Modern smart device-based concept of sensoric networks. Eurasip Journal Wireless Communication and Networking 2013(1(155)) ISSN 1687-1499
5. Honkavirta, V., Perälä, T., Ali-Löytty, S., Piché, R.: A comparative survey of WLAN location fingerprinting methods. In: 6th Workshop on Positioning, Navigation and Communication, WPNC 2009, pp, 243–251 (2009)
6. Li, B., Salter, J., Dempster, A.G., Rizos, C.: Indoor positioning techniques based on wireless LAN. Tech. report School of Surveying and Spatial Information Systems, UNSW, Sydney, Australia (2006)

7. Saha, S., Chauhuri, K., Sanghi, D., Bhagwat, P.: Location determination of a mobile device using IEEE 802.11b access point signals. In: Wireless Communications and Networking, WCNC 2003, vol. 3, pp. 1987–1992 (2003)
8. Drane, C., Macnaughtan, M., Scott, C.: Positioning GSM telephones. IEEE Commun. Mag. 36(4), 46–54, 59 (1998)
9. Gustafsson, F., Gunnarsson, F.: Mobile Positioning Using Wireless Networks. IEEE Signal Processing Magazine 22(4), 41–53 (2005)
10. Shen, J., Oda, Y.: Direction estimation for cellular enhanced cell-id positioning using multiple sector observations. In: 2010 International Conference on Indoor Positioning and Indoor Navigation (2010)
11. Varshavsky, A., de Lara, E., LaMarca, A., Hightower, J., Otsason, V.: GSM Indoor Localization. Pervasive and Mobile Computing Journal (PMC) 3(6), 698–720 (2007)
12. Cizmar, A., Papaj, J., Dobos, L.: Security and QoS integration model for MANETS. Computing and Informatics 31(5), 1025-1044 (2012)
13. Wong, C., Klukas, R., Messier, G.: Using WLAN infrastructure for angle-of-arrival indoor user location. In: IEEE Vehicular Technology Conference, VTC 2008-Fall, pp. 1–5 (2008)
14. Muthukrishnan, K., Koprinkov, G., Meratnia, N., Lijding, M.: Using Time-of-Flight for WLAN Localization: Feasibility Study. Technical Report no. TR-CTIT-06–28, Centre for Telematics and Information Technology (CTIT), University of Twente (2006)
15. Golden, S., Bateman, S.: Sensor Measurements for Wi-Fi Location with Emphasis on Time-of-Arrival Ranging. IEEE Transactions on Mobile Computing 6(10), 1185–1198 (2007)
16. Günther, A., Hoene, C.: Measuring Round Trip Times to Determine the Distance between WLAN Nodes. Technical Report no. TKN-04–16, Telecommunication Networks Group, Technical University Berlin, p. 43 (2004)
17. Mazuelas, S., Bahillo, A., Lorenzo, R.M., Fernandez, P., Lago, F.A., Garcia, E., Blas, J., Abril, J.: Robust Indoor Positioning Provided by Real-Time RSSI Values in Unmodified WLAN Networks. IEEE Journal of Selected Topics in Signal Processing 3(5), 821–831 (2009)
18. Kriz, P., Maly, F.: Agent-based approach to community wireless network management. In: Proceedings of the 17th International Conference Computers (Part of CSCC 2013), pp. 183–188 (2013) ISBN 978-960-474-311-7
19. Machaj, J., Brida, P.: Performance comparison of similarity measurements for database correlation localization method. In: Nguyen, N.T., Kim, C.-G., Janiak, A. (eds.) ACIIDS 2011, Part II. LNCS, vol. 6592, pp. 452–461. Springer, Heidelberg (2011)
20. Brida, P., Machaj, J., Benikovsky, J.: A Modular Localization System as a Positioning Service for Road Transport. Sensors 2014(14), 20274–20296 (2014)
21. Machaj, J., Brida, P.: Impact of Wi-Fi Access Points on performance of RBF localization algorithm. ELEKTRO, 70–74 (2012) ISBN: 978-1-4673-1180-9

HPC Cloud Technologies for Virtual Screening in Drug Discovery

Rafael Dolezal, Vladimir Sobeslav[✉], Ondrej Hornig, Ladislav Balik,
Jan Korabecny, and Kamil Kuca

Faculty of Informatics and Management, University of Hradec Kralove,
Rokitanskeho 62, 50003 Hradec Králové, Czech Republic
{rafael.dolezal,vladimir.sobeslav,ondrej.hornig,
ladislav.balik,jan.korabecny,kamil.kuca}@uhk.cz

Abstract. Increasing development in computer technologies, storage capabilities, networking, and multithreading as well as accelerated progression of computational chemistry methods in past decades have allowed performing advanced virtual screening (VS) methods to assist drug discovery. By employing virtualized computer resources and suitable middleware it is possible to support growing computer demands of VS with flexible cloud services adapted for High Performance Computing (HPC). This paper resumes the possibilities of cloud services for VS applications in drug discovery and shows possible track to move VS computing software to cloud environment.

Keywords: Clouds · HPC · Virtual screening · Drug discovery · OpenStack

1 Introduction

Rapid advances in current computing technologies have offered higher effectiveness and cost-savings to many areas dealing with large-scale and data-intensive processes. Thanks to globally growing availability and decreasing price-performance ratio of computer resources, a novel paradigm of distributed task processing has actually formed as cloud computing. Besides the most widespread software cloud applications like Google Apps, Dropbox, Office 365, etc., important contribution is expected in bioinformatics, computational chemistry and similar fields where such virtualized technologies are able to deliver scalable computer performance for demanding scientific calculations. Despite attractive flexibility of cloud services, a challenging point will be to put cloud computing to the same footing with High Performance Computing (HPC) which has become a prominent tool to carry out large bioinformatic tasks. One of the most essential utilization of cloud services may be seen in rational drug discovery by virtual screening (VS) scenarios where extensive calculations gain steadily increasing proportion. Since all VS methods demand for efficient computational resources, their progress and success depend on development of computer technologies tailored to high performance computing. Amongst several lately developed cloud middleware, we focus in the present paper particularly on OpenStack implementation suitable for VS software methods.

© Springer International Publishing Switzerland 2015
N.T. Nguyen et al. (Eds.): ACIIDS 2015, Part II, LNAI 9012, pp. 440–449, 2015.
DOI: 10.1007/978-3-319-15705-4_43

Cloud solution provides to its users several benefits - it is more scalable, flexible and cost effective. It is possible to distinguish between several approaches in delivering and using cloud services. Most complex variant is Software as a Service (SaaS) approach. Cloud service delivers applications directly to users through Internet browser. That means that every application is from the user perspective used in the same environment. Nowadays there are a lot of examples of such services: Google Mail, Microsoft Exchange or Youtube video portal. Less complex, Platform as a Service (PaaS) approach is used for developer purposes, for example to code, test and deploy applications. Behind PaaS term Github or Google Code services can be imagined. Delivering of infrastructure platform, which only means to virtualize server hardware, data storage and networking equipment is usually abbreviated as IaaS (Infrastructure as a Service). This approach brings usually web-browser based orchestration of hardware sources [1].

Although more and more scientists trust to cloud providers (it is necessary to take into consideration data control and privacy), today there is negligible group that deny sharing private data with third party provider of public cloud environment. Every association that does not want to rely on cloud outside of its own administration can deploy more secure cloud environment for its own use that is what is called private cloud [2]. Ethical aspects of data storing, mining and distributing have to be accepted [3].

2 Virtual Screening Approach

In almost all fields related to drug design and discovery, computational instruments have enabled to significantly streamline processing of accumulated bio-information in order to disclose, explain and utilize the biological potential hidden in various chemical compounds. With previous experimental methods based on high-throughput screening of combinatorial chemistry libraries containing several thousand organic compounds, analogical *in silico* screening methods may currently compete to identify new drugs [5]. Provided sufficient computer resources are available and convincing theoretical background relating chemical structure and biological activity has been elaborated, VS methods can evaluate millions of molecular models in affordable time-span without need of any chemicals, microorganisms, tissues or animals (Fig. 1.).

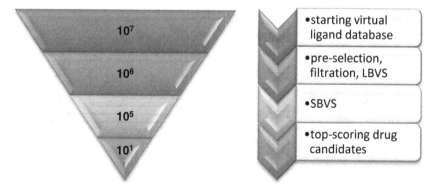

Fig. 1. Major steps involved in VS workflow. Initial database containing millions of potential drugs is systematically reduced to several promising candidates.

There are basically two approaches which may be employed to rationalize the research and development of novel drugs [6]. The first type methods try to deduce more or less general model of drug efficacy by data-mining of extent virtual databases containing information on biological activities and chemical structures of potential drugs. Particularly, inclusion of quantitative information on the biological activities is a distinctive feature of this approach. After completing supervised training process, successfully validated model can be used to estimate unknown biological activities of novel chemical structures. Consequently, the prediction of biological activity may serve to indicate the most promising chemical compound in the given therapeutical context. In case the model of structure-activity relationship is applied to screen virtual ligand databases involving tens thousands or more chemical compounds, the method is referred as ligand-based virtual screening (LBVS).

The second approach is aimed at finding such molecules which exhibit relatively strong *in silico* binding affinity to the target protein active site. This calculation method is often implemented as high-throughput molecular docking of large virtual ligand databases in powerful computer clusters. Contrary to LBVS, molecular docking does not exploit biological activities but requires knowledge of 3D structure of the target protein. Due to the necessity of the structure knowledge on the target, the approach is called structure-based virtual screening (SBVS).

Both LBVS and SBVS have been lately applied in numerous drug discovery studies. Examples of successful drug discovery projects are: three novel submicromolar antagonsits of TLR4-MD2 receptor found by a combined LBVS-SBVS approach [7], potent inhibitors of trihydroxynaphtalene reductase with antifungal activity identified by ligand-based 3D similarity searching and molecular docking to a homology-built enzyme model [8], several high affinity compounds for human β2-adrenergic receptor discovered by high-throughput pharmacophore VS and extensive molecular docking [9], and many others.

2.1   Classical HPC Solution

In past as well as in present, demanding calculations underlying VS in drug discovery have been managed especially by traditional computer clusters (Fig. 2).

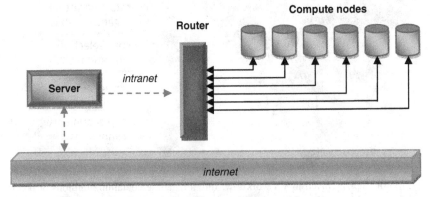

Fig. 2. Traditional computer cluster architecture

Analogical architecture has been adopted by modernized HPC resources which mostly consist of a powerful compute node domain controlled by a front-server for task scheduling. The compute nodes in HPC structure are mutually interconnected by high-speed network like Gigabit Ethernet, Myrinet or Infiniband [10]. In addition, the infrastructure is usually linked to a high performance parallel file system (e.g. Lustre, IBM's General Parallel File System) for handling data [11]. Task distribution, scheduling, queuing, controlling and monitoring over HPC clusters is managed by batch systems such as TORQUE, PBS Pro, LSF or SGE [12].

Commonly, virtualization is often introduced to HPC grid infrastructures to flexibly split individual physical computers into two or more virtual machines. According to computer resource requirements submitted to scheduler, several virtual machines deployed on physical cluster nodes may be grouped into virtual clusters to perform entered tasks as an organized unit. This scheme of virtualized clusters is mainly necessary if the submitted jobs are mutually dependent like in parallelized calculations over different nodes. Allocation of virtual clusters may be carried out by requesting standard batch system interface or through injection of node images. Until the resources asked from scheduler are not available, a task or multi-task is kept in a global queue. Further extension of HPC concept can be reached by preemption. In this case, even full machine appropriation by high priority tasks is enabled to temporarily overtake control over all local resources while suppressing actual low priority processes. Naturally, modern HPC clusters seem to be very suitable for VS jobs which are ordinarily performed distributively and/or in parallel.

Typically, many thousands of molecular docking jobs executed by programs like AutoDock Vina, FlexX, PhDock, Surflex or Glide are performed within a medium-size SBVS. To exemplify the issue, we outline how SBVS may works by a simple example. Starting point is to prepare virtual ligand database and a model of the biological target. Since large drug databases are available online (e.g. zinc.docking.org), a custom virtual ligand database can be prepared simply by downloading various molecular subsets (e.g. lead-like, fragment-like, drug-like). Thereafter, suitable format conversion and preliminary chemical calculations are due to obtain correct input for a docking program. Similarly, a target protein downloaded from online databases (e.g. www.pdb.org) is adjusted for molecular docking. Thus, sufficient input data for high-performance molecular docking involve a folder of ligand files (i.e. virtual ligand library) and a file with the target structure (e.g. enzyme, receptor, protein). Essentially, a configuration file containing parameters (size and location of gridbox, path to input files, etc.) for docking program must be designed before SBVS can start. If AutoDock Vina is utilized, the config file also assigns how many CPUs can be used to parallelize the calculation [13]. Finally, the task is translated by a script into a batch and then submitted to HPC scheduler for processing.

Simple example of a bash script for processing virtual ligand database by high-throughput molecular docking in AutoDock Vina is given below:

```
#!/bin/bash

VINA=~/vina/vina        # Executable program for molecular docking
RESULTS=~/vina/results/ # Path for program outputs
LIGANDS=~/vina/ligands/ # Virtual ligand database
ENZYME=~/vina/enzyme    # Model of the biological target
CONF=~/vina/conf.t      # Configuration file for Vina program

$RESOURCES="-l walltime=40h -l mem=10gb -l scratch=500mb -l
nodes=1:ppn=16"         # Requirements on computer resources

i=1
for a in $LIGANDS
do
    echo "$VINA --config $CONF --ligand $a --receptor $ENZYME --out
    $RESULTS" > ./job_$i.sh # Preparing job script for scheduler
    chmod +x job_$i.sh
    qsub $RESOURCES ./job_$i.sh # Submitting job to PBS interface
    i++
done
```

From the technological point of view, the most important work is entrusted to scheduler which must secure that all submitted jobs will complete on compute nodes. In the above example, a shared file storage system is assumed. Once the calculations are finished, the resulting program outputs contain the valuable information on which compound might be a promising drug.

3   Open Source Driven Cloud Computing Platform

As mentioned earlier, HPC platforms are used for exhaustive calculations submitted by scientist and engineers. HPC clusters stands for a lot of computer nodes which are usually regular computer machines networked together with one or more cluster management servers [2]. Speed of calculations depends on number of computer nodes connected to HPC environment, communication channels used, amount of data transferred via them and a capacity of management computer nodes (that means an ability to split computational task and delegate separate jobs to nodes). Every HPC cluster needs for its run a lot of resources. Firstly, organization has to hire people that take care of infrastructure software or hardware and its updates and upgrades. Secondly, it has to pay electricity bills. Electric power is consumed by machines that compute HPC jobs and also supports systems that ensure cooling, networking, etc. Another cost is also a price for purchase of computer hardware [14].

For large government medical institutions or medical companies, HPC is a suitable way. Smaller teams and so popular start-ups, which are shorter on financial sources, need to find an alternative way offering them comparable computational power without additional license, utility or labor costs. Using clouds is a good way to meet these requirements [15]. The usual approach for invoicing clients of cloud environments is called pay-as-you-go. That means user pays only for what he consumed. Formula for final invoice amount may look as in this example: number of CPU cores, amount of memory and disk storage and software licenses multiplied by hour tax and number of hours in which this infrastructure was used. This approach is fully predictable and cost effective for client organization [16].

Contemporaneous popularity of cloud computing is given by changed strategy in utilization of computer resources and software transition. While classical HPC is based more or less on rigid computer architecture which scalability is limited, clouds may integrate thousands of different physical computers and distribute complex software settings and configurations via one interface distributed by virtualization platform or cloud orchestration software.

Because VS methods are very source demanding and a lot of organizations have implemented data policy criteria which deny data of organization to leave its infrastructure (or administrative domain), administrators may take into consideration private cloud solution. In this work, the term private cloud is used to describe an on-demand infrastructure that is run and operated by organization itself.

Every virtualized platform has its own administration overhead that demand additional sources which have to be compensated by benefits from using virtualized platform, respectively cloud solution on local infrastructure. This advantage is mainly user-friendliness of cloud served services. Typical users or scientists cannot afford to take care of operating systems, firmware updates and networking at all. And that is what cloud solution does for such type of organization. A narrow circle of people takes care of infrastructure while a lot of others use the systems and software on demand and on-click.

Migrating projects between platforms (mean hardware, operating system or computational engine) usually bring challenging errors and non-standard situations. That's the place where standardization is very important. Many of commercial projects offer loads of features that no other project has (such as Microsoft Azure, Amazon Web Services or VMware vCloud Suite), but mutual compatibility is a serious problem. If organization chose provider or platform one time, it is hard to switch to another one. But on the other hand, there are some open-source platform projects that are based on standardized and open software, usually developed by user community. Recently, several open-source cloud middleware implementations have been developed (e.g. OpenStack, OpenNebula, Nimbus, Eucalyptus). OpenStack represents a robust and scalable solution deployed in datacenters worldwide [4].

3.1 Virtualized Computing Platform in OpenStack

In previous chapter we introduced usual approach to manage VS jobs on the top of high-performance computer cluster. This approach is also used in research performing institutions, such as Centre for Biomedical Research of University Hospital Hradec Králové. These institutions usually use computational resources provided to them by an organization running some kind of supercomputer or a computing grid. MetaCentrum NGI (part of Czech national academic association called CESNET) provides computational resources to number of Czech academic research institutions [17]. In the environment of MetaCentrum's computing grid, researchers are able to use the AutoDock

Vina software for VS combined with PBS Pro. AutoDock Vina software is an open-source program designed for molecular-docking simulation and PBS Pro is a variation of shell scripting environment for distributing computational jobs.

Virtualizitaion of server infrastructure (as cloud is an advanced virtualization technology with management and billing) is nowadays often used also for this type of task. The decision to move AutoDock Vina and PBS Pro to the OpenStack cloud infrastructure was made. As transformation method was chosen a virtualization on CentOS 6.5 x64 operating system with above mentioned software installed. The comparison to the same software installed and operated on real hardware was issued. VS task performed on the Metacentrum NGI computer cluster was approximately between 1.1 and 1.15 times faster than on the as much as possible same hardware configured virtual machine. Some little difference is generated by the use of different CPUs, memory and other components, but there is a significant influence of virtualization overhead which has to be taken into consideration.

For these purposes, it is not useful to virtualize only one computer. Cloud computing is based on virtualization concept connecting multiple computers. Main purpose of this approach is to provide a dynamically scalable platform and to bring a user friendly environment (Fig. 3) to deliver infrastructure services – consisting of mainly preconfigured templates for compute nodes and therefore provide a service oriented environment to the researchers.

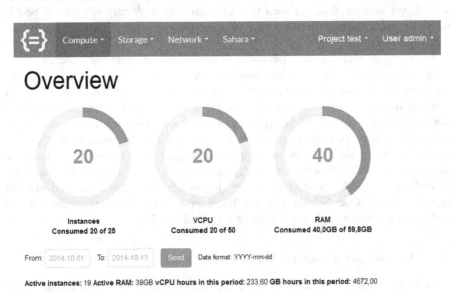

Fig. 3. OpenStack dashboard implementation in laboratory

In cloud virtualized platform, there is not a problem to generate more computing nodes connected via shared virtual network. This type of a cluster can be used as a real hardware cluster connected via classical network media and devices.

3.2   OpenStack with Sahara Plugin to Apache Hadoop Management

Future works will be focused on OpenStack implementation which includes Apache Hadoop, an ideal open-source solution for distributed computing. It uses MapReduce algorithm to split and distribute computational jobs across connected compute nodes. It also uses uncommon distributed file system, HDFS, to support large data sets usually used in this type of computation.

In our Laboratory of Computer Networks and Operating Systems we developed OpenStack (version Icehouse on Ubuntu 14.04 LTS) private cloud solution with Sahara plug-in (formerly Savanna) specially developed for Apache Hadoop management (Fig. 4). This environment is now in beta testing phase in cooperation with CBR. They prepare their first compute job to compare AutoDock Vina and PBS Pro to Hadoop and real hardware to virtualized platform of private cloud.

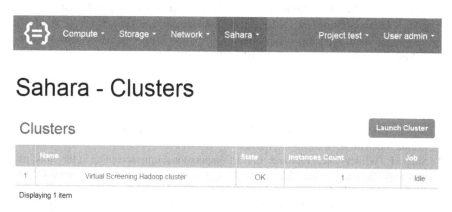

Fig. 4. OpenStack dashboard with Sahara plug-in for Apache Hadoop

Our goal is to achieve fully functional and scalable Hadoop system running on the top of OpenStack IaaS cloud solution that will be accessible to large group of biomedical scientist for their virtual screening (and also other computational) jobs. We would like to provide to them a platform where they can receive computational platform as simple and scalable as possible. Multiuser oriented environment is also our mission.

4   Conclusion

Drug discovery remains to be a risky, time-consuming and very expensive endeavor though great progress in pharmaceutical technology has been achieved in recent years. Developing of novel drug roughly takes 15 years and consumes investments about 1 billion USD. Advanced computer-aided drug discovery methods might significantly decrease the costs and save the time necessary to develop new applicable drugs. One of the practically applied methods is a virtual screening, innovative computing discipline with large requirements which can be satisfied by HPC or cloud computing.

This paper brings preview of possibilities of conversion frequently used high-performance computational tasks into cloud environment, which can bring better management tools and source dedication. Private or public cloud can be developed from open-source technologies. Usage of open-source software and its components may result in multivendor or multiprovider compatibility that is the main reason why OpenStack was chosen.

From the evaluation of beta testing and first computational tasks performed in this new environment it is obvious that user experience for bioscientists may be comparable as in MetaCentrum's computing grid environment. This aspect connected with cloud optimal workload brings advantages for both sides of the project.

Acknowledgements. Access to computing and storage facilities owned by parties and projects contributing to the National Grid Infrastructure MetaCentrum, provided under the programme "Projects of Large Infrastructure for Research, Development, and Innovations" (LM2010005), is greatly appreciated. The paper is supported by the project of specific science Smart networking & cloud computing solutions (FIM, UHK, SPEV 2015).

References

1. Kupfer, D.M.: Cloud Computing in Biomedical Research. Aviation, Space, and Environmental Medicine **83**(2), 152–153 (2012)
2. Sitter, T., Willick, D.L., Floriano, W.B.: Computational drug screening in the cloud using HierVLS/PSVLS. In: The 2013 World Congress in Computer Science, Computer Engineering, and Applied Computing, Las Vegas, p. 7 (2013)
3. Chen, Y., Randy, V.P., Katz, H.: What's New About Cloud Computing Security? University of California at Berkeley, p. 8 (2010)
4. Truksha, V.: Cloud platform comparison: CloudStack, Eucalyptus, vCloud Director and OpenStack (2012) (October 13, 2014). http://www.networkworld.com/article/2189981/tech-primers/cloud-platform-comparison–cloudstack–eucalyptus–vcloud-director-and-openstack.html
5. Martell, R.E., et al.: Discovery of novel drugs for promising targets. Clin. Ther. **35**(39), 1271–1281 (2013)
6. Lavecchia, A., Di Giovanni, C.: Virtual screening strategies in drug discovery: a critical review. Curr. Med. Chem. **20**(23), 2839–2860 (2013)
7. Svajger, U., et al.: Novel toll-like receptor 4 (TLR4) antagonists identified by structure and ligand-based virtual screening. Eur. J. Med. Chem. **70**, 393–399 (2013)
8. Brunskole Svegelj, M., et al.: Novel inhibitors of trihydroxynaphthalene reductase with antifungal activity identified by ligand-based and structure-based virtual screening. J. Chem. Inf. Model. **51**(7), 1716–1724 (2011)
9. Yakar, R., Akten, E.D.: Discovery of high affinity ligands for beta2-adrenergic receptor through pharmacophore-based high-throughput virtual screening and docking. J. Mol. Graph. Model. **53**, 148–160 (2014)
10. Rashti, M.J., Afsahi, A.: 10-Gigabit iWARP Ethernet: comparative performance analysis with InfiniBand and Myrinet-10G. IEEE

11. Krishnan, S., Tatineni, M., Baru, C.: myHadoop-Hadoop-on-Demand on Traditional HPC Resources. San Diego Supercomputer Center Technical Report TR-2011-2. University of California, San Diego (2011)
12. Wolski, R., et al.: Using Batch Controlled Resources to Support Urgent Computing
13. Trott, O., Olson, A.J.: AutoDock Vina: improving the speed and accuracy of docking with a new scoring function, efficient optimization, and multithreading. J. Comput. Chem. **31**(2), 455–461 (2010)
14. Browne, J.C., et al.: Comprehensive, open-source resource usage measurement and analysis for HPC systems. Concurrency and Computation: Practice and Experience **26**(13), 2191–2209 (2014)
15. AT&T. Medical Imaging in the Cloud (2012) (October 12, 2014). https://www.corp.att.com/healthcare/docs/medical_imaging_cloud.pdf
16. Fusaro, V.A., et al.: Biomedical Cloud Computing With Amazon Web Services. PLoS Comput. Biol. **7**(8), e1002147 (2011)
17. CESNET, z.s.p.o. Metacentrum NGI (2014) (September 14, 2014). http://www.metacentrum.cz/cs/

Internet of Things, Big Data
and Cloud Computing

Delivery of e-Health Services in Next Generation Networks

Paweł Świątek[1], Halina Tarasiuk[2(✉)], and Marek Natkaniec[3]

[1] Wrocław University of Technology, 27 Wybrzeże Wyspiańskiego Street,
50-370 Wrocław, Poland
pawel.swiatek@pwr.edu.pl
[2] Institute of Telecommunications, Warsaw University of Technology,
15/19 Nowowiejska Street, 00-665 Warsaw, Poland
halina@tele.pw.edu.pl
[3] AGH University of Science and Technology, al. Mickiewicza 30,
30-059 Krakow, Poland
natkanie@kt.agh.edu.pl

Abstract. The main contribution of this paper is the proposal of delivery of e-health services which fully draws from recent achievements in the areas of networking, signaling and distributed application development. Moreover, we present a fully functional prototype implementation of a system in which the proposed solution was evaluated. Our prototype includes QoS-aware e-health services supported by a signaling system of Next Generation Networks (NGN) with the new NGN e-health service component. Experimental evaluation of the proposed signaling scheme that supports QoS guarantees of e-health services is also presented in this paper.

Keywords: e-Health services · Quality of Service · NGN · Signaling system

1 Introduction

The rapid growth of telecommunications allows for introducing completely new technologies and services in medicine. In recent years technological development in computing and networking has largely made the delivery of electronic health (e-health) services, including medical diagnosis and patient care possible from a distance. The main goal of e-health services is to suitably supervise the monitoring of the patient's health condition. The evolution of technology has enabled the development of advanced systems based on more powerful, portable, and easy-to-use terminals and applications, such as hand-held devices, video telephones or small computer systems. Therefore, e-health applications and smart health are one of the main groups of applications among Internet of Things (IoT). Recent research and activities of standardization bodies (e.g.: [5], [8]) show that delivery of future e-health services will require among others: 'anywhere and anytime' connectivity [5], end-to-end cross-domain quality of service assurance [8] and on-demand composition of personalized context-aware services [10]. This in turn requires the integration of ICT (Information and Communication Technologies) systems on multiple (i.e.: networking, signaling, application and business model) levels [9], [10]. Recent

© Springer International Publishing Switzerland 2015
N.T. Nguyen et al. (Eds.): ACIIDS 2015, Part II, LNAI 9012, pp. 453–462, 2015.
DOI: 10.1007/978-3-319-15705-4_44

trends show that the most suitable solution for such an integration is convergence of all-IP networking architectures controlled by NGN signaling on one hand (networking and signaling levels) and service-oriented architecture approach on the other (application and business levels). Such an integration received much attention from industry and academia in past few years (e.g.: [6], [7]). Taking into account the above related works, we propose an evolutionary approach to offer e-health services in IPv6 QoS network based on NGN and DiffServ architectures [1], [2], but implemented as one of three Parallel Internets in virtualized network infrastructure [11]. Therefore, the solution exploits the availability of IPv6 features in health monitoring devices, as well as Future Internet research.

The remainder of this paper is organized as follows. Section 2 describes two new exemplary e-health services: SmartFit and Asthma. Section 3 provides a proposal of the integration of the above-mentioned e-health services with NGN and details about a new e-health service component. Experimental evaluation of this service component is shown in Section 4. Section 5 concludes the paper.

2   Exemplary e-Health Services

For the purpose of presentation of the e-health services delivery three exemplary use-cases of two e-health applications are presented in this section. The first two use-cases concern the SmartFit application. Its main functionality is connected to a decision support and remote monitoring of a sportsman during technical training [13], [14]. The third use-case concerns remote monitoring and treatment of an asthmatic with the Asthma application.

SmartFit – routine monitoring: in this use-case it is assumed that sportsman requests a monitoring service which consists of: automatic monitoring of vital signals, archiving in data repository, sending notification and decisions concerning their current workout. There are four actors in this scenario: (1) Sportsman – equipped with sensors and a mobile device gathering vital signals from sensors; (2) SmartFit services server (SFSS) – which upon requests, acts on behalf of the sportsman by configuring necessary services; (3) Monitoring service – receives raw signals from sportsman, monitors them and sends notifications to the sportsman when necessary; (4) Data repository service – stores workout data received from monitoring service.

SmartFit – emergency monitoring: in addition to routine monitoring use-case the emergency monitoring requires the following: (1) voice consultation with a trainer – accomplished by setting a VoIP session with trainer; (2) additional vital signals are monitored, previously monitored signals may be required to be monitored with denser resolution – monitoring and repository services are reconfigured with XML-RPC messages. Moreover, in this use-case two additional actors are introduced: Trainer – who handles an emergency event, SIP servers – responsible for setting up a voice connection between trainer and sportsman.

Asthma monitoring: the treatment of asthma is highly related to proper patient monitoring and restricted to informing the patient about the risk of symptom intensification or drug recommendation. Three different tests were defined for asthma: on-demand test, morning test, and evening test. According to the official patient's diary, the patient should answer first some normalized questions, which are specific for each test. The lung efficiency (PEF/FEV1) and wheezes should be measured as a next step. It should be also pointed out that the values of PEF/FEV1 may vary in these tests because of the part of the day. The measurements are sent via Bluetooth to the patient's mobile device and further via any of the available wireless technology to the e-health asthma server. The server forwards all required data to the medical doctor. On the basis of those data, the medical doctor decides the diagnosis, and as a consequence, the patient promptly receives information about his health status and recommended medicines. It is assumed that in the proposed asthma system, the information is simplified to four grades of patient illness. The medical doctor can also trigger via e-health asthma server voice or video consultation, if needed.

Since health is the live-critical area of the IoT, the Quality of Service (QoS) should be a major requirement of e-health services. Therefore, e-health services and their applications, which require low latency, real time or high throughput, should be supported by QoS mechanisms in the network [3]. A summary of QoS requirements of SmartFit and Asthma applications are presented in Table 1.

Table 1. QoS requirements of exemplary e-health services

E-health service/ Requirements	SmartFit	Asthma
Low latency	Routine monitoring (EMG, ECG, Acceleration); Emergency monitoring (ECG, Heart Rate, Pressure); Signaling;	Heart Rate; Lung efficiency measurement; Signaling;
Real time	Voice consultation; Video consultation;	Voice consultation; Video consultation LQ; Video consultation HQ;
High throughput	File transfer;	-
Best effort	-	Air quality monitoring; GPS;

We believe that a promising solution to guarantee the above-mentioned QoS requirements for e-health services is the integration of these services with NGN architecture.

3   Integration of e-Health Services with NGN

Fig. 1 presents the NGN architecture with the new integrated e-health functional elements: e-health service component, e-health applications, and e-health customer-premises equipment.

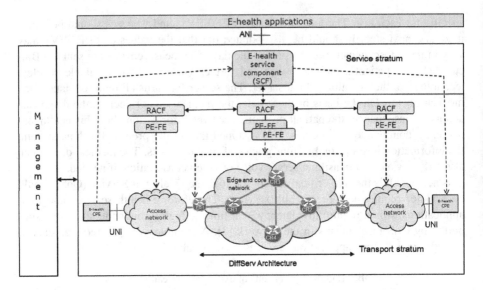

Fig. 1. NGN architecture with integrated e-health functional elements and interfaces of the signaling system. SCF – Service Control Function; RACF – Resource and Admission Control Function; PE-FE – Policy Enforcement Functional Entity; ANI – Application-Network Interface; UNI – User-Network Interface; CPE – Customer-premises Equipment; ER – Edge router; CR – Core router.

3.1   NGN Architecture

According to the ITU-T recommendations [1], [2] the NGN architecture is divided into the following blocks: transport stratum, service stratum, management, and applications. Transport stratum covers the underlying networks and signaling system. In our proposal, it consists of e-health customer-premises equipment (CPE), IEEE 802.11 access networks, and single IPv6 QoS domain with edge and core routers (ER, CR). The heart of the transport stratum is the signaling system [12]. The main role of the signaling system is to perform call setup/release procedures. As a consequence, to perform admission control algorithms to limit admissible traffic load in the network and in that way to guarantee packet transfer according to the target values of QoS parameters (packet delay, jitter, packet losses, throughput) [3]. For this purpose, the signaling system contains two main functional modules: resource and admission control function (RACF) and policy enforcement functional entities (PE-FEs). As a novelty of the proposed solution, connections can be established as on-demand or planned connections [12].

To integrate e-health service with NGN architecture, new e-health service control function (SCF) component enhances NGN service stratum functionalities compared to available NGN recommendations [1], [2]. In the remaining part of the paper, the e-health service component will be referred to as SCF. In order to control application specific communication and to satisfy requirements for data transfer QoS requests for systems resources, a reservation must be passed from the e-health service component to the transport stratum. In the NGN architecture this is accomplished by the mapping of data flows into end-to-end classes of service (CoSs) and by the reservation of proper amount of network resources (RACF and PE-FE). In this proposal we also take into account the experiences of research related to QoS guarantees over heterogeneous networks as presented in [3]. Following this research we could also map end-to-end CoSs into CoSs specific for considered network technologies (e.g. IP, IEEE 802.11e [7]).

3.2   New e-Health Service Component (SCF) - Exemplary Scenario

An exemplary scenario of delivery of certain service in an exemplary application is presented in Fig. 2a.

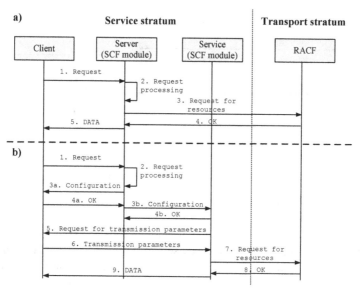

Fig. 2. a) Communication service request through SCF-RACF interface and b) signaling messages exchange for an exemplary service

In this scenario a client (user/sensor) sends a request to an application server (1) for certain resources (e.g. medical data). This request is received and processed (2) by an SCF module. Based on the information included in the request (1) and on the results of request processing (2), the SCF module generates a request for an end-to-end CoS and sends it to the transport stratum through the SCF-RACF interface. Upon arrival of the acknowledgement (4) of communication, a resources reservation SCF module transmits the information necessary to the server to begin communication with the client. Afterwards, the server delivers requested data to the client (5). Much more complex communication scenarios for various types of applications are possible (e.g. Fig. 2b), nevertheless, in general each complex communication scenario can be decomposed into sub-scenarios requiring an end-to-end communication between a pair of nodes. An important advantage of separation of the service stratum in the NGN architecture is that the method for requesting an end-to-end communication service is independent from the application type. Such a decoupling is achieved with the use of the SCF module, which stands on the border of service and transport strata, and maps application's service requests into a uniform for all applications SCF-RACF interface. This means, that requests sent to a transport stratum are independent from application-specific negotiation and signaling protocols.

SmartFit Routine Monitoring Scenario. The flow of messages and sequence diagram in an exemplary SmartFit routine monitoring scenario are presented in Fig. 3.

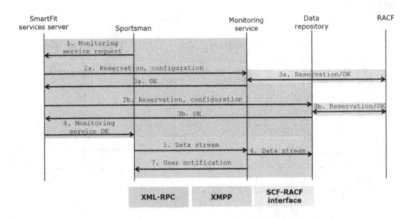

Fig. 3. The flow of messages and sequence diagram in a routine monitoring scenario

A Routine monitoring scenario begins when the sportsman sends a monitoring service request (1) to the SmartFit services server. Then SFSS configures monitoring service and data repository on behalf of a sportsman (2a, 2b). When configuration of services (monitoring and repository) is confirmed (3a, 3b), the SFSS confirms preparation of requested routine monitoring service to the sportsman by sending acknowledgement (4). Afterwards transmission of vital signals may begin. The sportsman sends their vital signals to a monitoring service (5), which after signal processing passes them to data repository service for storage (6). Signal processing at the monitoring service may result in notifications concerning sportsman current state (7). The first phase of routine monitoring service delivery (1-4) is done with the use of XML-RPC protocol. In the second phase, negotiation between each pair of actors exchanging data is accomplished with the use of XMPP protocol. This involves setting up connections for data transmission (5, 6). Depending on the actual workout scenario (especially on type of transmitted data and QoS requirements) different types of connections may be established. The transmission of notifications (7) will mostly be done by sending simple XMPP messages.

Due to the limitation of space, only the SmartFit routine monitoring scenario is presented in this paper.

4   Experimental Evaluation of the e-Health Service Component

It is obvious that the proposed e-health service component, with its signaling scheme introduces certain contribution to the overall complex service setup delay. This delay is crucial for all e-health applications (e.g. audio-visual monitoring of a patient in an emergency situation). Therefore, in this section we present the results of experimental evaluation of the performance of the proposed signaling scheme of SCF. The performance of the signaling scheme is measured as the overall delay between sending the complex service setup request and receipt of connections setup acknowledgement. Signaling performances at transport stratum were evaluated e.g. in [4].

4.1   Experiments Outline

The performance evaluation of the proposed signaling scheme was conducted with the use of the topology presented in Fig. 4, which included: three layer 3 switches SW_1 through SW_3 (Juniper EX4200), three servers S_1 through S_3 (Intel Core i7 960, 12GB RAM) and client application C_1 which acted as a requests generator.

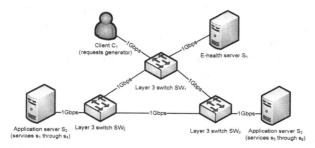

Fig. 4. Topology of the test-bed for signaling performance evaluation

Since the sole aim of the experiments was to evaluate the performance of a signaling scheme we used 1Gbps connections between devices in order for the network not to be a bottleneck. Requests generated by the client's application C_1 where sent to e-health server application located on server S_1. For each received request e-health server initiated a multi-level signaling process consisting in: configuration of necessary services located on servers S_2 (services s_1 through s_4) and S_3 (services s_5 through s_8), negotiations between required services (from set of all available services $S=\{s_1,...,s_8\}$) and communication resources reservation. Each i-th request req_i consisted of a connected directed acyclic graph $G_i=\langle N_i, E_i\rangle$, which nodes $N_i \subset S$ represented services taking part in delivery of requested complex service and edges $E_i \subset S \times S$ represented the direction of data transmission. In other words, the set of edges E_i of graph G_i defined which connections between services from set $S=\{s_1,...,s_8\}$ must be established in order to deliver the requested complex service given by request req_i.

Requests generated by client C_1 consisted of graphs having k edges, where $k=\{1,...,4\}$. Number of edges k represented the complexity of requested service in terms of the number of connections necessary to be established to deliver required functionality. For each generated request a k-edged graph $G_{i,k}$ was chosen uniformly at random from a set of all graphs $\Gamma_k = \{G_k = \langle N_k, E_k \rangle : N_k \subset S, card(E_k) = k\}$. Next, for each node of graph $G_{i,k}$ a service from set S was randomly chosen in such a way, that resulting graph $G_{i,k}$ formed an acyclic connected graph. In the conducted experiment, twenty one tests were performed. Each test run consisted of generation and execution of a series of twenty two thousand requests for different types of requested complex services (increasing number of edges in request graphs $k = \{1,...,4\}$) and for increasing requests arrival rate ($\lambda = \{1,...,5\}$) where requests interarrival times were exponentially distributed. In the additional twenty first run requests graph sizes k and requests arrival rate λ varied throughout the test. Requests were generated uniformly at random from the set of all possible request graphs with the number of edges equal to four or less ($G_i \in \Gamma_1 \cup \Gamma_2 \cup \Gamma_3 \cup \Gamma_4$).

4.2   Results

Results of the experiments runs concerning completion times of 1-edged requests are presented in Fig. 5. Due to the limitation of space we present only results obtained for 1-edged requests. The plot from Fig. 5a presented histograms of the frequencies of appearance of requests completion times in consecutive time intervals of length 50ms for increasing requests arrival rate $\lambda = \{1,...,5\}$ for 1-edged requests. In corresponding Fig. 5b distribution of requests completion times for increasing request arrival rate is presented. This distribution can be treated as rough estimates of requests completion time probability distribution functions.

(a)   (b)

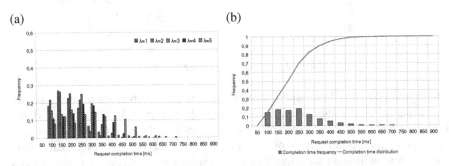

Fig. 5. a) Frequencies, and b) Distribution of 1-edged requests completion times for increasing requests arrival rates

Additionally in Fig. 6, frequencies of appearance histograms and distributions of requests completion times cumulated over all examined requests arrival rates are shown. Results of the conducted tests are summarized on plots presented in the Fig. 7. Histogram of frequencies and distribution of requests completion times for the twenty first experiment run, which model an exemplary usage of the proposed system, are presented in Fig. 7.

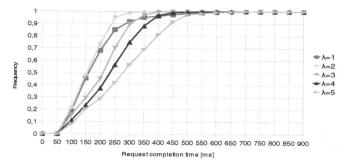

Fig. 6. Cumulated frequencies and distribution of 1-edged requests completion times for all requests arrival rates

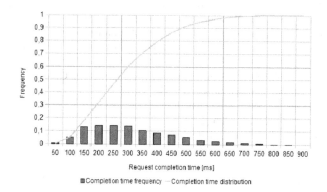

Fig. 7. Histogram of frequencies and distribution of completion times for all request sizes k={1,...,4} and requests arrival rates λ={1,...,5}

The results of our investigation show that the proposed signaling system performs quite well for the assumed conditions. The average time of single connection setup varies from around 200ms for the least loaded system to 300ms for high requests arrival rate. Moreover, more than 95% of requests (for all arrival rates) are serviced in less than 400ms (see Fig. 5 and Fig. 6). Notice, that in the real-life scenario when requests sizes and arrival intensities vary in time, the average request completion time equals to 300ms, and 95% of requests are serviced in less than 550ms (see Fig. 7).

5   Conclusions

In this work, it was shown that signaling system is a crucial part of e-health service. Its performance was thoroughly evaluated in a number of experiments. We have also presented illustrative use-case scenario of SmartFit e-health application. The system has been tested in the PL-LAB experimental infrastructure [11]. An important observation is that all examined characteristics of the requests completion times scale linearly with both: requests size k and requests arrival rate λ. This fact allows us to claim, that the proposed signaling scheme is efficient and scalable at least for the conditions investigated in this research.

Acknowledgment. This work was partially funded by the European Union within the European Regional Fund project number POIG.01.01.02-00-045/09-00, "Future Internet Engineering" and within the European Social Fund.

References

1. ITU-T Rec. Y.2012. Functional requirements and architecture of next generation networks (April 2010)
2. ITU-T Rec. Y.2111. Resource and Admission Control Functions in Next Generation Networks (November 2008)
3. Burakowski, W., et al.: Provision of End-to-End QoS in Heterogeneous Multi-Domain Networks. Annals of Telecommunications **63**(11) (2008)
4. Tarasiuk, H., et al.: Performance Evaluation of Signalling in the IP QoS System. Journal of Telecommunications and Information Technology **3**, 12–20 (2011)
5. Yelmo, J.C., et al.: A user-centric service creation approach for next generation networks. In: Innovations in NGN: Future Network and Services, First ITU-T Kaleidoscope Conf., pp. 211–218 (2008)
6. Gobernado, J., et al.: Management of Service Sessions in an NGN-SOA Execution Environment. IEEE Comm. Magazine **48**(8), 103–109 (2010)
7. Natkaniec, M., et al.: Supporting QoS in Integrated Ad-Hoc Networks. Wireless Personal Communications **56**(2), 183–206 (2011)
8. Aragues, A., et al.: Trends and challenges of the emerging technologies toward interoperability and standardization in e-health communications. IEEE Comm. Magazine **49**(11) (2011)
9. Świątek, P., et al.: Service composition in knowledge-based SOA systems. New Generation Computing **30**(2/3), 165–188 (2012)
10. Grzech, A., Świątek, P., Rygielski, P.: Dynamic resources allocation for delivery of personalized services. In: I3E 2010, IFIP AICT 341, pp. 17–28. Springer (2010)
11. Tarasiuk, H., et al.: A proposal of the IPv6 QoS system implementation in virtual infrastructure. In: Proc. of Networks 2012, Rome, Italy (October 2012)
12. Tarasiuk, H., Rogowski, J.: On the signaling system in the IPv6 QoS parallel internet. In: Proc. of 8th IEEE, IET Int. Symposium on Communication Systems, Networks and Digital Signal Processing, CSNDSP 2012, Poznań, Poland (2012)
13. Świątek, P., et al.: Application of wearable smart system to support physical activity. In: Advances in Knowledge-based and Intelligent Information and Engineering Systems. IOS Press, pp. 1418–1427 (2012)
14. Brzostowski, K., et al.: Adaptive decision support system for automatic physical effort plan generation - data-driven approach. Cybernetics and Systems **44**(2–3), 204–221 (2012)

Implementation and Performance Testing of ID Layer Nodes for Hierarchized IoT Network

Jordi Mongay Batalla[1(✉)], Mariusz Gajewski[2],
Waldemar Latoszek[2], and Piotr Krawiec[2]

[1] Warsaw University of Technology, Nowowiejska Street 15/19, 00-665 Warsaw, Poland
jordim@interfree.it
[2] National Institute of Telecommunications, Szachowa Street 1, 04-894 Warsaw, Poland
{m.gajewski,w.latoszek,p.krawiec}@itl.waw.pl

Abstract. Recent advances in technologies for smart devices are having a significant impact in IoT (Internet of Things) scenarios as, e.g., intelligent buildings. Sensor/actuator networks use small and non-intrusive devices consuming reasonable amount of energy and offering improved performance. On the other hand, highly specialized devices providing high reliability are interconnected by dedicated network infrastructure because of safety reasons. This article discusses early stage of the implementation of an innovative hierarchical network infrastructure for connecting IoT objects and services where the location of the nodes is closely related to the structure of the environment as it occurs in intelligent buildings/enterprises.

1   Introduction

Networks consisting of specialized sensors and actuators play a crucial role in currently under development intelligent buildings. There are two observable areas where the networked IoT objects are successfully used: energy saving and security. Both require sensors (i.e., passive infra-red, fume detectors, etc.) and actuators (i.e. light switches, window actuators, etc.) located in selected areas of a building. Their location is strictly dependent on the structure of the building and connections between them, which creates a hierarchical network that can be modeled as a tree topology, as shown in Fig. 1.

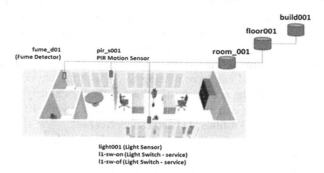

Fig. 1. Example of hierarchical network for data transmission in intelligent building

© Springer International Publishing Switzerland 2015
N.T. Nguyen et al. (Eds.): ACIIDS 2015, Part II, LNAI 9012, pp. 463–472, 2015.
DOI: 10.1007/978-3-319-15705-4_45

This article discusses implementation details together with test results of the deployed hierarchical network for connecting IoT objects where node location may be defined by the same environment structure, as it occurs in the case of intelligent buildings with fix nodes and fix sensors/actuators. The presented implementation is based on the ID Layer concept that we developed and presented in [1]. The discussed network node has been mostly developed in Linux kernel module and extends the solution proposed in Flexible Packet Forwarding (FPF) method [2].

2   Context

Until recently, sensor networks were mainly the hermetic solutions based on specialized devices, dedicated network and proprietary protocols developed by suppliers. Over time, popular solutions have become factory standards widely used in the industry control [3] (e.g. Controller Area Network), but also in other areas [4]. For example, more and more smart devices have been equipped with standard wired or wireless Ethernet interfaces [5], thus they may share the same network infrastructure as other applications. Therefore, it is desirable to implement such nodes using layer-2 network that is backwards compatible with Ethernet.

In order to develop the location-oriented network topology, the key idea is to embed the physical network connectivity structure into a (logical) topological space (e.g. introducing a metric or Euclidean space). This approach was presented among others in VIRO (Virtual Id ROuting) [6] to illustrate how the novel topological perspective enables the development of the scalable resilient network routing algorithms. Another example of this approach is SEATTLE [7], which introduces OSPF-style shortest routing in layer 2 for inter-connecting objects and Ethernet switches. Such solutions are aimed at reducing network-wide flooding – often typical for Ethernet switches needed to forward packets whose locations are yet to be learned, especially in the case of wide layer-2 networks, which encompass small LAN networks.

Other approaches aimed at replacing the current global IP address space by flat identifiers, have been adopted by VRR [8], UIP [9] and ROFL [10]. They make use of several methods of hashed id assignment (mostly based on DHT), which produces an id-space completely independent of the underlying network topology. As a result, these methods perform routing based on logical distance to the id of the destination.

Real Time Control Systems consider not only topological addressing, but also transmission parameters such as packet delay and predictability of the response time. In such systems, the transmission is moderated by a controller/supervisor, which grants permissions to specific devices (i.e. sensors/actuators) by sending appropriate tokens. The order of polling is fixed according to the address table (with flat structure) stored in the controller and can have nothing to do with the physical placement of devices. The device with the next highest address is the logical neighbor, even when they can be located at the extreme ends of a physical network. Example solutions encompass, among others, Profibus [11] and DeviceNet [12].

In turn, the ID Layer concept [1] assumes hierarchical addressing scheme for the same purpose, i.e. each level of address hierarchy is represented by one address segment. Forwarding process is based on analysis of particular address segments, however, this approach does not require physical node on each level of hierarchy since forwarding functionality may be performed also by virtualized nodes.

3   Implementation of ID Layer Node

The main objective of the implementation of the ID_Layer node is to develop the ID layer in the network level instead of building an overlay network. The architecture proposed in [1] has a hierarchical structure, in which each node has a human-readable identifier related to its location. These nodes, called further as ID_Layer nodes, include connected objects (sensors/actuators) and also address the services offered by the objects. It is assumed that the name of each node, object and service is formed as an 8-ASCII extended character (word). The naming scheme uses hierarchical ID-based addressing scheme, which is created and managed in conjunction with the location of the IoT object. The human-readable names of nodes, object and services are also used for packet routing across the network.

All included functional blocks of ID-Layer node were created as software modules in user and kernel space of Linux operating system. Fig. 2 shows main building blocks of ID_Layer node and functional dependencies between them. The main modules are: (1) **Forwarding Administration Tool**, which is the configuration module. This module gives the possibility for an administrator to configure the Forwarding Table and to assign a node name; (2) **Forwarding Module**, which is responsible for sending a frame to the required node, regardless of whether the frame is a *data frame*, a *registration message* or a *resolution message*. More detailed definition of different types of frames is given in [1]. This module communicates with Registration Module (registration process), Resolution Module (resolution process) and the Forwarding Administration Tool during the initial node configuration procedure; (3) **Registration Module**, which is responsible for the registration of new objects/services in the node, to which the object/service are connected, while the (4) **Resolution Module** allows to obtain the information about all the objects/services registered in the specified node.

The implemented ID-Layer node performs functionalities of forwarding, registration of objects/services and resolution of services.

For forwarding frames, the ID address is included in the header of the ID frame together with the information about header length. Moreover, each node has assigned its own address by the administrator [1]. This allows to perform forwarding actions in the node only by comparing the ID with the address of the node without the necessity of running routing protocols.

Registration is the process by which objects (and basic/composed services offered by them) inform about the own characteristics to the closest network node. The network nodes will maintain information about the connected objects and offered services.

At last, resolution is in charge of discovering the services offered by the objects and presenting them to the users (IoT applications).

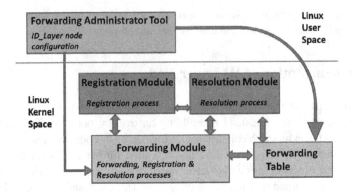

Fig. 2. General architecture of the ID_Layer node

Details of the functional processes performed by the modules are given in the next sub-sections, starting from the configuration of the ID-Layer node.

3.1   Configuration of ID_Layer Node

The Forwarding Administration Tool located at the user space communicates with the Forwarding Module (kernel space) for basic node configuration in the following areas: Forwarding Table configuration and node name configuration.

The Forwarding Table configuration is performed by adding new entries in an appropriate data structure maintained by the ID_Layer node in the kernel space. The following sample command is performed to configure one entry:

> ./cf_tool add_ethernet *room001 eth4 aa:11:b0:c0:00:01*
> where:
> *room001* – name of the next node
> *eth4* – name of the interface through which the frame will be sent
> *aa:11:b0:c0:00:01* – destination MAC address

In order to perform a complete configuration of the node, it is necessary to assign a node name. It is performed by issuing the following command:

> ./cf_tool add_name   *floor111.build111.room111*

3.2   Forwarding Process

The data forwarding process is performed by the Forwarding Module using the Forwarding Table, which is a data structure that stores necessary information about routes to the adjacent nodes. The forwarding process applies only to data frames and *resolution messages,* in which the destination address means the domain name of the node, while the *registration messages,* for which the fixed destination address (*.locathst*) is established, are forwarded without querying the Forwarding Table.

Data and resolution frames contain similar formats, as far as addressing concerns (ID frame header is presented in Fig. 3). The format of the ID frame header contains destination and source addresses as well as Message info (more information about data format can be found in [1]). Both, the source and the destination addresses, contain different levels (corresponding to different hierarchical levels) separated by dots (e.g., *build001. floor001.room0001*). Each level contains 8 bytes. In the 2-byte *message info* field, the first four bits define the message type, the next bit identifies whether the message is multicast or unicast. Finally, 11 bits indicate the length of the message in bytes.

Fig. 3. Exemplary ID_Layer frame

When a frame arrives to the node, this compares the destination address with the entries of the Forwarding Table and forwards the frame, following the rule inserted into the Forwarding Table. The forwarding operation is preceded by a validation of the destination address. The aim of this step is to check whether the appropriate part of the destination address contained in the frame is consistent with the node name assigned by the administrator. For example, a frame with address floor001.room0001 should not arrive from the parent node interface (interface where the parent node is connected) to a node with address .floor002, but it may arrive from the child node interface (this case would be the case when the frame should be directed to the destination through the parent node). Each level of the destination address is compared with the corresponding level of the own node name set previously by the administrator. In the case of a failed name validation, the frame is forwarded to the node at a higher level of hierarchy (parent node). In the case of positive validation, the frame is forwarded according to the forwarding rule set in the Forwarding Table. From the implementation's point of view , there are two options for performing such a validation. The first option consists of converting the destination address of the frame as well as the address of the node to integer type and comparing the integers. In the second option, both the destination address and the own node name are stored and compared as character variables. It is supposed that the first option is quicker since the number of comparisons is proportional to the number of levels, whereas the second option requires a number of comparisons proportional to the number of characters (which is equal to 8 times the number of levels). In the test experiments presented in the next section, we will compare the performance of the two options.

If the validation process finishes positively, then the forwarding process goes ahead by searching the relevant part of the domain name that will be used during the forwarding procedure. The information about the own node name allows the algorithm to find the relevant part of the name. If the node name is *floor001.build001* and the destination address set in the frame header is *floor001.build001.room0001*, then the forwarding will be based on the last part of the address (*room0001*).

The next step of the algorithm is to find an appropriate entry in the Forwarding Table for the relevant part of the domain name.

The Forwarding Table consists of entries with 3 fields: *destination_MAC*, *dev_name* and *next_node*, as shown in Table 1.

Table 1. Data structure of Forwarding Table entries

Name of variable	Type of variable	Description
destination_MAC	uint8_t [48]	Destination MAC address of the node interface to which the frame is sent
dev_name	char [5]	Name of the outgoing interface
next_node	char [8]	Domain address of the next node

The information about the *next_node* (destination node address) is used to calculate a specific *index* of the entry in the Forwarding Table. In order to add the appropriate entry to the table, the algorithm converts the 8-byte long *next_node* name to 1-byte numerical value according to the following iteration algorithm (1):

$$index = ((37 * index) + ch \rightarrow name[i] \& 0xff) \qquad (1)$$

where *index* is the value of the converted *next_node* (1 byte), *i* is an iteration variable and *ch→name[i]* is the i-level of the domain node name.

Then, the *find_entry* function queries the Forwarding Table about the interface connected to the value *index*. On the basis of the information contained in this entry, the frame is sent to the next node.

Fig. 4 shows the sequence diagram of the forwarding process.

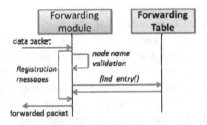

Fig. 4. Sequence diagram of the forwarding process

3.3   The Registration Process

The registration process illustrated in the sequence diagram presented in Fig. 5 is initiated upon the receipt of a specific *register message* [1] by the Forwarding Module. Then, the Forwarding Module reads the appropriate *message info* field [1] placed in the message header and redirects the process to the Registration Module without querying the Forwarding Table. In the Registration Module, the function *register_handle* is called. This function maintains main data structure with information about currently registered objects or services (object/service identifier –*id* and full address of the object/service - *ObjectAddress*). If the data structure does not yet store the entry with the demanded *id* of the object/service, then a new entry is created with the data contained in the *register message*. Finally, the Registration Module passes to the Forwarding Module the necessary information used for sending *response message* to the object/service (that sent the *register message*) in order to confirm the registration process.

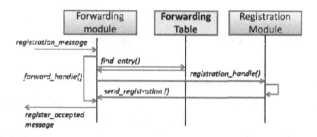

Fig. 5. Sequence diagram of the registration process

3.4 Resolution Process

The resolution process illustrated in the sequence diagram of Fig. 6 is initiated by the user (IoT application) in order to retrieve information about registered objects/services.

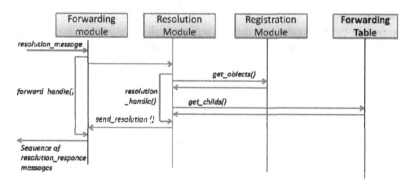

Fig. 6. Sequence diagram of the resolution process

For this purpose, the user's application sends a resolution message, which is forwarded in accordance with the forwarding algorithm until the destination node. When the message reaches the destination node, the Forwarding Module of this node extracts the message info and, based on this information, redirects the process to the Resolution Module where the function resolution_handle is executed. In the next step, the Resolution Module retrieves information about registered objects/services of the queried node from the Register Module (get_object function) and retrieves information about all child nodes from the Forwarding Table (get_child function), which enables the identification of the objects/services registered in the nodes of lower hierarchy. In the final phase of the resolution_handle function, the Resolution Module sends the appropriate information to the Forwarding Module. The latter builds a resolution response message, in which the source and the destination addresses are interchanged. Moreover, information about registered objects/services and names of the child nodes are placed into the information field of the resolution response message. Finally, these messages are sequentially forwarded back to the requester.

4   Performance Tests of Forwarding Process

Even if the advantages of ID addressing and ID Layer forwarding are numerous for IoT applications [1], the proposed solution risks fail in the case when the implementation does not fulfill the requirements of performance necessary for forwarding a large amount of packets. The aim of the presented here performance tests is to show that the deployed solution is efficient enough to be used in IoT scenarios. More precisely, we will demonstrate that the ID-Layer node performance is comparable to IP router implemented on Linux OS (software development). Moreover, the test deal with scalability issues show the behavior of the ID_Layer node for increasing up to 8 number of the domain levels, i.e., for increasing hierarchy atomization.

The testbed /consists of one System Under Test (SUT) and one tester. The SUT is the ID_Layer node installed on HP ProLiant DL360G6 server, which runs Linux Operating System. The tester is the Spirent TestCenter (equipped with CM-1G-D4 card). The tester and the SUT are connected by two 1 Gbps Ethernet links in ring topology, as proposed in the benchmarking for testing network interconnect devices presented in RFC 2544 [13]. We performed tests for the following frame size: 96B, 112B, 128B, 160B, 256B, 384B, 512B, 1024B and 1518B, and the stream was the maximum allowed by the interfaces, i.e., 1 Gbps. In these conditions, we measured the frame losses observed in the SUT due to overload of the server.

The results presented below shows the Frame Loss Ratio for different frame sizes and increasing number of domains. First of all, let us remark that the software IP router implemented on Linux OS was installed in the same hardware and the Frame Loss Ratio of the IP router was, at least, 20 times higher than the ID_Layer node (for all frame sizes), even in scenarios with 8 domain levels. Note that the IP router performance is not affected by level complexity because of the same nature of IP addresses. For clarity purposes, we did not present the values of Frame Loss Ratio in the figure.

As one of the major features of the forwarding process is the validation of the destination address in the ID_Layer frame, we compared two approaches for implementation, which is described above. As stated above, this address is composed of the levels of the domain name separated by dots. The introduction of this functionality results in the need for additional computational effort caused by parsing the destination address.

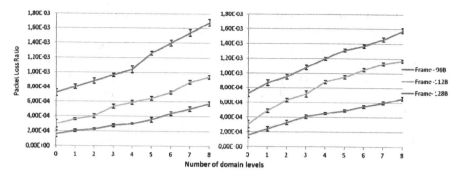

Fig. 7. Test results for frames 96B, 112B, 128B frames (both options of forwarder implementation): *a)* name validation with conversion of variables. *b)* name validation without conversion of variables (0 number of domain levels means forwarding without validation name).

The results revealed that the increasing number of levels of the domain name in the destination address causes a higher Frame Loss Ratio for both options of the ID_Layer forwarder (i.e., name validation with and without conversion of variables). The increase of Frame Loss Ratio is approximately linear for each test scenario. In addition to this, it can be noted that, for the same tests (frames with the same number of domain levels), there is no significant difference between the values of Frame Loss Ratio for both implementations of the forwarder. .

In Fig. 7, we presented results only for frames not bigger than 128B, but the tests were performed for other frame sizes according to the test assumptions. The results of the other tests confirmed that the forwarder transmits frames with bit rate equal to the maximum link bitrate - 1Gb/s for frames larger than 160 Bytes independently of the test scenario (number of domain levels). For the frame size equal to 160B, the Frame Loss Ratio parameter does not exceed the level of 10^{-6}. This means that the main limitation of the forwarder performance is the performance of network interface as frame loss occurs only for very small frames.

It can be concluded that the size of the address contained in the frame header has a small impact on the value of the Frame Loss Ratio parameter and, on the other hand, the performance of the ID_Layer node is satisfactory (at the same level as software IP router).

5   Conclusions

More and more smart devices used in current developments are equipped with Ethernet interfaces, which allow sharing the same network infrastructure between different applications. This approach allows to reduce implementation costs and does not adversely affect the functionality of the intelligent building solutions.

The based on the ID layer concept implementation was developed over the Linux operating system. It engages mainly kernel resources to improve effectiveness of the solution.

We conducted performance tests of the prototype aimed at checking the effectiveness of the implemented solution for different lengths of the address field. In result, we calculated the limit performance of the node, which could be located on different levels in hierarchy tree. The test results presented above confirm the usefulness of kernel based approach for ID layer implementation. In particular, it is able to serve high frame rates and can be implemented as additional functionality of Linux based network nodes.

Further works will cover, besides the completion of routing functionality according to assumptions in [1], development of system for centralized domain names management.

References

1. Mongay Batalla, J., Krawiec, P.: Conception of ID layer performance at the network level for Internet of Things. Springer Personal and Ubiguitous Computing (2013)
2. Bęben, A., Mongay Batalla, J., Wiśniewski, P., Xilouris, G.: A Scalable and Flexible Packet Forwarding Method for Future Internet Networks. In: IEEE Globecom 2014. The source code for Flexible Packet Forwarding (FPF) method may be found in: http://tnt.tele.pw.edu.pl/software_fpf.php
3. Website of CAN in Automation (CiA). http://www.can-cia.de/
4. Świątek, P., Klukowski, P., Brzostowski, K., Drapała, J.: Application of wearable smart system to support physical activity. In: Advances in Knowledge-Based and Intelligent Information and Engineering Systems, pp. 1418–1427. IOS Press (2012)
5. Hunt J.: Building automation migrates towards Ethernet and wireless, Industrial Wireless Book, Issue 69, April 2012. http://www.iebmedia.com/wireless.php?id=8593&parentid=74&themeid=255&hft=69&showdetail=true&bb=1
6. Sourabh, J., Yingying, C., Zhi-Li, Z.: VEIL: A "Plug-&-Play" virtual (ethernet) Id layer for below IP networking. In: 1st IEEE workshop on below IP Networking 2009 (In conjunction with IEEE Globecom 2009) (2009)
7. Kim, C., Caesar, M., Rexford, J.: Floodless in SEATTLE: a scalable ethernet architecture for large enterprises. In: SIGCOMM (2008)
8. Caesar, M., Castro, M., Nightingale, E.B., O'Shea, G., Rowstron, A.: Virtual ring routing: network routing inspired by DHTS. SIGCOMM Computer Communication Rev. (2006)
9. Caesar, M., Condie, T., Kannan, J., Lakshminarayanan, K., Stoica, I.: Rofl: routing on flat labels. In: SIGCOMM (2006)
10. Ford, B.: Unmanaged internet protocol: taming the edge network management crisis. SIGCOMM Computer Communication Rev. (2004)
11. Website of PROFIBUS and PROFINET International (PI). http://www.profibus.com
12. Website of global standards development and trade association for Common Industrial Protocol or "CIP" – and the network adaptations of CIP – EtherNet/IP, DeviceNet, CompoNet and ControlNet. http://www.odva.org/
13. Bradner, S., McQuaid, J.: Benchmarking Methodology for Network Interconnect Devices. Requests For Comments RFC 2544 (1999)

Modeling Radio Resource Allocation Scheme with Fixed Transmission Zones for Multiservice M2M Communications in Wireless IoT Infrastructure

Sergey Shorgin[1], Konstantin Samouylov[2(✉)], Yuliya Gaidamaka[2], Alexey Chukarin[2], Ivan Buturlin[2], and Vyacheslav Begishev[2]

[1] Institute of Informatics Problems of RAS, Moscow, Russia
sshorgin@ipiran.ru
[2] Peoples' Friendship University of Russia, Moscow, Russia
{ksam,ygaidamaka}@sci.pfu.edu.ru, chukarin@yandex.ru,
ivan.buturlin@gmail.com, begishevu@mail.ru

Abstract. This paper focuses on the paradigm of Internet of Things (IoT), which gives rise to communications between a large number of different technological devices, e.g. sensors and controllers. The increasing demand for various services without human intervention motivates service providers to apply machine-to-machine (M2M) communications. In this paper, a LTE cellular system where M2M devices and human-to-human (H2H) users transmit their data into the wireless network is considered. In particular, a radio resource allocation scheme for M2M communications in IoT infrastructure is proposed and analyzed. The scheme is based on fixed transmission zones at which M2M traffic is served according to the Processor Sharing (PS) discipline. Also, the Markovian model to evaluate main performance measures, i.e. data transmission delays and blocking probabilities, is proposed. To carry out the numerical analysis the recursive algorithm for computing the stationary probability distribution is developed.

Keywords: Internet of things (IoT) · Machine-to-machine (M2M) · Human-to-human (H2H) · Radio resource allocation · Markovian model · Blocking probability · Data transmission delay

1 Introduction

Today, increasingly capable mobile devices, represented by advanced smartphones and tablets, are employed to aid people in their daily routines, from communication and social interaction to storing and processing their important private information. Current wireless systems are struggling to meet the anticipated acceleration in user traffic demand on future internet services and applications [1] aggravated by the rapid proliferation of M2M communications. With the expected 13-fold growth of M2M data over the next five years [2], mobile network operators are challenged with the need to significantly improve capacity and coverage across their wireless deployments. To augment the existing cellular technology, mobile industry is taking decisive

© Springer International Publishing Switzerland 2015
N.T. Nguyen et al. (Eds.): ACIIDS 2015, Part II, LNAI 9012, pp. 473–483, 2015.
DOI: 10.1007/978-3-319-15705-4_46

steps in many aspects of fifth generation (5G) wireless system design. Nevertheless 5G wireless systems will not be a universal one-size-fits-all solution, but rather become a converged set of various radio access technologies, integrated under the control of the operator's cellular network. The paradigm of heterogeneous networks (HetNets) has been introduced as a next-generation networking architecture enabling aggressive capacity and coverage improvements towards future 5G networks [3, 4]. An important recent trend in HetNets is the increasing co-existence between cellular (e.g., LTE) and local area networks (e.g., WiFi) [5, 6], which in turn gives rise to the co-existence of M2M and H2H (Human-to-Human) traffic.

As numerous unattended wireless M2M devices (sensors, actuators, smart meters, etc.) connect to the LTE network, preventive measures are needed to ensure that their uncontrolled transmissions do not disrupt conventional communication [7, 8]. Along these lines, wireless industry has been designing overload control mechanisms to protect priority human-centric communication. With respective procedures standardized previously for Release 11 of 3GPP LTE, the research community has now moved forward with the goal to enable efficient IoT operation [9, 10]. It is widely known that the characteristics of M2M communications are significantly different from those of H2H traffic. With small and infrequent data chunks typical for M2M, the network needs additional mechanisms to carry such traffic with low blocking probabilities and with minimum data transmission delays. This need is becoming especially urgent in cellular systems, such as LTE, which have been optimized for streaming session-based traffic. Therefore, it is necessary to build mathematical models that allow for the preliminary evaluation measures of M2M traffic in coexistence with H2H users over wireless IoT infrastructure.

This paper investigates the influence of M2M communications on dynamic resource allocation in wireless IoT infrastructure, based on 3GPP LTE standards. In particular, the analytical formulas for the key performance characteristics of M2M and H2H communications, such as data transmission times and blocking probabilities, are developed. Unlike [11, 12], where the simplified model of wireless IoT infrastructure for uniform M2M traffic was considered, in this paper the stochastic model with multiservice M2M traffic is constructed. Our framework allows optimizing radio resource allocation procedures in the LTE cellular network [13, 14 ,15] and achieving understanding of resulting system performance to reach good balance between M2M and H2H communications. The rest of the paper is organized as follows. Section 2 details the mathematical model for multiservice M2M communications and introduces its core assumptions. Also, the main performance measures are defined and the algorithm for computing the stationary probabilities distribution is developed. In Section 3, performance analysis and some case study is performed. The conclusion of this paper is presented in Section 4.

2   System Model of LTE Cell with H2H and M2M Traffic

We consider single LTE cell, which receives call setup requests from H2H users and M2M devices. Let's assume that all H2H users and M2M devices have the same Signal to Noise Ratio (SNR), and that all user devices do not change their position relative to the cell for the period under consideration. Thus, established radio channels will have the same characteristics, and the data rate will depend only on the number of allocated frequency resources - Units of Radio Resource (URR). The system capacity is C URR. All users employ identical H2H-service, such as voice telephony or video on demand. Additionally, the cell supports transmission of M2M data chunks of K types from a plurality of M2M devices, i.e. unlike the monoservice case presented in [11, 12] the multiservice case for M2M communications is considered. The system reserves C_{H2H} URR to provide H2H-services, and $C_{M2M} := C - C_{H2H}$ URR are available for M2M devices. The scheme of a cell serving multiservice M2M and H2H traffic is shown in Fig. 1.

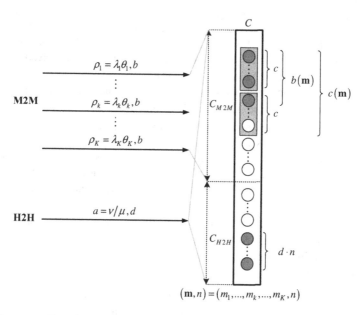

$$(\mathbf{m}, n) = (m_1, \ldots, m_k, \ldots, m_K, n)$$

Fig. 1. Resource allocation model with fixed transmission zones for multiservice M2M communications

A minimum of b URR is required to transmit a M2M data chunk of any type. In order to transmit data chunks, URR are grouped into zones, each measuring c URR. Hence $M := \lfloor c/b \rfloor = \max\{y \in \mathbb{N}: y \le c/b\}$ is the maximum number of data chunks which can be transmitted in a fixed zone of size c URR. Let us introduce another parameter $S := \lfloor C_{M2M}/c \rfloor$ - the number of transmission zones of size c URR to serve traffic from M2M devices.

The arrival flow of k-th type requests from M2M devices is assumed to be Poisson with the rate of λ_k [s^{-1}], whereas the length of each data chunk is exponentially distributed with the mean θ_k [bit]. Denote $\rho_k := \lambda_k \theta_k$ the offered load of k-th type from M2M devices. Suppose a H2H request for service requires d URR. We consider the arrival flow of requests of H2H requests from the users to be Poisson with the rate of v [s^{-1}], while the service time of H2H request to be exponential with the mean of $1/\mu$ [s]. Denote $a := v/\mu$ the offered load of H2H requests.

This model represents a combination of models with unicast and elastic traffic [16]. Elastic traffic is transmitted in accordance with the following principle: after a new request arriving the transmission rate of all already served by the system requests decrease and the available for elastic traffic resource is divided equally between all the accepted requests in the system. We assume that H2H requests and M2M requests operate independently of each other.

In the presented model, three different scenarios are possible when a new M2M request generated by a M2M device of any type arrives to the system.

1. The M2M request is accepted for service and no additional resources for M2M traffic are allocated. This scenario corresponds to the situation when the rescheduling of URR within already allocated band allows to transmit all already served by the system data chunks and the newly arrived data chunk with the transmission rate not less than b .

2. The M2M request is accepted for service and a new fixed transmission zone of size c URR within C_{M2M} URR is allocated for its service. This scenario corresponds to the situation when at the instant of the request arriving the number of already served data chunks does not allow rescheduling of URR within already allocated band to maintain the transmission rate not less than b . At the same time, there should be at least c URR of free (unallocated) resource available for M2M traffic to allocate a new transmission zone within C_{M2M} URR.

3. The M2M request is blocked due to lack of resources without any impact on the rate of the input Poisson process.

Similarly, two different scenarios are possible when a new service request is generated by a H2H device:

1. The H2H request is accepted for service when at the instant of its arriving there are at least d URR free within C_{H2H} URR.

2. Otherwise the H2H request is blocked due to lack of resources.

Let $m_k(t)$ be the number of M2M data chunks transmitted at the instant t and $n(t)$ be the number of users which at the instant t are receiving H2H-service, $t \geq 0$. Then the functioning of the considered LTE cell with H2H and M2M traffic can be described by the $(K+1)$ -dimensional stochastic process

$$\{(\mathbf{m}(t), n(t)), \, t > 0\} := \{(m_1(t), \ldots, m_K(t), n(t)), \, t > 0\}$$

over the state space

$$X := \{\mathbf{m} \geq 0, n \geq 0 : n \cdot d \leq C - c(\mathbf{m}), \, c(\mathbf{m}) \leq C_{M2M}\}, \tag{1}$$

where $c(\mathbf{m}) := c \cdot \lceil b(\mathbf{m})/M \rceil = c \cdot \min\{y \in \mathbb{N}, y \geq b(\mathbf{m})/M\}$ is the number of URR

allocated for the transmission of $m_\bullet := \sum_{k=1}^{K} m_k$ M2M data chunks, and $b(\mathbf{m}) := b \cdot m_\bullet$ is

a minimum number of URR necessary for $m_\bullet$ data chunks transmission. Under the assumptions about Poisson arrival and exponential service the process $\{(\mathbf{m}(t), n(t)), t > 0\}$ is the Markovian process (MP).

Let us introduce blocking sets B_k for the k-th type M2M data chunks and B_{H2H} for H2H users as follows:

$$B_k := \{(\mathbf{m}, n) \in X : n \cdot d > C - c(\mathbf{m} + \mathbf{e}_k) \vee c(\mathbf{m} + \mathbf{e}_k) > C_{M2M}\}, \quad k = \overline{1, K}, \tag{2}$$

$$B_{H2H} := \{(\mathbf{m}, \mathrm{n}) \in X : (n+1)d > C - c(\mathbf{m})\}. \tag{3}$$

Taking into account the minimum transmission rate of b URR for M2M data chunks of any type the blocking sets B_k can be determined by the following formula:

$$B_k = \{(\mathbf{m}, n) \in X : n \cdot d > C - c\lceil (m_\bullet + 1)/c \rceil \vee c\lceil (m_\bullet + 1)/c \rceil > C_{M2M}\}, \quad k = \overline{1, K}. \tag{4}$$

According to [17] and using (2) the blocking set B_{M2M} can be represented as follows:

$$B_{M2M} = \{(\mathbf{m}, n) \in X : n \cdot d > C - c(m_\bullet + 1) \vee c(m_\bullet + 1) > C_{M2M}\}. \tag{5}$$

Under the assumptions above there is a stationary distribution of MP $\{(\mathbf{m}(t), n(t)), t > 0\}$, and steady state probabilities $p(\mathbf{m}, n)$, $(\mathbf{m}, n) \in X$, satisfy the following system of the balance equations:

$$p(\mathbf{m}, n)\left[\sum_{k=1}^{K}\left(\lambda_k \cdot 1\{(\mathbf{m}, n) \notin B_{M2M}\} + (c(m_\bullet)/\theta_k) \cdot 1\{m_k > 0\}\right) + \right.$$

$$+ v \cdot 1\{(\mathbf{m}, n) \notin B_{H2H}\} + \mu n\Big] = \sum_{k=1}^{K} p(\mathbf{m} - \mathbf{e}_k, n) \cdot \lambda_k \cdot 1\{m_k > 0\} +$$

$$+ \sum_{k=1}^{K} p(\mathbf{m} + \mathbf{e}_k, n)(c(m_\bullet + 1)/\theta_k) \cdot 1\{(\mathbf{m}, n) \notin B_{M2M}\} + \tag{6}$$

$$+ p(\mathbf{m}, n+1)\mu(n+1) \cdot 1\{n > 0\} + p(\mathbf{m}, n-1) \cdot v \cdot 1\{n > 0\},$$

$$k = \overline{1, K}, \quad (\mathbf{m}, n) \in X. .$$

The Fig. 2 shows that Kolmogorov's criterion is satisfied; therefore MP $\{(\mathbf{m}(t), n(t)), t > 0\}$ is reversible Markovian process. Then its stationary probabilities satisfy to system of the detailed balance equations:

$$\begin{cases} p(\mathbf{m}, n) \cdot (c(m_\bullet)/\theta_k) = p(\mathbf{m} - \mathbf{e}_k, n) \cdot \lambda_k, \ m_k > 0, \ k = \overline{1, K}, \ (\mathbf{m}, n) \in X \\ p(\mathbf{m}, n) \cdot \mu n = p(\mathbf{m}, n-1) \cdot v, \ n > 0, \ (\mathbf{m}, \mathrm{n}) \in X. \end{cases} \tag{7}$$

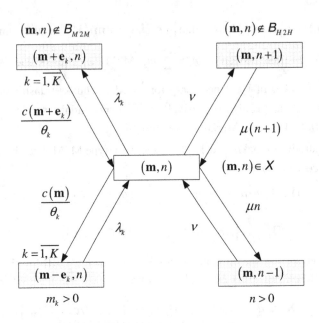

Fig. 2. Fragment of state transitions diagram for MP $\{(\mathbf{m}(t), n(t)),\ t > 0\}$

From the system of equations (7) stationary distribution of the MP $\{(\mathbf{m}(t), n(t)),\ t > 0\}$ can be found in an explicit product form:

$$p(\mathbf{m}, n) = G^{-1}(X) c^{m_\bullet} \left(\prod_{i=1}^{m_\bullet} \lceil b \cdot i/M \rceil \right)^{-1} \cdot \left(\prod_{k=1}^{K} \rho_k^{m_k} \right) \cdot (a^n/n!),\ (\mathbf{m}, n) \in X\ , \qquad (8)$$

where $G(X) = p^{-1}(\mathbf{0}, 0)$ is a normalizing constant.

From the system of equations (7) we have also obtained an important result, which can be written in the form of given below recursive algorithm.

The algorithm for computing the stationary distribution

Input data: $C, C_{H2H}, K, c, b, v, \mu, d;\ \lambda_k, \theta_k, k = \overline{1, K}$.

Step 1. Calculation of the system parameters

 1.1. $\rho_k = \lambda_k \cdot \theta_k, a := v/\mu$.

 1.2. $C_{M2M} = C - C_{H2H}, S = \lfloor C_{M2M}/c \rfloor, M = \lfloor c/b \rfloor$.

Step 2. Calculation of the non-normalized probability $g(\mathbf{m}, n)$ and normalizing constant $G(X)$

 2.1. $g(\mathbf{0}, 0) = 1, G(X) = 1$.

 2.2. $c(m_\bullet) = c \cdot \lceil b(m_\bullet)/c \rceil, b(m_\bullet) = b \cdot m_\bullet,\ m_\bullet = 0, ..., \lfloor S \cdot c/b \rfloor$.

2.3. Calculation of $g(\mathbf{m},n)$:

2.3.1. $g(m_1,...,0,...,0,0) = g(m_1 - 1,...,0,...,0,0)(\rho_1/c(m_\bullet))$,

$G(\mathbf{X}) = G(\mathbf{X}) + g(m_1,...,0,...,0,0)$, $m_1 = 1,...,\lfloor S \cdot c/b \rfloor$.

2.3.2. $g(m_1,m_2,...,0,...,0,0) = g(m_1,m_2 - 1,...,0,...,0,0)(\rho_2/c(m_\bullet))$,

$G(\mathbf{X}) = G(\mathbf{X}) + g(m_1,m_2,...,0,...,0,0)$, $m_2 = 1,...,\lfloor (S \cdot c - m_\bullet \cdot b)/b \rfloor$,

$m_\bullet = 0,...,\lfloor S \cdot c/b \rfloor$.

2.3.3. $g(m_1,m_2,...,m_k,...,m_K,0) = g(m_1,m_2,...,m_k,...,m_K - 1,0)(\rho_K/c(m_\bullet))$,

$G(\mathbf{X}) = G(\mathbf{X}) + g(m_1,m_2,...,m_k,...,m_K,0)$, $m_K = 1,...,\lfloor (S \cdot c - m_\bullet \cdot b)/b \rfloor$,

$m_\bullet = 0,...,\lfloor S \cdot c/b \rfloor$.

2.4 $g(m_1,m_2,...,m_k,...,m_K,n) = g(m_1,m_2,...,m_k,...,m_K,n-1)\dfrac{a}{n}$,

$n = 1,...,\lfloor (C - c(m_\bullet))/d \rfloor$, $m_\bullet = 0,...,\lfloor S \cdot c/b \rfloor$.

Step 3. Calculation of stationary distribution $p(\mathbf{m},n)$.

3.1. $p(m_1,m_2,...,m_k,...,m_K,n) = g(m_1,m_2,...,m_k,...,m_K,n)/G(\mathbf{X})$,

$n = 0,...,\lfloor (C - c(m_\bullet))/d \rfloor$, $m_\bullet = 0,...,\lfloor S \cdot c/b \rfloor$.

Knowing the stationary probability distribution of $p(\mathbf{m},n)$, it is possible to calculate blocking probabilities of requests from M2M devices and H2H users:

$$B_{M2M} := \sum_{(\mathbf{m},n) \in B_{M2M}} p(\mathbf{m},n),$$

$$B_{H2H} := \sum_{(\mathbf{m},n) \in B_{H2H}} p(\mathbf{m},n)$$

The formula for the mean M2M data chunk transmission time may be given as:

$$T_k := \sum_{(\mathbf{m},n) \in X} m_k \cdot p(\mathbf{m},n) / \lambda_k (1 - B_{M2M}), \ k = \overline{1,K}. \tag{9}$$

Further, we continue by numerically analyzing the probability measures of the considered resource distribution model with the fixed transmission zone for M2M traffic in LTE cell with H2H users.

3   Numerical Analysis

As an example, we consider a stand-alone LTE cell with the peak capacity of $C = 50$ Mbps, which is distributed between H2H users and M2M devices. For the H2H user service, the system reserves $C_{H2H} = 10$ Mbps of its capacity. We consider that all M2M devices transmit two types of data chunks, i.e. $K = 2$. Assume that $\theta_1 = 100$ Kbyte and $\lambda_1 \in [2,...,98]$; $\theta_2 = 200$ Kbyte and $\lambda_2 \in [1,...,49]$. In this case the arrival flows of requests are equal $\rho_1 = \rho_2$. As an illustration of a H2H service, we consider streaming video, which has a requirement on the minimum throughput as $d = 3$ Mbps. Assume the H2H offered load rate to be $a = 5$. Let up to $S = 2$ fixed transmission zones be allocated for M2M data chunks transmission, each of which comprises $c = 20$ Mbps. We also determine 1 URR = 880 Kbps. Fig. 3 introduces plots illustrating H2H request blocking probabilities, M2M data chunk blocking probabilities, and mean chunk transmission time on increasing M2M offered load.

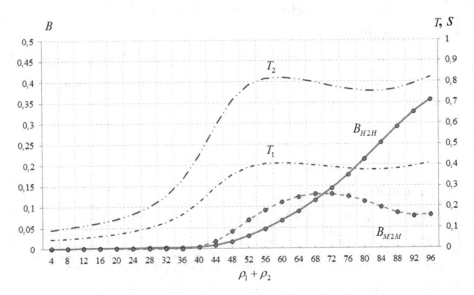

Fig. 3. Blocking probabilities and mean data chunk transmission time

The figure indicates that the mean chunk transmission time varies significantly with the changing offered load. With the increasing of the offered M2M traffic load $\rho_1 + \rho_2$ the number of available URR of the first allocated zone is decreasing, while mean transmission time is increasing. When the transmission rate reaches its minimum value of b URR, the second fixed zone is allocated wherein the mean transmission time is slightly reduced. With further growth of the offered M2M load the number of available URR is decreasing so the mean transmission time tends to its maximum.

To select the optimal strategy of radio resources allocation, we recommend to use a set of criteria that take into account the features of M2M-service traffic. In this paper the problem of finding the optimal size c of a fixed transmission zone allocated for M2M data chunks transmission is formulated as multicriterion optimization problem for the objective function vector $\mathbf{U}(c) = (U_1(c), U_2(c))$, where scalar objective functions are the mean M2M data chunk transmission time $U_1(c)$ and the mean number of occupied URR $U_2(c)$:

$$U_1(c) = \sum_{k=1}^{K} T_k(c), \tag{10}$$

$$U_2(c) = \sum_{(\mathbf{m},n)\in X} \left(c \cdot \left\lceil \frac{m_*}{M} \right\rceil + d \cdot n \right) \cdot p(\mathbf{m},n). \tag{11}$$

As illustration we use one of commonly used approaches, i.e. a weighted sum method [18]:

$$\min_c U(c) = \min_c \left(\varpi_1 U_1(c) + \varpi_2 U_2(c) \right) \tag{12}$$

subject to: $\varpi_1 + \varpi_2 = 1$, $0 \le \varpi_i \le 1$, $i = 1,2$;

$$c \cdot S \le C_{M2M};$$

$$B_{H2H} \le B^*_{H2H};$$

$$B_{M2M} \le B^*_{M2M}.$$

Numerical analysis of the multicriterion optimization problem was carried out for the network with two types of M2M data chunks $S = 2$, the offered M2M traffic load $\rho_1 + \rho_2$, and the offered H2H traffic load $a = 5$. As constraints on blocking probabilities of requests from M2M devices and H2H users the following values were used: $B^*_{H2H} = 10^{-3}$ and $B^*_{M2M} = 10^{-2}$. Fig. 4 shows the results of multicriterion optimization problem solution with the weighted sum method with weights $\varpi_1 = 0,8$ and $\varpi_2 = 0,2$. For a given set of weighting coefficients the optimization problem has a unique solution at the point $c = 15$ URR. This optimal size of the fixed transmission zone permits to provide the necessary quality of service for M2M-traffic and increases the resistance of a single LTE cell to overload due to the presence of free (unoccupied) radio resources.

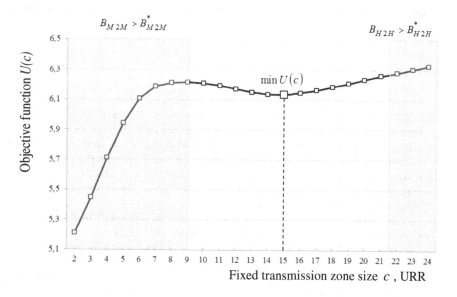

Fig. 4. Numerical solution of multicriterion optimization problem

4   Conclusion

On the mobile communications side, further progress is necessary in enabling higher-bandwidth for IoT infrastructure. Service quality and availability (connectivity, latency, mobility, energy-efficiency, etc.) need to be improved as well by offering more adequate mechanisms to handle heterogeneity in mobile devices, clouds, and wireless networks. In this paper, we addressed a resource sharing problem for M2M traffic in 3GPP LTE stand-alone cell as a part of future research for wireless IoT infrastructure. We give an analytical model with heterogeneous elastic traffic from M2M devices and minimum rate guarantees. The resource allocation scheme is based on fixed transmission zones at which traffic from the M2M devices is being served according to the PS discipline. We propose an analytical solution to calculate the model performance measures under the assumption of simplified physical model.

In the future we expect to continue the analysis of the model by removing the obvious simplifications. The case of general distribution of data chunks length, MMPP arrivals, different SNR values for H2H and M2M users will be investigated. It is evident that these assumptions do not expect the analytical solution of closed form and one has nothing to do but to use simulation techniques to analyze the performance measures of the above resource allocation scheme.

An interesting task for future study is the optimization of various admission control schemes for wireless IoT infrastructure. In prospect, we aim to find a way of simulating the initial data for analytical modeling because, for the time being, we chose their values in a theoretical way. Values based on a simulation approach are more interested, and can enhance our proposal.

Acknowledgements. The reported study was partially supported by the RFBR, research projects No. 13-07-00953, 14-07-00090.

References

1. Świątek, P., Juszczyszyn, K., Brzostowski, K., Drapała, J., Grzech, A.: Supporting content, context and user awareness in future internet applications. In: Álvarez, F., Cleary, F., Daras, P., Domingue, J., Galis, A., Garcia, A., Gavras, A., Karnourskos, S., Krco, S., Li, M.-S., Lotz, V., Müller, H., Salvadori, E., Sassen, A.-M., Schaffers, H., Stiller, B., Tselentis, G., Turkama, P., Zahariadis, T. (eds.) FIA 2012. LNCS, vol. 7281, pp. 154–165. Springer, Heidelberg (2012)
2. Cisco Visual Networking Index: Global Mobile Data Traffic Forecast Update. http://www.cisco.com/c/en/us/solutions/collateral/service-provider/visual-networking-index-vni/white_paper_c11-520862.html
3. Andrews, J.G.: Seven ways that HetNets are a cellular paradigm shift. In: IEEE Communications Magazine 51, pp. 136–144. IEEE Press, New York (2013)
4. Bangerter, B., Talwar, S., Arefi, R., Stewart, K.: Networks and devices for the 5G era. In: IEEE Communications Magazine 52, pp. 90–96. IEEE Press, New York (2014)
5. Bennis, M., Simsek, M., Saad, W., Valentin, S., Debbah, M.: When cellular meets WiFi in wireless small cell networks. In: IEEE Communications Magazine 51, pp. 44–50. IEEE Press, New York (2013)

6. Galinina, O., Andreev, O., Gerasimenko, S., Koucheryavy, M., Himayat, Y., Shu-Ping Yeh, N., Talwar, S.: Capturing spatial randomness of heterogeneous cellular/WLAN deployments with dynamic traffic. In: IEEE Journal on Selected Areas in Communications 32(6), pp. 1083–1099. IEEE Press, New York (2014)

7. Wang, L., Kuo, G.S.: Mathematical modeling for network selection in heterogeneous wireless networks – a tutorial. In: IEEE Communications Surveys & Tutorials 15, pp. 271–292. IEEE Press, New York (2013)

8. Gerasimenko, M., Petrov, V., Galinina, O., Andreev, S., Koucheryavy, Y.: Impact of MTC on Energy and Delay Performance of Random-Access Channel in LTE-Advanced. Wiley Transactions on Emerging Telecommunications Technologies, Special Issue: Machine-to-Machine: An Emerging Communication Paradigm 24(4), 366–377 (2013)

9. Beale, M.: Future challenges in efficiently supporting M2M in the LTE standards. In: WCNC Workshops, pp. 186–190. IEEE (2012)

10. Hasan, M., Hossain, E., Niyato, D.: Random access for machine-to-machine communication in LTE-advanced networks: issues and approaches. In: IEEE Communications Magazine 51, pp. 86–93. IEEE Press, New York (2013)

11. Zheng, K., Hu, F., Wang, W., Xiang, W., Dohler, M.: Radio resource allocation in LTE-advanced cellular net-works with M2M communications. In: IEEE Communications Magazine 50(7), pp. 184–192. IEEE Press, New York (2012)

12. Borodakiy, V.Y., Buturlin, I.A., Gudkova, I.A., Samouylov, K.E.: Modelling and Analysing a Dynamic Resource Allocation Scheme for M2M Traffic in LTE Networks. In: Balandin, S., Andreev, S., Koucheryavy, Y. (eds.) NEW2AN 2013 and ruSMART 2013. LNCS, vol. 8121, pp. 420–426. Springer, Heidelberg (2013)

13. Gudkova, I., Samouylov, K., Buturlin, I., Borodakiy, V., Gerasimenko, M., Galinina, O., Andreev, S.: Analyzing Impacts of Coexistence between M2M and H2H Communication on 3GPP LTE System. In: Mellouk, A., Fowler, S., Hoceini, S., Daachi, B. (eds.) WWIC 2014. LNCS, vol. 8458, pp. 162–174. Springer, Heidelberg (2014)

14. GPP specifications: 22.368, Service requirements for Machine-Type Communications (MTC). http://www.3gpp.org/DynaReport/22368.htm

15. GPP specifications: 37.888, Study on provision of low-cost Machine-Type Communications (MTC) User Equipments (UEs) based on LTE (Release 12). http://www.3gpp.org/DynaReport/36888.htm

16. GPP specifications: 37.869, Study on enhancements to Machine-Type Communications (MTC) and other mobile data applications. http://www.3gpp.org/DynaReport/37869.htm

17. Gudkova, I.A., Samouylov, K.E.: Analysis of an admission model in a fourth generation mobile network with triple play traffic. Automatic Control and Computer Sciences 47(4), 202–210 (2013)

18. Cohon, J.L.: Multiobjective programming and planning: Academic Press, New York (1978)

Towards On-Demand Resource Provisioning for IoT Environments

Andreas Kliem[✉] and Thomas Renner

Department of Telecommunication Systems, Complex and Distributed IT Systems,
Technische Universität Berlin, Einsteinufer 17, 10587 Berlin, Germany
{Andreas.Kliem,Thomas.Renner}@tu-berlin.de

Abstract. The set of connected embedded devices surrounding and providing resources to us will constantly grow in the future. Currently these devices are often treated as pure data sources and there is no notion of providing and provisioning the compute and storage resources they offer to the users. Paradigms like Infrastructure as a Service, Platform as a Service or, Pay as you Go are very popular and successful in the Cloud Computing domain. We will discuss whether these paradigms can be applied to the Internet of Things domain in order to create a Cloud of Things that is surrounding us and provides resources in an ubiquitous fashion.

1 Introduction

Following the Ubiquitous Computing [1] and Internet of Things (IoT) [2] visions, connected and embedded devices like sensors, smart phones, or entertainment devices build the foundation for many applications we use every day. Examples are E-Health [3], Smart Homes [4], or Transportation [5]. A significant amount of infrastrcutre needs to be deployed and managed in order to take advantage of all the interconnected devices. Many approaches that investigate in providing such an infrastructure follow a Gateway based solution, which can be briefly described as a single system (e.g. a router), that integrates available sensors and dispatches the resulting data streams between sources (e.g. sensors) and corresponding sinks (e.g. end-user applications hosted at a compute center) [6] [7]. The Gateway therefore acts as a bridge between data acquisition and data processing.

Gateway based solutions often rely on the assumption that devices can be treated as pure data sources. Instead, the upcoming category of smart devices (e.g. Smart TVs, Smart Phones) additionally offers compute- and storage resources. This leads to a resource oriented view, where each node participating in the IoT infrastructure can offer one or more of the three resources data, compute or, storage as a service to the users. Compute- and storage resources can be used for in-network processing for instance, which reduces the demand for communication and improves energy efficiency [8]. Compared to Gateway based solutions, the gap between data acquisition and processing is mitigated because both tasks are handled by the same infrastructure. Similar to Mesh networks [9], pure data resources

© Springer International Publishing Switzerland 2015
N.T. Nguyen et al. (Eds.): ACIIDS 2015, Part II, LNAI 9012, pp. 484–493, 2015.
DOI: 10.1007/978-3-319-15705-4_47

(i.e. sensors) are not longer bound to a specific Gateway, but rather integrated where they are needed by any kind of smart device capable of doing so. A typical use case can be found in apartment-sharing communities, where each resident brings its own devices and at least some of them are shared by the community (e.g. in the living room/kitchen).

Consuming resources as a service, such as Infrastructure as a Service (IaaS) and Platform as a Service (PaaS) has become popular along with Cloud Computing. By applying the efficient, scalable, and easy to use Everything-as-a-service [10] model to the IoT, multiple users can explore and utilize the offered resources at the same time, while their devices become an active part of the infrastructure. With this motivation, we introduce our proposal of the Device Cloud. We will identify participating entities and required components. An architecture draft of a distributed and resource oriented middleware platform, called the Device Cloud Middleware, will be introduced in Section 2. Arising challenges and corresponding solution approaches will be discussed based on a review of related application domains and concepts in Section 3. Finally, Section 4 will conclude the paper.

2   Device Cloud Architecture

This section introduces our proposal of the Device Cloud architecture shown in Figure 1. Smart devices establish a distributed middleware platform, the Device Cloud Middleware, which offers a shared resource pool that is composed of the physical resources offered by the devices. On top of that resource pool, the middleware provides a homogeneous service execution environment to efficiently utilize the resources (i.e. data, compute and, storage). The architecture is separated into three layers, the *Physical Space*, the *Runtime Space* and the *Social Space*. The main goal of this abstraction is to hide the complexity, in order that neither sensors or devices nor applications or users should have to take care of the heterogeneity of the corresponding spaces (e.g. which node integrates another node or executes an application).

The **Physical Space** includes all collaborating physical devices, systems and networks. A segmentation into three node types is defined: *Device Nodes (DN)* refer to data sources (i.e. sensors), *Aggregation Nodes (AN)* refer to smart devices that additionally provide compute- or storage resources and *Backend Nodes (BN)* refer to data stream sinks and management components hosted on dedicated servers. ANs are the core nodes of the Device Cloud approach. The Device Cloud Middleware facilitates the ANs to expose all resources made available by the *Physical Space* to the upper layers. As shown in Figure 1, ANs bridge the gap between the *Physical* and the *Runtime Space*. The physical AN device itself and its offered resources belong to the *Physical Space* while the Device Cloud Middleware software hosted by the ANs belongs to the *Runtime Space*.

The **Runtime Space** manages and provisions available data, compute and, storage resources by means of a shared resource pool. The main challenge is to

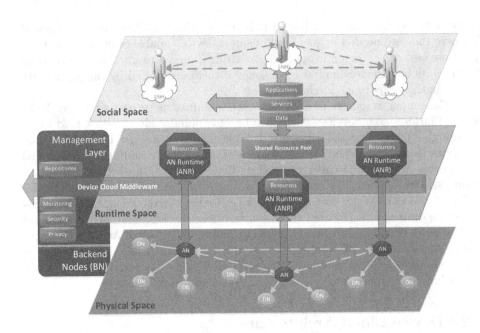

Fig. 1. High level Device Cloud architecture with space abstraction

keep transparency regarding heterogeneity and availability, because nodes in the *Physical Space* can come and go, rely on different communication and transport protocols as well as data formats. Besides on demand resource provisioning, the Device Cloud Middleware, which basically constitutes the *Runtime Space*, decouples the *Physical* from the *Social Space*. Following core components are defined within the *Runtime Space*:

- **Aggregation Node Runtime (ANR)** is a software node running on ANs. All nodes together establish the distributed Device Cloud Middleware. The distributed fashion is hidden from the other layers. Besides resource integration and provisioning, ANRs provide a uniform service execution environment that abstracts from the underlying device specific characteristics. In general, ANRs provide and manage the utilization of the shared resource pool similar to the Cloud Computing PaaS paradigm.
- The **Management Layer** monitors the state of the Device Cloud Middleware nodes and hosts knowledge repositories used by the ANRs to adapt to the current environment. Knowledge can refer to the state of other nodes or the network, but especially targets software modules and configurations, that are required to handle certain devices (e.g. device driver and data transformation modules needed to integrate a discovered DN). In addition, the Management Layer has to provide monitoring and management features to cope with issues like privacy, security, Quality of Service (QoS) or, access control.

- The **Shared Resource Pool** can be envisioned as the set of data, compute and, storage resources provided by the *Physical Space*. The Device Cloud Middleware manages and provisions these resources to users and applications.

In order to provide a Cloud-like on demand resource provisioning among multiple users, the Device Cloud Middleware needs to map available resources to applications. Therefore many conditions like data locality, characteristics of the transport (e.g. bandwidth), characteristics of the ANs themselves (e.g. disappearing nodes) or, QoS have to be considered, which is discussed in more detail in Section 3.

The **Social Space** reflects all collaborating users and their applications in the Device Cloud. Based on the *Runtime Space*, it provides an environment for ubiquitous applications and services, in which resources are not tightly bound to one user but rather can be shared among each other. From the perspective of a user, the platform that serves his needs is no longer a set of statically bound physical devices. Instead, it is a set of shared resources and ANRs that are deployed on top of the overall physical infrastructure. Therefore, several user-centric "virtualized" platforms can coexist on the same physical nodes. Moreover, a user's personal platform becomes highly adaptive, scalable and mobile, since resources appearing close to the user can be shared and integrated on the fly.

3   Device Cloud Challenges

With emphasis on the difficulties that arise out of the introduced Device Cloud architecture, selected IoT related middleware challenges [6] [12] will be extended and discussed in this section.

3.1   Runtime Device Integration and Abstraction

One challenge is to seamlessly integrate and configure the heterogeneous resources of the *Physical Space*. The Device Cloud needs to integrate and handle mobile and stationary devices from different application domains with different characteristics (e.g. communication protocols, data formats, operating systems). Due to the constantly changing environment, approaches that integrate and configure devices manually are not feasible [11]. Because DNs and ANs are shared, belong to different users/authorities and each AN can have different DN integration capabilities (e.g. supported communication protocols), it becomes unlikely that administrators or application developers can take responsibility for integrating and managing all devices manually. Moreover, it is necessary to embrace this physical dynamics in a unified way on a low layer and to hide the complexity from the upper layers (e.g. the *Social Space*) [13]. Abstraction can be achieved by providing an automatic device integration and management process. Therefore, the Device Cloud needs to understand the devices (e.g. capabilities, data structures they produce, device configurations) or at least needs to be able to gather required integration knowledge on demand (e.g. appearance of a new DN).

Various approaches investigated in automatic or semi-automatic device integration at runtime, for instance OpenHAB [14], DOG [15] and Hydra [16]. They follow a gateway based approach, as already discussed in Section 1. Usually the gateway contains all required device specific control and protocol logic. Due to the scalability and the amount of available devices, it is not feasible to store all required device specific logic (e.g. device drivers) on each single AN in the Device Cloud. Instead, a runtime integration mechanism is required, that takes the resource constraints of the ANs into account. Moreover, none of the approaches gives a clear account on how new devices can be discovered at runtime and how the platform can be able to handle new devices that were not foreseen during system development [17].

3.2   Interoperability, Data-Models and Nomenclatures

Once devices are connected and data resources appear, sufficient knowledge about the characteristics and the content of the resulting data streams is required to properly link data sources and sinks (i.e. devices to applications). Moreover, efficient routing and QoS need to be realized, while preserving transparency introduced with the space abstraction and without having a-priori knowledge about the collaborating entities. Especially semantic interoperability [19], that refers to a uniform understanding of the generated and exchanged information, is important to enable functionalities like content-based data routing, context-awareness, QoS or, dependency resolution between application layer components. In general, interoperability is difficult to achieve because devices are heterogeneous and follow different or no standards. Therefore, apart from global standardization efforts [20], generic and adaptable core interoperability models are required. According to other interoperability related standardization activities, such as the ISO/IEEE 11073 [21] known from the medical application domain, three core models can be defined:

- **Information Model:** An information model characterizes and describes the information that represents the current state of a device and the knowledge it provides for other devices (e.g. nomeclatures, data models).
- **Service Model:** The service model describes basic primitives that are sent between devices in order to exchange knowledge defined by the information model. Information and service model are tightly coupled to the topic device abstraction.
- **Communication Model:** The communication model defines abstract requirements for the transport (e.g. a reliable channel), how the abstract syntax of the information model has to be transferred into a transfer syntax, or a state model, which specifies the communication process or life cycle of a connection between devices.

The core models need to be specified in a generic and adaptable way. Though, it is important, that the core models are abstract to ensure longevity and sufficient coverage of the heterogeneous resources. Concrete implementations of the core models (e.g. Bluetooth and ZigBee communication models) have to be injectable into the Device Cloud Middleware at runtime.

3.3 Service Execution Environment

Besides proper utilization of data resources, another key aspect of the Device Cloud is on demand provisioning of compute- and storage resources. Therefore, a service execution environment, that allows applications from the *Social Space* to utilize available resources from the *Physical Space*, is required. Again, the transparency constraint is important. Neither users nor application developers should take care of the distributed fashion of the Device Cloud Middleware, the heterogeneous resources or, disappearing parts of the execution environment (i.e. if an ANR disappears/fails). Accordingly, the main challenges of the service execution environment are to efficiently provision and distribute the available resources, hide heterogeneity, provide service migration and data replication metrics, scalability in case of insufficient resources in the shared pool and, allow for isolation between the applications.

Hiding heterogeneity is related to the interoperability challenge discussed in the previous section. Service migration and data replication metrics are essential to optimize the behavior of the distributed Device Cloud Middleware and to cope with appearing and disappearing nodes, that can significantly change the topology and capabilities of the physical resources the Device Cloud Middleware is based on. With the help of the management layer, the behavior and state of the Device Cloud Middleware and the participating nodes needs to be monitored, which allows to provide knowledge required for proper migration and replication decisions. Related approaches that target service migration and distributed execution environments can be found in the area of multi agent systems [22]. Due to the distributed fashion of the middleware, the scalability challenge is rather easy to tackle by allocating ANRs on BNs (i.e. dedicated servers). Providing isolation is important in terms of privacy and security, if several applications and data of different users are moved between and executed on shared resources. Therefore, the service execution environment is required to provide proper isolation capabilities like virtualization or sandboxes, while paying respect to the resource constraints of ANs. Upcoming concepts like Cyber-Physical Clouds [23] [24], Vehicular Clouds or Mobile Cloud Computing [25] show future directions towards tackling this challenge. In these approaches, devices basically act as servers that move in space and execute virtual devices. Virtual devices can migrate between physical ones, which is referred to as cyber-mobility (i.e. moving between devices hosts). Additionally, virtual devices can move with their hosts, which is referred to as physical mobility [24]. Similar to regular Cloud Computing, this allows for efficient resource utilization and enables robust and safe execution of virtual devices, since each virtual devices can be isolated from each other.

3.4 Multitenancy

On demand resource provisioning and collaboration requires concepts to share the access to resources among multiple users. Resources can be categorized as exclusive or non-exclusive. Exclusive compute- and storage resource are private to their owners and will not become part of the shared resource pool. Exclusive data resources instead, can become part of the shared resource pool if necessary. From the *Physical Space* point of view, a DN, allowing only one AN to connect to it, is exclusive for instance. However, by applying techniques such as sensor virtualization [26] [27], several virtual instances can be exposed within the *Runtime Space*. More generally, this refers to the ability of the Device Cloud Middleware to provide a single data resource (i.e. its data stream) to several applications possibly using different data formats. In addition to sharing data resources, sharing compute- and storage resources as already discussed in Section 3.3 has to be made available by deploying a homogeneous execution environment on top of the heterogeneous ANs.

Besides utilizing the shared resources, the Device Cloud Middleware has to provide provisioning capabilities, which includes access control, accounting and dynamic granting and withdrawal of device access tokens. In contrast to regular IaaS clouds for instance, the Device Cloud has to manage a peer to peer collaboration of entities. Each participating user owning some resources (i.e. devices) can act as a resource provider and consumer at the same time. Therefore it is necessary, that users are able to authenticate against each other and that identities of resources and users can be proofed. In order to allow for on demand provisioning and reduce the amount of user interaction, state models, decision metrics and, corresponding rule sets are required.

3.5 Security and Privacy

Apart from related issues already mentioned in the previous challenges (e.g. isolation, access control), security and privacy is a key challenge that needs to be addressed on a large scale due to the huge amount of participating devices and users. The Device Cloud needs to protect against and recover from security attacks, that arise out of the devices physical accessibility [28], their reduced ability to host own security modules and, the platform openness [18]. Moreover, sharing resources requires to establish secure end-to-end links between participants. Additionally, because of the online deployment features discussed in Section 3.1, a notion of trust for exchangeable software modules (i.e. identity, access permissions) needs to be provided. These challenges are related to mobile agents security concepts [29].

Another important challenge in IoT is the definition of privacy policies [18]. The huge amount of policies, required to ensure user-centric control of personal data and devices, needs to be managed and consolidated in an automated fashion. On the Internet, protocols like OAuth 2.0 [30] and OpenID [31] have been defined to enable for privacy-respecting and user consent authorization models. Applying these protocols to the Device Cloud can be used to increase privacy.

3.6 Non-functional Requirements and QoS

The integration of the various IoT devices and the collaboration of the participating entities lead to a higher amount of data and increased competition regarding the shared resources (e.g. network, compute). Since the Device Cloud will allow to host applications from both non-critical (e.g. entertainment) and critical (e.g. e-Health) application domains on the same infrastructure, it is necessary to provide mechanisms that allow to express QoS constraints and non-functional requirements such as reliability or latency. Due to the distributed design of the Device Cloud Middleware, it is crucial that the required data is available at the required location within the required time frame and can be processed properly.

Regarding routing of data streams, we have to deal with many data streams that share the same commodity transport (i.e. Ethernet) and highly differ regarding their characteristics (e.g. bandwidth, latency requirements) [20]. When talking about devices and data resources in the *Physical Space*, we also have to consider actuators, that often introduce non-functional requirements like determinism or reliability. Therefore it is necessary to introduce respective capabilities and models, that allow to easily apply, monitor and, adjust these QoS policies in an automated fashion. Optimization and on-line adjustment is important because of the dynamic set of nodes participating. Promising approaches and technologies like Software Defined Networks [32] or Data-centric infrastructures (i.e. OMG DDS [33]) can be integrated into the Device Cloud Middleware for this purpose.

4 Conclusion

The presented architecture approach tries to facilitate the upcoming and growing category of smart devices in order to apply the popular IaaS, PaaS and, on-demand resource provisioning paradigms known from Cloud Computing to the IoT domain. The approach establishes the notion of a shared resource pool, that not only covers resources in terms of data (i.e. sensors) but also considers compute- and storage resources provided by the growing set of smart devices (e.g. smart TVs, smart phones, routers). This not only targets resource utilization or in-network processing of data, but also offers new ways of people to people and people to thing collaboration by sharing physical resources that surround us. We discussed challenges that arise from merging Cloud Computing features with IoT infrastructures and provided possible solutions by reviewing approaches from related research initiatives. Although a lot of effort will be required to integrate and adapt these approaches, especially because of the disappearing boundary between resource consumers and providers, we showed that the theoretical and technical foundations required were partly already investigated. Therefore, an application of IaaS and PaaS to IoT seems reasonable and is underlined by the increasing amount of resources smart devices will offer in the future.

References

1. Weiser, M.: The computer for the 21st century. Scientific American **265**(3), 94104 (1991). http://dx.doi.org/10.1038/scientificamerican0991-94
2. Ashton, K.: That 'Internet of Things' Thing. RFID Journal (2011)

3. Wang, Y.-W., Yu, H.-L., Li, Y.: Notice of retraction internet of things technology applied in medical information, pp. 430–433 (April 2011)
4. Chong, G., Zhihao, L., Yifeng, Y.: The research and implement of smart home system based on internet of things, pp. 2944–2947 (September 2011)
5. Yuqiang, C., Jianlan, G., Xuanzi, H.: The research of internet of things' supporting technologies which face the logistics industry, pp. 659–663 (December 2010)
6. Bandyopadhyay, S., Sengupta, M., Maiti, S., Dutta, S.: Role of middleware for internet of things: A study. International Journal of Computer Science and Engineering Survey 2(3), 94–105 (2011)
7. Alamri, A., Ansari, W.S., Hassan, M.M., Hossain, M.S., Alelaiwi, A., Hossain, M.A.: A survey on sensor-cloud: Architecture, applications, and approaches. International Journal of Distributed Sensor Networks 2013, 118 (2013). http://dx.doi.org/10.1155/2013/917923
8. Laukkarinen, T., Suhonen, J., Hännikäinen, M.: An embedded cloud design for internet-of-things. International Journal of Distributed Sensor Networks 2013 (2013)
9. Akyildiz, I.F., Wang, X., Wang, W.: Wireless mesh networks: a survey. Computer Networks 47(4), 445–487 (2005). http://www.sciencedirect.com/science/article/pii/S1389128604003457
10. Banerjee, P., Friedrich, R., Bash, C., Goldsack, P., Huberman, B., Manley, J., Patel, C., Ranganathan, P., Veitch, A.: Everything as a service: Powering the new information economy. Computer 44(3), 36–43 (2011)
11. Perera, C., Jayaraman, P., Zaslavsky, A., Christen, P., Georgakopoulos, D.: Dynamic configuration of sensors using mobile sensor hub in internet of things paradigm. In: 2013 IEEE Eighth International Conference on Intelligent Sensors, Sensor Networks and Information Processing, pp. 473–478 (April 2013)
12. Nagy, M., Katasonov, A., Khriyenko, O., Nikitin, S., Szydlowski, M., Terziyan, V.: Challenges of middleware for the internet of things. University of Jyvaskyla, Tech. Rep. (2009)
13. Lee, E.A.: Cyber physical systems: design challenges. In: 2008 11th IEEE International Symposium on Object Oriented Real-Time Distributed Computing, ISORC, pp. 363–369 (2008)
14. openHAB. http://www.openhab.org (Online; accessed August 15, 2014)
15. Bonino, D., Castellina, E., Corno, F.: The dog gateway: enabling ontology-based intelligent domotic environments. IEEE Transactions on Consumer Electronics 54(4), 1656–1664 (2008)
16. Eisenhauer, M., Rosengren, P., Antolin, P.: A development platform for integrating wireless devices and sensors into ambient intelligence systems. In: 6th Annual IEEE Communications Society Conference on Sensor, Mesh and Ad Hoc Communications and Networks Workshops, SECON Workshops 2009, pp. 1–3 (2009)
17. Kim, J.E., Boulos, G., Yackovich, J., Barth, T., Beckel, C., Mosse, D.: Seamless integration of heterogeneous devices and access control in smart homes. In: 2012 8th International Conference on Intelligent Environments (IE) (June 2012)
18. Stankovic, J.: Research directions for the internet of things. IEEE Internet of Things Journal 1(1), 3–9 (2014)
19. Heiler, S.: Semantic interoperability. ACM Computing Surveys 27(2), 271–273 (1995). http://dx.doi.org/10.1145/210376.210392
20. Atzori, L., Iera, A., Morabito, G.: The internet of things: A survey. Computer Networks 54(15), 2787–2805 (2010). http://www.sciencedirect.com/science/article/pii/S1389128610001568

21. ISO/IEC/IEEE Health Informatics-Personal Health Device Communication-Part 20601: Application Profile-Optimized Exchange Protocol. ISO/IEEE 11073-20601:2010(E), pp. 1–208 (January 2010)
22. Hirsch, B., Konnerth, T., Heßler, A.: Merging agents and services the jiac agent platform. In: Multi-Agent Programming. Springer, pp. 159–185 (2009)
23. Craciunas, S.S., Haas, A., Kirsch, C.M., Payer, H., Röck, H., Rottmann, A., Sokolova, A., Trummer, R., Love, J., Sengupta, R.: Information-acquisition-as-a-service for cyber-physical cloud computing. In: Proceedings of the 2nd USENIX Conference on Hot Topics in Cloud Computing, p. 14 (2010)
24. Kirsch, C., Pereira, E., Sengupta, R., Chen, H., Hansen, R., Huang, J., Landolt, F., Lippautz, M., Rottmann, A., Swick, R., et al.: Cyber-physical cloud computing: the binding and migration problem. In: Proceedings of the Conference on Design, Automation and Test in Europe, pp. 1425–1428 (2012)
25. Gerla, M.: Vehicular cloud computing. In: 2012 The 11th Annual Mediterranean Ad Hoc Networking Workshop (Med-Hoc-Net), pp. 152–155 (2012)
26. Alam, S., Chowdhury, M.M.R., Noll, J.: Senaas: an event-driven sensor virtualization approach for internet of things cloud. In: 2010 IEEE International Conference on Networked Embedded Systems for Enterprise Applications (November 2010). http://dx.doi.org/10.1109/NESEA.2010.5678060
27. Yuriyama, M., Kushida, T.: Sensor-cloud infrastructure-physical sensor management with virtualized sensors on cloud computing. In: 2010 13th International Conference on Network-Based Information Systems (NBiS) (2010)
28. Ravi, S., Raghunathan, A., Chakradhar, S.: Tamper resistance mechanisms for secure embedded systems. In: Proceedings of the 17th International Conference on VLSI Design, pp. 605–611 (2004)
29. Hohl, F.: Time limited blackbox security: protecting mobile agents from malicious hosts. In: Vigna, G. (ed.) Mobile Agents and Security. LNCS, vol. 1419, pp. 92–113. Springer, Heidelberg (1998). http://dl.acm.org/citation.cfm?id=648051.746183
30. OAuth, OAuth 2.0. http://oauth.net/ (Online; accessed August 15, 2014)
31. Recordon, D., Reed, D.: Openid 2.0. In: Proceedings of the Second ACM Workshop on Digital Identity Management, DIM 2006 (2006). http://dx.doi.org/10.1145/1179529.1179532
32. Lantz, B., Heller, B., McKeown, N.: A network in a laptop: rapid prototyping for software-defined networks. In: Proceedings of the 9th ACM SIGCOMM Workshop on Hot Topics in Networks, p. 19. ACM (2010)
33. Pardo-Castellote, G.: Omg data-distribution service: architectural overview. In: 2003 Proceedings of the 23rd International Conference on Distributed Computing Systems Workshops (2003). http://dx.doi.org/10.1109/ICDCSW.2003.1203555

ComSS – Platform for Composition and Execution of Streams Processing Services

Paweł Świątek[✉]

Chair of Computer Communication Systems, Wroclaw University of Technology,
Wybrzeze Wyspianskiego 27, Wroclaw 50-370, Poland
pawel.swiatek@pwr.edu.pl

Abstract. This work addresses specific research challenges imposed by
contemporary IoT (Internet of Things) applications, which are modular,
distributed, utilize sensor networks and are typically implemented as services, according to the Service Oriented Architecture (SOA) paradigm.
They also require maintaining and processing of streaming data, which
leads to the need of effective management of composite streaming services (often with quality of service (QoS) restrictions). We propose a
prototype solution supporting development and management of such IoT
applications, a ComSS Platform acting as middleware for running composite streams' processing services. The results of the tests, conducted
in order to assess the performance of various service selection and negotiation methods are discussed and analyzed in order to identify the best
solutions for future development of distributed IoT software platforms.

Keywords: Streaming services · Stream computing · IoT platform ·
Services composition · Services execution · Platform as a service (PaaS)

1 Introduction

Internet of Things paradigm and data stream processing is rapidly gaining popularity due to consequent increase in number of IoT enabled devices and their
capabilities. This creates new opportunities for development of data processing
and analysis methods, which will be a welcome amenity for various applications
which can benefit from the use of IoT devices. This can be especially important for e-Health domain, where such solutions can be used in remote patient
monitoring and diagnostics (e.g.: [3,4,17]).

The greatest barrier which inhibits the development of such solutions is the
lack of standards which could enable simple and robust connectivity and composability of IoT devices and data streams they generate [13]. Hence, the main
task for today is the development of proper standards which will provide the
required level of connectivity and composability.

In this paper we describe a complete solution for automatic negotiation and
quality maintenance of the distributed, service-based IoT applications by discussing our prototype middleware software – ComSS Platform – dedicated to

© Springer International Publishing Switzerland 2015
N.T. Nguyen et al. (Eds.): ACIIDS 2015, Part II, LNAI 9012, pp. 494–505, 2015.
DOI: 10.1007/978-3-319-15705-4_48

running composite stream processing services. Models assumed for batch and streaming services are described in the next section. Section three is dedicated to the composite stream processing services. In section 4 a general architecture of the ComSS Platform is presented. Finally in section 5 the results of ComSS Platform efficiency test are presented along with the discussion and conclusions about the applicability of chosen service selection and composition mechanisms in such an environment. Last section summarizes the platform efficiency tests and outlines our plans for the future work.

2   Stream Processing

Service Oriented Architecture (SOA) becomes increasingly popular as a base for modern distributed processing systems [7,8,16]. In SOA, system functionalities are provided through web services, described by both functional and non-functional, parameters [14]. Functional description contains data types and structures used in interaction with the web service. Non-functional parameters describe the performance and service reliability. For instance, a fall detection service's functional description contains information that service processes the data stream from movement sensors, that fall would be detect in data stream and immediately send an alert to monitored person's caretakers. Non-functional description of such a service might describe service's data processing latency and the detection efficiency.

Distributed processing is inherently related to network communication. In classical SOA approach, services can be interpreted as batch services, that are invoked with standard request-response pattern (fig. 1). In this communication pattern, user sends all data required as a single package in the service request. The response from the service is also single data portion.

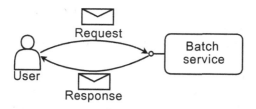

Fig. 1. Batch service communication model.

The described communication pattern implies some properties of such services. First of all, requests should include a complete and integral portion of data for processing. Also, due to the request-response communication pattern, batch services are stateless, therefore, the requests are processed independently from each other and even though consequent requests will not be related and will be processed by different processing instances.

Despite the fact that this approach is useful for many applications, there are use cases where request-response approach appears to be incorrect or hard to implement. An example of such a use case is data stream processing. Streaming services that handle data streams operate on continuously incoming stream of data [1]. The main characteristic is that they cannot be easily partitioned to sensible data packages, therefore it should be processed as a whole. Data stream processing algorithms usually operate on data samples, however majority of algorithms includes consequent data samples and relations between them. Another feature of streaming services is that a system has no impact on sample order. After processing each of the data portion can be deleted or stored in the cache, which is usually smaller than stream.

We propose a model of streaming service communication presented in Fig. 2. The main difference from the batch service is the separation of control and data processing interfaces.

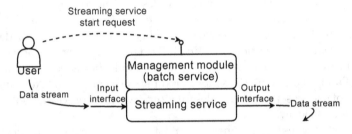

Fig. 2. Streaming service communication model.

In the proposed model, streaming service consists of two modules: the management module and the streaming service that handles data processing. The management module is a batch service which manages streaming service parameters and its state. Throughout its control interface service instances can be started, updated and stopped. The interface offers also operations for monitoring the service instances and setting execution parameters of the streaming service.

The second module is the streaming service. As mentioned before, execution parameters affect the configuration of this module and the data stream processing algorithms. In our model we can specify separate input and output interfaces in services. Depending on the services, the number of inputs and outputs may vary. The data stream processing starts when data streams are sent to the defined interface for processing. It is important to notice that the interfaces should only process information and synchronise data. Any control messages should be send to dedicated management interface.

3   Composite Stream Processing Services

In SOA-based systems, basic functionalities are in capacity of delivering atomic web services. In order to provide, more complex functionalities atomic web services are composed into composite web services [9]. Classical composition methods used in SOA-based systems, adopt batch services so that they cannot be properly implemented as methods for streaming service composition. Having that in mind, a dedicated approach for streaming services composition, is required.

Composite streaming service can be defined as a directed, acyclic graph. The graph nodes contain information about atomic streaming service, stream sources and stream destinations. The edges represent connections between atomic streaming services and stream directions. Additionally, we distinguish services in composite streaming service into two types: internal and external services. Internal services are those which set complex functionality and are implemented to the model of streaming service. External services are abstract services used for providing input and output communication interfaces of the composite streaming service. They do not process any data streams and are not mapped to any specific streaming services.

Service composition is a process aimed at fulfilling both functional and non-functional requirements defined by the user [15,18]. The composition process starts with connecting the required functionalities into data flow graph structure. Next, a set of services implementing the required functionality is found for each of the functional requirements. In the last step, a composite service, which fulfils non-functional requirements, is selected from the set of all possible composite services [10].

According to our streaming service model, it is important to notice that a service composed of internal services has the same input and output interface configuration as those on the border of the subgraph. The border services are integrated with the external services. For this reason, the streaming service management module allows communication with both the internal and external services.

Finally, the composition process yields a cross-compliance between atomic services. This requirement is strictly bonded with two parameters describing data stream and data interfaces: stream type and stream format. Stream type describes the type of data to be sent from data source (e.g. sensor, data base, etc.), while the stream format determines the way of the data organization. Both of the parameters should match with the interfaces of streaming services in order to make them connectable.

4   ComSS Platform

Composition of streaming services and execution of composite streaming services requires a dedicated infrastructure which supports these processes and ensures their correctness. In order to achieve this goal we have built a distributed software platform which enables streaming service composition and execution. The composition process supported by this platform consists of three main stages: (i) composite streaming service structure generation, commonly named as composition; (ii) atomic services selection and (iii) service instances initialization.

The proposed platform, called ComSS Platform (COMposition of Streaming Services) has been built based on the SOA paradigm. It means that all components of the platform provide web service interfaces that have been designed and implemented according to the SOA standards. Figure 3 shows a structure of typical platform workflow.

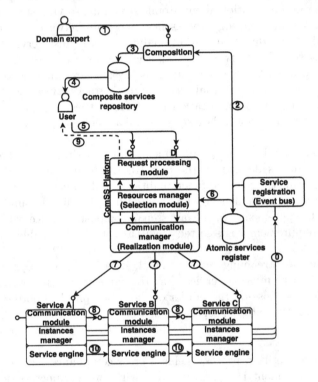

Fig. 3. ComSS Platform typical workflow.

4.1 Composite Streaming Service Structure Generation

First step of composite service building is the structure generation [5,15,19]. As it was previously mentioned, composite service functionality can be described as a graph. Each node in this graph represents a single functionality and edges represent data flow directions between these functionalities. Such graph is called the functionalities graph and it is the result of the composition phase described in this section. At the initial stage of composition process (step 1 on Fig. 3) domain expert define service functional requirements [6]. Functional requirements are defined by description of input stream which should be processed and expected output stream, which are described by types and formats.

Composition service uses atomic services repository to get list of available single functionalities (step 2). Each functionality is described by type and number of input and output interfaces. This description contains also the non-functional description which is specified in the next section. Based on atomic services description and composite service requirements the composition has to create structure of functionalities. The outcome of this phase is stored in composite services repository (step 3).

Composite service functionality graph contains description of atomic services classes with connections between them, along information about source and destination of data stream. Composition process used in our prototype relies on matching data types between sources and destinations. It is worth mentioning that there are also streaming services that do not operate within our platform, however they could still become data sources or data destinations for composite services, e.g. health sensors or video cameras. The only requirement for this is the correct description of the data streams that they produce/receive. The correct description should contain stream data type, stream format and, in case of destinations, device address.

Composite services repository contain functionality graphs that can be provided to users. Each structure can be used as many times as it is required (step 4). However, it is worth remembering that at this stage users are not choosing graphs with atomic services but only with their descriptions. It means that created service structure can be used as long as there are available services of classes defined in structure description.

4.2   Non-Functional Optimization – Service Selection

Next step of building composite streaming service is the atomic services selection [2]. The main aim of this stage is to find atomic services which assure requirements defined in the request. The input data for this stage are functionality graph prepared in the previous stage and user-defined non-functional requirements (step 5). Non-functional requirements may concern various QoS parameters. Such as, composite streaming service's processing time, cost or maximum number of service instances [11] [12].

Such request is forwarded by Request processing module to Selection module. In our prototype Selection module is a component of Resources manager. The Resource manager is responsible for guarding the maximum number of instances and assuring services quality level. Resources manager uses information stored in Atomic services register to fulfil this tasks during the service selection (step 6).

The prototype selection module implements three selection algorithms: Full search, Branch & Bound and Genetic algorithm. These algorithms are responsible for non-functional parameters optimization. The outcome of the service selection phase is the Services graph which contains selected services that fulfil both functional and non-functional requirements for composite streaming service [9].

4.3   Service Instances Initialization

The last step of streaming service composition is the service instances initialization. The main goal of this process is to prepare service instances and start services and thereby start composite service. This process is mainly based on communication between platform and services. There are two types of communication links that can be specified: signalization and negotiation. First one is used for communication between platform and services (step 7), second one is responsible for direct communication between services (step 8).

Communication on this link can be divided into three steps: resources initialization, negotiations between services and services instance starting. During resources initialization, services announce their readiness for operation. The second step is the negotiations stage. This step is responsible for service instance interfaces configuration, especially for determining data streams formats between neighbouring services and for configuration of connection interfaces.

Due to some problems regarding data stream format negotiations like blockade when service cannot convert formats we propose three negotiation methods. All of them have been implemented in ComSS Platform. First of them is the Ad-hoc negotiations method. After resources reservation, each service communicate through negotiation interface with neighbouring services negotiates mutual stream format. When the service cannot convert stream formats, it waits for its input negotiations to finish and then it decreases available output formats list according to the formats selected on the input and initializes new negotiations with the next service. In this case, the same format must be used on the input and on the output of streaming service.

Ad-hoc method cannot prevent negotiation deadlocks. In order to overcome this limitation, we have prepared the Sequential negotiation method. It is similar to the Ad-hoc method, but, in contrary to the Ad-hoc method, negotiations are carried on according to the sequence of services in the composite service structure graph. In case of failure on each of nodes on the previous node format can be changed. This modifications ensures that the negotiations will be successful if a feasible solution exists.

Both of methods described above aim in finding described matching formats between services, however, they are not considering the influence of different formats on quality. For data stream of one type, there can be many different formats that represent the same data in different way. This variety of formats may lead to the situation when the overall performance of composite streaming service is significantly decreased because of format conversions. In order to solve this problem, we developed the third method of negotiations which is called Planning negotiations. In this approach data stream formats are determined centrally in Communication manager. The Planning process is not only looking for mutual format for each pair of services but it also compares different solutions in order to select the one with the highest quality level.

The last step is the instances start-up, summarizes whole composite service building process. When *communication manager* creates instances graph by extending service graph with information about instances, the instances graph is sent back to *request processing module*, which returns it to the user (step 9). From this moment user can start data stream sending to newly deployed composite streaming service (step 10). ComSS platform also provides an interface for services deletion. This process releases the resources, because after this operation all instances associated with selected composite streaming service are being stopped. Communication with services during this operation is similar to the last step of instances initialization - the communication manager is sending stop message to each service instance associated with composite streaming service.

5   ComSS Platform Efficiency

ComSS Platform offers services which provide users with functionality to build composite streaming services. In order to evaluate the platform, several tests have been conducted, which mainly aimed in measuring the efficiency of the platform. The efficiency was measured as the overall delay impressed by the whole composite service composition process as well as the delays of each of composition phases contributing to the overall delay.

5.1   Testing Environment

Test environment was set up on three virtual machines. Each of these virtual machines was dedicated for deployment of one of the following platform components: *Atomic Service Register*, *Client Application* and *ComSS Platform*. As a test scenarios two different functionality graphs were prepared. First with three requested service types linked serially. Second one with six service types linked partly in parallel way.

All performed tests focused on measuring the composition execution time. The test measured processing times of *Resources Manager*, *Communication Manager* and whole CREATE request.

5.2   Measurement Results

Figures 4a – 5c show changes of CREATE request execution time in relation to the number of services and the number of formats. This figure shows that execution time in case of Ad-hoc and sequential negotiation does not increase with the increase in the number of formats (fig. 4a – 4d).

Another observation is that Full search and Branch & Bound algorithms behave similarly. Time of processing increases with the increase in number of services in service register. Also, as predicted, Full search algorithm is considerably slower than Branch & Bound.

It seems that selection algorithms should not be related to the number of formats, but in case of Branch & Bound algorithm there is an increase in performance when the number of formats is lower.

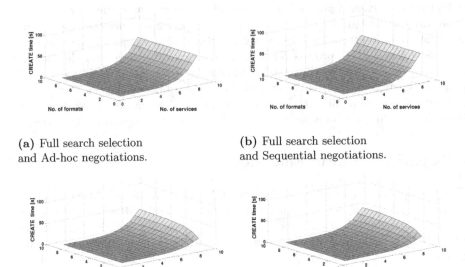

(a) Full search selection
and Ad-hoc negotiations.

(b) Full search selection
and Sequential negotiations.

(c) Branch & Bound selection
and Ad-hoc negotiations.

(d) Branch & Bound selection
and Sequential negotiations.

Fig. 4. Processing time of CREATE request in relation to the numbers of formats and services

Figure 5a shows that Genetic selection algorithm execution time is not related to the number of services (in register). In our prototype the values of parameters of genetic algorithm were selected after preliminary tests, however the results of these tests are out of scope of this paper, thus they will not be described. The criterion for parameters selection considered the execution time of Genetic algorithm and the quality of solution found.

Figures 5a – 5c presents the characteristic of Planning-based negotiations method in combination with different selection algorithms. Intuitively planning process has to take more time than simple negotiations method, especially when the services has to communicate during negotiation, despite the fact that the formats have been chosen centrally. Times observed on this figures increase up to sixty seconds when we use fast genetic selection, over hundred seconds using Branch & Bound and almost hundred and twenty seconds during Full search.

An important issue that is worth considering is where and when should we use the exact algorithms which were proven to be inferior in terms of efficiency. The answer to this issue is directly connected to the required quality. Only by using Full search or Branch & Bound selection and Planning negotiations composition process can find the solution with the highest possible quality level.

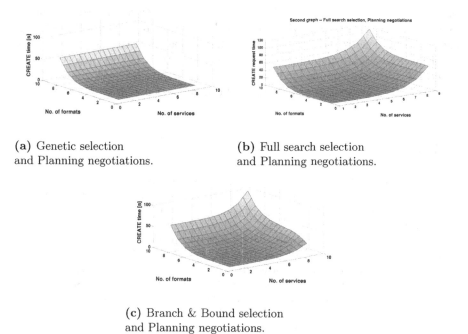

(a) Genetic selection
and Planning negotiations.

(b) Full search selection
and Planning negotiations.

(c) Branch & Bound selection
and Planning negotiations.

Fig. 5. Processing time of CREATE request in relation to the numbers of formats and services

6 Conclusion

ComSS platform is designed to be a middleware for running and management of composite streaming services and its performance depends on many factors, including the number of available services, their instances and data and stream formats used. Composite service queries may be simple or may concern complex service applications involving many services. In this context the time of composition execution is a key performance indicator of such a platform which determines how fast it can respond and deploy the service application required by the user. During the tests *ComSS Platform* processing time was measured. Result analysis indicate that for simple service structures there is no significant impact of service selection algorithm and negotiations method used on the required service composition and execution. This implies, that in such cases any algorithm may be used. Important fact, shown by experiments is that little extension of functionality graph (in terms of the number of nodes and connections) can increase execution time from seconds to almost two minutes. Fortunately in such cases, there are algorithms which calculate results as fast as for simple functionality graphs, genetic selection algorithm and Sequential negotiations are examples of them. However, if quality assurance is needed, more precise methods, e.g. Branch & Bound selection algorithm and Planning negotiations method should be used -

service quality assurance takes time. The compromise between fast service building and quality of service provisioning could be offered by dynamically selected algorithms and methods – which means that proper heuristics should be proposed and carefully evaluated. In case where quality is not so important and structures are more complicated fast algorithms should be used. Simple functionality graphs structures processing could be processed with better but slower algorithms.

References

1. Babcock, B., Babu, S., Datar, M., Motwani, R., Widom, J.: Models and issues in data stream systems. In: Proceedings of the Twenty-First ACM SIGMOD-SIGACT-SIGART Symposium on Principles of Database Systems, PODS 2002, pp. 1–16. ACM, New York (2002)
2. Bao, H., Dou, W.: A qos-aware service selection method for cloud service composition. In: 2012 IEEE 26th International Parallel and Distributed Processing Symposium Workshops PhD Forum (IPDPSW), pp. 2254–2261 (2012)
3. Brzostowski, K., Drapała, J., Grzech, A., Świątek, P.: Adaptive decision support system for automatic physical effort plan generation-data-driven approach. Cybernetics and Systems **44**(2–3), 204–221 (2013)
4. Brzostowski, K., Drapała, J., Świątek, J.: System analysis techniques in ehealth systems: a case study. In: Pan, J.-S., Chen, S.-M., Nguyen, N.T. (eds.) ACIIDS 2012, Part I. LNCS, vol. 7196, pp. 74–85. Springer, Heidelberg (2012)
5. Charif, Y., Sabouret, N.: An overview of semantic web services composition approaches. Electron. Notes Theor. Comput. Sci. **146**(1), 33–41 (2006)
6. Falas, Ł., Stelmach, P.: Web service composition with uncertain non-functional parameters. In: Camarinha-Matos, L.M., Tomic, S., Graça, P. (eds.) DoCEIS 2013. IFIP AICT, vol. 394, pp. 45–52. Springer, Heidelberg (2013)
7. Grzech, A., Juszczyszyn, K., Kołaczek, G., Kwiatkowski, J., Sobecki, J., Świąek, P., Wasilewski, A.: Specifications and deployment of soa business applications within a configurable framework provided as a service. In: Ambroszkiewicz, S., Brzezinski, J., Cellary, W., Grzech, A., Zielinski, K. (eds.) Advanced SOA Tools and Applications. SCI, vol. 499, pp. 7–71. Springer, Heidelberg (2014). http://dx.doi.org/10.1007/978-3-642-38957-3_2
8. Grzech, A., Juszczyszyn, K., Świątek, P.: Methodology and platform for business process optimization. In: Selvaraj, H., Zydek, D., Chmaj, G. (eds.) Progress in Systems Engineering. AISC, vol. 330, pp. 377–382. Springer, Heidelberg (2015). http://dx.doi.org/10.1007/978-3-319-08422-0_56
9. Grzech, A. Świątek, P.: Modeling and optimization of complex services in service-based systems. Cybernetics and Systems **40**(8), 706–723 (2009)
10. Grzech, A., Świątek, P., Rygielski, P.: Dynamic resources allocation for delivery of personalized services. In: Cellary, W., Estevez, E. (eds.) Software Services for e-World. IFIP AICT, vol. 341, pp. 17–28. Springer, Heidelberg (2010)
11. Huang, A.F.M., Lan, C.W., Yang, S.J.H.: An optimal qos-based web service selection scheme. Inf. Sci. **179**(19), 3309–3322 (2009)
12. Ko, J.M., Kim, C.O., Kwon, I.H.: Quality-of-service oriented web service composition algorithm and planning architecture. J. Syst. Softw. **81**(11), 2079–2090 (2008)

13. Mongay Batalla, J., Krawiec, P.: Conception of id layer performance at the network level for internet of things. Personal and Ubiquitous Computing **18**(2), 465–480 (2014). http://dx.doi.org/10.1007/s00779-013-0664-0
14. Stelmach, P., Schauer, P., Kokot, A., Demkiewicz, M.: Universalplatform for composite data stream processing services management. In: Zamojski, W., Mazurkiewicz, J., Sugier, J., Walkowiak, T., Kacprzyk, J. (eds.) New Results in Dependability and Computer Systems. AISC, vol. 224, pp. 399–407. Springer, Heidelberg (2013)
15. Stelmach, P., Świątek, P., Falas, Ł., Schauer, P., Kokot, A., Demkiewicz, M.: Planning-Based method for communication protocol negotiation in a composition of data stream processing services. In: Kwiecień, A., Gaj, P., Stera, P. (eds.) CN 2013. CCIS, vol. 370, pp. 531–540. Springer, Heidelberg (2013)
16. Świątek, P., Brzostowski, K., Drapała, J., Juszczyszyn, K., Grzech, A.: Development of intelligent ehealth systems in the future internet architecture. In: Klempous, R., Nikodem, J. (eds.) Innovative Technologies in Management and Science. TIEI, vol. 10, pp. 73–94. Springer, Heidelberg (2015)
17. Świątek, P., Klukowski, P., Brzostowski, K., Drapała, J.: Application of wearable smart system to support physical activity. Advances in Knowledge-based and Intelligent Information and Engineering Systems, 1418–1427 (2012)
18. Świątek, P., Stelmach, P., Prusiewicz, A., Juszczyszyn, K.: Service composition in knowledge-based soa systems. New Generation Computing **30**(2–3), 165–188 (2012)
19. Zeng, L., Benatallah, B., Ngu, A.H.H., Dumas, M., Kalagnanam, J., Chang, H.: Qos-aware middleware for web services composition. IEEE Transactions on Software Engineering **30**(5), 311–327 (2004)

Experience-Oriented Enhancement of Smartness for Internet of Things

Haoxi Zhang[1(✉)], Cesar Sanin[2], and Edward Szczerbicki[3]

[1] Chengdu University of Information Technology, No. 24 Block 1, Xuefu Road,
Chengdu 610225, China
Haoxi@cuit.edu.cn
[2] Faculty of Engineering and Built Environment, School of Engineering,
The University of Newcastle, Callaghan, NSW 2308, Australia
Cesar.Sanin@newcastle.edu.au
[3] Gdansk University of Technology, Gdansk, Poland
Edward.Szczerbicki@zie.pg.gda.pl

Abstract. In this paper, we propose a novel approach, the Experience-Oriented Smart Things that allows experiential knowledge discovery, storage, involving, and sharing for Internet of Things. The main features, architecture, and initial experiments of this approach are introduced. Rather than take all the data produced by Internet of Things, this approach focuses on acquiring only interesting data for its knowledge discovery process. By catching decision events, this approach gathers its own daily operation experience, which is the interesting data, and uses such experience for knowledge discovery. An initial experiment was made at the end of this paper, by applying this approach to a sensors-equipped bicycle, the bicycle is able to learn user's physical features and recognize its user out of other riders. Customized version of Decisional DNA is used in this approach as the knowledge representation technique. Decisional DNA is a domain-independent, and flexible, and standard experiential knowledge repository solution that allows knowledge to be acquired, reused, evolved and shared easily. The presented conceptual approach demonstrates how knowledge can be discovered through its domain's experiences and stored as Decisional DNA.

Keywords: Knowledge representation · Decisional DNA · Machine learning · Intelligent systems · Internet of things

1 Introduction

Thanks to advances in fields such as wireless communication, sensing, automatic identification, and could computing, that a sensing, networked, and intelligent world has been building up. The concept of the Internet of Things (IoT) [1][2][3] is to connect all things in the world to the Internet, and eventually build an intelligent world around us, where things can communicate with each other, make decisions by themselves, and act accordingly without explicit instructions, and even know what we need, what we want, and what we like [3][5]. More and more governments, academics, and researchers are taking part in constructing such an intelligent environment

© Springer International Publishing Switzerland 2015
N.T. Nguyen et al. (Eds.): ACIIDS 2015, Part II, LNAI 9012, pp. 506–515, 2015.
DOI: 10.1007/978-3-319-15705-4_49

that is composed of various computing systems, such as smart health care, intelligent transportation, smart home, and global supply chain logistics [6][7][8]. As a result, one of the most critical problems arises now: how do we transform the data generated and captured by IoT into knowledge to make an intelligent world for human beings? In this paper, we introduce a novel approach that uses experiences of IoT as the main source for knowledge discovery, storage, involving, and sharing.

This paper is organized as follows: section two describes an academic background on basic concepts related to our work; section three presents the features, architecture and experiments for the experience-oriented knowledge discovery approach, called the Experience-Oriented Smart Things. Finally, in section four, concluding remarks are drawn.

2   Background

2.1   Data from IoT

Basically, every single thing of IoT might produce data containing various kinds of information. According to the work [27], data produced by IoT can be divided into two classes: the data about things and the data generated by things. The former refers to data that describe IoT themselves, like identity, state, and location, etc., whereas the latter refers to data captured or created by things. The data about things usually contain information that can be used to improve the performance of the systems, infrastructures, and things of IoT. The data generated by things usually carry information that are the results of operation or interaction with humans, among systems, and between human and systems; which can be used to enhance the services provided by IoT.

In recent years, the total amount of data produced worldwide every year has exceeded one zettabyte [3], and the data generated by IoT per day have increased fast beyond limits of available data process tools today. Hence the term "big data" were introduced to describe this data-deluge situation [28]. Although a range of traditional tools [29] are used to solve or ease the issues of handling the big data problem, such as data condensation [30], divide and conquer [31], incremental learning, and random sampling [32], theses tools are generally not powerful enough to deal with such amount of data from IoT [33][34].

Consequently, a number of research proposals and attempts have been made. Among them, a new trend to solve the big data problem is to reduce the complexity of input data [35][36][37]. Different from reducing the complexity of input data, distributed computing, feature selection, and cloud computing are some other promising directions for dealing with the issue [6][38][39].

2.2   Knowledge Discovery on IoT

Finding valuable information from data produced by IoT is the most challenging part and the essential goal of big data processing. By utilizing knowledge discovery technologies, such as data mining and machine learning, intelligence and smartness can be added to IoT.

Several IoT smartness researches and theories can be found in literature. López et al. [13] proposed an architecture that integrates fundamental technologies for realizing the IoT into a single platform and examined them. The architecture introduces the use of the Smart Object framework [15][16] to encapsulate sensor technologies, radio-frequency identification (RFID), object ad-hoc networking, embedded object logic, and Internet-based information infrastructure. They evaluated the architecture against a number of energy-based performance measures, and showed that their work outperforms existing industry standards in metrics such as delivery ratio, network throughput, or routing distance. Finally, a prototype implementation for the real-time monitoring of goods flowing through a supply chain was presented in detail to demonstrate the feasibility and flexibility of the architecture. Key observations showed that the proposed architecture has good performance in terms of scalability, network lifetime, and overhead, as well as producing low latencies in the various processes of the network operation. Li et al. [11] introduced the Smart Community as a new Internet of Things application, which used wireless communications and networking technologies to enable networked smart homes and various useful and promising services in a local community environment. The smart community architecture was defined in their paper, then solutions for robust and secure networking among different homes were described, at the end, two smart community applications, Neighborhood Watch and Pervasive Healthcare, were presented. In [12], a cognitive management framework that will empower the Internet of Things to better support sustainable smart city development was presented. The framework introduced the virtual object (VO) concept as a dynamic virtual representation of objects and proposed the composite VO (CVO) concept as a means to automatically aggregate VOs in order to meet users' requirements in a resilient way. In addition, it illustrated the envisaged role of service-level functionality needed to achieve the necessary compliance between applications and VOs/CVOs, while hiding complexity from end users. The envisioned cognition at each level and the use of proximity were de-scribed in detail, while some of these aspects are instantiated by means of building blocks. A case study, which presented how the framework could be useful in a smart city scenario that horizontally spans several application domains, was also described. In [14], Lee et al. applied human learning principles to user-centered IoT systems. Their work showed that IoT systems could benefit from a process model based on principles derived from the psychology and neuroscience of human behavior that emulates how humans acquire task knowledge and learn to adapt to changing context.

According to the survey of [3], after a comprehensive comparison of different data mining technologies, and their applications for IoT, a promising direction was found that by using knowledge discovery technologies the IoT can be made smarter or even more intelligent.

2.3   Decisional DNA and Set of Experience Knowledge Structure

The Decisional DNA is a novel knowledge representation theory that carries, organizes, and manages experiential knowledge stored in the Set of Experience Knowledge Structure [20]. The Set of Experience Knowledge Structure (SOEKS or shortly SOE) has been developed to capture and store formal decision events in an explicit way [18]. It is a flexible, standard, and domain-independent knowledge representation structure [17]. And is a model based upon available and existing knowledge, which must adapt to the decision event it was built from (i.e. it is a dynamic structure that depends on the information provided by a formal decision event) [21]; moreover, SOEKS can be stored in XML or OWL files as ontology in order to make it transportable and shareable [19] [22].

SOEKS consists of variables, functions, constraints and rules associated in a DNA shape enabling the integration of the Decisional DNA of an organization [21]. Variables normally implicate representing knowledge using an attribute-value language (i.e. by a vector of variables and values) [23], and they are the centre root and the starting point of SOEKS. Functions represent relationships between a set of input variables and a dependent variable; besides, functions can be applied for reasoning optimal states. Constraints are another way of associations among the variables. They are restrictions of the feasible solutions, limitations of possibilities in a decision event, and factors that restrict the performance of a system. Finally, rules are relationships between a consequence and a condition linked by the statements IF-THEN-ELSE. They are conditional relationships that control the universe of variables [21].

Additionally, SOEKS is designed similarly to DNA at some important features. First, the combination of the four components of the SOE gives uniqueness, just as the combination of four nucleotides of DNA does. Secondly, the elements of SOEKS are connected with each other in order to imitate a gene, and each SOE can be classified, and acts like a gene in DNA [21]. As the gene produces phenotypes, the SOE brings values of decisions according to the combined elements. Then, a decisional chromosome storing decisional "strategies" for a category is formed by a group of SOE of the same category. Finally, a diverse group of SOE chromosomes comprise what is called the Decisional DNA [18].

3   The Experience-Oriented Smart Things

Based on the research background and technology review given in section 2, the Experience-Oriented Smart Things (EOST) approach is proposed for knowledge discovery and reuse on IoT. This section presents the main features, architecture, and initial experiments of this research.

3.1   Main Features

The EOST is designed and proposed to allow experiential knowledge discovery, storage, involving, and sharing for IoT. Moreover, the EOST shall be able to handle the big data issue, and compatible with a range of different things, and most importantly, the EOST shall be smart. In order to achieve these goals, the three key features of EOST, namely experience-oriented, cloud-base, and learning, as given below.

a) Experience-oriented: one of the good ways to deal with the big data issue is to capture only the interesting data instead of all the data. By mimicking man learning from experience, the EOST abstracts experiences from data of things, and uses experience instead of all data; just like man remembers experience but all details of how the certain experience was haven. Which turns big data into small data.

b) Cloud-based: technically, the EOST is designed as an open platform for all things. To allow that, could computing and open application program interface (API) are important technologies to have in this approach.

c) Learning: the EOST is supposed to learn from data of things so that better performance, smarter behavior, and more efficient operation can be achieved on IoT. This is where knowledge discovery technologies come into play.

3.2   System Architecture

The four-layer architecture is designed for our conceptual EOST platform, which consists of Physical Layer, Operating System Layer, Application Layer, and EOST Layer (see Fig. 1).

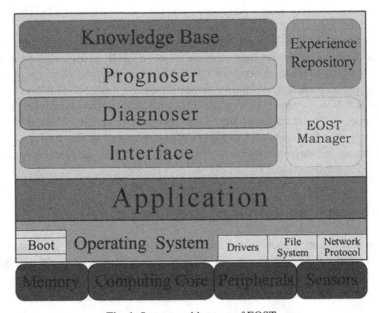

Fig. 1. System architecture of EOST

At the bottom, there is the physical layer that consists of the memory, computing units, peripherals, networking hardware, and most importantly, the sensing entities of IoT. It is the fundamental layer underlying the logical data structures of higher-level functions in the system.

At the second level, there is the operating system layer where the operating system of IoT runs and manages the computing hardware of IoT, and provides data transfer services among the EOST layer, the application layer, and the underlying physical layer.

Upon the operating system layer, there is the application layer running applications developed to fulfill different tasks and offer various functionalities to the end-user; and with the help of the EOST layer, these applications can access knowledge-based services to make the whole system intelligent and being capable of acquiring, reusing, improving and sharing knowledge.

Finally, the EOST layer is at the top. It is the central core of our research, and is designed to work as the "brain" in order to bring smartness to IoT applications: It analyses and routes data, extracts experiences from data, learns from experiences, manages experiences and knowledge, cooperates with other mechanisms, and interacts with the IoT application. The EOST layer is composed of a set of computer software, namely: Interface, Diagnoser, Prognoser, EOST Manager, Experience Repository and Knowledge Base. The Interface connects the EOST with its outer environment, and provides knowledge-based services and functionalities to the IoT applications. The Diagnoser is the place where the IoT scenario data are gathered and organized; in our case, we link each experience with a certain scenario describing the circumstance under which the experience was acquired. Scenario data are essential for learning as it gives the clue for mapping data generated by things and data about things. The Prognoser is in charge of analyzing scenario data, and creating experiences and updating knowledge based on machine learning and data mining algorithms. The EOST Manager works as a dispatcher that manages all inner logical of EOST, and runs the whole EOST system. The Experience Repository and the Knowledge Base are where experiences and knowledge stored and managed respectively.

3.3   Initial Experiments

In order to exam our concept, we designed an IoT application, a sensor-equipped bicycle. By using wireless communication technologies, such as Bluetooth and NFC, the bicycle sends sensor data to the smart phone APP; afterwards, these data are sent to the EOST via 3G or WIFI for knowledge discovery. Finally, the EOST sends suggestions back to assist the bicycle user for decision making.

The main hardware components of the bicycle consist of a NXP LPC1769 board [24], a HC-06 Bluetooth module, and two MD-PS002 pressure sensors. The NXP LPC1769 is an ARM 32-bit Cortex-M3 Microcontroller with MPU, CPU clock up to 120MHz, 64kB RAM, 512kB on-chip Flash ROM with enhanced Flash Memory Accelerator. It supports In-Application Programming (IAP) and In-System Programming (ISP), has eight channel general purpose DMA controller, nested vectored interrupt controller, AHB Matrix, APB, Ethernet 10/100 MAC with RMII interface and dedicated DMA, USB 2.0 full-speed Device controller and Host/OTG controller with DMA, CAN 2.0B with two channels, four UARTs, one with full modem interface, three I2C serial interfaces, three SPI/SSP serial interfaces, I2S interface, General purpose I/O pins, 12-bit ADC with 8 channels, 10-bit DAC, and four 32-bit timers with capture/compare. The NXP LPC1769 board is easy to use, low power, and very handy to have different peripherals and sensors working together. Through the HC-06 Bluetooth module, the board is able to communicate with other devices, such as a smart phone, so that the captured data can be sent for further processing.

The initial experiment was designed to evaluate the usability of the EOST to IoT. First, as a whole new exploring in combination of knowledge representation and IoT applications, whether the Decisional DNA can be adapted for IoT must be examined. Second, the capability of knowledge capturing of Decisional DNA on IoT needs to be tested. Finally, the application is required to remember its user to examine the smartness of the EOST.

In terms of the adaptability examination of Decisional DNA, we converted the file format of SOEKS from XML to plain text so that the captured data can be organized and stored on the NXP LPC1769 board. Every minute while user's riding, pressure sensors collected the two tires' real-time tire pressure. Table 1 illustrates a fragment of the tire pressure collected at a time. The ID is used to indicate two tires: number one stands for the front tire, while number two stands for the rear. Besides pressure, date and time are captured at the same time too: they are collected for future use, such as learning the riding routine of user.

Table 1. A fragment of the captured tire pressure data

ID	Pressure (bar)	Date	Time
1	1.59	2014-09-15	15:42:27
2	1.54	2014-09-15	15:42:27

By organizing and sending captured data to the APP running on an Android phone via Bluetooth connection, tire pressure information was collected. Then, the APP sends the information to the EOST, and the EOST will store it as experience base on the principals of Decisional DNA. Finally, the EOST analyzes experiences and extracts knowledge from them. In this initial experiment, we introduced the Farthest-First [24] algorithm to help learn the user's normal weight distribution based on tire pressure information, and eventually distinguish its user from other riders (i.e. user clustering): we collect tire pressures when the user is riding in order to train the system, after training, we can change the rider, and the bicycle is able to detect the change from the tire pressure differences. The Fig. 2 shows the result of the user clustering in Weka [25] by using real-time data of tire pressures: the system clusters the riders correctly; the Cluster 1 (marked as cross) stands for the user, and the other riders is clustered as the Cluster 2 (marked as solid dot).

As we can see from the initial experiment, by using the Decisional DNA and some machine learning algorithms, for example, the FarthestFirst algorithm in this case, we created a knowledge-based platform, the EOST, for IoT. Through its open API, the IoT applications can connect to EOST platform, and access the smartness services provided by EOST.

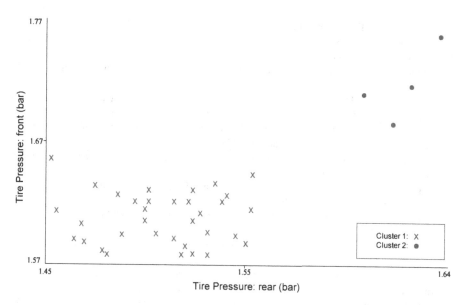

Fig. 2. The result of the user clustering on EOST

4   Conclusions and Future Work

In this paper, we introduced a conceptual platform, the EOST, providing cloud-based smartness services to IoT. Instead of taking all data from IoT, this approach focuses on acquiring only interesting data for its knowledge discovery process. By catching decision events, this approach gathers IoT application's own daily operation experience (i.e. the interesting data), and uses such experience for knowledge discovery process. An initial experiment was made at the end of this paper, a sensors-equipped bicycle is used to test our research idea. The initial experiment shows that the bicycle is able to learn user's physical features and recognize its user out of other riders by accessing the smartness services from EOST.

Making IoT intelligent is a very challenging goal to achieve, and our conceptual research is just at its first stage. There are quite a lot improvements, refinements, and further work remaining to be done, some of them are:

- More detailed requirements and specifications of the EOST need to be done.
- Evaluate and compare different knowledge discovery technologies in order to design and optimize the knowledge discovery strategy inside the platform.
- Further design and development of open APIs to support different IoT applications to access the smartness services provided by EOST.

References

1. Atzori, L.: Antonio Iera, and Giacomo Morabito.: The internet of things: A survey. Comput. Netw. **54**(15), 2787–2805 (2010)
2. Ashton, K.: That 'Internet of Things' Thing. RFID Journal.
   http://www.rfidjournal.com/article/print/4986

3. Tsai, C., et al.: Data Mining for Internet of Things: A Survey. 1-21 (2013)
4. Kortuem, G., et al.: Smart objects as building blocks for the internet of things. IEEE Internet Comput. **14**(1), 44–51 (2010)
5. Perera, C., et al.: Context aware computing for the internet of things: A survey, 1-41 (2013)
6. Bandyopadhyay, D., Jaydip, S.: Internet of things: Applications and challenges in technology and standardization. Wireless Pers. Commun. **58**(1), 49–69 (2011)
7. Domingo, M.C.: An overview of the Internet of Things for people with disabilities. Journal of Network and Computer Applications **35**(2), 584–596 (2012)
8. Miorandi, D., et al.: Internet of things: Vision, applications and research challenges. Ad Hoc Netw. **10**(7), 1497–1516 (2012)
9. López, T.S., et al.: Taxonomy, technology and applications of smart objects. Information Systems Frontiers **13**(2), 281–300 (2011)
10. López, T.S., et al.: Adding sense to the Internet of Things. Pers. Ubiquit. Comput. **16**(3), 291–308 (2012)
11. Li, Xu, et al.: Smart community: an internet of things application. Communications Magazine, IEEE **49**(11), 68–75 (2011)
12. Vlacheas, P., et al.: Enabling smart cities through a cognitive management framework for the internet of things. IEEE Communications Magazine **51**(6), (2013)
13. López, T.S., et al.: Adding sense to the Internet of Things. Pers. Ubiquit. Comput. **16**(3), 291–308 (2012)
14. Lee, S.W., Oliver, P., Zeungnam, B.: Applying human learning principles to user-centered IoT systems. Computer **46**(2), 46–52 (2013)
15. Vasseur, J.-P., Adam D.: Interconnecting smart objects with ip: The next internet. Morgan Kaufmann (2010)
16. The IPSO Alliance. http://www.ipso-alliance.org
17. Maldonado Sanin, C.A.: Smart Knowledge Management System. PhD Thesis, Faculty of Engineering and Built Environment - School of Mechanical Engineering, University of Newcastle, E. Szczerbicki, Doctor of Philosophy Degree, Newcastle (2007)
18. Sanin, C., Szczerbicki, E.: Experience-based Knowledge Representation SOEKS. Cybernetics and Systems **40**(2), 99–122 (2009)
19. Sanin, C., Szczerbicki, E.: An OWL Ontology of Set of Experience Knowledge Structure. Journal of Universal Computer Science **13**, 209–223 (2007)
20. Zhang, H.: Cesar Sanín, and Edward Szczerbicki.: Implementing Fuzzy Logic to Generate User Profile in Decisional DNA Television: The Concept and Initial Case Study. Cybernetics and Systems **44**(2–3), 275–283 (2013)
21. Sanin, C., Mancilla-Amaya, L., Szczerbicki, E., CayfordHowell, P.: Application of a Multi-domain Knowledge Structure: The Decisional DNA. Intel. Sys. For Know. Management, SCI **252**, 65–86 (2009)
22. Sanín, C., Toro, C., Sanchez, E., Mancilla-Amaya, L., Zhang, H., Szczerbicki, E., Crasco, E., Peng, W.: Decisional DNA: A Multi-technology Shareable Knowledge Structure for Decisional Experience. Neurocomputing **88**, 42–53 (2012)
23. Lloyd, J.W.: Logic for Learning: Learning Comprehensible Theories from Structure Data. Springer, Berlin (2003)
24. The NXP LPCXpresso Board for LPC1769. http://www.nxp.com/demoboard/OM13000.html
25. Hochbaum, D.S., Shmoys, D.B.: A best possible heuristic for the k-center problem. Mathematics of operations research **10**(2), 180–184 (1985)

26. Witten, I. H., Frank, E.: Data Mining: Practical machine learning tools and techniques. Morgan Kaufmann (2005)
27. Ali, N., Abu-Elkheir, M.: Data management for the internet of things: green directions. In: Proc. IEEE Globecom Workshops. pp. 386–390 (2012)
28. Russom, P.: Big data analytics. TDWI Best Practices Report, Fourth Quarter (2011)
29. Xu, R., Wunsch, D.: Clustering. vol. 10. John Wiley & Sons (2008)
30. Zhang, T., Ramakrishnan, R., Livny, M.: BIRCH: an efficient data clustering method for very large databases. In: ACM SIGMOD Record, vol. 25(2), pp. 103–114. ACM (1996)
31. Guha, S., Meyerson, A., Mishra, N., Motwani, R., O'Callaghan, L.: Clustering data streams: Theory and practice. IEEE Trans. Knowl. Data Eng. 15(3), 515–528 (2003)
32. Ng, R.T., Han, J.: CLARANS: A method for clustering objects for spatial data mining. IEEE Trans. Knowl. Data Eng. 14(5), 1003–1016 (2002)
33. Madden, S.: From databases to big data. IEEE Internet Comput. 16(3), 0004–6 (2012)
34. Cantoni, V., Lombardi, L., Lombardi, P.: Challenges for data mining in distributed sensor networks. In: IEEE 18th International Conference on Pattern Recognition, ICPR 2006, vol. 1, pp. 1000–1007 (2006)
35. Baraniuk, R.G.: More is less: signal processing and the data deluge. Science 331(6018), 717–719 (2011)
36. Ding, C., He, X.: K-means clustering via principal component analysis. In: Proceedings of the twenty-first international conference on Machine learning, p. 29. ACM (2004)
37. Chiang, M.C., Tsai, C.W., Yang, C.S.: A time-efficient pattern reduction algorithm for k-means clustering. Inf. Sci. 181(4), 716–731 (2011)
38. Gubbi, J., Buyya, R., Marusic, S., Palaniswami, M.: Internet of Things (IoT): A vision, architectural elements, and future directions. Future Generation Computer Systems 29(7), 1645–1660 (2013)
39. Aggarwal, C.C., Ashish, N., Sheth, A.: The internet of things: a survey from the data-centric perspective. In: Managing and mining sensor data, pp. 383–428. Springer US (2013)

Artificial Intelligent Techniques
and Their Application in Engineering
and Operational Research

A Cloud Computing Platform for Automatic Blotch Detection in Large Scale Old Media Archives

Kiok Ahn[1], Mingi Kim[1], Md. Monirul Hoque[2]([✉]), and Oksam Chae[2]

[1] ITA Technology, Seoul, South Korea
{kiokahn,mingi}@itatech.co.kr
[2] Department of Computer Engineering, Kyung Hee University, Seoul, South Korea
{monirul,oschae}@khu.ac.kr

Abstract. In this paper, we present an adaptive detection technique for blotch error, evident in old archived media. Traditional pixel based blotch detection methods, due to sensitivity of threshold, fail to detect blotch, if present, at identical location in successive frames. Furthermore, as the amount of archive data is quite large, processing time needs to be considered. To alleviate problems associated with traditional methods and speed up the process, in this paper, we have proposed a cloud computing solution where the blotch detector is a five frame based Rank order difference (ROD) detectors. False alarm is reduced by integrating adapting refinement based on local neighborhood statistics of candidate blotch regions in spatio-temporal domain. Experiment is performed on real archives to evaluate the efficacy of proposed solution.

Keywords: Blotch Detection · Adaptive Refinement · Cloud Computing · Parallel Processing · Old Archive Media

1 Introduction

Old archive preservation is of great importance in terms of historical record as well as the means to quality improvement for reproduction purposes. In order to preserve the archive materials by converting these contents to digital file, it is possible that noise or errors contained in the film or tape is either maintained or displayed in other forms; thus reduce the quality of the generated files. It is important to realize that successful treatment of any missing/degraded data problem must involve detection of the missing/corrupted regions. This would enable the reconstruction algorithm to concentrate on these areas and so the reconstruction errors at non-corrupted sites can be reduced [1]. Archived films often suffer damage and quality degradation through inappropriate storage and wear and tear etc. These defects were categorized in the project BRAVA [2] and include blotch, line scratches, brightness variation, and frame vibration, amongst many others. The most common type of defect found in old video archive is the

© Springer International Publishing Switzerland 2015
N.T. Nguyen et al. (Eds.): ACIIDS 2015, Part II, LNAI 9012, pp. 519–528, 2015.
DOI: 10.1007/978-3-319-15705-4_50

Fig. 1. Frame containing blotch [3]

presence of blotch (shown in Fig. 1), which usually appears as black, white or semitransparent regions with different degree of degradation, shape, and size.

Cloud computing is an established model of business computing that delivers computing as a service rather than a product. The cloud computing services falls into three basic types such as, IaaS, PaaS and SaaS [4] showed as a visual model in Fig. 2. As the size of archived media is usually big and with limited resources i.e. processing speed and storage for HD videos, parallel execution of error detection method is needed in order to minimize checking time and efforts required to keep ahead of other conventional quality check system.

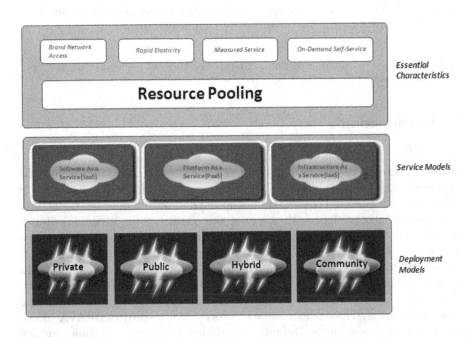

Fig. 2. Illustration of Cloud Computing

Until this recent, several works have been put forward in detecting blotches [1,3,5–10]. Temporal detectors such as SDIa and SDIp [1] are based on the computation of the DFD, which is the difference between two consecutive motion compensated (MC) frames with additional constraints. In spite of simplicity, these methods produce false alarms or miss blotch depending on the defined threshold, i.e. for low and high threshold respectively. Spatial filtering methods [5–7] assume spatial inconsistency of defects but provide false alarms on sharp and textured regions, i.e. when image spatial patterns look like defects patterns, and fail to detect blotches exceeding the filter size. Spatio-temporal methods such as MOS [8], ML3Dex [9], ROD [10], sROD [3] extend spatial filtering to the temporal domain, often using motion compensated (MC) frames and achieve better performance than spatial or temporal methods alone. Traditional methods assume that blotches cause discontinuities simultaneously in both temporal directions [3] i.e. blotches seldom appear at identical location in consecutive frames. Due to this assumption, performances of these methods highly depend on threshold selection. In order to detect blotches at same position over consecutive frames, threshold needs to be set at very low for these methods; leads to high degree of false alarm. The improved five frame based ROD method [11], can be useful in this situation but still suffers from false alarm. Probabilistic methods [12,13] perform well in real situations, but have a high computational cost, which become intractable when neighborhood order exceeds first or second order [12].

To address those aforementioned problems along with the processing resource constraints for large scale archive content, we have proposed an adaptive refinement technique which works on blotch pixels, initially detected by the five frame based ROD detector[11]. Refinement technique, based on local neighborhood image statistics of candidate blotch region, can effectively reduce the sensitivity to threshold setting. Furthermore, a cloud computing platform is proposed to fulfill the constraints when dealing with large scale archive data.

The remainder of this paper is organized as follows. Section 2 describes our blotch detection system. Section 3 illustrates the architecture of our cloud computing platform along with the arrangement of blotch detection task. Section 4 shows the experimental results and related analysis. Finally we conclude our contribution in section 5.

2   Proposed Detection System

A classical way to tackle the blotch detection problem is to proceed by estimating motion first between consecutive frames and then perform blotch detection on these motion compensated frames. Each of these steps has been tackled in the literature with many different approaches. As for the motion estimation step, a detailed comparison can be found in [14], where the authors propose a new objective that formalizes median filtering heuristics. This new objective includes a non local term of flow and image boundaries information that robustly integrates flow estimates over large spatial neighborhoods. This method outperforms state of the art methods for motion computation and consequently adopted in our work.

Here first we recall the formulation of three [10] and five [11] MC frame based ROD and describe them briefly. ROD is generally more robust to motion estimation errors than any of the SDI detectors although it requires the setting of three thresholds. The essence of the detector is the premise that blotched pixels are outliers in the local distribution of intensity. In ROD, they defined a list of pixel in the motion compensated previous and next frame (as in Fig. 3). The next step is to sort p_1 to p_6 in to the list of $r(k) = [r_1(k), r_2(k), r_3(k), r_4(k), r_5(k), r_6(k)]$ where $r_1(k)$ is the minimum. The median of these pixels is then calculated as $m(k) = (r_3(k) + r_4(k))/2$. The rank-ordered differences are defined as $d(k) = [d_1(k), d_2(k), d_3(k)]$, where:

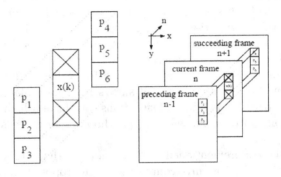

Fig. 3. Pixel position in ROD [10]

$$d_i(k) = \begin{cases} r_i(k) - x(k) & \text{if } x(k) \leq m(k) \\ x(k) - r_{7-i}(k) & \text{if } x(k) > m(k) \end{cases} \qquad (1)$$

Then the pixel position at k is detected as corrupted if any of the following is true

$$d_i(k) > T_i . \qquad (2)$$

where $i = 1, 2, 3$ and T_1, T_2, T_3 are user defined threshold and $T_1 < T_2 < T_3$. The most important selection is the appropriate value of T_1. The extended five frame based ROD detector is the combination of three classical ROD detectors. Here first(R_1), second(R_2), and third(R_3) detector work on $(n-1), n, (n+1)$, $(n-1), n, (n+1)$, $(n-1), n, (n+1)$ MC frames, respectively. The pixel $x(k)$ in current frame n, is detected as corrupted/blotched if at least one of the rank order differences exceeds some preselected thresholds i.e. T_1, T_2, T_3. If the corruption is found using detector R_1, the blotch is on the current frame n. If it is found by using R_2 or R_3 detector, there are occluded blotches and there exists blotches at the same position in successive frames along the motion trajectories between the nth and $(n-1)th$ frames or between the nth and $(n+1)th$ frames, respectively. In case of consecutive blotch, five frame based ROD detector [11] can detect blotch due to the inclusion of R_2 or R_3 detector while for original ROD [10], the threshold T_1 is needed to set at very low($\backsim 5$); resulting in large false alarm.

Original[10] and five frame based ROD[11] detectors did not propose any post processing to reduce false alarm, if any. In this paper, we propose an adaptive refinement technique which works on the initially detected blotch region i.e. first blotch is detected by five frame based ROD [11] and then adaptive refinement is performed on that detection result. It is adaptive in the sense that it takes decision based on the local image statistics, and therefore vary spatially. It is spatially coherent in the sense that gray level differences are observed in a whole neighborhood of the current pixel and furthermore, taken into consideration in temporal domain. The assumption follow here is that for each pixel $x(k)$, the goal is to decide whether the differences between the neighborhood of $x(k)$ in current frame and the neighborhood of $x(k)$ in previous and next frames are meaningful, i.e. due to a blotch, or if they can be explained by the local statistics of the region. The idea is to accept detection results if patches following a refinement model H_0 will not be detected as containing blotches. To determine whether a detected blotch pixel belongs to a specific region, regions consisting of pixels with similar properties will have to be extracted from the available data. Adjacent pixels within a blotch tend to have similar intensities. A pair of pixels is considered to be similar if they belong to a neighborhood of 3×3 and their intensity difference is smaller than some predefined value (set as 5). Therefore, adjacent pixels flagged by the blotch detector are considered to be part of the same candidate blotch. To differentiate between the various candidate blotches, a unique label is assigned to each of them. After determining candidate blotch, the formulation of refinement model is as follows:

Proposition 1. *Let $|N_0|$ be the size of a candidate blotch region i.e. no of candidate blotch pixel in that region. For a pixel x in candidate blotch at frame n, consider a patch u_n of size $(|N_0|/2) \times (|N_0|/2)$ centered at pixel x and its corresponding patches u_{n-1}, u_{n+1} in considered previous and next frames. The refinement model hypothesis states that the point differences between considered successive frames are realizations of i.i.d random variables following a centered normal distribution $N(0, \sigma_x^2)$.*

Should the refinement model H_0 satisfies, then the term D_x^p (motion compensated patch difference between $(n-1)th$ and nth frames) is a realization of the distribution $N(0, \sigma_x^2/(|N_0|/2))$.

$$D_x^p = \frac{2}{N_0} \sum_{y \epsilon (|N_0|/2)\times(|N_0|/2)} u_n(x,y) - u_{n-1}(x,y) \ . \tag{3}$$

and the same result hold for D_x^n. If the observed differences D_x^p, D_x^n are too large (set at 5), we can conclude that these differences can hardly be explained by the model hypothesis. In this case, the model hypothesis is rejected, which means a blotch is confirmed in x.

3   Cloud Computing Platform

In order to speed up the detection process by parallel processing, we propose a cloud computing platform similar to [15], OVCE (Online Virtual Computing Environment) test bed which is designed by Media Chorus Inc. This application independent test bed is composed of four layers: client layer (CL), admin server (AS), task management server (TMS) and virtualized check server (VCS) cluster. The architecture of the platform is shown in Fig. 4. CL uploads task configuration XML, video to be checked along with shell scripts and executable programs to AS. The XML describes the resource requirement such as task ID, application ID, and hardware resources for executing the task. That task ID define which kind of error detection method client want to use and application ID define the specific criteria function for the detection method. The AS consists of three functions. First based on the received XML file and user requirements, it determines the numbers of task management server needed to finish the task in parallel. The FTP server receives uploaded videos, extracts the frames from the video and passes the frames and shell scripts and programs to TMS. The third function of AS is to accumulate the output of VCS clusters to generate the final blotch detection in the video. Task management server (TMS) contains two functions; based on the task XML file received from AS, TMS assigns the task to the lower end of virtual machine for execution. For management purpose, TMS knows the status and cost of resources of each virtual machine and the task running on cloud platform needs to be registered previously. And the second one is to check extracted frames and scripts sent from the FTP server for saving the corresponding resources of each task. Virtual machine cluster is responsible for the specific implementation of tasks. First of all, each virtualized check server creates a separate operating environment and the resources required for the tasks is allocated according to the configuration XML file assigned by TMS. Then, shell scripts and programs corresponding to the task are also downloaded automatically, and the execution begins. Meanwhile, each virtual machine periodically sends status signal or message to TMS to notify its usability and availability. After finishing the execution, the detection result is passed to AS. In order to ensure the reliability, an error detection task may also be send to multiple virtual machines for execution. AS accepts the results according to the ID numbers and filters out the repeated results to speed up the output of the correct one.

4   Experimental Results

At first we will show the experimental results for blotch detection and then we describe parallel processing performance for OVCE.

4.1   Blotch Detection

This section confronts our proposed i.e. five frame based ROD with adaptive refinement method with SDIa[1], SDIp[1], ROD[10], sROD[3], and five frame

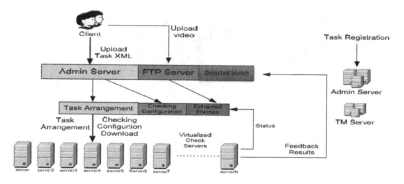

Fig. 4. Online Virtual Computing Environment (OVCE) for blotch detection

based ROD [11] on video archives provided by National Archives of Korea (NAK). The provided digital archives are made from film and tape-based videos with a wide variety of content as reference videos containing different types of archive related errors. Total size of the archive is almost 20 TByte with 15000 digital SD contents. A set of 30 blotch containing video sequences were selected from these reference videos and 300 error frames were extracted from those video sequences. In our first experiment, threshold is set to 15 for SDIa and SDIp while parameter s is set to 10 for sROD. For ROD and five frame based ROD, three thresholds are set as 15, 20 and 25. In these parameter setting, among all methods, only five frame based ROD can detect consecutive blotch pixels with lots of false alarm, which is further reduced by our refinement process (as illustrated in Fig. 5h). Next we set the thresholds for ROD [10] detectors as 5, 10 and 15. Threshold for SDIa and SDIp is set to 5 and sROD parameter s is set to 5. The corresponding result is shown in Fig. 6. It can be seen that due to the sensitivity of detectors, high number of false alarm occurred. From this result, we can conclude that threshold setting for consecutive blotch detection at same location is the most critical task for these methods. For five frame based ROD, threshold is not required to set at very low and with threshold range, between 15 to 20, is good enough for single and consecutive frame blotch detection.

4.2 OVCE Performance

The AS and TMS are PCs Intel Xeon CPU 2.66 GHz, and 12GB RAM with Turbo Linux 10.5 as operating system. Each host server has four virtual machines, and each one is Intel Quad Core CPU 2.66GHz and 1GB RAM. The operating systems are also the same as AS. All the servers are connected via 100 MB Ethernet. This section presents the blotch detection execution time comparison on regular PC and our cloud computing platform. On our platform, blotch detection program can be executed in parallel, which is truly different from regular PC. New detection task can be assigned to the virtual machines with the lowest workload for execution. As shown in Fig. 7, with the increase of the scale

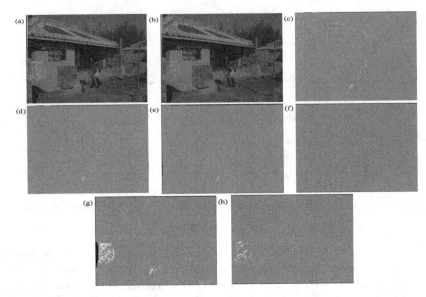

Fig. 5. Blotch detection result: Consecutive frames containing blotch (a-b); Detection by SDIa(c); Detection by SDIp (d); Detection by ROD (e); Detection by sROD (f); Detection by Five frame ROD (g); Detection result after adaptive refinement (h) [Courtesy: NAK]

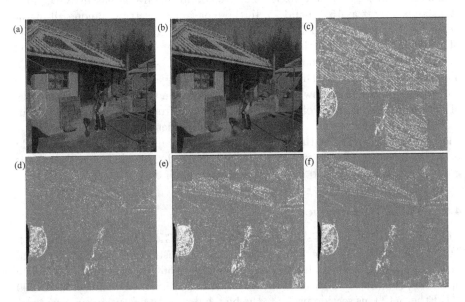

Fig. 6. Effect of low threshold on blotch detection: Consecutive frames containing blotch (a-b); Detection by SDIa (c); Detection by SDIp (d); Detection by ROD (e); Detection by sROD (f). It can be seen that low threshold results in large number of false alarm in the whole frame.

of virtual machine cluster, the throughput increases dramatically. Although the detection speed drops with the increase of test images, which is expected and can be handled by increasing no of virtual machines.

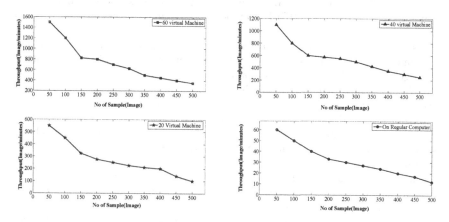

Fig. 7. Performance comparison of analyzing frames for blotch presence detection for different no. of virtual machines

5   Conclusion

In this paper, we have presented a cloud computing solution of blotch detection in large scale data. Manual thresholding plays an important role in traditional pixel based blotch detection. Determination of optimal thresholding can not be done manually. The adaptive refinement proposed in this paper overcomes the effect of manual thresholding and plays significant role in correct blotch detection. Cloud based solution ensures fast and timely delivery of results for large scale archive. Combined together with other archive error detection techniques, this platform can be the future of digital archive restoration.

Acknowledgments. This research was supported by "Archives Preservation Technology R & D Program" funded by the National Archives of Korea (NAK).

References

1. Kokaram, A.C., Morris, R.D., Fitzgerald, W.J., Rayner, P.J.: Detection of missing data in image sequences. IEEE Transactions on Image Processing **4**, 1496–1508 (1995)
2. BRAVA: Broadcast Restoration of Archives Through Video Analysis. http://brava. ina.fr
3. Van Roosmalen, P.M.B.: Restoration of archived film and video. Universal Press (1999)

4. Borko, F., Armando, E.: Handbook of cloud computing (2010)
5. Nieminen, A., Heinonen, P., Neuvo, Y.: A new class of detail-preserving filters for image processing. IEEE Transactions on Pattern Analysis and Machine Intelligence, pp. 74–90 (1987)
6. Hardie, R.C., Boncelet, C.: Lum filters: a class of rank-order-based filters for smoothing and sharpening. IEEE Transactions on Signal Processing 41, 1061–1076 (1993)
7. Buisson, O., Besserer, B., Boukir, S., Helt, F.: Deterioration detection for digital film restoration. In: 2013 IEEE Conference on Computer Vision and Pattern Recognition, pp. 78–78. IEEE Computer Society (1997)
8. Arce, G.R.: Multistage order statistic filters for image sequence processing. IEEE Transactions on Signal Processing 39, 1146–1163 (1991)
9. Kokaram, A.C.: Motion picture restoration. PhD thesis, Citeseer (1993)
10. Nadenau, M.J., Mitra, S.K.: Blotch and scratch detection in image sequences based on rank ordered differences. In: Proc. of 5th Int. Workshop on Time-Varying Image Processing, Citeseer, pp. 27–35 (1996)
11. Gangal, A., Kayikçioglu, T., Dizdaroglu, B.: An improved motion-compensated restoration method for damaged color motion picture films. Signal Processing: Image Communication 19, 353–368 (2004)
12. Bornard, R.: Probabilistic approaches for the digital restoration of television archives. Ecole Centrale Paris, Paris (2002)
13. Kokaram, A.C.: Motion picture restoration: digital algorithms for artefact suppression in degraded motion picture film and video. Springer (1998)
14. Sun, D., Roth, S., Black, M.J.: Secrets of optical flow estimation and their principles. In: 2010 IEEE Conference on Computer Vision and Pattern Recognition (CVPR), pp. 2432–2439. IEEE (2010)
15. Yan, H.B., Li, P., Zhu, C.G., Du, Y.J., Cui, G.: Accelerating image local invariant features matching using cloud computing platform. In: 2011 IEEE International Conference on Computer Science and Automation Engineering (CSAE), vol. 3., pp. 21–25. IEEE (2011)

High-Accuracy Phase-Equalizer
for Communication-Channel Compensation

Noboru Ito[✉]

Faculty of Science, Toho University, Miyama 2-2-1, Funabashi, Chiba 274-8510, Japan
050488in@is.sci.toho-u.ac.jp

Abstract. In digital communications, a nonlinear-phase communication channel needs to be equalized by using a phase-equalizer in such a way that the nonlinear-phase can be compensated and thus the whole communication channel has nearly linear-phase. As a result, the transmitted signal waveform will not be distorted. This paper proposes an improved iterative second-order-cone-programming (iterSOCP) scheme for designing an allpass digital phase-equalizer in the minimax sense. That is, the maximum absolute error of the phase response (maximum phase-error) is to be minimized. This iterSOCP scheme simplifies the minimax design problem as an iterative SOCP problem, and this minimax design problem is solved by utilizing an SOCP solver to find the coefficients of the allpass phase-equalizer. It should be noted that the solution is always a sub-optimal solution due to the non-linear feature of the design problem itself. We will use an example to demonstrate that the improved iterSOCP minimax design results in more accurate design results than other design techniques such as the linear-programming design and other iterSOCP designs.

Keywords: Signal processing · Digital communications · Phase-equalization · Phase-equalizer · IterSOCP · Optimization

1 Introduction

In digital communications, a nonlinear-phase communication-channel distorts the waveform of the transmitted signal. To keep the signal waveform unchanged after signal transmission, the nonlinear-phase characteristic must be compensated by using a phase-equalizer. That is, a phase-equalizer is cascaded with the nonlinear-phase communication-channel such that the nonlinear-phase is compensated and thus the whole communication-channel has almost linear-phase.

The allpass mathematical model is attractive for designing a digital phase-equalizer because the allpass model has unit-amplitude in the whole frequency-band. Thus, one simply needs to consider the approximation of its phase-response [1–4]. This unit-amplitude feature has also been successfully employed in designing other types of digital systems like variable-fractional-delay digital filters [5–8]. Such unique variable filters are needed in many practical applications [9]-[15]. This paper is concerned with the phase-equalizer design with allpass model.

© Springer International Publishing Switzerland 2015
N.T. Nguyen et al. (Eds.): ACIIDS 2015, Part II, LNAI 9012, pp. 529–539, 2015.
DOI: 10.1007/978-3-319-15705-4_51

Although allpass model has unit-amplitude feature, its phase is a highly non-linear function of its coefficients, which makes the phase-equalizer design (phase-fitting problem) difficult to solve. In [1], the allpass phase-equalizer design is approximated as a linear-programming problem, and other optimization techniques can also be developed for solving the design problem. In [2–4], we have shown that the design problem can also be approximated as iterative second-order-cone-programming (iterSOCP) problems, where different iterative procedures have been developed for getting higher and higher design accuracy.

Along the lines of the iterSOCP designs, this paper presents a further improved iterSOCP scheme for designing an allpass phase-equalizer. Generally speaking, the minimax design of an allpass phase-equalizer is not a true SOCP problem. Like the basic ideas presented in [2–4], we approximate this non-SOCP problem as an iterSOCP problem. The basic differences between this new iterSOCP scheme and those in [2–4] are the ways in which the unknown denominators of the error function is iteratively treated as knowns. Moreover, the iterative procedures are terminated through using different stopping criterions. As a result, the design accuracy can be further improved. We will use the same design example as that used in [2–4] for comparing the different iterSOCP schemes and show that the improved iterSOCP scheme proposed in this paper outperforms other design techniques including the linear programming design proposed in [1] and the existing iterSOCP designs developed in [2–4].

2   IterSOCP Design

Let us first consider the phase-equalizer design problem using allpass model. That is, a digital system that approximates a given desired phase-response $\theta_d(\omega)$ needs to be designed. After designing such a phase-equalizer, we can cascade it into the practical communication-channel for compensating the nonlinear-phase of the digital communication-channel.

The phase-equalizer is designed to approximate the given $\theta_d(\omega)$ by using the allpass digital system with transfer function

$$
\begin{aligned}
H(z) &= \frac{a_K + a_{K-1}z^{-1} + a_{K-2}z^{-2} + \cdots + z^{-K}}{1 + a_1 z^{-1} + a_2 z^{-2} + \cdots + a_K z^{-K}} \\
&= \frac{z^{-K} \cdot D(z^{-1})}{D(z)}
\end{aligned}
\tag{1}
$$

with

$$
\begin{aligned}
D(z) &= 1 + a_1 z^{-1} + a_2 z^{-2} + \cdots + a_K z^{-K} \\
D(z^{-1}) &= 1 + a_1 z^1 + a_2 z^2 + \cdots + a_K z^K.
\end{aligned}
\tag{2}
$$

The frequency-response of the allpass phase-equalizer in (1) is

$$H(\omega) = \frac{e^{-jK\omega} \cdot \left(1 + \sum_{k=1}^{K} a_k e^{jk\omega}\right)}{1 + \sum_{k=1}^{K} a_k e^{-jk\omega}} = \frac{N(\omega)}{D(\omega)} \tag{3}$$

whose numerator frequency-response is

$$N(\omega) = e^{-jK\omega}\left(1 + \sum_{k=1}^{K} a_k e^{jk\omega}\right)$$
$$= e^{-jK\omega} + \sum_{k=1}^{K} a_k e^{j(k-K)\omega} \tag{4}$$

and its denominator frequency-response is

$$D(\omega) = 1 + \sum_{k=1}^{K} a_k e^{-jk\omega}. \tag{5}$$

We can determine the desired frequency-response from the given phase-response $\theta_d(\omega)$ as

$$H_d(\omega) = e^{j\theta_d(\omega)}.$$

As a result, the frequency-response error is the difference between $H(\omega)$ and $H_d(\omega)$, i.e.,

$$e_H(\omega) = H(\omega) - H_d(\omega)$$
$$= \frac{\hat{e}_H(\omega)}{D(\omega)} \tag{6}$$

where

$$\hat{e}_H(\omega) = \hat{e}_R(\omega) + j\hat{e}_I(\omega) \tag{7}$$

and $\hat{e}_R(\omega)$ and $\hat{e}_I(\omega)$ are the real part and imaginary part, respectively [2]-[4], which can be detailed as

$$\hat{e}_R(\omega) = \cos(K\omega) - \cos\theta_d(\omega) - \sum_{k=1}^{K} a_k \left[\cos(\theta_d(\omega) - k\omega) - \cos(K-k)\omega\right]$$
$$\hat{e}_I(\omega) = -\sin(K\omega) - \sin\theta_d(\omega) - \sum_{k=1}^{K} a_k \left[\sin(\theta_d(\omega) - k\omega) + \sin(K-k)\omega\right]. \tag{8}$$

Consequently, the frequency-response error $e_H(\omega)$ in (6) includes both numerator and denominator as

$$e_H(\omega) = \frac{\hat{e}_R(\omega) + j\hat{e}_I(\omega)}{D(\omega)}. \tag{9}$$

Clearly, $\hat{e}_R(\omega)$ and $\hat{e}_I(\omega)$ are linear functions of the phase-equalizer coefficients $\{a_1, a_2, \cdots, a_K\}$, but $\hat{e}_H(\omega)$ itself is a non-linear function of the coefficients $\{a_1, a_2, \cdots, a_K\}$ due to the involved denominator $D(\omega)$, as is shown also in [2–4].

Theoretically, the minimax design aims to minimize the maximum absolute error (maximum-error) of $|e_H(\omega)|$, and the design can be formulated as

$$\begin{aligned} \text{minimize} \quad & \epsilon \\ \text{subject to} \quad & |e_H(\omega)| \leq \epsilon \end{aligned} \tag{10}$$

with

$$|e_H(\omega)| = \left| \frac{\hat{e}_H(\omega)}{D(\omega)} \right| = \frac{\sqrt{\hat{e}_R^2(\omega) + \hat{e}_I^2(\omega)}}{|D(\omega)|}. \tag{11}$$

Hence, the minimax design in (10) can be described in the form of

$$\begin{aligned} \text{minimize} \quad & \epsilon \\ \text{subject to} \quad & |\hat{e}_H(\omega)| \leq |D(\omega)| \cdot \epsilon. \end{aligned} \tag{12}$$

Here, it should be mentioned that the above design constraint is highly nonlinear due to $|D(\omega)|$, and the most difficult problem is how to deal with the denominator $|D(\omega)|$. If $|D(\omega)|$ is known, the constraint (12) becomes a second-order-cone (SOC) constraint. In [2–4], different techniques have been tried to assume the unknown $|D(\omega)|$ as a known.

As long as $|D(\omega)|$ is known, the above minimax design becomes simpler to solve. This is because the following vector belongs to the SOC. That is,

$$\begin{bmatrix} |D(\omega)| \cdot \epsilon \\ \hat{e}_R(\omega) \\ \hat{e}_I(\omega) \end{bmatrix} \in \mathcal{K}_q \tag{13}$$

where $\mathcal{K}_q$ denotes the SOC. Let

$$\begin{aligned} \boldsymbol{a} &= \begin{bmatrix} a_1 \ a_2 \cdots a_K \end{bmatrix}^T \\ \boldsymbol{y} &= \begin{bmatrix} \epsilon \ \boldsymbol{a} \end{bmatrix}^T \\ \boldsymbol{b}^T &= \begin{bmatrix} -1 \ 0 \cdots 0 \end{bmatrix} \end{aligned} \tag{14}$$

the SOCP design (12) can be formulated as

$$\begin{aligned} \text{maximize} \quad & \boldsymbol{b}^T \boldsymbol{y} \\ \text{subject to} \quad & \begin{bmatrix} |D(\omega)| \cdot \epsilon \\ \hat{e}_R(\omega) \\ \hat{e}_I(\omega) \end{bmatrix} \in \mathcal{K}_q \end{aligned} \tag{15}$$

or more concretely,

$$\begin{aligned} \text{maximize} \quad & \boldsymbol{b}^T \boldsymbol{y} \\ \text{subject to} \quad & \boldsymbol{c} - \boldsymbol{A}^T \boldsymbol{y} \in \mathcal{K}_q \end{aligned} \tag{16}$$

where

$$c = \begin{bmatrix} 0 \\ c_1 \\ c_2 \end{bmatrix}$$

where

$$c_1 = \cos(K\omega) - \cos\theta_d(\omega)$$
$$c_2 = -\sin(K\omega) - \sin\theta_d(\omega)$$

are the functions of the normalized frequency ω,

$$A^T = \begin{bmatrix} -|D(\omega)| & 0 & 0 & \cdots & 0 \\ 0 & f_1(\omega) & f_2(\omega) & \cdots & f_K(\omega) \\ 0 & g_1(\omega) & g_2(\omega) & \cdots & g_K(\omega) \end{bmatrix}$$

is also the function of normalized frequency ω, and

$$f_k(\omega) = \cos\left(\theta_d(\omega) - k\omega\right) - \cos(K - k)\omega$$

$$g_k(\omega) = \sin\left(\theta_d(\omega) - k\omega\right) + \sin(K - k)\omega.$$

Here, we present an improved iterSOCP scheme for solving the minimax design in (12), where the denominator $|D(\omega)|$ is involved.

1) Set the iteration $i = 0$ and a large initial maximum-error $\epsilon^{(0)}$. Then, randomly generate an initial coefficient-vector

$$\boldsymbol{a}^{(i)} = \left[a_1^{(i)} \; a_2^{(i)} \; \cdots \; a_K^{(i)} \right]^T$$
$$= \left[a_1^{(0)} \; a_2^{(0)} \; \cdots \; a_K^{(0)} \right]^T$$

which is used at the first iteration.

2) Update the iteration as $i = i + 1$. Use the coefficients $\boldsymbol{a}^{(i-1)}$ from the preceding iteration to compute the denominator

$$D^{(i-1)}(\omega) = 1 + \sum_{k=1}^{K} a_k^{(i-1)} e^{-jk\omega}. \tag{17}$$

This computation gets $D^{(i-1)}(\omega)$. Once $D^{(i-1)}(\omega)$ is known, we solve the SOCP problem

$$\text{maximize} \quad \boldsymbol{b}^T \boldsymbol{y}$$

$$\text{subject to} \quad \begin{bmatrix} |D^{(i-1)}(\omega)| \cdot \epsilon \\ \hat{e}_R(\omega) \\ \hat{e}_I(\omega) \end{bmatrix} \in \mathcal{K}_q. \tag{18}$$

This produces a new coefficient-vector $\boldsymbol{a}^{(i)}$ and the maximum-error $\epsilon^{(i)}$.

3) If $\epsilon^{(i-1)} > \epsilon^{(i)}$, then go back to 2) to repeat the above SOCP design. Otherwise, stop the iteration and take $\boldsymbol{a}^{(i)}$ as the final coefficient-vector

$$\boldsymbol{a} = \boldsymbol{a}^{(i)}.$$

It should be noted here that this stopping criterion is different from those in [2–4]. The new stopping criterion leads to higher design accuracy.

To solve the iterSOCP design, we need to sample the frequency $\omega \in [0, \pi]$ with step-size $\pi/(L-1)$ first. Sampling ω leads to uniform discrete points (L points). Then, the design constraints on the L points are imposed and the iterSOCP design in (18) is carried out.

3   Illustrative Example

The desired phase $\theta_d(\omega)$ is given by

$$\theta_d(\omega) = \begin{cases} -12\omega, & \text{for } \omega \in [0, \pi/2] \\ -8\omega - 2\pi, & \text{for } \omega \in [\pi/2, \pi]. \end{cases} \tag{19}$$

To approximate $\theta_d(\omega)$, we set the phase-equalizer order $K = 10$ and point-number $L = 1001$. With the initial coefficients listed in Table 1, which are generated randomly, the iterSOCP design method yields the sub-optimal coefficients also listed in Table 1.

Table 1. Initial and Final Coefficients

a_k	Initial Coefficients	Final Coefficients
1	-0.408038103891631	-1.273008065959045
2	0.318174651105881	0.812248342687515
3	-1.324512664216800	-0.202030792819760
4	-1.455923489180074	-0.072657347296024
5	0.734057733652676	0.032261588906438
6	-0.861814983406448	0.039438081185572
7	0.535336733703611	-0.009993972515208
8	0.456845405750621	-0.041002412083123
9	-1.337164877160273	0.032196937407078
10	1.835682522115138	0.000941850738569

The design accuracy is evaluated by using the four types of design errors, which are the normalized root-mean-squared (NRMS) error of the phase-response

$$\epsilon_{\theta 2} = \left[\frac{\displaystyle\int_0^\pi |e_\theta(\omega)|^2 d\omega}{\displaystyle\int_0^\pi |\theta_d(\omega)|^2 d\omega} \right]^{1/2} \times 100\% \tag{20}$$

the maximum phase-error

$$\epsilon_{\theta\,\max} = \max\{|e_\theta(\omega)|, \omega \in [0,\pi]\} \tag{21}$$

the NRMS frequency-response error

$$\epsilon_{H2} = \left[\frac{\displaystyle\int_0^\pi |e_H(\omega)|^2 d\omega}{\displaystyle\int_0^\pi |H_d(\omega)|^2 d\omega}\right]^{1/2} \times 100\% \tag{22}$$

and the maximum frequency-response error

$$\epsilon_{H\,\max} = \max\{|e_H(\omega)|, \omega \in [0,\pi]\} \tag{23}$$

where $e_\theta(\omega)$ is the phase-error

$$e_\theta(\omega) = \theta(\omega) - \theta_d(\omega) \tag{24}$$

and $\theta(\omega)$ is the actual phase.

Fig. 1 depicts the desired phase $\theta_d(\omega)$, and Fig. 2 plots the actual phase $\theta(\omega)$. To check the difference between Fig. 1 and Fig. 2, we plot the absolute phase-errors $|e_\theta(\omega)|$ in Fig. 3. Furthermore, Fig. 4 plots the absolute frequency-response errors $|e_H(\omega)|$ in decibel (dB) for checking the frequency-response errors.

Table 2 tabulates the design errors. To compare the design results with those from other existing designs, we also performed those designs by using the same

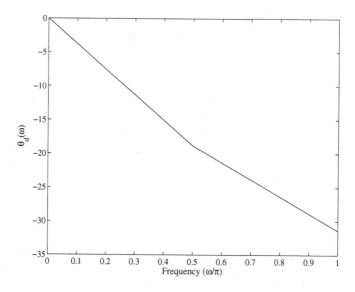

Fig. 1. Desired phase (phase specification)

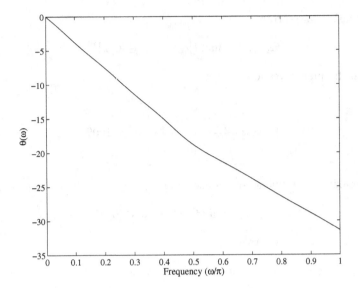

Fig. 2. Actual phase

Table 2. Four Types of Design Errors

Method	$\epsilon_{\theta2}(\%)$	$\epsilon_{\theta\,max}$	$\epsilon_{H2}(\%)$	$\epsilon_{H\,max}$ (dB)
LP [1]	0.3721	0.1779	7.2651	-15.01
iterSOCP [2]	0.1597	0.0922	3.1193	-20.71
iterSOCP [3]	0.1872	0.0879	3.6569	-21.12
iterSOCP [4]	0.1930	0.0755	3.7706	-22.45
New iterSOCP	0.1843	0.0716	3.6013	-22.90

order $K = 10$ and the same point-number $L = 1001$. All the design errors are also given in Table 2. It is clear from Table 2 that the presented iterSOCP design leads to much smaller maximum errors than other designs. Since the final objective of the minimax design is to minimize the maximum-error, we have achieved this final goal because we have further reduced the maximum-error through using the improved iterSOCP scheme.

4   Concluding Remarks

In this paper, we have presented an improved iterSOCP scheme for designing an allpass phase-equalizer, which is required in digital communications for compensating the nonlinear-phase characteristic of a digital communication-channel so that the waveform of the transmitted communication signal can be retained

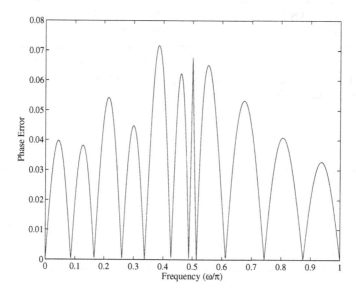

Fig. 3. Phase errors

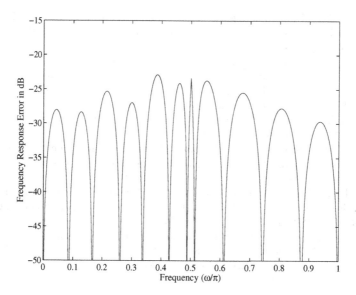

Fig. 4. Frequency-response errors

without distortion. The minimax design of an allpass phase-equalizer is approximated as an iterSOCP problem, and then the iterSOCP problem is iteratively solved. An illustrative example has been given to verify that the improved iter-SOCP design can get a much better design than other design methods including the existing iterSOCP designs. As a result, the improved iterSOCP scheme can be applied to the design of high-accuracy allpass phase-equalizers for compensating the nonlinear-phase of a digital communication channel.

References

1. Sugahara, K.: Linear Programming Design of IIR Digital Phase Networks. IECE Trans. **J68–A**(5), 444–450 (1985)
2. Ito, N., Deng, T.-L.: Generalized SOCP scheme for designing all-pass digital phase circuits. In: Proc. IEEE MICC 2013, pp. 255–258. IEEE Press, New York (2013)
3. Ito, N.: Phase-Correction-Network Design Using the SOCP Optimization Scheme. In: Proc. IEEE ISSNIP 2014, pp. 1–4. IEEE Press, New York (2014)
4. Deng, T.-B.: Novel Iterative Second-Order-Cone-Programming Scheme for Designing High-Accuracy Phase-Circuits. Journal of Circuits, Systems, and Computers **23**(5), 1–14 (2014). World Scientific, Singapore
5. Deng, T.-B.: Noniterative WLS Design of Allpass Variable Fractional-Delay Digital Filters. IEEE Trans. Circuits Syst. I, Reg. Papers **53**(2), 358–371 (2006). IEEE Press, New York
6. Deng, T.-B.: Generalized WLS Method for Designing Allpass Variable Fractional-Delay Digital Filters. IEEE Trans. Circuits Syst. I, Reg. Papers **56**(10), 2207–2220 (2009). IEEE Press, New York
7. Deng, T.-B.: Minimax Design of Low-Complexity Allpass Variable Fractional-Delay Digital Filters. IEEE Trans. Circuits Syst. I, Reg. Papers **57**(8), 2075–2086 (2010). IEEE Press, New York
8. Deng, T.-B.: Closed-Form Mixed Design of High-Accuracy All-Pass Variable Fractional-Delay Digital Filters. IEEE Trans. Circuits Syst. I, Reg. Papers **58**(5), 1008–1019 (2011). IEEE Press, New York
9. Deng, T.-B.: Discretization-Free Design of Variable Fractional-Delay FIR Digital Filters. IEEE Trans. Circuits Syst. II, Analog Digit. Signal Processing **48**(6), 637–644 (2001). IEEE Press, New York
10. Deng, T.-B., Nakagawa, Y.: SVD-Based Design and New Structures for Variable Fractional-Delay Digital Filters. IEEE Trans. Signal Processing **52**(9), 2513–2527 (2004). IEEE Press, New York
11. Deng, T.-B., Lian, Y.: Weighted-Least-Squares Design of Variable Fractional-Delay FIR Filters Using Coefficient-Symmetry. IEEE Trans. Signal Processing **54**(8), 3023–3038 (2006). IEEE Press, New York
12. Deng, T.-B.: Coefficient-Symmetries for Implementing Arbitrary-Order Lagrange-Type Variable Fractional-Delay Digital Filters. IEEE Trans. Signal Processing **55**(8), 4078–4090 (2007). IEEE Press, New York

13. Deng, T.-B.: Decoupling Minimax Design of Low-Complexity Variable Fractional-Delay FIR Digital Filters. IEEE Trans. Circuits Syst. I, Reg. Papers **58**(10), 2398–2408 (2011). IEEE Press, New York
14. Deng, T.-B.: Minimax Design of Low-Complexity Even-Order Variable Fractional-Delay Filters Using Second-Order Cone Programming. IEEE Trans. Circuits Syst. II, Exp. Briefs **58**(10), 692–696 (2011). IEEE Press, New York
15. Deng, T.-B., Chivapreecha, S., Dejhan, K.: Bi-minimax Design of Even-Order Variable Fractional-Delay FIR Digital Filters. IEEE Trans. Circuits Syst. I: Regular Papers **59**(8), 1766–1774 (2012). IEEE Press, New York

Using Different Norms in Packing Circular Objects

Igor Litvinchev$^{(\boxtimes)}$, Luis Infante, and Lucero Ozuna

Faculty of Mechanical and Electrical Engineering, UANL
Pedro de Alba S/N, 66450 San Nicolas de los Garza, NL, Mexico
`igorlitvinchev@gmail.com`

Abstract. A problem of packing unequal circles in a fixed size rectangular container is considered. The circle is considered in a general sense, as a set of points that are all the same distance (not necessary Euclidean) from a given point. An integer formulation is proposed using a grid approximating the container and considering the nodes of the grid as potential positions for assigning centers of the circles. The packing problem is then stated as a large scale linear 0-1 optimization problem. Valid inequalities are proposed to strengthening the original formulation. Nesting circles inside one another is considered tacking into account the thickness of the circles. Numerical results on packing circles, ellipses, rhombuses and octagons are presented to demonstrate the efficiency of the proposed approach.

1 Introduction

Packing problems generally consist of packing a set of items of known dimensions into one or more large objects or containers to minimize a certain objective (e.g. the unused part of the container or waste). Packing problems constitute a family of natural combinatorial optimization problems applied in computer science, industrial engineering, logistics, manufacturing and production processes (see e.g. [2, 5, 7, 10, 11] and the references therein).

Along with industrial applications one may find packing problems, e.g. in healthcare issues. In [22] automated radiosurgical treatment planning for treating brain and sinus tumours was considered. Radiosurgery uses the gamma knife to deliver a set of extremely high dose ionizing radiation, called "shots" to the target tumour area. For large target regions multiple shots of different intensity are used to cover different parts of the tumour. However, this procedure may result in large doses due to overlap of the different shots. Optimizing the number, positions and individual sizes of the shots can reduce the dose to normal tissue and achieve the required coverage.

Packing problems for regular shapes (circles and rectangles) of objects and/or containers are well studied [12]. In circle packing problem the aim is to place a certain number of circles, each one with a fixed known radius inside a container. The circles must be totally placed in the container without overlapping. The shape of the container may vary from a circle, a square, a rectangular, etc. Many variants of packing circular objects have been formulated as nonconvex (continuous) optimization problems with decision variables being coordinates of the centres [12]. Non-overlapping typically is assured by nonconvex constraints representing that the Euclidean distance separating

N.T. Nguyen et al. (Eds.): ACIIDS 2015, Part II, LNAI 9012, pp. 540–548, 2015.
DOI: 10.1007/978-3-319-15705-4_52

the centres of the circles is greater than a sum of their radii. The nonconvex problems can be tackled by available nonlinear programming (NLP) solvers, however most NLP solvers fail to identify global optima and global optimization techniques have to be used [5]. The nonconvex formulations of circular packing problem give rise to a large variety of algorithms which mix local searches with heuristic procedures in order to widely explore the search space. We will refer the reader to review papers presenting the scope of techniques and applications for regular packing problem [1, 4, 6, 18, 19, 20].

Irregular packing problems involve non standard shapes of objects and/or containers. Irregular shapes are those that require non-trivial handling of the geometry. One of the most common representations for irregular shape is a polyhedral domain which may by nonconvex o multi-connected. Heuristic and metaheuristic algorithms are the basis for the solution approaches [7, 21].

In this paper we study approximate packing of circular-like objects using a regular grid to approximate the container. The circular-like object is considered in a general sense, as a set of points that are all the same distance (not necessary Euclidean) from a given point. Thus different shapes, such as ellipses, rhombuses, rectangles, octagons can be treated the same way by simply changing the norm used to define the distance. In a sense, we demostrate that packing some irregular objects is as simple as packing circles. The nodes of the grid are considered as potential positions for assigning centers of the circles. The packing problem is then stated as a large scale linear 0-1 optimization problem. Valid inequalities are proposed to strengthening the original formulation. Nesting circles inside one another is considered tacking into account the thickness of the circles. Numerical results on packing circles, ellipses, rhombuses and octagons are presented to demonstrate the efficiency of the proposed approach.

To the best of our knowledge, the idea to use a grid was first implemented in [3] in the context of cutting problems. This approach was recently applied in [9, 14-17, 21] for packing problems. This work is a continuation of [14]. The rest of the paper is structured as follows. In Section 2 the main integer programming model for packing problem is presented. Section 3 is related to the experimental results on packing circles, ellipses, rhombuses and octagons to show that our methodology is efficient. A final section concludes this work.

2   The Principal Model

Suppose we have non-identical circles C_k of known radius R_k, $k \in K = \{1, 2, ... K\}$ which have to be packed in a container G. It is assumed that no two objects overlap with each other and each packed object lies entirely in the container. Here we consider the circle as a set of points that are all the same distance R_k (not necessary Euclidean) from a given point. In what follows we will use the same notation C_k for the figure bounded by the circle, $C_k = \{z \in \mathbb{R}^2 : \|z - z_{0k}\| \le R_k\}$, assuming that it is easy to understand from the context whether we mean the curve or the figure. Denote by S_k the area of C_k.

Let at most M_k circles C_k are available for packing and at least m_k of them have to be packed. Denote by $i \in I = \{1, 2..., n\}$ the node points of a regular grid covering the rectangular container. Denote by d_{ij} the distance (in the sence of norm used to define the circle) between points i and j of the grid. Define binary variables $x_i^k = 1$ if centre of a circle C_k is assigned to the point i; $x_i^k = 0$ otherwise. In what follows we will say that the object is assigned to the node i if the corresponding reference point is assigned to that node and will denote this as C_k^i.

In order to the circle C_k assigned to the point i be non-overlapping with other circles being packed, it is necessary that $x_j^l = 0$ for $j \in I$, $l \in K$, such that $d_{ij} < R_k + R_l$. For fixed i, k let $N_{ik} = \{j, l : i \neq j, d_{ij} < R_k + R_l\}$. Let n_{ik} be the cardinality of N_{ik}: $n_{ik} = |N_{ik}|$. Then the problem of maximizing the area covered by the circles can be stated as follows:

$$\max \sum_{i \in I} \sum_{k \in K} S_k^2 x_i^k \tag{1}$$

subject to

$$m_k \leq \sum_{i \in I} x_i^k \leq M_k, \quad k \in K, \tag{2}$$

$$\sum_{k \in K} x_i^k \leq 1, \quad i \in I, \tag{3}$$

$$x_i^k = 0 \text{ for } C_k^i \setminus (G \cap C_k^i) \neq \varnothing \text{ for } i \in I, k \in K, \tag{4}$$

$$x_i^k + x_j^l \leq 1, \text{ for } i \in I; k \in K; (j,l) \in N_{ik}, \tag{5}$$

$$x_i^k \in \{0,1\}, \quad i \in I, k \in K. \tag{6}$$

Constraints (2) ensure that the number of circles packed is between m_k and M_k; constraints (3) that at most one centre is assigned to any grid point; constraints (4) that C_k can not be assigned to the node i if C_k^i is not totally placed inside G; pair-wise constraints (5) guarantee that there is no overlapping between the circles; constraints (6) represent the binary nature of variables.

Similar to plant location problems [22] we can state non-overlapping conditions in a more compact form. Summing up pair-wise constraints (5) over $(j,l) \in N_{ik}$ we get

$$n_{ik} x_i^k + \sum_{j,l \in N_{ik}} x_j^l \leq n_{ik} \text{ for } i \in I, k \in K \tag{7}$$

Note that constraints similar to (7) were used in [9] for packing equal circles ($K = 1$).

Proposition 1. [15,17]. Constraints (5), (6) are equivalent to constraints (6), (7).

Thus the problem (1)-(6) is equivalent to the problem (1)-(4), (6), (7). To compare two equivalent formulations, let

$$P_1 = \{x \ge 0 : x_i^k + x_j^l \le 1, \text{ for } i \in I, \ k \in K, \ (j,l) \in N_{ik}\},$$

$$P_2 = \{x \ge 0 : n_{ik} x_i^k + \sum_{j,l \in N_{ik}} x_j^l \le n_{ik} \text{ for } i \in I, \ k \in K\}.$$

Proposition 2. [15, 17]. $P_1 \subset P_2$.

As follows from Proposition 2, the pair-wise formulation (1)-(6) is stronger [23] than the compact one. Numerical experiments presented in [17] demonstrate that the pair-wise formulation is also computationally more attractive since it provides a tighter LP-bound. Bearing in mind these reasons we restrict ourselves by considering below only pair-wise formulations.

By the definition, $N_{ik} = \{j,l : i \ne j, d_{ij} < R_k + R_l\}$ and hence if $(j,l) \in N_{ik}$, then $(i,k) \in N_{jl}$. Thus a half of the constraints in (5) are redundant since we have

$$x_i^k + x_j^l \le 1, \text{ for } i \in I, \ k \in K, \ (j,l) \in N_{ik},$$

$$x_j^l + x_i^k \le 1, \text{ for } j \in I, \ l \in K, \ (i,k) \in N_{jl}.$$

The redundant constraints can be eliminated without changing the quality of LP-bound giving a reduced pair-wise non overlapping formulation.

In many applied problems nesting circles inside one another is permitted. For example, in [8, 10, 11] nesting is considered in the context of packing pipes of different diameters into a shipping container. Nesting is also essential for storing different cylinders one over another in the form of cylindrical towers.

To consider nesting circles inside one another, we only need to modify the non-overlapping constraints. In order to the circle C_k assigned to the point i be non-overlapping with other circles being packed (including circles placed inside this circle), it is necessary that $x_j^l = 0$ for $j \in I$, $l \in K$, such the $R_k - R_l < d_{ij} < R_k + R_l$ for $R_k > R_l$. Let

$$\Omega_{ik} = \{j,l : i \ne j, R_k - R_l < d_{ij} < R_k + R_l, R_k > R_l\}$$

Then the non-overlapping constraints for packing circles with nesting can be stated in the form

$$x_i^k + x_j^l \le 1, \text{ for } i \in I; k \in K; (j,l) \in \Omega_{ik} \qquad (8)$$

Constraints (3) have to be relaxed in case of nesting.

If nesting is permitted, e.g., in the case of packing plastic pipes [8,10,11], it may be necessary to take into account the thickness of the pipe, i.e. the difference between external and internal size of the object. To consider nesting-subject-to-thickness we need only to redefine the set Ω_{ik}. Let g_k be the thickness of the circle C_k. For $R_k - g_k > R_l$ For Ω_{ik} defined as

$$\Omega_{ik} = \{j,l : i \ne j, R_k - g_k - R_l < d_{ij} < R_k + R_l, R_k - g_k > R_l\}, \qquad (9)$$

we get non-overlapping constraints (8) for the case of "nesting-subject-to-thickness". The rest of the optimization model stated above remains unchanged.

Note that all constructions proposed above, including Propositions 1,2, remain valid for any norm used to define the circular-like object. In fact, changing the norm affects only the distance d_{ij} used in the definitions of the sets N_{ik}, Ω_{ik} in the non-overlapping constraints (5), (8). That is, by simple pre-processing we can use the basic model (1)-(6) for packing different geometrical objects of the same shape. It is important to note that the non-overlapping condition has the form $d_{ij} \geq R_k + R_l$ no matter which norm is used. For example, a circular object in the maximum norm $\|z\|_\infty := \max_i \{|z_i|\}$ is represented by a square, taxicab norm $\|y\|_1 := \sum_r |y_r|$ yields a rhombus. In a similar way we may manage rectangles, ellipses, etc. Using a superposition of norms, we can consider more complex circular objects. For $\|y\| := \max_r \{|y_r|, \gamma \sum_r |y_r|\}$ and a suitable $0.5 < \gamma < 1$ we get an octagon, an intersection of a square and a rhombus.

We may expect that the linear programming relaxation of the problem (1)-(6) provides a poor upper bound for the optimal objective. For example, for $K = 1$ and suitable M_k, m_k the point $x_i^k = 0.5$ for all $i \in I$ may be feasible to the relaxed problem with the corresponding objective growing linearly with respect to the number of grid points.

To tightening the LP-relaxation for (1)-(6) without nesting we consider valid inequalities aimed to ensure that no grid point is covered by two circles. Define matrix $\left[\alpha_{ij}^k \right]$ as follows. Let $\alpha_{ij}^k = 1$ for $d_{ij} < R_k$, $\alpha_{ij}^k = 0$ otherwise. By this definition, $\alpha_{ij}^k = 1$ if the circle C_k centred at i covers point j. The following constraints ensure that no points of the grid can be covered by two circles:

$$\sum_{k \in K} \sum_{j \in I} \alpha_{ij}^k x_j^k \leq 1, \quad i \in I. \tag{10}$$

Note that (10) is not equivalent to non-overlapping constraints (5). Constraints (10) ensure that there is no overlapping in grid points, while (5) guarantee that there is no overlapping at all. The valid inequality (10) holds for any norm used to define the circular object.

3   Numerical Results

In this section we present a numerical study on packing equal circles, ellipses, rhombuses and octagons by varying the definition of the norm in (1)-(6). It is assumed that the supply of the objects is unlimited. The standard Euclidean and taxicab norms were used to define circles and rhombuses, $\|z\| := (2z_1^2 + z_2^2)^{1/2}$ and $\|z\| := \max\{|z_1|, |z_2|,$ $(1/\sqrt{2})(|z_1| + |z_2|)\}$ were used for ellipses and octagons. A rectangular uniform grid of size Δ along both sides of the container was implemented. It is not hard to verify that for these particular shapes constraints (4) constraints (4) can be relaxed by reducing correspondingly the size of the container. The test bed set of 9 instances was used for packing maximal number of equal circular objects into a rectangular container of width 3 and height 6. The values of radii and grid size are the same as in [9, Table 3]. All optimization problems were solved by the system CPLEX 12.6. The runs were executed on a desktop computer with CPU AMD FX 8350 8-core processor 4 Ghz and 32Gb RAM.

The results of the numerical experiment are given in Table 1. Here the first four columns present instance number, radius R of the circular object, size of the grid Δ and the number of binary variables (dim). The following columns give the number of objects packed and corresponding CPU time (in seconds) for circles (C), ellipses (E), rhombuses (R) and octagons (O). For all problem instances $mipgap = 0$ was set as a stopping criterion for running CPLEX. The asterisk indicates that the computation was interrupted after the computation time exceeded 12-hours CPU time. For problem instances where optimality was not achieved within time limit the number of objects corresponds to the best integer solution and the number in parenthesis indicates the value of $mipgap$ in % obtained to the moment of interruption. Packings for instance 7 are presented in Fig. 1.

Table 1. Packing circles, ellipses, rhombuses and octagons

No.	R	Δ	dim	O	CPU	R	CPU	C	CPU	E	CPU
1	0.5	0.125	697	18	1	28	1	18	1	34	11
2	0.625	0.15625	1403	9	52	15	11	10	41	21	25
3	0.5625	0.0703125	2449	12	202	20	312	13	186	27	288
4	0.375	0.046875	1425	26	49	39	399	32	4	59	2
5	0.3125	0.078125	2139	41	6850	76	6	45	114	99	3
6	0.4375	0.0546875	3666	20	1430	35	2829	21	17654	44 (4)	*
7	0.25	0.0625	3649	72	22	127	17	74 (5)	*	137 (7)	*
8	0.275	0.06875	2880	50	20495	75 (4.6)	*	61	177	108 (6)	*
9	0.1875	0.046875	6897	106 (12.5)	*	167 (9)	*	140 (5)	*	261 (4)	*

As can be seen from Table 1 for all types of the objects the large instances were not solved to optimality within the time limit. The computation was interrupted obtaining a feasible solution within 5-10% of proven relative suboptimality.

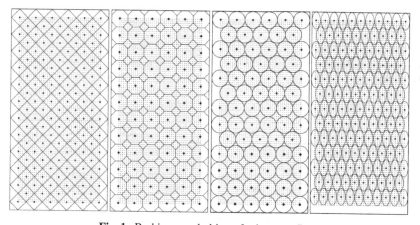

Fig. 1. Packing equal objects for instance 7

However, the most of the computation time was used to tightening the dual bound to prove optimality/suboptimality of the feasible solution obtained on early stages of CPLEX. For example, for the case of octagons and instance 9 a feasible solution with 105 objects packed was obtained within 300 sec., for instance 5 - 41 objects in 30 sec, for instance 6 – 19 objects in 30 sec. A similar behaviour was noticed for other objects and instances where the computation time exceeds 1000 sec. For all these instances at most 600 sec. was necessary to get a feasible solution with relative difference 1-2% from the best feasible or optimal solution. Thus we may consider CPLEX as a sufficiently fast tool to find good feasible solutions and in this sense it is comparable with heuristics proposed, e.g., for circle packing problems in [9].

Table 2 presents an effect of introducing valid inequalities (10) in the problem formulation (1)-(6). The columns here indicate the type of the object (C, E, O, R), the value of the LP-relaxation before and after introducing valid inequalities, LP and LPC, correspondingly. The value of LP-relaxation without valid inequalities (second column in Table 2) is just the same for all types of circular objects and equals to the half of the instance dimension (all variables are 0.5). We see that introducing valid inequalities (10) improves significantly the quality of LP bound for all shapes of the objects. The detailed study of this subject for the case of circles one can find in [17] for the same test bed instances.

Table 2. LP-relaxations

No.	LP	O	LPC	R	LPC	C	LPC	E	LPC
1	348.5	18	19	28	33.43	18	19	34	36
2	701.5	9	10	15	16.87	10	10	21	25
3	1224.5	12	14.0743	20	22.25	13	14.07	27	29.91
4	712.5	26	30.9485	39	41.37	32	36.33	59	68.86
5	1069.5	41	53.4043	76	94.76	45	53.4	99	110
6	1833.5	20	22.5537	35	39.72	21	23.86	43	49.787
7	1824.5	72	90.9767	127	157.96	74	90.98	137	182
8	1440	50	59.014	75	79.53	61	72	108	134.56
9	3448.5	106	134.342	167	182.28	140	162	261	273.61

The results of a small computational experiment for packing two octagons in a square 30x30 container maximizing the total area of the objects are presented in Table 3.

Table 3. Packing 2 different octagons

No.	R_1, R_2	dim	N-	CPU	N+	CPU	N+T	CPU
1	0.6, 6.3	441	627.48	1	842.21	1	804.37	1
2	0.6, 6.3	961	699.06	6	971.05	3	910.209	5
3	1, 5.3	441	699.35	1	952.82	1	922.99	1
4	1, 5.3	961	750.09	57	1158.27	129	1019.1	49

Here the first three columns give instance number, radii, and a number of grid points (integer variables). The last columns give the total area without nesting (N-), with nesting (N+) and with nesting and thickness (N+T), as well as corresponding CPU time in sec. The thickness g_k in (1.9) was defined as $0.1R_k$. The packings obtained for instance 4 are presented in Fig. 2.

Fig. 2. Packing two octagons for instance 4

4   Conclusions

Integer formulations were proposed for approximate packing circular-like objects with nesting and taking into account the thickness of objects (the difference between external and internal size of the object). It was demonstrated that by simply changing the definition of the distance (preprocessing) it is possible to use the same basic models for packing different circular-like objects such as circles, ellipses, rhombuses, octagons, etc.

Valid inequalities were considered to strengthening the original formulation. The results of our numerical experiment indicate that the valid inequalities improve significantly the LP-bound. Note that these inequalities can be used for packing different shapes. An interesting area for future research is the generalization of valid inequalities for the case of nesting. To cope with large dimension of arising problems it is interesting to study the use of Lagrangian relaxation and corresponding heuristics [13, 23].

This work was partially supported by grants from RFBR (12-01-00893-a) and CONACYT (167019).

References

1. Akeb, H., Hifi, M.: Solving the circular open dimension problem using separate beams and look-ahead strategies. Computers & Operations Research **40**, 1243–1255 (2013)
2. Baltacioglu, E., Moore, J.T., Hill, R.R.: The distributor's three-dimensional pallet-packing problem: a human-based heuristical approach. International Journal of Operations Research **1**, 249–266 (2006)
3. Beasley, J.E.: An exact two-dimensional non-guillotine cutting tree search procedure. Operations Research **33**, 49–64 (1985)
4. Birgin, E.G., Gentil, J.M.: New and improved results for packing identical unitary radius circles within triangles, rectangles and strips. Computers & Operations Research **37**, 1318–1327 (2010)

5. Castillo, I., Kampas, F.J., Pinter, J.D.: Solving circle packing problems by global optimization: Numerical results and industrial applications. European Journal of Operational Research **191**, 786–802 (2008)

6. Correia, M.H., Oliveira, J.F., Ferreira, J.S.: Cylinder packing by simulated annealing. Pesquisa Operacional **20**, 269–286 (2000)

7. Fasano, G.: Solving Non-standard Packing Problems by Global Optimization and Heuristics. Springer (2014)

8. Frazer, H.J., George, J.A.: Integrated container loading software for pulp and paper industry. European Journal of Operational Research **77**, 466–474 (1994)

9. Galiev, S.I., Lisafina, M.S.: Linear models for the approximate solution of the problem of packing equal circles into a given domain. European Journal of Operational Research **230**, 505–514 (2013)

10. George, J.A., George, J.M., Lamar, B.W.: Packing different–sized circles into a rectangular container. European Journal of Operational Research **84**, 693–712 (1995)

11. George, J.A.: Multiple container packing: a case study of pipe packing. Journal of the Operational Research Society **47**, 1098–1109 (1996)

12. Hifi, M., M'Hallah, R.: A literature review on circle and sphere packing problems: models and methodologies. Advances in Operations Research, Article ID 150624 (2009). doi:10.1155/2009/150624

13. Litvinchev, I., Rangel, S., Mata, M., Saucedo, J.: Studying properties of Lagrangian bounds for many-to-many assignment problems. Journal of Computer and Systems Sciences International **48**, 363–369 (2009)

14. Litvinchev, I., Ozuna, L.: Packing circles in a rectangular container. In: Proc. Intl. Congr. on Logistics and Supply Chain, Queretaro, Mexico, pp. 24–30, October 2013

15. Litvinchev, I., Ozuna, L.: Integer programming formulations for approximate packing circles in a rectangular container. Mathematical Problems in Engineering, Article ID 317697 (2014). doi:10.1155/2014/317697

16. Litvinchev, I., Ozuna, L.: Approximate packing circles in a rectangular container: valid inequalities and nesting. Journal of Applied Research and Technologies **12**, 716–723 (2014)

17. Litvinchev, I., Infante, L., Ozuna Espinosa, E.L.: Approximate circle packing in a rectangular container: integer programming formulations and valid inequalities. In: González-Ramírez, R.G., Schulte, F., Voß, S., Ceroni Díiaz, J.A. (eds.) ICCL 2014. LNCS, vol. 8760, pp. 47–60. Springer, Heidelberg (2014)

18. Lopez, C.O., Beasley, J.E.: A heuristic for the circle packing problem with a variety of containers. European Journal of Operational Research **214**, 512–525 (2011)

19. Lopez, C.O., Beasley, J.E.: Packing unequal circles using formulation space search. Computers & Operations Research **40**, 1276–1288 (2013)

20. Stoyan, Y.G., Yaskov, G.N.: Packing congruent spheres into a multi-connected polyhedral domain. International Transactions in Operational Research **20**, 79–99 (2013)

21. Toledo, F.M.B., Carravilla, M.A., Ribero, C., Oliveira, J.F., Gomes, A.M.: The Dotted-Board Model: A new MIP model for nesting irregular shapes. Int. J. Production Economics **145**, 478–487 (2013)

22. Wang, J.: Packing of unequal spheres and automated radiosurgical treatment planning. Journal of Combinatorial Optimization **3**, 453–463 (1999)

23. Wolsey, L.A.: Integer Programming. Wiley, New York (1999)

SCADA Based Operator Support System for Power Plant Fault Diagnosis

N. Mayadevi[1]([✉]), S.S. Vinodchandra[2], and Somarajan Ushakumari[1]

[1] Department of Electrical Engineering, College of Engineering Trivandrum,
Trivandrum, India
mayamohana@hotmail.com
[2] Computer Centre, University of Kerala, Trivandrum, India

Abstract. Supervisory Control and Data Acquisition Systems (SCADA) play an important role in the monitoring and control of power generation plants. Whenever faults occur in any part of the power plant, critical alarms are generated by SCADA system. This paper discusses on the assessment of such faults using a set of rules encoded with human expert knowledge. Proposed system uses a search algorithm on SCADA history alarm database which is driven by the knowledge base rules. The expert system components are developed for fault diagnosis and for operator guidance. Rules and Meta rules for knowledge induction and backward chaining inference engine are used for fault diagnosis. The encoded knowledge helps in searching the SCADA alarm database to arrive at the root cause of the fault occurred. The decision support system provides step by step procedure to be followed by an operator to restore the system to normal operating conditions. The tedious manual processing involved in analysing SCADA history alarm database containing vast amount of information can be reduced by using the proposed operator support system. The explanation facility and structured methodology followed by the hybrid intelligent system makes fault diagnosis an effortless and efficient process. The system is implemented using Java Expert System Shell(JESS) in Eclipse platform.

Keywords: Fault diagnosis · Expert system · SCADA · Operator support system

1 Introduction

The supervision and monitoring of power plant is a complex task due to numerous parameters involved in power generation process and the complex interactions existing between them. With the fast progress in computer aided techniques, various tools have been developed to monitor, control and automate plant related activities. Still the most important task of fault diagnosis and abnormal event detection in process plants remains largely as a manual activity [1]. If well experienced plant operators are available, faults can be identified and corrected in time.

© Springer International Publishing Switzerland 2015
N.T. Nguyen et al. (Eds.): ACIIDS 2015, Part II, LNAI 9012, pp. 549–558, 2015.
DOI: 10.1007/978-3-319-15705-4_53

But scarcity of experienced human experts, who can make effective decisions during stressful fault conditions, is one of the major problems faced by power industries. The fault analysis strategy is further complicated by the increased size and complexity of modern power generation units. Fault diagnosis for complex, non-linear, dynamic systems has become an important topic of research over the past few decades. As a result, there is a wide range of fault analysis strategies developed, depending on the nature of system under consideration. These techniques include model based approaches, knowledge based approaches, data driven techniques, simulation based approaches, neural and fuzzy based approaches and statistical techniques [2].

Recent work has focused on knowledge based expert system technology as a method for processing complex combinations of decision conditions using the experts knowledge. An expert system is a set of program that manipulate encoded knowledge to solve problems in a specialized domain, which normally requires human expertise [3]. An expert system's knowledge is obtained from expert sources and coded in suitable form for the system to use in its inference or reasoning processes [4]. Main advantages of developing an expert systems for diagnostic problem-solving are transparent reasoning, ability to reason like human experts and the ability to provide explanations for the solutions provided [2].

Power plant fault diagnosis that uses expert system concept has been researched extensively and often reported[5-6]. Current trend in the development of diagnostic expert system is to bring together knowledge base and database concepts with hybrid modelling techniques for root cause diagnosis and automatic data analysis [7-8].

Embedding expert system components in real time plant monitoring system such as SCADA systems mainly try to help the operators to improve their performance during emergency situations. Modern SCADA systems are integrated with intelligent alarm processors which reduce the number of alarms generated during fault conditions [9]. Most of the existing SCADA based fault diagnosis expert systems primarily focused on important machinery like turbine, generator set etc. [10-11]. Extending these systems for entire plant fault diagnosis is difficult and costly.

This paper proposes a fault diagnosis strategy for the entire plant, based on the database of plant history. It uses a hybrid technique for fault diagnosis integrating procedure oriented and knowledge oriented techniques. SCADA alarm database together with knowledge base act as the back bone of the proposed system. SCADA alarm history is explored with the help of stored knowledge to diagnose the root cause of already occurred black out situation. It also provides the steps to be followed for restoring the system back to normal condition. Generally plant operators manually analyse plant conditions and historical SCADA data to locate and isolate fault. This tedious manual processing involved in plant fault diagnosis can be reduced with the help of the proposed operator support system.

2   SCADA System of Power Plant

SCADA system supervises controls, optimizes and manages power generation and transmission systems. SCADA system components comprises of master terminal units, Remote Terminal Units (RTUs), I/O Subsystems, software that run the control algorithms and generate control outputs. It is through the operator interface, Human Machine Interface (HMI) that humans interact and controls the process.

RTUs gathers information from various remote sensing input devices like valves, pumps, motors, meters, etc. Essentially, data is either analog (real numbers), digital (on/off) or pulse data.The RTU receives a binary data stream from the protocol that the communication equipment uses [12]. The data is then interpreted and the CPU directs the appropriate action at the site. The communication infrastructure provides connection between the supervisory system and to the RTUs [13].

On each fraction of a second signals are obtained from the RTUs. Each of these signals, values and flags are logged into the SCADA system and also backup to the SCADA database. A supervisory system provides good alarms (well defined), commented and specific to different running modes [14]. An important part of the SCADA implementation is such alarms. Conventionally, the operator's attention is directed towards a fault by the alarm system. Operators have to play a vital role in detecting the causes of abnormal conditions and taking corrective measures to prevent system failure [15]. These tasks can be very complex, especially during emergency situations, when control centre operators receive a huge flow of real-time information and is forced to make decisions under great stress, like, absence of the most experienced operators [16]. Only highly skilled and experienced personnel is able to identify the most important alarms and root causes in such an alarm flood.

SCADA act as supervisory level only and has no ability to analyse the plant fault condition. The system proposed in this work integrate an intelligent program with existing SCADA system, that can advise the operator about the root cause of the fault/alarm and the steps needed to clear it, to improve overall system performance.

3   Architecture of Operator Support System

Architecture of operator support system developed for a hydroelectric power plant is shown in fig.1. It consists of an expert system component and a diagnosis engine. Basic components of the expert system includes knowledge base, inference engine and knowledge base editor. Diagnosis engine is interfaced with SCADA system through open protocol client interfaces and consists mainly of pre-processor, alarm filter and search engine.

3.1   Expert System

Power plant is considered as an electro-mechanical system consisting of both electrical and mechanical components. The electrical fault in the plant is isolated

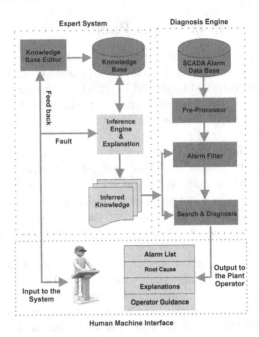

Fig. 1. Architecture of Fault Diagnosis System

by relay protection schemes. But faults in mechanical part of the system is mostly identified and corrected by maintenance team of the power plant. This study has mainly considered the mechanical faults in the system.

The main hurdle to create a new expert system application for problem diagnosis or fault analysis is the development of the knowledge base, which demands a huge interaction between the knowledge engineer and human expert [17].

The knowledge engineering phase of this research involved the identification of the different components and corresponding fault/alarm for the main components of the system. Major mechanical components of the system include spherical valve, turbine, governing system, governor, governor oil pump motor, generators, exciter, exciter transformer, generator transformer, isolators, circuit breakers and cables. The main mechanical problems that occur in a generating unit are usually caused by the high temperature of the bearings, vibration of the rotor and turbine, governor problems, governor bearing oil level problem, governor oil circuit problems, and inadvertent movements of the spherical valve and so on. The persistence of any such problem can seriously impair the machine, its accessories, or both if it is not relived from service immediately. The existing SCADA software in the plant is programmed to display alarm messages on HMI on occurrence of any of the fault conditions to warn operator. Once the cause of alarm is solved, the message will be cleared. If corrective actions are delayed, the highest fault level of mechanical trip occurs. In this study, all important alarms

Table 1. List of Alarms Selected for Study

Generator NDE Bearing Vibration	Generator NDE Bearing Low Level Oil
Turbine Bearing High Temp	Turbine Bearing High Oil Level
Generator DE Bearing Low Level Oil	Generator DE Bearing High Temp
Generator NDE Bearing High Temp	Generator DE Bearing Vibration
Stator Core High Temp	HPU Low Level Pressure

related to mechanical trip conditions are identified and causes of each alarm are analysed with the help of plant operator and manuals. Knowledge about the problem domain is extracted by conducting interviews with plant experts and SCADA developers. Plant manuals, books and operator diary also helped in knowledge base creation. The alarm list and consequent alarm list are prepared based on acquired knowledge. Ten critical alarms of the plant are considered in this study. Table 1 gives the list of selected alarms. Most of these alarms are caused on account of mechanical fault in the system.

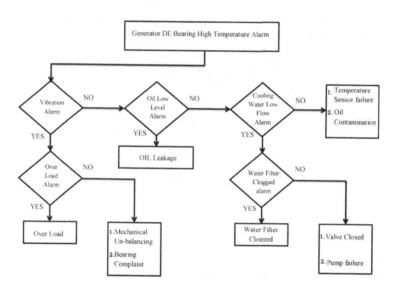

Fig. 2. Reasons of GEDE Bearing High Temp Alarm

For knowledge representation rules and meta rules are used. Meta rules basically follows, *IF* Complaint *THEN* Hypothesis form. It is mainly used to control the inference process. Rules having, *If* Condition *Then* Action form is used in diagnosis/chaninng process. With the rule editor part of Graphical User Interface (GUI), rules are created and stored in knowledge base. For example, Fig.2 shows the causes for a critical alarm such as *Generator DE Bearing High Temp Trip*. The corresponding rules can be made by successive analysis of the procedure(Fig.3). Knowledge base builder can then make set of SCADA Tag based rules for *Generator DE Bearing High Temp Trip*. Rules contain SCADA based Tags and also other reason codes, which are defined while developing expert system. For example, causes such as oil contamination has no sensors or SCADA tags. Such reasons are coded using defined reason codes as OILC. The knowledge is kept in the knowledge base of expert system for making future decisions. The input of the decision maker is a set of signals from the sensors recorded by the SCADA database.

Meta Rules :   IF (Complaint) Then (Hypothesis 1) (Hypothesis 2) (Hypothesis 3) ....
IF (MECHANICAL TRIP) THEN
                    (SATOR CORE HIGH TEMPERATURE TRIP)
                    OR (TURBINE BEARING HIGH OIL LEVEL TRIP)
                    OR (GENERATOR NDE BEARING LOW LEVEL OIL TRIP)
                    OR (TURBINE BEARING HIGH TEMP TRIP)
                    OR(GENERATOR DE BEARING LOW LEVEL OIL TRIP)
                    OR (GENERATOR DE BEARING HIGH TEMP TRIP)
                    OR (GENERATOR DE BEARING VIBRATION TRIP)
                    OR (GENERATOR NDE BEARING VIBRATION TRIP)
                    OR (HPU LOW LEVEL PRESSURE TRIP)
Rules :   IF(condition 1) (condition 2)(condition 3) (condition)..... THEN (Action)
    R1 : IF(VIBRATIONA) OR (OIL LOWLEVELA) OR
          COOLING WATER LOW FLOW A) OR
          (OIL ONTAMINATION) OR (SENSOR FAILURE) THEN
          (GENERATOR DE BEARING HIGH TEMP)
    R2 :   IF(OVERLOADA) THEN (VIBRATIONA)
    R3 :   IF(OIL LEAKAGE) THEN (OIL LOWLEVELA)
    R4 :   IF (WATER FILTER CLOGGED A) OR (VALVE CLOSED)
             OR (PUNB FAILURE) THEN (COOLING WATER LOW FLOW A)
    R5 :   IF (WATER FILTER CLOGGED A) THEN (WATER FILTER CLOGGED)

Fig. 3. Rules and Meta Rules

3.2   Diagnosis Engine

The diagnosis engine takes as input a log of SCADA alarms. An alarm message normally contains a time stamp, tag name,eventid, tagid, unitnumber and action field.Automatic fault analysis based on this historical SCADA alarm database involves various steps. Algorithm 1 illustrates various steps involved in root cause diagnosis.

The alarm data base needs to be converted into suitable format before applying search algorithm. The pre-processing involves various steps, such as data cleaning, data reduction and data filtering. Data cleaning involves the removal of data with less quality and handling the missing data with format conversion,if any needed. After consulting plant experts some of the attributes are removed to reduce file size. The database is then sorted based on each unit. Repeated and insignificant alarms are also removed in the pre-processing stage.

Algorithm 1. SCADA Root Cause Analysis

Input: AlarmName,Date,Time ;
Output: Fault Reasons
Procedure Root Cause Diagnosis :
1: $Processed\_Data \leftarrow SCADA\_PreProcessing(Unit\_No, Date)$
2: $Tag\_List\_Tree \leftarrow Infrence(Rulebase, AlarmName)$
3: $Filtered\_Data \leftarrow Filter\_SCADA(Processed\_Data, Tag\_List\_Tree)$
4: $Level\_No \leftarrow 1$
5: $First\_Level\_Tag\_List \leftarrow Get\_TAGS(Level\_No, Tag\_List\_Tree)$
6: Diagnose_Process (First _Level _Tag _List, Level_No, Tag _List _Tree)

Algorithm 2. Function SCADA Search

1: **procedure** DIAGNOSE_PROCESS(First _Level _Tag _List, Level_No, Tag_List_Tree)
2:     $k \leftarrow 1$
3:     **while** $First\_Level\_Tag\_List[k] \neq EMPTY$ **do**
4:         $FLAG \leftarrow Flag\_Value(First\_Level\_Tag\_List[k], Filtered\_Data)$
5:         **if** $FLAG == 1$ **then**
6:             $PRINT(First\_Level\_Tag\_List[k])$
7:             $Level\_No \leftarrow Level\_No + 1$
8:             $Diagnose\_Process(First\_Level\_Tag\_List[k], Level\_No, Tag\_List\_Tree)$
9:         else$k \leftarrow k + 1$
10:             **if** $First\_Level\_Tag\_List[k] -¿Type \neq SCADA\_Tag$ **then**
11:                 PRINT _ALL _Remaining _Tags (First _Level _Tag _List[k])
12:             **end if**
13:         **end if**
14:     **end while**
15: **end procedure**

The inference engine accepts user input at the beginning that specifies alarm name, date, unit number and time. The production system inference cycle queries and responses to the questions through the input-output interface.It then uses this dynamic information along with the state knowledge stored in the knowledge base by backward chaining process. When consistent matches are found, corresponding rules are placed in a conflict set. To find an appropriate and consistent match substitutions are required. The inference process is carried out recursively in three stages viz; match, select and execute. During the match stage, the contents of work memory are compared to fact and rules contained in the knowledge base. When consistent matches are found the corresponding rules are placed in the conflict set. Based on inference process, output tag list is prepared which act as the basic data structure for search procedure.

The tag list is implemented as a tree with link list having fields like tags and an operator which can be either AND or OR. AND-OR operator is required, as rule bases can contain AND OR parts. List contains another field indicating whether the rule contain information that has a SCADA tag or not, for example reason such as Sensor Malfunction has no TAG.

The pre-processed SCADA database is then filtered based on tag list contents. The database search algorithm is implemented as a recursive technique at various levels as given in algorithm 2. If SCADA alarm database contains root cause tag, it is displayed on the Graphical User Interface (GUI) along with path followed by diagnosis procedure which explains the process. In certain cases, malfunctions can occur due to other reasons. If knowledge base contains these reason codes, that also can be visualised through GUI. If no information is available operator guidance information will be displayed on the GUI.

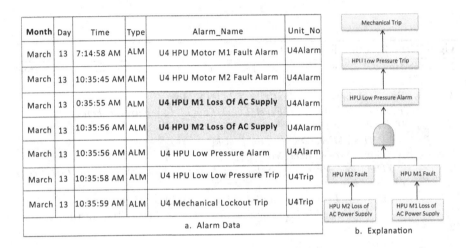

Month	Day	Time	Type	Alarm_Name	Unit_No
March	13	7:14:58 AM	ALM	U4 HPU Motor M1 Fault Alarm	U4Alarm
March	13	10:35:45 AM	ALM	U4 HPU Motor M2 Fault Alarm	U4Alarm
March	13	0:35:55 AM	ALM	**U4 HPU M1 Loss Of AC Supply**	U4Alarm
March	13	10:35:56 AM	ALM	**U4 HPU M2 Loss Of AC Supply**	U4Alarm
March	13	10:35:56 AM	ALM	U4 HPU Low Pressure Alarm	U4Alarm
March	13	10:35:58 AM	ALM	U4 HPU Low Low Pressure Trip	U4Trip
March	13	10:35:59 AM	ALM	U4 Mechanical Lockout Trip	U4Trip

a. Alarm Data

b. Explanation

Fig. 4. A Sample Experimental Result

4 Evaluation

A prototype system that can handle above mentioned ten critical alarms related to mechanical trip is developed using JESS expert system shell. JESS rules are created for storing alarm reasons and built in Rete algorithm for reasoning engine [18]. User interfaces are created with the help of Java programs in Eclipse plat form.

This study is conducted by collecting SCADA data from January 2010 to October 2013. To test the developed system, operator diary during the period was referred and test cases were selected. Fig.4 illustrates test results of one cases study. This test case was reported as a mechanical trip error in previous year. The user input included trip time and date. The SCADA database for a period of twenty four hours was processed with the proposed algorithm. More than thousand two hundred alarms were listed during the period. Manual anlayis has taken more than one hour as per the case diary reports. The search algorithm was fast enough to process all alarm cascades in less than a minute. Search algorithm oriented filtering techniques has reduced the alarm list to only seven entries as illustrated in Fig.4(a).

Here *High Pressure Unit (HPU) Motor Loss Of AC Supply* was identified as root cause of mechanical trip. Fig. 4(a)illustrates the resultant alarm table with root cause highlighted. The flow of data analysis and knowledge that have helped to arrive at root cause is displayed in fig. 4(b).

If operator require, the advisory system provides instructions for clearing the fault conditions, based on rules stored in the knowledge base.

5   Conclusion

This paper presents the design and implementation of a hybrid intelligent system for power plant fault diagnosis based on the available SCADA history alarm database. Existing SCADA system is a monitoring system that act at supervisory level only. It produces too many alarms and events during fault conditions. Manually analysing this database to locate and identify the reasons of fault consumes time and needs the presence of expert plant operator. The expert's knowledge is the key factor in manual analysis. The proposed system attempts to encode this expert knowledge and uses it to automate SCADA alarm analysis strategy. Knowledge base system developed, convert the valuable operational knowledge of plant operators into explicit knowledge to be used in future.

The tool developed has an object-oriented environment that allows creation of hybrid knowledge base expert system, by encoding an experts knowledge as a set of frames and rules and does knowledge inferencing using either built-in backward or forward chaining inference engine. A backward chaining technique is used for obtaining all reason codes that can have a SCADA tag in many cases. If SCADA tag based fault has occurred a recursive search algorithm implemented on available SCADA history database displays root cause. The proposed system will decrease the stress currently imposed on operators during fault conditions and is expected to reduce outage time and consequent financial losses. It is expected that the technology developed can be applied to any complex plant that is controlled by SCADA monitoring system.

References

1. Venkatasubramanian, V., Rengaswamy, R., Yin, K., Kavuri, S.N.: A Review of Process Fault Detection and Diagnosis Part I: Quantitative Model-Based Methods. Computers and Chemical Engineering **27**, 313–326 (2003)
2. Venkatasubramanian, V., Rengaswamy, R., Kavuri, S.N.: A Review of Process Fault Detection and Diagnosis Part II: Process History Based Methods. Computers and Chemical Engineering **27**, 327–346 (2003)
3. Russel, S., Norvig, P.: Artificial Intelligence: A Modern Approach. Prentice Hall, USA (2002)

4. Roth, R.M., Wood, W.C.I., Delphi, A.: Approach to acquiring knowledge from single and multiple experts. In: Conference on Trends and Directions in Expert Systems (1990)
5. Arroyo Figueroa, G., Solis, E., Villavicencio, A.: SADEP a Fuzzy Diagnostic System Shell- an Application to Fossil Power Plant Operation. Expert System with Applications **14**, 43–52 (1998)
6. Kang, S.J., Moon, J.C., et al.: A Distributed and Intelligent System Approach for the Automatic Inspection of Steam-Generator Tubes in Nuclear Power Plants. IEEE Transaction on Nuclear Science **3**(3), 1713–1722 (2002)
7. DaSilva, V., Linden, R.: A framework for expert systems development integrated to a SCADA/EMS environment. In: Proceedings of the 14th International Conference on Intelligent System Applications to Power Systems, China, pp. 608–613 (2007)
8. Amaya, E.J., Alvares, A.J.: SIMPREBAL: An expert system for real-time fault diagnosis of hydro generator machinery. In: Proceedings of IEEE Conference on Emerging Technologies and Factory Automation, Bilbao, pp. 1–8 (2010)
9. Zhao, W., Bai, X., Wang, W.: A novel alarm processing and fault diagnosis expert system based on BNF rules. In: Proceedings of IEEE/PES Transmission and Distribution Conference and Exhibition: Asia and Pacific, China, pp. 1–6 (2005)
10. Yang, S.K.: A Condition-Based Failure-Prediction and Processing for Preventive Maintenance. IEEE Transactions on Reliability **52**(3), 373–380 (2003)
11. Nabeshima, K.: Nuclear Reactor Monitoring with the Combination of Neural Network and Expert System. Mathematics and Computers in Simulation **60**(5), 233–244 (2002)
12. Mak, K.-H., Holland, B.L.: Migrating Electrical Power Network SCADA Systems to TCP/IP and Ethernet Networking. Power Engineering Journal **16**(6), 305–311 (2002)
13. Warcuse, J., Menz, B., Payne, J.R.: Servers in SCADA Applications. IEEE Transaction on Industrial Application **9**(2), 1295–1334 (1997)
14. Bann, J., Irisarri, G., Kirschen, D.: Integration of Artificial Intelligence Applications in the EMS: Issues and Solutions. IEEE Transactions on Power Systems **11**(1), 475–482 (1996)
15. Bransby, M., Jenkinson, J.: Alarming Performance. Computing and Control Engineering Journal **9**(2), 61–67 (1998)
16. Vale, Z.A., Machade-Moura, A.: Sparse an Intelligent Alarm Processor and Operator Assistance, A.I in power Systems. IEEE Expert **12**(2), 86–93 (1996)
17. Teo, C.Y., Gooi, H.B.: Artificial intelligence in diagnosis and supply restoration for a distribution network. In: Proceedings of IEEE Generation Transmission and Distribution Conference, vol. 145, pp. 444–450 (1999)
18. Yang, S.J.H., Zhang, J., Chen, I.Y.L.: A JESS-Enabled Context Elicitation System for Providing Context-Aware Web Services. Expert Systems with Applications **34**, 2254–2266 (2008)

Lino – An Intelligent System for Detecting Malicious Web-Robots

Toni Gržinić[1]([✉]), Leo Mršić[2], and Josip Šaban[3]

[1] Croatian Academic and Research Network, Josipa Marohnica 5, Zagreb, Croatia
toni.grzinic@carnet.hr
[2] IN2data Ltd Data Science Company, Josipa Marohnica 1/1, Zagreb, Croatia
leo.mrsic@in2data.com
[3] Hypo Alpe-Adria Bank, Alpe Adria Platz 1, Klagenfurt, Austria
josip.saban@hypo-alpe-adria.com

Abstract. These days various robots are crawling the Internet, they are also called: bots, harvesters or spiders. Popular search engines use a similar technique to index web pages - they have an autonomous agent (called robot or bot) that is in charge of crawling various attributes of web sites. Lately, this crawler technique is exploited by malicious users, for example harvesters, which are used for scraping e-mail addresses from websites in order to build a spam list for spambots. Recently, robots are also misused to buy flight tickets or do fast bids in on-line auction system. In this paper we present an intelligent system called Lino which tries to solve the mentioned problem. Lino is a system that simulates a vulnerable web page and traps web crawlers. We collect various features and perform a feature selection procedure to learn which features mostly contribute to the classification of visitor behaviour. For the classification purpose we use state of the art machine learning methods like Support Vector Machine and decision tree C 4.5.

1 Introduction

Today's Internet has outgrown the original concepts of its creators. From a small academic network that facilitated research, it has grown into a diverse network that ranges from the business processes accelerator of international companies to the home user entertainment platform. Although it brings many positive changes into our daily lives, Internet also became a threat to its users. It now has its own black economy, where criminals sell stolen confidential information and where information trading can damage computer systems. The biggest challenge in the field of security are so called botnets or networks of infected computers controlled by a central node. Such networks are used for various purposes - from stealing bank numbers and credentials to denial of service attacks.

Bots are infected computers that, combined, make up the botnet, and are used to either hide the central node or facilitate the spreading of malicious content. It is often the case that bots run automated procedures that violate the security of the attacked Web sites. Therefore, one of the challenges in recent years

© Springer International Publishing Switzerland 2015
N.T. Nguyen et al. (Eds.): ACIIDS 2015, Part II, LNAI 9012, pp. 559–568, 2015.
DOI: 10.1007/978-3-319-15705-4_54

is to find a model for distinguishing legitimate or human users from automated ones like malicious bots.

The robots themselves are not dangerous or illegal. One such useful robot is GoogleBot, which is used to index the Internet content. Opposite to these kinds of robots we have others which are used for extraction of various illicit or illegal activities. Robots, controlled and created by malicious users, retrieve information from web sites using popular attacks in order to compromise selected web sites or to simply determine vulnerability to some kind of attack.

Researching these automated users raises two fundamental questions. The first question is how to monitor the development of robots that present a threat to web sites and are developing in parallel with new technologies on the Internet. The second question is related to the relevance of the dataset for analysis and comparison of algorithms for the detection and classification of these types of threats. In the real world, due to the development of technology, there are no standardized datasets because it is difficult to gather data that will be relevant for a longer period of time. Also, a dataset for research purpose is difficult to obtain from third parties, for example companies, because it creates a privacy issue for their users or interfere too much with the technical solution used. To solve this problem we built a specialized data collection system named Lino. In addition to human visitors, web sites are frequently visited by robots. Their names and purposes are different, but in the literature they are also known as crawlers, harvesters or spiders. It is worth to mention that, besides popular Web search engine robots that crawl the website and accelerate its visibility, also exist malicious bots. Their purpose is to check for the website vulnerabilities, and they do that by filling Web forms, spreading spam or sending fraud messages via e-mail. Spambots are specialized robots seeking e-mail addresses to which spam e-mail is sent. It is also often the case that robots periodically retrieve content from portals and later use it illegally, for example robots that retrieve product prices from popular e-commerce sites. Recently, robots wee used to perform various frauds like beating human competition in booking plane or concert tickets [1].

2   Literature Review and Similar Systems

Most previous works analyse user behaviour using web server log records [7], [9], [11] and [12]. For detection of malicious robots various machine learning methods are used: decision trees [5], [12], probabilistic models such as Bayesian networks [5] [11] [10] and Markov chains [7] [9], Ripper [11], SVM [11], neural networks [11], Naive Bayes [11], the k-nearest neighbors algorithm [11], time series [4] and clustering [5]. Research experiments like [7] and [9] can detect anomalies in user behaviour and analyse the characteristics of the text in requests sent to a web server. Using the request feature they created probabilistic models such as Markov chains and Bayesian networks that work in supervised mode until they learn to distinguish between normal and anomalous behaviour.

---

[1] A recent and popular example of a bot fraud is booking for dinner tables in Silicon Valley - http://bit.ly/1esBSxG

Accurate classification of malicious and benevolent users is the result of proper knowledge and careful selection of features that describe individual users. For example, Tan and Kumar [12] combined the features of previous studies that were supplemented with their new features, the new feature set represents the main contributions of their research. Their feature set is the following: access to the robots.txt file, checking a user agent string, checking the IP address of the user and sum of HEAD requests without Referer field. In recent works [4], [10], [5] and [11] feature set has expanded significantly, additional features are now used, for instance: clicks per session, number of unique IP address, session duration, the percentage of access to images, the percentage of access to PDF files, distribution of requests to individual sites, the percentage of responses that return an error (HTTP error codes 400-499), doubtful Referer fields, ignored cookies, IP addresses belonging to a particular web browser, not using the parameter GET in requests, access to child sites in sequential order (typically alphabetical), the ratio of sequential access to the same web link and standard deviation of depth requests through a hierarchy of web sites. Also, an important step in the analysis is to identify user sessions, with the help of which you can connect a set of actions with records in web server log [10] [11] [12]. Joining the session to the user is used primarily to monitor the user's movement pattern on the website, or to identify the behavioural characteristics of users which can greatly assist in the detection of automated visitors.

Lino is a variation of a honeypot system. Honeypot systems are security tools that are deliberately vulnerable to certain types of attacks and constitute a type of traps for malicious users. Their primary task is to be tested and exploited by a malicious user. The exploitation of such systems is used by researchers to study new techniques of attack and recognize potentially dangerous groups of users. A good overview of the state of modern honeypot systems is presented by the European Network and Information Security Agency [1]. Hayati and colleagues presented a HoneySpam robot profiling system that collects e-mail addresses and later uses them as a destination of their spam campaigns [3] - in this paper the architecture of the system is described and the collected data is analyzed. Provos [8] in his paper also presented the architecture of Honeyd system that is used to analyse the network layer attack - though it has been ten years since the creation of this system it still serves as an example and a basis for developing modern honeypots.

3 Methodology

3.1 Creating a System for Data Collection

For the purpose of collecting data we built a system called Lino that is used to record the available server and client variables, with the corresponding website used as a graphical interface. Lino consists of news and articles taken from other sources where the news is usually attractive and current in order to attract a larger amount of potential users as well as robots. The system collected data for three months, from September 2013 until December 2013. Lino has no specific functionality; instead it is being used to display the content, text fields, and other website elements that are used for the collection of user actions. In other words, from every user's requests Lino tries to elicit useful information that will be used in next steps. Since we are interested in malicious clients, we decided to hide in the page specific keywords, so called Google Dorks, used by attackers to find victims of a specific attack.

Example of a keyword listed as Google Dork:

```
inurl:"wp-download.php?dl_id="
```

Keywords are actually intended for the Google search engine queries, as in the example above where the attacker seeks vulnerable parts of the popular blog engine Wordpress[2]. Also, on the site we hide popular spammer addresses, in order to attract more email harvesters. To reduce the risk that we do not collect enough data in a given time interval we made multiple instances of the system. In total we used three instances that contained links to other instances.

For each user who accessed the system Lino generated a session identifier which had a duration of 60 min. From every user request Lino captured the following data:

- Request timestamp
- Client IP address
- Client port used for the connection (web browser takes a random port)
- Requested path to the Web server
- Query parameters used in the request
- Request method used (GET, POST, HEAD ...)
- Type of protocol that is used
- If the client sends data using the POST method, the content is written
- Previous references from which the user came to the currently requested page (HTTP Referrer field)
- User Agent
- Session identifier

3.2 Labelling the Collected Dataset

Before labelling a dataset, it was necessary to reduce and generalize the data. We decided to split into sessions all requests to the application Lino (over 12,500 of them). The session can be viewed as a unique behavioural profile of a client who joined the system in a given time interval (in this paper we used a period of 4 hours). Session identifies requests that were made by a unique client. The client's session grouping is necessary to select a specific time ranges in which the user has been active on the website.

Following are the criteria by which we add sessions to clients:

- Time ranges of the data collected - we have divided the time into buckets of 4 hours

Table 1. Stored features from the session

Feature description	Feature name and type
Requests made in specific session	*num_requests* (numeric)
Requests with HTTP HEAD parameter it was often the case that the robot checks the availability of website using a HEAD method	*head* (boolean)
Since we had hidden links in the application, we calculated the percentage of accesses to those links	*perc_hidden_links* (numeric)
Standard deviation in the intervals between queries in a single session	*std_dev_timedelta* (numeric)
Duration of the session	*session_duration* (numeric)
The user receives a session identifier which is valid for 60 minutes. Users accessing without a browser or with disabled JavaScript change session identifiers on every request	*session_change* (boolean)
Sent data via a web form	*post_data* (boolean)
Accessing a robots.txt file, which contains the rules of behaviour for robots	*robots* (boolean)

- All requests with the same IP address and user agent in the time range of four hours (counting from the first request) are considered as a new session

When creating each session we also store the features presented in Table 1. Based on the above mentioned criteria, we got about 3,500 sessions, which were used to build a learning set, also storing additional attributes which included: reverse DNS queries for client IP address, client country of origin, client's AS number of the service provider. We automated the labelling procedure using publicly available data from MaxMind's GeoIP database[3]. After the automatised labelling procedure we manually skipped through the records to correct the wrongly classified instances, for examples unknown robots. We used aforementioned client based and session based data as features to identify if a new session is a robot or a human. Unfortunately, due to limited number of robots collected we could further segment the dataset on malicious and normal robots. Lino fulfilled its primarily purpose and collected mostly robots' sessions. Sessions distribution by user types is the following:

- Human 9%
- Robot 90%
- Unknown 1%

---

[2] More about this vunerability - http://www.exploit-db.com/exploits/5326/
[3] GeoIP database, https://www.maxmind.com/en/geoip2-databases

3.3   Feature Selection Procedure

Before training the classification model it was necessary to choose the features that are relevant and useful for the classification process. To evaluate the features we used the ranking of features based on the following methods [13]:

1. Information gain that ranks features based on the calculated information gain relative to the classification class, numerical features are first discretized.
2. Gain ratio ranks features based on the calculated gain ratio. Gain ratio is calculated as Information gain divided by the entropy of the feature for which the ratio is computed.
3. Symmetrical uncertainty is a measure that eliminates redundant and meaningless features, which have no interconnectivity with other features.
4. Relief method was proposed by Kira and Rendell [6] and is used for the selection of statistically relevant features, it is resistant to noise in the data and the interdependence of features. Features are evaluated in a way that is randomly sampled from a given set of instances and take nearest neighbours that belong to the class. If the neighbours are aligned with instances the weighting factor increases, in contrast if the closest neighbours are different the weighting factor decreases.

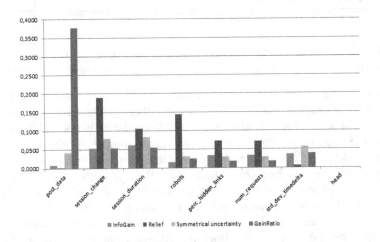

Fig. 1. Comparison of different methods for feature selection

If we look at the ranked features in Figure 1, we see that the features that dominate the dataset are:

– *post_data*, which shows us whether the client has filled/not filled the fake form in the Lino system
– *session_change*, which shows us if user, during the session, has changed the session identifier or not

- *session_duration*, duration of the session in seconds
- *robots*, which shows us whether the user accessed/not accessed robots.txt file, which defines the rules of robot conduct

Aforementioned features were selected manually, we ranked all features according to the score of the feature selection method. We selected most significant features for our classification models, in our case the top five features.

4   Classification Model Selection for Bot from Human Differentiation

A prerequisite for using supervised learning methods and selecting the optimal subset of features is a labelled dataset. Selected features should contribute to the generalization of some classes, ie. for each class they should be able to make a unique behavioural profile. To evaluate the performance of the classification method we used the K-fold cross validation method. For our purposes we used k = 10 parts - the relevant literature [13] states that k = 10 parts are a optimum number for estimating errors.

4.1   Decision Tree C 4.5

Firstly, for classification purposes, we evaluated a decision tree algorithm C 4.5, which is an upgrade of the classic algorithm ID3. Both algorithms are the result of research made by Ross Quinlan [13]. C 4.5 uses a dataset for learning to create a redundant tree. In the case of using similar data, in the learning and validation

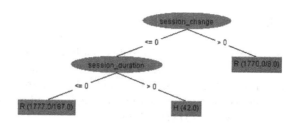

Fig. 2. Prunned tree, using the full set of features

set, classifier has good results but when we use an independent validation set the classifier usually produces bad results. After building a redundant tree, the tree is converted to the IF/THEN rules and the algorithm calculates the best conditions for classification accuracy we remove the IF conditions if they do not reduce the classification accuracy. Pruning is done from the leaves to the root of the tree and is based on the pessimistic estimation of errors; errors are related to the percentage of incorrectly classified cases in the training dataset. Based on the difference in accuracy of rules and standard deviation taken from the binomial distribution we define a certain upper limit of confidence that is usually 0.25, based on which the trees are pruned [14]. For building our models with C 4.5 we set the confidence threshold for pruning to 0.25 and the minimum number of instances per leaf is 2.

Table 2. Classification results for C 4.5 and SVM, experiment 1 uses only selected features. Experiment 2 uses selected features plus Country and ASN of client

	Class	TP Rate	FP Rate	F-measure	AUC
C 4.5 Experiment #1	Human	0.177	0	0.301	0.773
	Robot	1	0.823	0.972	0.773
C 4.5 Experiment #2	Human	0.793	0.002	0.872	0.985
	Robot	0.998	0.207	0.992	0.985
SVM Experiment #1	Human	0.265	0	0.419	0.801
	Robot	1	0.735	0.979	0.801
SVM Experiment #2	Human	0.962	0.006	0.942	0.976
	Robot	0.998	0.042	0.997	0.978

Prior to the classification we removed the class instance of unknown visitors because they represented human attempts to attack with manually entered vales or using non-existing browsers. Method C 4.5 has resulted in the pruned tree showed in Figure 2, which is the same with the optimal selection of features and using the full set of features. It is important to note that the algorithm C 4.5 is very good at choosing features by using heuristics in creating and deleting subtrees.

If we look at the results in Table 2, we can see that for the given features (Experiment 1) we have a classification accuracy of 94.5% and a perfect rate of correct positives for robots (TP rate). Classifier badly classifies human visitors (TPR = 0.177), and degrades the classification ability of the robot where the rate is false positive rate is high - 0.823. Looking at F-measure we can say that a good classifier detects correctly robots while wrongly classifies human visitors and commonly (> 80%) declares them robots.

We tested the C 4.5 classier with two additional features - client's country and ASN of service provider. This features were resolved from the IP address using the aforementioned GeoIP database. This subset (C 4.5 Experiment 2) is shown in Table 2. We reduced the number of false positives for the class Robot to 0.207, thus the result of classification for class Human was better 0.793.

4.2 Support Vector Machine

SVM is an algorithm that finds the maximum margin of separation between classes [2], while defining the margin as distance between the critical points which are closest to the surface of separation. Points nearest to the surface are called support vectors, the margin M can be seen as the width of the separation between the surfaces. Calculating the support vector is an optimization problem which can be solved using different optimization algorithms. The trick used in calculating SVM is using different kernel functions, which move unsolvable or inadequate problems to a higher dimension, where it can be solved. In our experiments we trained our SVM models with the sequential minimal optimization algorithm using a linear kernel $K(x,y) = <x,y>$ where $\epsilon = 1.0^{-12}$ and tolerance is set to 0.001, previously training data was normalised. For Experiment 1 features (Table 2) SVM performs better than C 4.5, the precision was 95.8%. Human visitors are still a problem, although SVM has much higher rate of true positives (26.5%). Increased detection rate of human visitor gives a lower rate of wrong detection of robots (73.5%). F measure is very good for robots and much better for human visitors (even better than method C 4.5) - but still too low to be used - (0.419). With additional features Country and ASN (Expertiment 2 in Table 2) we obtained a false rate for both classes under 5%. The true positive rate was also high for class Human 0.962 and for class Robot 0.998. We can conclude that with this subset of features and with regular retraining to avoid the concept drift, this model is feasible for every day use.

5   Conclusion

The main drawback with selected features is the detection of human visitors which as consequence gives a high false positive rate for robot detection. The root cause is probably the asymmetric dataset where there are many more robots than human visitors. A recommendation would be to expand the number of human visitors in order to get a more balanced dataset[4]. The reason for these shortcomings is too short time used for data collection and perhaps ambiguous texts that are found in Lino. The system should be set to parse a more popular domain and adapt these texts to target a smaller, more specialized, audience. Also, we should optimize vectors of attacks (dorks) incorporated in Lino.

In addition to the above-mentioned shortcomings this study consists of several contributions, keep in my mind that we did not want to present a perfect classifer. The first contribution is the way in which the data is collected, in most available literature datasets are used from activity logs of web servers we use a trap system for fooling and catching robots. There is no doubt that Lino should collect more client features and this is planned for future implementations and development. Also, features used in Lino are slightly different from previously used features. Researchers often look at error ratio, percentage of images in data and other similar features. We do not analyse these features because it is not relevant to Lino nor is it possible that we produce a query that will cause an error in Lino. Lino allows us to collect more useful data related to user behavior. For example, whether the user has posted something in a non-existing form,

---

[4] To obtain and use the collected dataset and the Lino collector system described in this research, please contact the authors.

if he clicked on hidden links or if he constantly changed identifier sessions. In relevant literature no one uses variables that are related to the time interval between queries. We use the standard deviation of the time elapsed between queries, which according to our feature selection does not contribute to the classification. Although it proved to be insignificant with the current dataset, automatised procedures should have lower standard deviations and a regular difference between queries.

References

1. Gorzelak, K., Grudziecki, T., Jacewicz, P., Jaroszewski, P., Juszczyk, L., Kijewski, P.: Proactive detection of network security incidents (2011)
2. Hastie, T., Tibshirani, R., Friedman, J., Hastie, T., Friedman, J., Tibshirani, R.: The elements of statistical learning, vol. 2. Springer (2009)
3. Hayati, P., Chai, K., Potdar, V., Talevski, A.: HoneySpam 2.0: profiling web spambot behaviour. In: Yang, J.-J., Yokoo, M., Ito, T., Jin, Z., Scerri, P. (eds.) PRIMA 2009. LNCS (LNAI), vol. 5925, pp. 335–344. Springer, Heidelberg (2009)
4. Jacob, G., Kirda, E., Kruegel, C., Vigna, G.: Pubcrawl: protecting users and businesses from crawlers. In: USENIX Security Symposium, pp. 507–522 (2012)
5. Kang, H., Wang, K., Soukal, D., Behr, F., Zheng, Z.: Large-scale bot detection for search engines. In: Proceedings of the 19th International Conference on World Wide Web, pp. 501–510. ACM (2010)
6. Kira, K., Rendell, L.A.: The feature selection problem: traditional methods and a new algorithm. In: AAAI, pp. 129–134 (1992)
7. Kruegel, C., Vigna, G., Robertson, W.: A multi-model approach to the detection of web-based attacks. Computer Networks 48(5), 717–738 (2005)
8. Provos, N.: Honeyd-a virtual honeypot daemon. In: 10th DFN-CERT Workshop, Hamburg, Germany, vol. 2 (2003)
9. Robertson, W., Vigna, G., Kruegel, C., Kemmerer, R.A., et al.: Using generalization and characterization techniques in the anomaly-based detection of web attacks. In: NDSS (2006)
10. Stassopoulou, A., Dikaiakos, M.D.: Web robot detection: A probabilistic reasoning approach. Computer Networks 53(3), 265–278 (2009)
11. Stevanovic, D., An, A., Vlajic, N.: Detecting web crawlers from web server access logs with data mining classifiers. In: Kryszkiewicz, M., Rybinski, H., Skowron, A., Raś, Z.W. (eds.) ISMIS 2011. LNCS (LNAI), vol. 6804, pp. 483–489. Springer, Heidelberg (2011)
12. Tan, P.N., Kumar, V.: Discovery of web robot sessions based on their navigational patterns. In: Intelligent Technologies for Information Analysis, pp. 193–222. Springer (2004)
13. Witten, I.H., Frank, E.: Data Mining: Practical machine learning tools and techniques. Morgan Kaufmann (2005)
14. Wu, X., Kumar, V., Quinlan, J.R., Ghosh, J., Yang, Q., Motoda, H., McLachlan, G.J., Ng, A., Liu, B., Philip, S.Y., et al.: Top 10 algorithms in data mining. Knowledge and Information Systems 14(1), 1–37 (2008)

Optimum Design of Finned Tube Heat Exchanger Using DOE

S.P. Praveen Kumar and Kwon-Hee Lee[✉]

Department of Mechanical Engineering, Dong-A University, Busan, South Korea
kumar.praveensp@gmail.com, leekh@dau.ac.kr

Abstract. Heat Exchanger is a device used to transfer heat from one medium to another. This paper discusses the use of Design of Experiments method in optimizing design parameters for design of finned tube heat exchanger (fins are neglected for computational purpose). The effects of four design parameters – tube thickness, tube inner diameter, horizontal and vertical distances between the tubes have been explored. It is investigated numerically using Ansys-CFX. The primary response under this study is the temperature difference and pressure drop. An L9 orthogonal array was used to accommodate the experiments. In this study, it is observed that the most effective parameter found to be the inner diameter and the thickness of the tube among other parameters having more influence at the heat transfer rate.

Keywords: Heat exchangers · Ansys-CFX · Numerical analysis · Design of Experiments (DOE) · ANOVA (Analysis of Variance)

1 Introduction

Heat Exchanger is used to transfer heat from one medium to another medium where this medium can either be separated to prevent mixing or be in direct contact depending on its application. Heat transfer rate of heat exchanger is maximized by maximizing the contact surface area between the media. Heating or cooling of a concerned fluid stream, condensation and evaporation of a single or multicomponent of a fluid stream and heat recovery or heat rejection from a system are the typical applications of heat exchanger. Finned tube heat exchangers are commonly used exchangers for the application of space conditioning systems, chemical industries, combined heat and power industries and so on.

Finned tube heat exchangers consists of a mechanically or hydraulically expanded round tubes in a block of parallel continuous fins. They are designed for maximum heat transfer associated with a minimum pressure drop associated with each fluid. In this study, we neglected the fins on the tubes for the computational purpose and were designed using Solidworks. The main purpose of this study is to determine the most significant parameter which affects the heat transfer rate with respect to the response characteristics. Numerical Analysis is carried out using Ansys-CFX. Temperature difference and Pressure drop are the two characteristics chosen as responses for this analysis [1].

© Springer International Publishing Switzerland 2015
N.T. Nguyen et al. (Eds.): ACIIDS 2015, Part II, LNAI 9012, pp. 569–576, 2015.
DOI: 10.1007/978-3-319-15705-4_55

DOE method is an optimal parameter design of experimental tool, which firstly involves in choosing several important parameters from a governing equation or relative characteristics of engineering and then inputs them into one appropriate plan table designed by Taguchi with plural levels for each parameter. By comparing those calculated results from the table, a set of optimal parameters with corresponding level can be found. Application of this methodology in the energy based engineering system has been scarce, however many tried and showed that this methodology can successfully applied on heat transfer studies [2]. They also analyzed the effects of various design parameters on the heat transfer and pressure drop of the heat exchanger with a slit using the DOE method. The optimum design value of each parameters was presented and the reproducibility of the results were discussed in this paper using this DOE method.

2   Numerical Solution

2.1   Model Description

The Finned tube heat exchanger considered for this study has bundle of tubes arranged in a straight having separated equally in horizontal and vertical direction from one another and enclosed in a rectangular tube. Total number of tubes is 36 and the fins on the tubes are eliminated in-order to make the computational part easier within short duration, also the memory required for this computation is procurable. The design parameters for this analysis are prescribed by a Z manufacturer. Also they wanted us to focus mainly on optimization of tubular arrangements and tube parameters of heat exchanger design. We already reported about the tubular arrangements in our previous study.

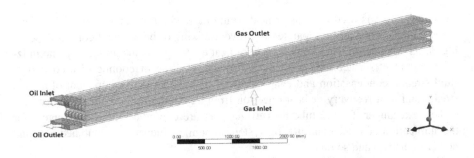

Fig. 1. Physical Model of Finned Tube Heat Exchanger

We utilized commercial CAD modeler SOLIDWORKS to design our preliminary model. The tube material is assumed to be steel and its total length is 8540mm. Initially, as per manufacturer design, it had 484 tubes, but later we reduced it to 36 pipes by keeping in consideration of required computational memory. Dowtherm oil at 215 °F is made to flow through the tubes and flue gas at 922.5 °F is made to pass over the tubes, heat exchange between these two fluids was studied. The inlet and outlet of the oil pipe diameter is four times of the tube diameter, this helps to accommodate large volume of oil thereby maintaining the flow continuity. [3]

2.2   Initial and Boundary Conditions

The numerical model is modelled with tetrahedral cells using of Ansys-CFX mesh. 672881 nodes and 1773254 elements were generated in the numerical model. This is a steady state analysis dealing with incompressible and turbulent fluid flows. For this case, the SST (Shear Stress Turbulent) model is used. This turbulent model is suitable for solving the flow near the wall and is most preferred model for bodies having curved structures. Unlike other available models, this model does not converge to solution quickly but the results generated are comparable with the experimental data. The governing equations for this analysis are navier-stokes equation, energy equation and continuity equation. [4]

The operating pressure of the oil is 150 psi and for the gas is 12.5 psi. The oil flows through the tubes from the oil inlet along longitudinal direction with a velocity of 200 feet per second and the gas flows over the tubes with a velocity of 175 feet per second along the transverse direction. To facilitate the heat transfer between the oil and gas, fluid-solid domain interface is engaged between oil (fluid domain) and tube (solid domain) and also with gas (fluid domain) and tube (solid domain). The fluid and solid properties and boundary conditions are depicted below in Table 1 and 2 respectively.

Table 1. Physical Properties of Fluids and Solid

	Material	Density [lb./ft³]	Specific Heat Capacity [Btu /lb. °F]	Thermal conductivity [Btu/hr.ft.°F]
Oil	Dowtherm Oil	56.41	0.4559	0.0643
Gas	Flue Gas	0.024	0.2735	0.0313
Tube	Steel	490.309	0.1037	34.96

Table 2. Boundary and Initial Conditions

Fluid	Velocity [ft./sec]	Inlet Temperature [F]	Operating Pressure [Psi]
Oil	200	215	150
Gas	175	922.5	12.5

2.3   Numerical Results

The response of the analysis are the temperature difference and pressure drop which helps in understanding clearly the heat exchange between these two fluids. The response is considered only at the oil outlets because there was not considerable change in temperature at the gas outlet.The results are extracted from the post processor of Ansys-CFX and are depicted below in Figure 2 and 3 respectively. The temerpature and pressure at the oil outlet is found to be 839.40 °F and 2.318 MPa respectively.

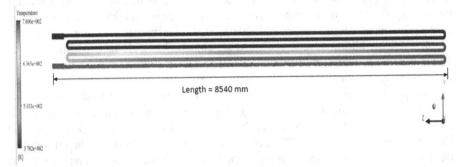

Fig. 2. Temperature Distribution of Numerical Analysis

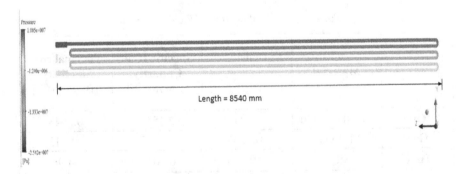

Fig. 3. Pressure Distribution of Numerical Analysis

3   Evaluation of the Design Using Design of Experiments

3.1   Design of Experiments

Design of Experiments (DOE) consists of a plan of experiments with the objective in controlled of acquiring data, executing these experiments and analyzing the data, in order to obtain the information about the behavior of a given process. The major steps of implementing the DOE are: (1) to identify the factors/interactions, (2) to identify the levels of each factor, and select an appropriate orthogonal array (OA), (3) to assign the factors to the columns of the OA, (4) to conduct the experiments, (5) to analyze the data and determine the optimum levels, and (6) to conduct the confirmation experiment. [5]

Definition of Design Variables

Initially the four independent design variables are chosen for this analysis, they are thickness of the tube (A), inner diameter of the tube (B), horizontal distance between two adjacent tubes (C) and vertical distance between two adjacent tubes (D).

DOE Implementation Using Orthogonal Array

In order to minimize the number of experiments required, Orthogonal Arrays (OA) were developed by Taguchi. OA uses only a portion of the total possible combinations to estimate the effect of the design variables with respect to the responses. Taguchi developed a family of OA matrices which can be used in various situations. These matrices are helpful in reducing the number of experiments and also provides reasonably rich information.

L9 Orthogonal Array

For our analysis we choose L9 Orthogonal Array. In L9 Orthogonal array, there are 9 experiments, 4 design variables and 3 levels for each design variable.

Table 3. L9 Orthoganal Array

Experiment Number	A	B	C	D
1	1	1	1	1
2	1	2	2	2
3	1	3	3	3
4	2	1	2	3
5	2	2	3	1
6	2	3	1	2
7	3	1	3	2
8	3	2	1	3
9	3	3	2	1

The levels are defined to each design variables. For L9 Orthogonal array, each design variables has 3 levels. The second level data for each design variable is obtained from the data used for the preliminary analysis. The first and third level datas are fixed by the lower and upper values around the initial data having equal intervals. The levels for our analysis is depicted below in Table 4.

Table 4. Levels for Design Variables

Levels	A (mm)	B (mm)	C (mm)	D (mm)
1	3.51	47.25	79.18	79.18
2	3.90	52.50	87.98	87.98
3	4.29	57.75	96.78	96.78

3.2 Experimental Results

The numerical analysis is carried out for the 9 experiments with the combination of design parameters mentioned in the L9 orthogonal array. The experimental result data are collected for the evaluation of effect of design variables with respect to the responses. The results of 9 experiments are depicted below in Table 5.

Table 5. Experimental Results of L9 Orthogonal Array

Experiment Number	Temperature Difference [F]	Pressure Drop [Mpa]
1	259.03	2.86
2	419.26	2.64
3	614.05	1.16
4	82.93	1.65
5	481.57	1.01
6	676.61	0.80
7	70.56	0.01
8	437.71	0.50
9	692.30	1.31

3.3 Discussion

The collected results has to be evaluated in order to find the significant variable among the design variables having more influence over the heat transfer rate.

Analysis of Variance is used to evaluate the collected data. ANOVA is a parametric procedure used to determine the statistical significance of the difference between the means of two or more groups of values. It is utilized in this analysis to determine the relative importance of each design variable with respect to temperature difference and pressure drop from the experimental results data. ANOVA for temperature difference and pressure drop is depicted below in Table 6 and 7 respectively. In the table, Dof is the number of degrees of freedom, S is the sum of squares of result data, V is the variance of the result data, and F is the variance ratio.

Table 6. ANOVA for Temperature difference

Factor	S	Dof	V	F
A	1410.10	2	705.11	
B	415452.41	2	207726.23	294.61
C	8416.39	2	4208.21	5.90
D	17874.18	2	8937.12	12.67

Table 7. ANOVA for Pressure Drop

Factor	S	Dof	V	F
A	4.10	2	2.01	14.43
B	0.27	2	0.13	
C	1.95	2	0.97	7.04
D	0.73	2	0.36	2.64

From the ANOVA tables, it can be seen that statistically only two design variables are significant. They are thickness (A) and inner diameter (B) of the tubes. Inner diameter of the tube (B) is significant for having high temperature differnce and thickness of the tube (A) is significant for having low pressure drop. The optimizing level for these significant variable is found using ANOVA and the numerical analysis is performed based on the combinations formed finally and an optimized result is obtained [6].

4   Conclusion

The design of experiments, the temperature differenece and pressure for our model is optimized in this study. An L9 orthogonal array was used to acdiffodate the experiments. The experimental results revealed that Temperature difference of our heat exchanger is mostly affected by the inner diameter of the tube and Pressure drop is affected mostly due to the thickness of the tube. Predicted value temperature difference and pressure drop from ANOVA were found to be 712.52 °F and -0.44891 MPa respectively. The optimal level for each design vairable is A3, B3, C2, and D2 are defined from ANOVA. Numerical Experiment is performed using the optimum level data and the response is extracted. The extracted experimental value of temperature difference and pressure drop values are 629.08 °F and 1.33 MPa respectively[7]. The other two variables horizontal and vertical distance between each tubes had only a slight effect on the responses but not much effective. With help of these results we have planned to conduct numerical analyis for the same model with fins in our future research. We have a plan to fabricate a prototype model of finned tube heat exhanger with the same parameters used for these analysis if the results of our analysis are pleasable and valid. Later we would like to compare our numerical analysis results with the thermal analysis of prototype model.

Acknowledgement. This research was financially supported by Valve Center from the *Regional Innovation Center (RIC)* Program of Ministry of Trade, Industry and Energy (MOTIE).

References

1. Praveen Kumar, S.P., Shin, B.S., Lee, K.H.: Numerical Studies on the Performance of Finned-Tube Heat Exchanger. World Academy of Science, Engineering and Technology **8**(2) (2014)
2. Lee, Y.H., Kim, S.Y., Park, M.K.: An Experimental Study of Shell and Tube Heat Exchanger Performance with Baffle Spacing. Translation of the Korean Society of Mechanical Engineering **25**(12), 1748–1755 (2001)
3. ANSYS, Inc.: Release 14.0 Documentation for the ANSYS Workbench (2013)
4. Ghori, M.V.: Numerical Analysis of Tube-Fin Heat Exchanger Using Fluent, vol. 1(2) (2012)
5. Shin, B.S., Lee, K.H.: Optimization of the cyclone separator for geothermal plant. In: International Conference on Manufacturing, Machine Design and Technology, p. 110, May 2013
6. Jang, B.H., Lee, K.H.: A Numerical Analysis for the Performance Improvement of a Channel Heat Exchanger. Translation of International Standard Book Number: 978-1-61804-082-4
7. Jeon, R.W., Shin, B.S., Praveen Kumar, S.P.: Comparison of Theoretical and Numerical Analysis of Shell and Tube Heat Exchanger. Applied Mechanics and Materials **607**, 329–332 (2014)
8. Incropera, F.P., Dewitt, D.P.: Heat and Mass Transfer

Extraction of Event Elements
Based on Event Ontology Reasoning

Wei Liu$^{(\boxtimes)}$, Feijing Liu$^{(\boxtimes)}$, Dong Wang, Ning Ding$^{(\boxtimes)}$, and Xu Wang

School of Computer Engineering and Science, Shanghai University,
Shanghai 200444, China
{liuw,liufeijing,Ces13721024}@shu.edu.cn

Abstract. This paper proposes an event elements extraction method based on event ontology reasoning by constructing an upper event ontology and event elements reasoning rules based on event non-taxonomic relations. Event elements extraction includes three steps: data preprocessing; complementing event elements initially; event elements reasoning. The experimental results show that this method can improve the accuracy of event elements extraction.

Keywords: Event ontology reasoning · Event elements · Event elements extraction

1 Introduction

In the field of NLP, event is a structured knowledge unit with bigger granularity than concept, which is in line with human cognition. Therefore, in the field of AI, researchers hope that event-related (including action, time, place and people) information can be automatically identified from text by machine, thus to achieve some automatic text processing tasks, such as text classification, topic detection and tracking and so on. Therefore, identification of event elements has become an important sub-task of event information extraction.

The machine learning method considers event extraction as a classification problem and has good robustness, but it requires large-scale corpus labeled as model training base, which results in very laborious manual annotation. For shortcomings of machine learning method, this paper proposes an event elements extraction method based on event ontology. This method enables machine to mimic users' reading habits, utilize event ontology to associate event information and reasons about event elements including place, time, subject and object.

2 Related Work

Machine learning method is more objective and does not require much human intervention and domain knowledge, which includes two key steps, classifier construction and feature selection. In [1], machine learning methods were utilized to identify

N.T. Nguyen et al. (Eds.): ACIIDS 2015, Part II, LNAI 9012, pp. 577–586, 2015.
DOI: 10.1007/978-3-319-15705-4_56

relevant semantic role of verb in Persian. In [2], machine learning methods were used to identify key events of Chinese news stories and extract event 5W1H elements. In [3] event elements identification was divided into two steps: identify named entities, time phrases and event elements words; use maximum entropy classifier to classify elements. A method was proposed in [4] to identify event type argument by combining event trigger expansion and a binary classifier and multi-class classification based on maximum entropy. A method was proposed in [5] to identify event elements based on semi-supervised clustering and feature weighting.

In addition to machine learning methods, pattern matching methods in event elements identification are also frequently used. The key idea of these methods is to create a series of models and match sentences with template to achieve a purpose of event identification and extraction. These methods are only suitable for specific areas and lack of versatility. In [6], a method of Web news-oriented event multi-elements retrieval is studied. In [7], multi-pattern matching method was utilized to identify event elements on the ACE Chinese corpus, but the rules adopted are limited and results are not satisfactory. In [8] and [9], a new task of cross-document event extraction method and a biomedical event extraction system were proposed respectively.

3   Construction of Event Ontology

In the field of information extraction, event is defined as "a refined retrieval-used theme". Topic Detection and Tracking (TDT) sponsored by DARPA defined event as "a thing happens in a certain time and place". The main definitions about event of this paper are from our previous work [10]. In this section, we review these concepts briefly.

3.1   Event Related Concepts

Definition 1 (*Event*). We define event as a thing happens in a certain time and environment, which some actors participate in and shows some action features. Event e can be defined as a 6-tuple formally: $e = (A, O, T, V, P, L)$, where A means an action set happened in an event; O means objects involved in the event; T means time; V means place; P denotes assertions; L means language expressions.

Definition 2 (*Event Class*). Event Class denotes a set of events with common features, defined as: $EC = (E, C_1, C_2, ..., C_6)$, $C_i = \{c_{i1}, c_{i2} ..., c_{im}, ...\}(1 \leq i \leq 6, m \geq 0)$. Where E is event set, called extension of event class. C_i is called intension of event class. It denotes common features set of certain event element (element i). c_{im} denotes one of common features of event element i.

Definition 3 (*Taxonomic Relations*). There exists subsumption relation between event class $EC_1 = (E_1, C_{11}, C_{12}, ..., C_{16})$ and event class $EC_2 = (E_2, C_{21}, C_{22}, ..., C_{26})$ if and only if $E_2 \subset E_1$, or $C_{2i} \subseteq C_{1i}(i = 1,2, ...,6)$. We call EC_1 hypernym event class and EC_2 hyponym event class, denoted as $EC_2 \subset EC_1$.

Definition 4 (*Non-taxonomic Relations*). ① Causal Relation: it is denoted as $(\lambda EC_1 \rightarrow EC_2)$. λ denotes the probability that events of event class EC_2 happen caused by events of event class EC_1. ②Follow Relation: Another event of EC_2 the subject participates in happens after an event of EC_1 specified subject participates in happens, denoted as$(EC_1 \rhd EC_2)$. ③Concurrence Relation: If EC_1 concur with EC_2 in a certain length of time, and the occurrence probability is above a specified threshold, there is a concurrence relation between EC_1 and EC_2, denoted as$(EC_1 \parallel EC_2)$. ④Composition Relation: If each event instance of event class EC_1 is composed of one event instance of event class EC_2 and other event classes, EC_2 is part of EC_1, denoted as$(EC_1 \lessdot EC_2)$.

Definition 5 (Event Ontology). Event ontology EO is defined as a 4-tuple ly: $EO = (UECS, ECS, R, Rules)$. $UECS$ is a set of upper event classes; ECS is a set of event classes; R means the relations between event classes; $Rules$ is set of rules be expressed in logic languages.

3.2   Upper Event Ontology

In support of event elements reasoning, this paper constructs an upper event ontology based on event ontology of reference [10]. Upper event ontology defines event taxonomic hierarchy model, as shown in Table 1.

Table 1. Upper event ontology structure

1 Class :HumanEvent	2 Class :NatureEvent
1.1 Class :SinglePersonEvent	2.1 Class :NonNatureForceEvent
1.1.1 Class :PersonObject_SinglePersonEvent	2.1.1 Class :PersonObject_ NonNatureForceEvent
1.1.1.1 Class :Continue_PO_SinglePersonEvent	2.1.1.1 Class :Continue_PO_NonNatureForceEvent
1.1.1.2 Class :Instant_PO_SinglePersonEvent	2.1.1.2 Class :Instant_PO_NonNatureForceEvent
1.1.2 Class :NonObject_SinglePersonEvent	2.1.2 Class : NonPersonObject _NonNatureForceEvent
1.1.2.1 Class :Continue_NO_SinglePersonEvent	2.1.2.1 Class :Continue_NPO_NonNatureForceEvent
1.1.2.2 Class :Instant_NO_SinglePersonEvent	2.1.2.2 Class :Instant_NPO_NonNatureForceEvent
1.1.3 Class :NonPersonObject_SinglePersonEvent	2.1.3 Class :NonObject_NonNatureForceEvent
1.1.3.1 Class :Continue_NPO_SinglePersonEvent	2.1.3.1 Class :Continue_NO_NonNatureForceEvent
1.1.3.2 Class :Instant_NPO_SinglePersonEvent	2.1.3.2 Class :Instant_NO_NonNatureForceEvent
1.2 Class :PublicEvent	2.2 Class :NatureForceEvent
1.2.1 Class :NonPersonObject_PublicEvent	2.2.1 Class :PersonObject_ NatureForceEvent
1.2.1.1 Class :Instant_NPO_PublicEvent	2.2.1.1 Class :Continue_NatureForceEvent
1.2.1.2 Class :Continue_NPO_PublicEvent	2.2.1.2 Class :Instant_NatureForceEvent
1.2.2 Class :PersonObject_PublicEvent	2.2.2 Class : NonPersonObject _ NatureForceEvent
1.2.2.1 Class :Continue_PO_PublicEvent	2.2.2.1 Class :Continue_NPO_NatureForceEvent
1.2.2.2 Class :Instant_PO_PublicEvent	2.2.2.2 Class :Instant_NPO_NatureForceEvent
1.2.3 Class :NonObject_PublicEvent	2.2.3 Class :NonObject_ NatureForceEvent
1.2.3.1 Class :Continue_NO_PublicEvent	2.2.3.1 Class :Continue_NO_NatureForceEvent
1.2.3.2 Class :Instant_NO_PublicEvent	2.2.3.2 Class :Instant_NO_NatureForceEvent

The first level of upper event ontology is divided into two categories: human event class and natural event Class, according to subject category of event class.

The second level is further sorted according subjects of event class. Human event class is divided into single person event class and public event class. Natural event class is divided into natural force events and non-natural force events.

The third level of upper event ontology is sorted according to objects of event class. Human event class is divided into person object event class, non-person object event class and non-object event class. Nature force event class is divided into person object nature force event class, non-person object nature force event class and non-object nature force event class while non-nature force event class can be divided in the same way.

The fourth level is sorted based on the third level according to time element. Time elements tend to be divided into time points and the time periods, and accordingly, events can be divided into transient events and continuous events.

4   Reasoning Rules and Reasoning Procedure of Event Elements

4.1   Reasoning Rules of Event Elements

Taxonomic relations and non-taxonomic relations have different effects on elements reasoning. In taxonomic relations, when the abstract event class an event belongs to is queried in upper event ontology, restrictions of the event elements can be obtained. For example, if an event class (such as thunderstorm) belongs to instant nature force event class, its start time and end time are same, and its object is empty.

Table 2. Event elements reasoning rules for event relations of $CPOPE \times CPOPE$

a. $(e_1 \in EC_1) \cap (e_2 \in EC_2) \cap (EC_1 \nless EC_1) \Rightarrow (ST(e_1) \geq ST(e_2)) \cap (ET(e_1) \leq ET(e_2))$

b. $(e_1 \in EC_1) \cap (e_2 \in EC_2) \cap (EC_1 \nless EC_1) \Rightarrow P(e_1) = P(e_2)$

c. $(e_1 \in EC_1) \cap (e_2 \in EC_2) \cap (EC_1 \nless EC_1) \Rightarrow Sub(e_1) = Sub(e_2)$

d. $(e_1 \in EC_1) \cap (e_2 \in EC_2) \cap (EC_1 \nless EC_1) \Rightarrow Obj(e_1) \subseteq Obj(e_2)$

e. $(e_1 \in EC_1) \cap (e_2 \in EC_2) \cap (EC_1 \rightarrow EC_1) \Rightarrow ST(e_1) < ST(e_2)$

f. $(e_1 \in EC_1) \cap (e_2 \in EC_2) \cap (EC_1 \rightarrow EC_1) \Rightarrow P(e_1) = P(e_2)$

g. $(e_1 \in EC_1) \cap (e_2 \in EC_2) \cap (EC_1 \rightarrow EC_1) \Rightarrow Obj(e_1) = Sub(e_2)$

h. $(e_1 \in EC_1) \cap (e_2 \in EC_2) \cap (EC_1 \rhd EC_1) \Rightarrow ET(e_1) < ST(e_2)$

i. $(e_1 \in EC_1) \cap (e_2 \in EC_2) \cap (EC_1 \rhd EC_1) \Rightarrow P(e_1) = P(e_2)$

j. $(e_1 \in EC_1) \cap (e_2 \in EC_2) \cap (EC_1 \rhd EC_1) \Rightarrow Sub(e_1) = Sub(e_2)$

k. $(e_1 \in EC_1) \cap (e_2 \in EC_2) \cap (EC_1 \parallel EC_1) \Rightarrow P(e_1) = P(e_2)$

l. $(e_1 \in EC_1) \cap (e_2 \in EC_2) \cap (EC_1 \parallel EC_1) \Rightarrow (ST(e_1) \leq ST(e_2)) \cap (ET(e_1) \geq ET(e_2))$

Non-taxonomic relations are main content of event elements reasoning in this paper, which can contact context. After studying features of all event types in the fourth level of upper event ontology and a large number of cases, a set of event elements reasoning rules are proposed for combination of every two event types according to relations between events. Table 2 shows twelve event elements reasoning rules for event relations of the two *Continue_PO_PublicEvent* event types(referred to as *CPOPE* type, namely two events existing relations belong to public continue events many people take part in, such as "*assistance*" and "*on-site rescue*"). Similarly, other combinations also have reasoning rules. In these rules, the $P(e_i)$ represents place, and $ST(e_i)$ represents start time, and $ET(e_i)$ represents end time, $Sub(e_i)$ represent subject, $Obj(e_i)$ represents object.

4.2   Process of Event Elements Identification

This paper mainly identify event elements appear in news reports, focusing on identifying and complementing four elements (time, place, subject and object). Segment for an article using segmentation tool, and get a set of events from texts using language performance of event classes in event ontology, and then extract all named entities from this sample, and create a two-dimensional matrix. In this matrix, columns are events; rows are event elements; 0 denotes it is not element of the event; 1 denotes place element; 2 denotes start time; 3 denotes end time; 4 denotes subject; 5 denotes object. Complementing event elements actually update the matrix. A_{ij} denotes a matrix consisting of all the events of an article.

$$
A_{ij} =
\begin{array}{c}
\\
e_1 \\
e_2 \\
e_3 \\
e_4 \\
e_5 \\
e_6 \\
e_7 \\
e_8
\end{array}
\begin{array}{cccccccccc}
w_1 & w_2 & w_3 & w_4 & w_5 & w_6 & w_7 & w_8 & w_9 & w_{10} \\
\left[\begin{array}{cccccccccc}
1 & 2 & 3 & 4 & 5 & 0 & 0 & 0 & 0 & 0 \\
1 & 0 & 3 & 4 & 5 & 0 & 0 & 2 & 0 & 0 \\
0 & 2 & 3 & 0 & 0 & 0 & 1 & 0 & 4 & 5 \\
0 & 0 & 3 & 4 & 5 & 2 & 1 & 0 & 0 & 0 \\
1 & 0 & 3 & 4 & 5 & 2 & 0 & 0 & 0 & 0 \\
0 & 2 & 3 & 5 & 0 & 0 & 1 & 0 & 4 & 0 \\
0 & 2 & 3 & 5 & 0 & 0 & 1 & 0 & 4 & 0 \\
1 & 2 & 3 & 4 & 0 & 0 & 0 & 0 & 0 & 5
\end{array}\right]
\end{array}
$$

Identification process of event elements are following three steps: data preprocessing, complementing event elements initially according to words position, event elements reasoning.

In data preprocessing stage, first, segment for an article and revise manually; then identify event triggers and corresponding event elements using event ontology. In order to calculate positional relations between words in subsequent steps, tag number of words in sentences using sentence as a unit.

In initial complementing stage, the distance between event triggers and elements can be calculated using paragraph number, sentence number and word number tagged in data processing. The nearest words are selected as initial complementing results. For an event a of the text, steps of initial complementing event elements are as follows:

Step 1: Get element word $argu1$ from elements words list; $confident\ (e, i)$ is confidence of the $i\text{-}th$ element of the event, which is initialized to 0, then go to step 2;

Step 2: Calculate confidence of the element: if (e and $argu1$ locate in the same sentence) $confidence= \alpha/|$word number of e - word number of $augu1|$; else if (e and $argu1$ locate in the same paragraph) $confidence = \beta/|$ word number of e - word number of $augu1|$; else $confidence = \gamma/|$ word number of e - word number of $augu1|$; then go to step 3;

Step 3: If the $confidence(e, i)$ is higher than the previous, it is updated to *the confidence*; or keep previous $confidence$ and element unchanged, then go to step 4;

Step 4: Go to step 1, continue to remove event elements from the list until all event elements are taken from the list.

Here α, β, and γ are used to calculate weight of confidence. Their values ensure that value of confidence decreases with increasing distance (in general, the confidence which elements and triggers are in a same sentence is larger than one in a different sentence). Therefore, values of α, β, γ can be respectively *100*, *10* and *1*.

In event elements reasoning stage, initial complementing results in the second stage are used to reason elements. First, get elements restrictions of the event through querying the upper event class the event belongs to. If an element of some events is default, it is not complemented. If subjects of some events are only people, they are complemented by named entities. Then use one of two associated events to reason related elements of another event through above reasoning rules. The event of the largest confidence is selected as a seed event. Input Seed events, and query events existing non-taxonomic relation with seed events from event ontology, and then query event types of two events associated with every relation, and then determine elements reasoning rules base according to event types of the two events and reason related information.

4.3　Case Study

The following is a semi-automated tagged news report: e_i denotes an event trigger, and l_i denotes place, and t_i denotes time, and p_i denotes participants (including subject and object).

新快报讯，8月20日早上6点(**t₁**)，　阿尔及利亚以东150公里的卜伊拉(**l₁**)发生汽车炸弹(**p₁**)爆炸(**e₁**)事件，　造成11人(**p₂**)死亡(**e₂**)。 (*According to Xinkuai Express, at 6:00 am on August 20th, car bomb was exploded 150 kilometers east of Algeria , killing 11 people dead.*)

当地媒体报道称，包括4名军事人员在内的31人(**p₃**)受伤(**e₃**)。　目前(**t₂**)，　当地(**l₂**)正对伤者(**p₄**)进行救治(**e₄**)。 (*Local media said that 31 people including 4 military officers were injured. So far, the wounded has been on treatment in the local.*)

Step 1, complemented elements are shown in table 3 according to words location distance, and *conf* is confidence:

Step 2, reason related information according to restrictions of event classes in event ontology:

e_1 *is_a Instant_NonNatureForceEvent* => e_1.*ST*=e_1.*ET*=t_1, e_1.*OBJECT*=null

The above formula shows that event type of e_1 mapped to upper event ontology is *Instant_NonNatureForceEvent*. Conclusions can be drawn from above: e_1 is an *InstantEvent*, so start time and end time are same; the event describes subject's own changes, so there is no object. In step 1, element word p_1 is complemented as object of e_1, which does not meet the restriction *Instant_NonNatureForceEvent* doesn't have object. So in step 2, object of e_1 is revised as empty.

Table 3. Event elements complementation through words location distance

Event	LOC	ST	ET	SUBJECT	OBJECT	conf
e_1	l_1	t_1	t_1	p_1	p_1	291.7
e_2	l_1	t_1	t_1	p_2	p_2	247.6
e_3	l_1	t_2	t_2	p_3	p_3	221

e_4	l_2	t_2	t_2	p_4	p_4	165

Similarly, we can draw:

e_2 *is_a Instant_NonObject_SinglePersonEvent*=> $e_2.ST=e_1.ET=t_1$, $e_2.OBJECT=null$

e_3 *is_a Continue_NonObject_SinglePersonEvent*=> $e_3.OBJECT =null$, $e_3.ET>e_3.ST=t_1$

e_4 *is_a Continue_PersonObject_PublicEvent*=> $e_4.OBJECT =$ 伤者, $e_4.ET>e_4.ST=t_2$

And $e_4.SUBJECT=$医疗人员, which is obtained through specific *treatment* event class.

Step 3, following relations are obtained from querying event ontology: e_1 *cause* e_2, e_1 *cause* e_3, e_2 *concur* e_3, e_3 *cause* e_4.

e_1 is used as seed event. According to event types of e_1 and e_2, corresponding reasoning rules are found from rules base, and then other event elements are reasoned:

e_1 *cause* e_2 =>$e_1.ST< e_2.ST$=> $e_2.ST = e_2.ET = t1+$; e_2 *concur* e_3 =>$e_2.ST=e_3.ST$=> $e_3.ST = e_3.ET = t_1+$

e_3 *cause* e_4 => $e_3.ST< e_4.ST$=> $e_4.ST=t_1++(t_2=t_1++)$; e_3 *cause* e_4 => $e_3.LOC= e_4.LOC(l_2=l_1)$ => $e_4. LOC = l_1$

Finally, results are shown in table 4.

Table 4. Results of elements reasoning

Event	LOC	ST	ET	SUBJECT	OBJECT
e_1	l_1	t_1	t_1	p_1	*null*
e_2	l_1	t_1+	t_1	p_2	*null*
e_3	l_1	t_1+	t_1++	p_3	*null*
e_4	l_1	t_1++	t_1+++	医务人员	p_4

So if an event itself doesn't have object element, its object value is null; the happen time of event is updated; some default elements are complemented according to event ontology; absolute time and place are reasoned from some relative time and place (such as "*current*", "*local*"). To a certain extent, event elements are complemented.

5   Experiment and Analysis

5.1   Experiment

The experimental data set is Chinese Emergency Corpus (CEC) [11], which contains earthquakes, fires, accidents and terrorist attacks five types, a total of 205. Place, subject, object and time elements are complemented for events of CEC. Experiment is divided into two parts.

Experiment 1: Elements are complemented according to positional relations between triggers and any other event elements.

Experiment 2: Reasoning results are used to complement event elements according to reasoning rules proposed in this paper. Based on experiment 1, Experiment 2 is divided into two parts: (1) event with the highest confidence and occurring in the first paragraph is used as a seed event; (2) event with the highest confidence and occurring in other paragraph is used as a seed event.

5.2   Analysis of Experiment Results

In experiment 1, statistic results of random selected 195 events from CEC are shown in table 5:

Table 5. Complemented results of neighboring position elements for different relations

Element	Precision	Recall	F1 measure
Place	65.8%	62.5%	64.1%
Time	60.3%	66.6%	63.3%
Subject	48.5%	72.3%	58.1%
Object	54.3%	52.7%	53.5%

Just as shown in table 5: recall, precision and F1 values of place and time elements have reached more than 60%, which indicates that extraction of the two elements can achieve a preliminary desired results using positional relation between triggers and elements, but subjects and objects are not. First, because number of subjects and objects words in the article appears much more than subject and object themselves, which easily interferes with each other in sentences; second, because annotations of subjects and objects are not as clear as time and place. Some subjects may be objects of other events and vice versa; third, because some events may have many subjects and objects, but this method can only complement one of them.

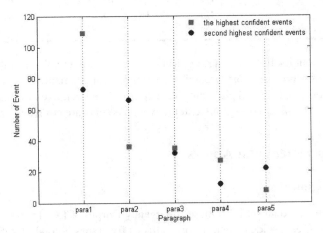

Fig. 1. Events distribution with the highest and second highest confidence

Experiment 1 only uses positional relations between words, but some factors like types of events and relations of context are not taken into account. So experiment 2 updates results of experiment 1 through reasoning rules. Event distributions with the highest confidence are shown in figure 1.

As shown in figure 1, events of the highest and second highest confidence generally appear in the first paragraph, or in the second paragraph, and some evenly distribute in other paragraphs. These events in the first paragraph are basically core events, but events

Table 6. Complement results based on element reasoning for different seed events

Element	FirstPara_Event			OtherPara_Event		
	P	R	F1	P	R	F1
Place	78.3%	82.3%	80.3%	73.5%	80.7%	76.9%
Time	80.9%	76.2%	78.5%	75.7%	73.6%	74.6%
Subject	69.3%	73.7%	71.4%	68.0%	66.9%	67.4%
Object	70.5%	68.6%	69.5%	65.6%	65.3%	64.5%

in other paragraph are almost not. So events with the highest confidence in the first paragraph and ones in other paragraph with the highest confidence are selected as seed events. Reasoning results of experiment 2 are shown in table 6.

Just as shown in table 6: P, R and F1 values are improved much more than experiment 1. In complementation of event elements, the interference of positional relations between words can be excluded. Identification of subject and object also avoid problems arising from experimental 1, especially for subject of many events. In addition, in the second experiment, events are mapped to event classes of event ontology, which can get restrictions of elements and take full account of categories and the default of event elements. For example, for some natural disaster events, their subject is nature through querying upper event ontology. In general, subject like nature is default, which causes that the real subject can't be obtained by complemented method of word distance in experiment 1. Event restrictions obtained from event ontology revise the results of experiment 1. For non-default elements, more suitable elements can be selected through element restrictions of event classes, which improves experimental effect. In selecting seed events, experimental results of the first part (with the highest confidence and in the first paragraph) are slightly better than the second part. The experiment results above show that event elements can be better identified by using event class relations and seed events in the first paragraph can achieve better experimental effect.

6   Conclusion

By construction of upper event ontology and event elements reasoning rules set, this paper proposed an event elements extraction method. This method reduces the dependence on scale of rules and corpus; Experimental results show that this proposed method can effectively improve identification performance of event elements. But there still need some improvements: the current accuracy of automatic identification of event indicators and event elements can't reach a more desired degree; the structure of event ontology affects the effect of identification; the reasoning rules need to be further improved. In our future research, event ontology will be enriched, including enriching and optimizing event types, restriction of event classes and reasoning rules of event elements. The event information identified can be used in text representation based on events, such as construction of text event networks.

W. Liu et al.

This paper is supported by the Natural Science Foundation of China, No.61305053 and No.61273328, and the Natural Science Foundation of Shanghai, No.12ZR1410900.

References

1. Saeedi, P., Faili. H.: Feature engineering using shallow parsing in argument classification of Persian verbs. In: Proc of the 16th CSI International Symposium on Artificial Intelligence and Signal Processing (AISP 2012), pp. 333–338 (2012)
2. Wang, W., Zhao, D.Y., Wang, D.: Chinese news event 5w1h elements extraction using semantic role labeling. In: Proc of the Third International Symposium on Information Processing (ISIP), pp. 484–489 (2010)
3. Chen, Z., Ji, H.: Language specific issue and feature exploration in Chinese event extraction. In: Proceedings of Human Language Technologies: The 2009 Annual Conference of the North American Chapter of the Association for Computational Linguistics, Companion Volume: Short Papers, pp. 209–212 (2009)
4. Zhao, Y., Qin, B., Che, W., et al.: Research on Chinese Event Extraction. Journal of Chinese Information Processing 22(1), 3–8 (2008). (in Chinese)
5. Fu, J., Liu, Z., Liu, W., et al.: Feature Weighting Based Event Argument Identification. Computer Science 37(3), 239–241 (2010). (in Chinese)
6. Zhong, Z.M., Li, C.H., Liu, Z.T., Dai, H.W.: Web news oriented event multi-elements retrieval. Journal of Software 24(10), 2366–2378 (2013). (in Chinese)
7. Tan, H.Y., Zhao, T.J., Zheng, J.H.: Identification of Chinese event and their argument roles. In: IEEE 8th International Conference on Computer and Information Technology Workshops, pp. 14–19 (2008)
8. Ji, H., Grishman, R., Chen, Z., Gupta, P.: Cross-document event extraction and tracking: task, evaluation, techniques and challenges. In: Proc of RANLP, pp. 166–172 (2009)
9. Miwa, M., Sætre, R., Kim, J.D., et al.: Event Extraction with Complex Event Classification using Rich Features. Journal of Bioinformatics and Computational Biology (JBCB) 8(1), 131–146 (2010)
10. Liu, Z., Huang, M., et al.: Research on event-oriented ontology. Computer Science 36(11), 189–192 (2009). (in Chinese)
11. Fu, J.: Research on Event-Oriented Knowledge Processing. Shanghai University, Shanghai (2010). (in Chinese)

Author Index

Printed in the United States
By Bookmasters